T4-AUB-747

HANDBOOK OF DIGITAL SYSTEM DESIGN

Second Edition

Wen C. Lin, Ph.D.

Professor
Department of Electrical Engineering
and Computer Science
University of California
Davis, California

CRC Press
Boca Raton Ann Arbor Boston

Library of Congress Cataloging-in-Publication Data

Lin, Wen C.
Handbook of digital system design / Wen C. Lin. — 2nd ed.
p. cm.
Includes bibliographical references.
ISBN 0-8493-4272-4
1. Digital electronics. 2. Electronic digital computers.
I. Title.
TK7868.D5L537 1990
621.381—dc20 90-1590
CIP

Direct all inquires to CRC Press, Inc., 2000 Corporate Blvd., N.W., Boca Raton, Florida, 33431.

International Standard Book Number 0-8493-4272-4

Library of Congress Card Number 90-1590

Acquiring Editor: Russ Hall
Production Director: Sandy Pearlman
Coordinating Editor: Andrea Raia
Cover Design: Chris Pearl
Internal Design: Andrea Raia
Copy Editor: Andrea Raia
Page Layout: Jonathan Pennell and Carlos Esser
Camera Work: Gary Bennett, Atlee Miller, Kevin Loung, and Helen Linna
Composition: CRC Press, Inc.
Indexing: Sharon Smith, Professional Indexing
Printer and Binder: Braun-Brumfield, Inc.

To My

Wife, Children,

and

Eldest Brother, Wen-Chan

PREFACE
TO THE FIRST EDITION

The objective of this handbook is to provide self-contained and sufficient design information for scientists and engineers who wish to undertake practical design of digital systems for their applications. It should also be suitable as a text or major reference book for seniors or graduate students in Electrical Engineering, Computer Engineering, or Biomedical/Clinical Engineering who are engaged in the design of digital systems, microcomputers, or instrumentation for their research and development projects. Without providing exercises or problems for drills or academic interest as is the format of conventional text books, this book covers principally three broad topics, namely, (1) fundamentals; it serves as a review to readers who are somewhat familiar with the basic theories but who may wish to refresh themselves on formulas, design equations, terminologies, symbols, and notations, (2) characteristics, properties and principles of operation of devices, modules and building blocks that are frequently used as components in digital system design, and (3) design procedures by means of examples, which provide the readers with some guidelines for system design.

The unique features of this handbook are that the necessary linear device characteristics and principles are included, and that a balance of circuit/system design is presented. The nontraditional composition of analog and digital materials for digital system design in this book is a response to the rapid advance in electronic technologies. Historically, digital engineers and logic designers knew nothing about linear circuitry and very little about digital electronic circuit design. If necessary, they could easily get help from a group within their department specializing in circuit design. Rapid advances in solid-state technology, such as large-scale-integrated (LSI) circuits, have reduced the demand for full-time circuit designers. As a result, the system designers are forced to acquire broader knowledge in linear and digital circuits, as well as logic and system design. Fortunately, this is not only possible but easily achievable with available technologies, as long as system designers are willing to learn some fundamentals of linear and digital circuit design. Assuming that the readers may be familiar with some but not all of the topics that have been covered, this book was written in somewhat of a tutorial style. Each subarea starts with fundamentals and then an attempt is made to clarify material which is traditionally confusing or overlooked. Finally, some design examples, if applicable, are presented.

The first two chapters basically deal with linear circuits, devices, and modules which are frequently encountered in digital system design. Operational amplifiers, linear active filters, phase-locked loops, and basic analog computing circuits are described and analyzed from a practical point of view. Chapters 3 to 5 are devoted to pulse, time, and logic circuits and devices. Here, the Binary State Analysis (BSA) and graphical analysis techniques are introduced and illustrated for TTL, ECL, CMOS, I^2L, and Schottky logic elements. A thorough treatment of R-S flip-flop, J-K flip-flop, D flip-flop, master-slave flip-flop, and other multivibrators, Schmitt trigger and medium scale integrated circuit devices is presented. In Chapters 6 to 8, the essential subsystems for a digital computer-based system, i.e., memory, arithmetic logic unit (ALU), and analog-digital-analog converters, are analyzed in detail. System noise problems are explored in Chapter 9. Finally, the topic of design of random (control) logic using programmable LSI is presented in Chapter 10.

The majority of the materials covered in this handbook were developed and tested successfully in the laboratories and classes for students in Engineering and Sciences during the author's tenure at Case Western Reserve University, Cleveland, OH. It is his pleasure to thank his former graduate students, especially Drs. Chi-Foon Chan and Mark Michael, and Mr. Wing-Kay Leung, who helped to develop the laboratory program. I am especially grateful to Professors Yaohan Chu, Harry W. Mergler, Wen H. Ko, E. L. Glaser, and Dr. William H. Ninke for their encouragement. My thanks goes to Miss Grace K. Lin and Mrs. Carol She for their skillful drawing and typing of the manuscript. Finally, I would like to thank Professor V. R. Algazi for his encouragement during the last stage of preparing the materials for this handbook.

PREFACE
TO THE SECOND EDITION

While preserving the broad coverage of the necessary materials for digital system designers in the first edition, we have made a number of improvements in this second edition as follows:

1. The organization of the book has been changed to a more conventional format; for example, 10.3.1 refers to Chapter 10, Section 3, Subsection 1.
2. For strengthening the reader's fundamentals in circuit analysis and synthesis, additional design examples have been included. For example, the graphical analysis technique for nonlinear devices has been detailed and its application in the signal bus has been illustrated with a practical design example. Similarly, a system design example has been used as a vehicle to demonstrate how through grounding and/or shielding, system noise problems can be prevented or minimized before the system is assembled.
3. Having witnessed the rapid growth in system design of Programmable Logic Devices (PLDs), such as PROM, PLA, PAL, and LCA in recent years, we have expanded the topic in sequential machine design with PLDs. In Chapter 10 we review the fundamentals of sequential machine design and their implementation with examples. Chapter 11 focuses on the design with PLDs. In Chapter 12, we introduce two important topics: computer-aided design for PLDs and a guideline for hardware debugging.

The author wishes to thank Mr. Russ Hall, Acquisition Editor, of CRC Press Inc., Boca Raton, FL, for his encouragement and trust in the developmental stage of this edition. A deep appreciation goes to Mr. Richard Darsie for his careful and thoughtful assistance, as well as some editorial suggestions, in the process of preparing and typing the manuscript for this second edition.

Wen C. Lin
Davis, California
1990

THE AUTHOR

Dr. Wen C. Lin, Ph.D., received the B.S. degree (top 5%) in Electrical Engineering from the National Taiwan University, Taiwan, Republic of China; M.S. and Ph.D. degrees in Electrical Engineering from Purdue University, Lafayette, Indiana.

He was an engineer with the Instrument Laboratory, Taiwan Power Company, engineer with General Electric Company, and senior engineer with Electronic Data Processing Division, Honeywell Corporation. In 1965, he joined Case Institute of Technology as an Assistant and then Associate Professor. Next, he served as a Professor of Electrical and Computer Engineering at Case Western Reserve University, Cleveland, Ohio, until 1978. Since 1978 he has been a Professor in the Department of Electrical Engineering and Computer Science at the University of California, Davis. Interested in pattern recognition, signal processing (speech, visual, etc.), biomedical engineering, autonomous robots, microcomputers, digital systems, and neural networks, Dr. Lin has published over 70 technical papers in those areas. He is the editor of the IEEE Press book, *Microprocessors: Fundamentals and Applications*, and the author of *Digital System Design Handbook for Scientists and Engineers* (1st and 2nd editions), CRC Press, and *Computer Organization and Assembly Language Programming for the PDP11 and VAX-11*, John Wiley & Sons. He is also a contributor to such books as the *International Encyclopedia of Robotics: Applications and Automation* (1988) John Wiley & Sons, and the *1991 Yearbook of the Encyclopedia of Physical Science and Technology*, Academic Press, Inc.

Dr. Lin is a member of Sigma Xi, Eta Kappa Nu, and a Senior Member of the IEEE. He has been listed in over 19 books of Who's Who in Society, Science and Technology compiled by the publishers in the U.S., Japan, and England.

CONTENTS

1 A REVIEW OF LINEAR CIRCUIT FUNDAMENTALS

The objective of this chapter is to review the circuit fundamentals which are essential to circuit designers. Some nontrivial examples will be used to illustrate how the fundamental theories or laws are applied in circuit and system analysis and design. It serves as a reminder to the readers and makes this book virtually self-contained.

1.1. BASIC LAWS AND THEORIES FOR LINEAR CIRCUITS

1.1.1. Ohm's Law

$$e = iR$$

where R is the resistance of a resistor, in ohms,
e is the voltage across the resistor, in volts, and
i is the current, in amperes.

1.1.2. Kirchoff's Laws

a. *Current Law:* The algebraic sum of all currents at a node must be zero at any time. If the sign of a current flowing into the node is assigned to be positive (negative), then the current flowing out of the node must be negative (positive).
b. *Voltage Law:* The sum of all voltage across each circuit element around a loop must be zero at all times. In reference to clockwise direction, if voltage-rise (from low to high) is assigned as positive, then the voltage-drop (from high to low) must be negative.

1.1.3. Thevenin's Theory

Across a pair of nodes, a linear network can be replaced by a voltage source with open-circuit voltage, E_g, in series with a resistor with a value of R_g, where R_g is the equivalent resistance between the nodes

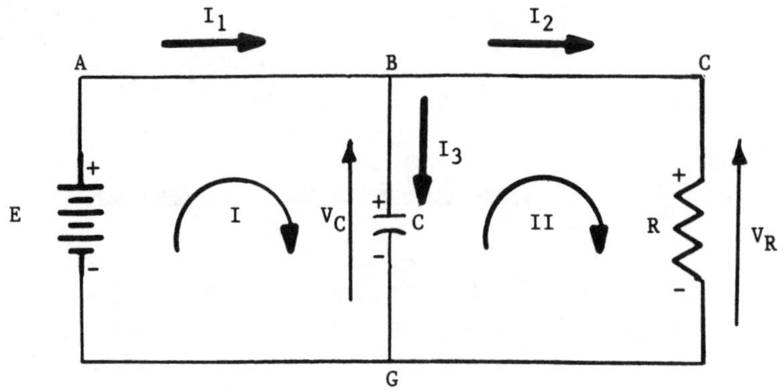

FIGURE1.1 Circuit contains one independent voltage source.

with all independent voltage and current sources, short-circuited and open-circuited, respectively.

1.2. EXAMPLES

Example 1

Problem: For the circuit shown in Figure 1.1, derive current and voltage equations.

Solution: Here we have one independent voltage source, the battery, E, whose polarity is given, so no polarity assignment is needed. By the following steps, we can label the current direction and voltage polarity for each of the other branches and the equations can then be derived. As shown in Figure 1.1,

a. Arbitrarily assume the direction of each branch current and the direction of the two voltage loops.
b. Assign symbolic names to each branch current and voltage.
c. Assign the polarity of the voltage across the circuit components, resistor R, and capacitor C, by the rule that all passive circuit components, such as resistors, capacitors, and inductors have their current entering terminals marked positive and the exiting terminals negative.
d. Write current equations according to Kirchoff's Current Law as follows: Node B: $I_1 - I_2 - I_3 = 0$, where I_1 is assigned flowing into B, and thus it is positive, while I_2 and I_3 flow out of B and thus are negative.
e. Write the voltage equations according to Kirchoff's Voltage Law as follows: Loop I: Starting at G and walking along the branches clockwise as assigned in Step a, assign the voltage sign for each branch by the rule that *positive* implies voltage-rise and *negative*

implies voltage-drop. If for any branch, the entering point was marked negative and exiting point positive in Step c, we define it as voltage-rise, otherwise we define it as voltage-drop. For example, at branch E we have "rise", thus positive; the voltage across C is "drop", thus negative. Therefore, by writing the voltage equation for Loop I we have $E - V_C = 0$. Loop II: As with Loop I we begin at G and walk along the branches clockwise as assigned in Step a. Note that this time the voltage across C is "rise", so that it must be assigned as positive, and the voltage across R is "drop", thus negative. Therefore, the voltage equation for Loop II is $V_C - V_R = 0$.

Example 2

Problem: For the circuit shown in Figure 1.2, determine I_i, I_2, I_3, I_0, and V_0, if $V_i = 10$ V, $R_1 = 100\ \Omega$, $R_2 = 100\ \Omega$, $R_3 = 500\ \Omega$, $R_4 = 100\ \Omega$, and $\beta = 10$. Note that V_i is an independent voltage source, while βI_i is a dependent current source. That is, the magnitude of the latter is a function of I_i which is not a constant although the current gain, β, is.

Solution: Label the current direction and voltage polarity for each branch as shown in the circuit by the procedure outlined in Example 1. Remember that once the current direction and voltage polarity are assigned, we must not change them until the end of the analysis.

Node A: $$I_i + I_0 - I_2 = 0 \quad (1.1)$$
Node B: $$\beta I_i + I_3 - I_0 = 0 \quad (1.2)$$

By Kirchoff's Voltage Law,

Loop I: $$V_i - I_i R_1 - I_2 R_2 = 0 \quad (1.3)$$
Loop II: $$I_2 R_2 + I_3 R_3 + I_0 R_4 = 0 \quad (1.4)$$

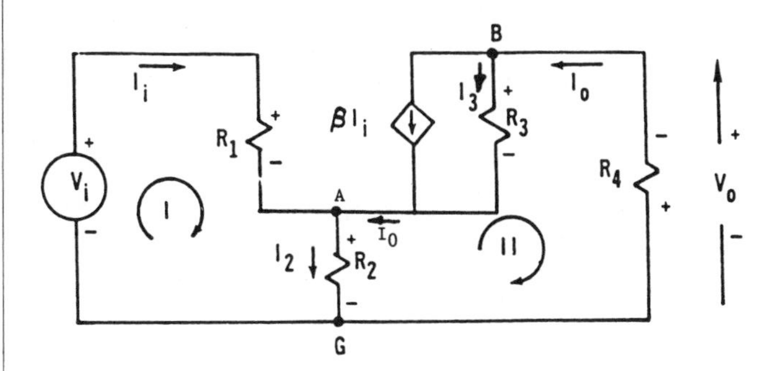

FIGURE 1.2 Circuit contains both dependent and independent sources.

Substituting the numerical values into Equations (1.1) to (1.4), we have

$$I_i + I_0 - I_2 = 0 \quad (1.5)$$
$$10I_i + I_3 - I_0 = 0 \quad (1.6)$$
$$10 - 100I_i - 100I_2 = 0 \quad (1.7)$$
$$100I_2 + 500I_3 + 100I_0 = 0 \quad (1.8)$$

Solving Equations (1.5) to (1.8),

$$I_i = 11.1 \text{ mA} \qquad I_2 = 88.88 \text{ mA}$$
$$I_3 = -33.33 \text{ mA} \qquad I_0 = 77.77 \text{ mA}$$
$$V_0 = -I_0 R_4 = -7.77 \text{ V}$$

Comment: Note that the voltage across R_2 for Loop I, in reference to the loop arrow of Loop I, is a voltage-drop, thus the sign for I_2R is negative; however, in Loop II it is a voltage-rise, in reference to the loop arrow of Loop II, thus its sign is positive. In this example, a review of the applications of the basic circuit laws was carried out through a numerical example. It is important to point out that determination of the signs of the currents and voltages based on the current flow directions is usually the major problem in deriving equations, if care is not taken.

Example 3

Problem: For the network shown in Figure 1.3(a), derive the Thevenin equivalent circuit at the terminals a,a′.

Solution: By Kirchoff's Current Law,

Node A: $$I_1 + \beta I_1 = I_2 \quad (1.9)$$

By Kirchoff's Voltage Law,

Loop I: $$E = I_1R_1 + I_2R_2 \quad (1.10)$$

From Equation (1.9),

$$I_1 = I_2/(1 + \beta) \quad (1.11)$$

Substitute I_1 in Equation (1.11) into Equation (1.10) to get

$$E = \frac{R_1}{1 + \beta} I_2 + I_2R_2 = \left(\frac{R_1}{1 + \beta} + R_2\right) I_2 \quad (1.12)$$

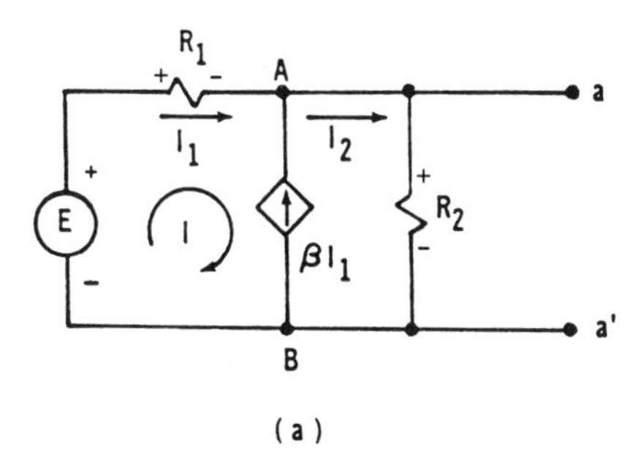

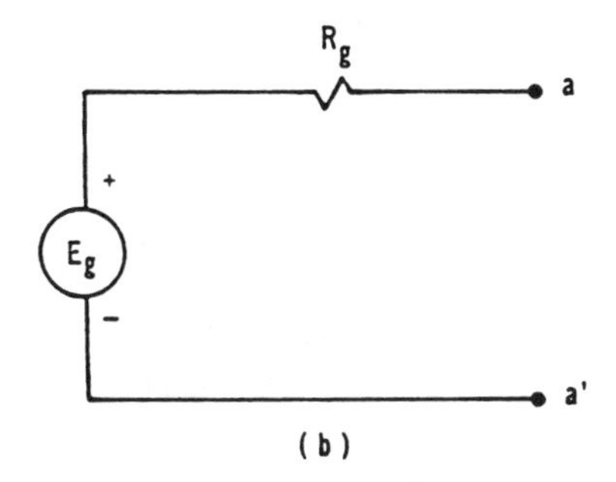

$$E_g = \frac{(1+\beta)R_2}{R_1 + (1+\beta)R_2} E$$

$$R_g = \frac{R_1 R_2}{R_1 + (1+\beta)R_2}$$

FIGURE 1.3 Thevenin equivalent circuit.

then

$$I_2 = \frac{E}{\left(\frac{R_1}{1+\beta} + R_2\right)}$$

The Thevenin Voltage is thus

$$E_g = I_2 R_2 = R_2 \frac{E}{\left(\frac{R_1}{1+\beta} + R_2\right)} = \frac{R_2(1+\beta)}{R_1 + R_2(1+\beta)} E \quad (1.13)$$

Let I_{sc} = the short circuit current of terminals aa′. In other words, I_{sc} is I_2 with aa′ shorted or $R_2 = 0$, i.e., from Equation (1.12),

$$I_{sc} = I_2\big]_{R_2=0} = (1+\beta)\frac{E}{R_1}$$

By Thevenin theory, and Equation (1.13),

$$R_g = \frac{E_g}{I_{sc}} = \frac{R_2(1+\beta)E}{R_1 + R_2(1+\beta)} \times \frac{R_1}{(1+\beta)E} = \frac{R_1 R_2}{R_1 + (1+\beta)R_2}$$

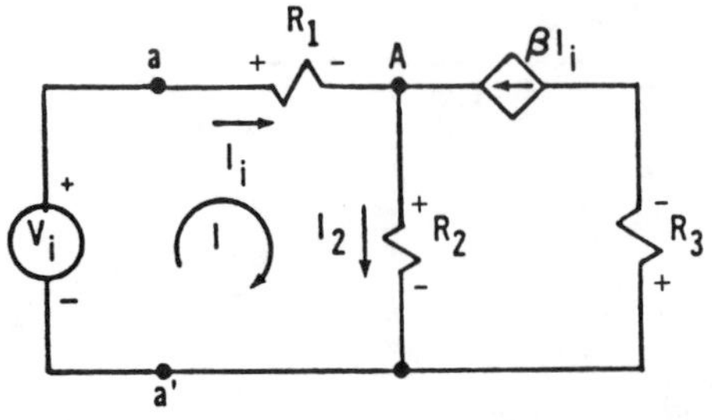

FIGURE 1.4 Circuit for solving input resistance.

The Thevenin's equivalent circuit is shown in Figure 1.3(b).

Comment: Since the network contains a dependent current source which cannot be open-circuited, the derivation of R_g is more complicated, and the ratio of E_g to I_{sc} is used for finding R_g. It is important to note that the Thevenin's equivalent resistance at aa′ is also known as the *output resistance* of the network at aa′. Often, a digital system designer needs to know, especially in interface circuit design, the input/output resistances or impedances of the given networks or black box, therefore, one would use this technique from time to time. The next example will show how the input resistance of a network can be found.

Example 4

Problem: Determine the input resistance at terminals aa′ of the network shown in Figure 1.4.

Solution:

Node A: $$I_2 = I_i + \beta I_i = (1+\beta)I_i \tag{1.14}$$

Loop I: $$V_i = I_i R_1 + I_2 R_2 \tag{1.15}$$

Substitute I_2 in Equation (1.14) into Equation (1.15) to get

$$V_i = I_i R_1 + I_i(1+\beta)R_2 = I_i\left[R_1 + (1+\beta)R_2\right] \tag{1.16}$$

By definition, the input resistance at terminal aa′, i.e., $R_i \triangleq \frac{V_i}{I_i}$, from Equation (1.16) is

$$R_i = \frac{V_i}{I_i} = R_1 + (1+\beta)R_2 \tag{1.17}$$

REFERENCES

1. Hayt, W. H., Jr. and Kemmerly, J. E., *Engineering Circuit Analysis*, 4th ed., McGraw-Hill, New York, 1987.
2. Fitzgerald, A. E., Higginbottham, D. E., and Grabel, A., *Basic Electrical Engineering*, 5th ed., McGraw-Hill, New York, 1981.

2 BASIC MODULES (BUILDING BLOCKS) FOR LINEAR SYSTEMS

The objective of this chapter is to present the fundamentals and characteristics of the most popular analog building blocks which are often used in digital system design. With emphasis on practical applications, unnecessary rigorous circuit analysis will not be presented, and approximation techniques will be used whenever applicable.

2.1. SINGLE-STAGE DIFFERENTIAL AMPLIFIER

The circuit shown in Figure 2.1(a) is a single-stage differential amplifier which often is the basic element of more complex circuitries, both analog, such as operational amplifier, and digital, such as ECL (Emitter-Coupled-Logic), and transmission line driver/receiver circuits. Due to its symmetrical circuit configuration, it has the desirable properties of low dc drift, high common mode rejection, and split-phase outputs. The small ac signal and dc signal analyses for the circuit are described as follows.

2.1.1. ac Small Signal Analysis

a. Common Mode Analysis

Due to integrated circuit technology, it is feasible to assume that the circuit shown in Figure 2.1(a) can be designed to have the following electrical properties: $R_{C1} = R_{C2} = R_C$; $Q_1 = Q_2$ or Q_1 and Q_2 have identical h-parameters. Figure 2.1(a) can be redrawn as shown in Figure 2.1(b). Since the left and right circuits are identical, or $V_{E1} = V_{E2}$, there is no current flow in the wire connecting the emitters of Q_1 and Q_2. Therefore the wire can be cut and the circuit can be split into two halves as shown in Figure 2.1(c). Figure 2.1(d), then, shows the small signal equivalent circuit of the left-hand side. For simplicity, let us assume that the h_{oe} and h_{re} of the transistors can be neglected; then the circuit can be reduced further as shown in Figure 2.1(e). Based on Figure 2.1(e), we have

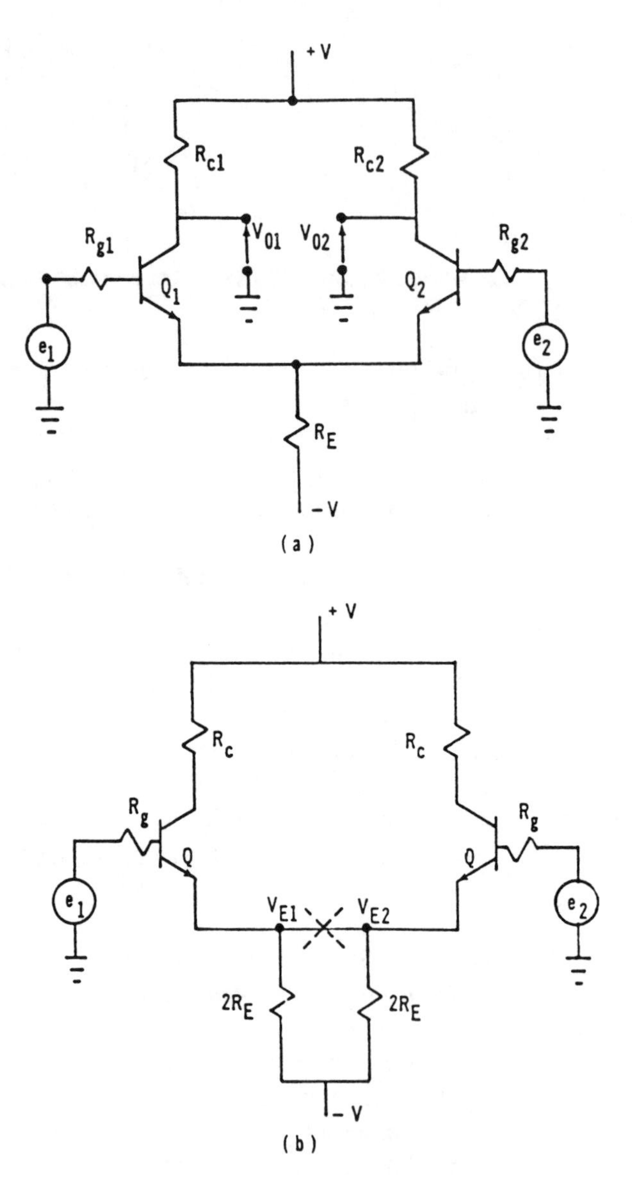

FIGURE 2.1 Single-stage differential amplifier.

$$e_1 = i_b(R_g + h_{ie}) + i_e 2R_E \tag{2.1}$$

$$V_{01} = -h_{fe} i_b R_C \tag{2.2}$$

$$i_e = (1 + h_{fe}) i_b \tag{2.3}$$

From Equations (2.1) to (2.3),

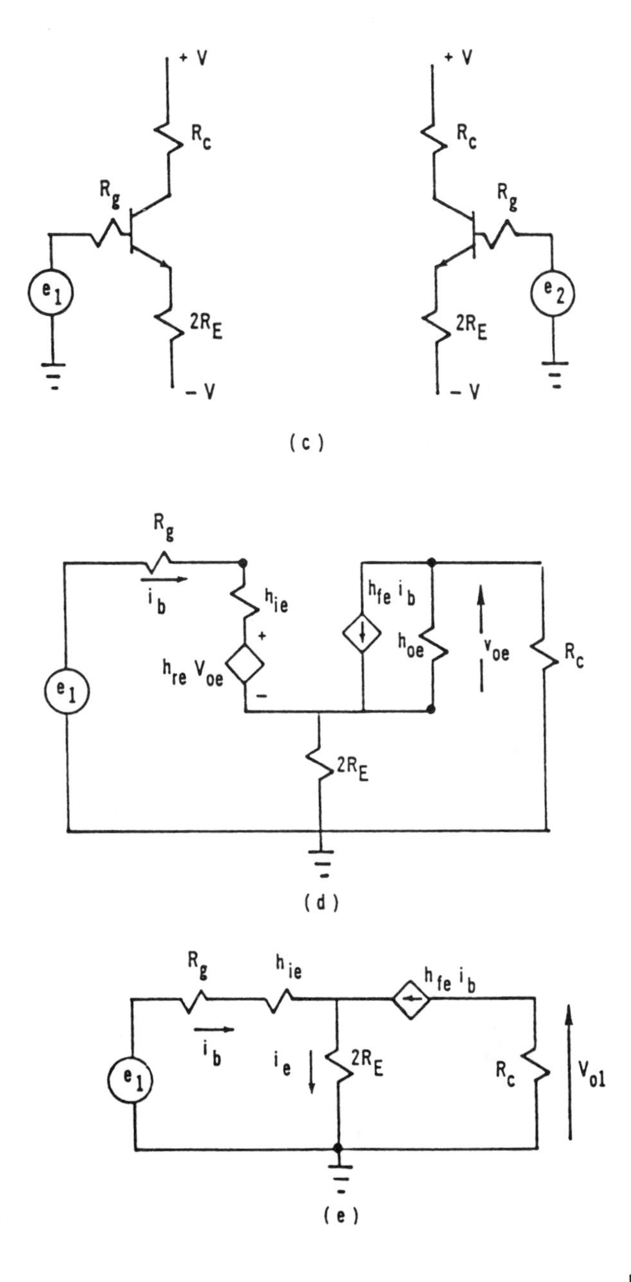

FIGURE 2.1 continued. Single-stage differential amplifier.

$$\frac{V_{01}}{e_1} = -\frac{h_{fe}R_C}{(R_g + h_{ie}) + (1 + h_{fe})2R_E} = -\frac{\dfrac{h_{fe}}{1 + h_{fe}}R_C}{2R_E + \dfrac{R_g + h_{ie}}{1 + h_{fe}}}$$

Since normally $h_{fe} >> 1$, then

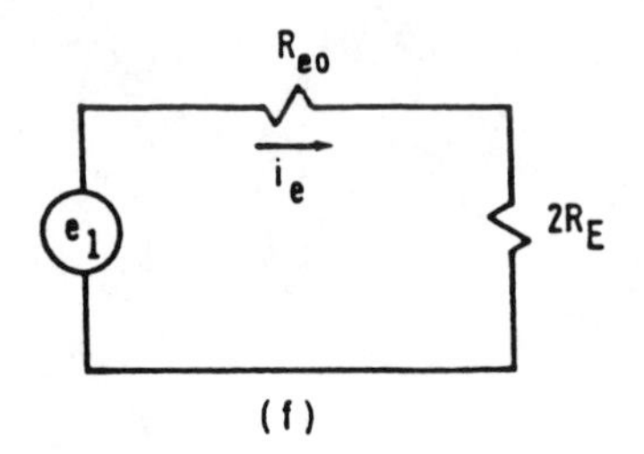

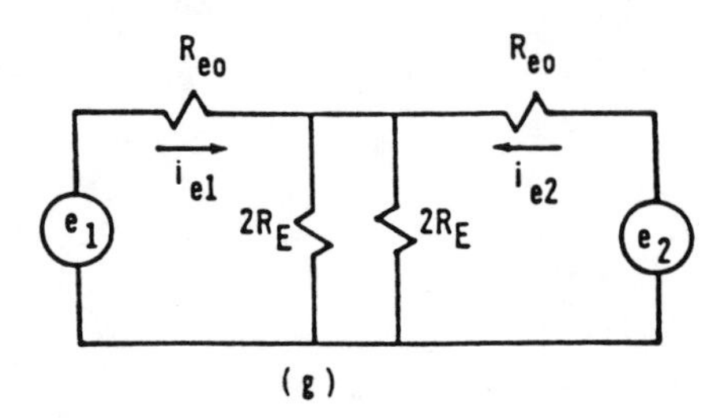

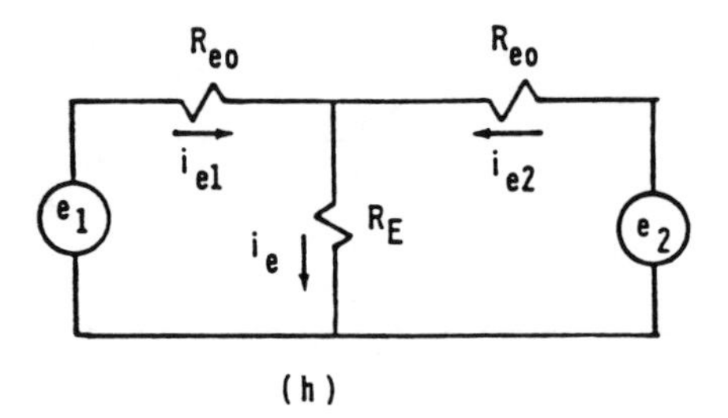

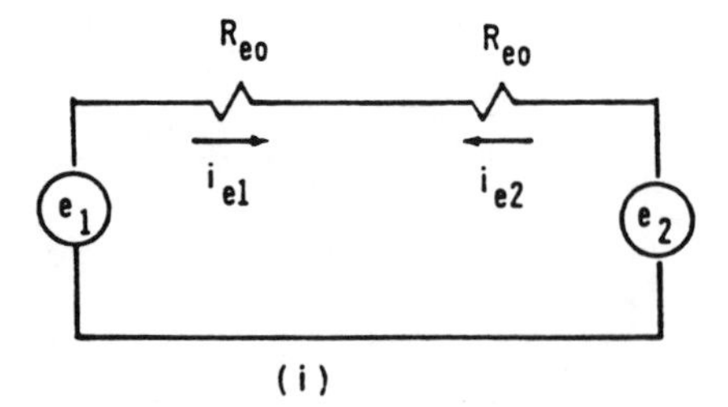

FIGURE 2.1 continued. Single-stage differential amplifier.

$$\frac{h_{fe}}{1 + h_{fe}} \simeq 1$$

So we have

$$\frac{V_{01}}{e_1} = -\frac{R_C}{2R_E + R_{eo}} \tag{2.4}$$

where R_{eo} is the equivalent output resistance at the emitter, i.e.,

$$R_{eo} \triangleq \frac{R_g + h_{ie}}{1 + h_{fe}} \tag{2.5}$$

where "$\triangleq$" means "defined as". Similarly,

$$\frac{V_{02}}{e_2} = -\frac{R_C}{2R_E + R_{eo}} \tag{2.6}$$

From Equations (2.4) and (2.6),

$$V_{01} + V_{02} = -\frac{R_C}{2R_E + R_{eo}}(e_1 + e_2)$$

$$\frac{V_{01} + V_{02}}{e_1 + e_2} = -\frac{R_C}{2R_E + R_{eo}} \tag{2.7}$$

Let us define *common mode gain* as

$$CMG \triangleq \frac{\frac{1}{2}(V_{01} + V_{02})}{\frac{1}{2}(e_1 + e_2)}$$

Then we have

$$CMG = -\frac{R_C}{2R_E + R_{eo}} \tag{2.8}$$

b. Differential Mode Analysis

From Equations (2.1), (2.3), and (2.5),

$$e_1 = i_b(R_g + h_{ie}) + i_e 2R_E$$

$$= i_e\left[\frac{R_g + h_{ie}}{1 + h_{fe}} + 2R_E\right]$$

$$= i_e\left[R_{eo} + 2R_E\right] \tag{2.9}$$

According to Equation (2.9), an equivalent circuit can be shown in Figure 2.1(f). Similarly,

$$e_2 = i_e\left[R_{eo} + 2R_E\right] \tag{2.10}$$

Figures 2.1(g) and (h) depict the composite equivalent circuit based on Equations (2.9) and (2.10).

R_E is normally very large. In fact, most of the differential amplifiers replace R_E by a current generator supplying I_e, which has high output resistance, so that it is reasonable to assume that

$$R_E >> R_{eo},\ i_{e1} >> i_e,\ i_{e2} >> i_e$$

or

$$i_{e1} + i_{e2} = i_e \simeq 0$$

which gives $i_{e1} = -i_{e2}$. If an equivalent circuit is shown in Figure 2.1(i), then

$$e_1 - i_{e1}R_{eo} = e_2 - i_{e2}R_{eo}$$

$$e_1 - e_2 = (i_{e1} - i_{e2})R_{eo} = 2i_{e1}R_{eo} \tag{2.11}$$

But from Equations (2.2) and (2.3), by eliminating i_b, we have

$$V_{01} = -h_{fb}R_C i_{e1} \simeq -R_C i_{e1} \tag{2.12}$$

$$V_{02} = -h_{fb}R_C i_{e2} = h_{fb}R_C i_{e1} \simeq R_C i_{e1} \tag{2.13}$$

where

$$h_{fb} = \frac{h_{fe}}{1 + h_{fe}} \simeq 1$$

By subtracting Equation (2.13) from Equation (2.12), we get

$$V_{01} - V_{02} = -2h_{fb}i_{e1}R_C \simeq -2i_{e1}R_C \tag{2.14}$$

Then, dividing Equation (2.14) by Equation (2.11),

$$\frac{V_{01} - V_{02}}{e_1 - e_2} \simeq -\frac{2i_{e1}R_C}{2i_{e1}R_{eo}} = -\frac{R_C}{R_{eo}} \tag{2.15}$$

Let us define the *differential mode gain*:

$$\text{DMG} \triangleq \frac{\frac{1}{2}(V_{01} - V_{02})}{\frac{1}{2}(e_1 - e_2)} \simeq -\frac{R_C}{R_{eo}} \tag{2.16}$$

For single-end output, we can have Equation (2.12) divided by Equation (2.11):

$$\frac{V_{01}}{e_1 - e_2} = -\frac{h_{fb}R_C}{2R_{eo}} \simeq -\frac{R_C}{2R_{eo}} \tag{2.17}$$

and Equation (2.13) divided by Equation (2.11):

$$\frac{V_{02}}{e_1 - e_2} = \frac{h_{fb}R_C}{2R_{eo}} \simeq \frac{R_C}{2R_{eo}} \tag{2.18}$$

Equations (2.17) and (2.18) are defined as *single-end* to *differential-end voltage gain*. Note that there is phase inversion in Equation (2.17), while there is no inversion in Equation (2.18). The circuit, therefore, has the phase-splitting property. Equations (2.17), (2.18), and (2.16) reveal that the voltage gain for the single-end output is one half of the DMG.

c. Common Mode Rejection Property

Let us define *common mode rejection,*

$$\text{CMR} \triangleq \frac{\text{DIFFERENTIAL MODE GAIN}}{\text{COMMON MODE GAIN}} = \frac{\text{DMG}}{\text{CMG}}$$

From Equations (2.16) and (2.8) we have

$$\begin{aligned} \text{CMR} &\simeq \left(-\frac{R_C}{R_{eo}}\right) \div \left(-\frac{R_C}{2R_E + R_{eo}}\right) \\ &= 1 + \frac{2R_E}{R_{eo}} \end{aligned} \tag{2.19}$$

From Equation (2.19), we see that as R_E increases, the CMR will increase accordingly. Normally, R_E is the output resistance of a constant current generator and can be assumed to have a large value. Therefore, the CMR of a differential amplifier is normally very high. In practice, as the amplifier is being used with the differential mode configuration, high CMR is a very desirable feature. As shown in Figure 2.2, e_s is the input signal to be amplified. The input lines 1 and 2, however, would also act as antennas picking up noise, e_1 and e_2. As a result, e_1 and e_2 will be amplified by the CMG of Equation (2.8), while e_s will be amplified by the DMG of Equation (2.16). It is evident that the higher the CMR of Equation (2.19), the better the signal-to-noise ratio of the amplifier.

The CMR of a given amplifier can be experimentally determined. Figure 2.3 shows the circuit diagram for determining CMR, by respectively measuring the ratios

$$\frac{V_{do}}{e_s} = \text{DMG}$$

and

$$\frac{\frac{1}{2}(V_{01} + V_{02})}{e_{CM}} = \text{CMG}$$

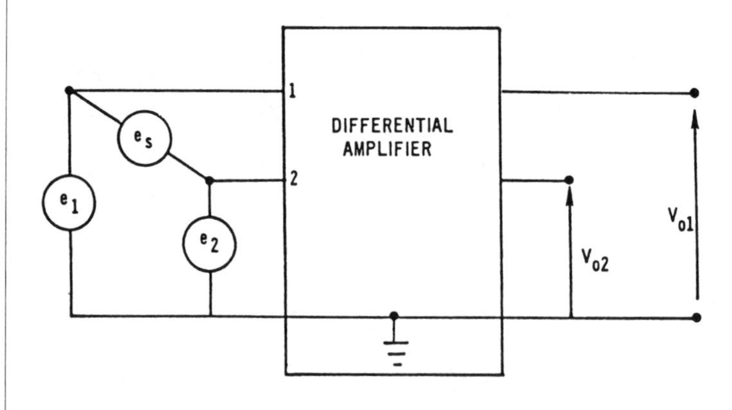

FIGURE 2.2 Common mode rejection.

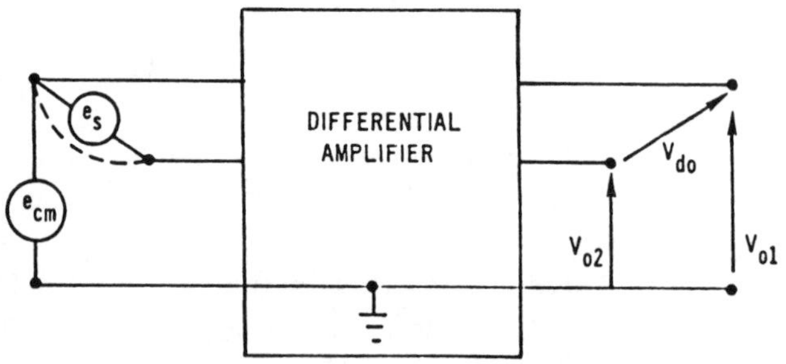

FIGURE 2.3 Common mode rejection measurement.

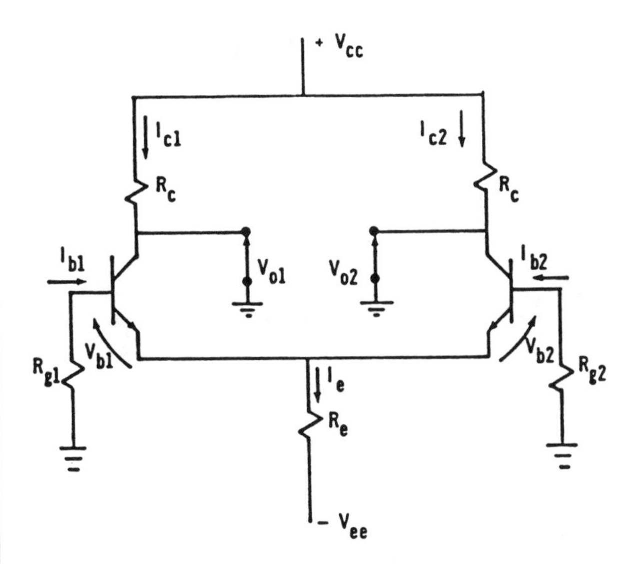

FIGURE 2.4 dc analysis.

(with e_s = o, or short-circuited as the dotted line shown), where e_s is a signal generator used as the differential input, while e_{CM} is a signal generator used to generate a common mode input signal.

2.1.2. dc Analysis

In reference to Figure 2.4,

$$I_{b1}R_{g1} + V_{b1} = I_{b2}R_{g2} + V_{b2} \tag{2.20}$$

$$V_{01} = V_{CC} - I_{C1}R_C$$

$$V_{02} = V_{CC} - I_{C2}R_C$$

$$V_{01} - V_{02} = -I_{C1}R_C + I_{C2}R_C \tag{2.21}$$

Let

$$\beta_1 = \frac{I_{C1}}{I_{b1}},\ \beta_2 = \frac{I_{C2}}{I_{b2}}$$

Then Equation (2.21) becomes

$$\begin{aligned} V_{01} - V_{02} &= -\beta_1 I_{b1} R_C + \beta_2 I_{b2} R_C \\ &= R_C \beta_1 I_{b1} \left(\frac{\beta_2 I_{b2}}{\beta_1 I_{b2}} - 1 \right) \end{aligned} \tag{2.22}$$

From Equation (2.20),

$$I_{b1}R_{g1}\left(\frac{I_{b2}R_{g2}}{I_{b1}R_{g1}}-1\right)=\left(V_{b1}-V_{b2}\right) \tag{2.23}$$

and from Equations (2.22) and (2.23), we have

$$V_{01}-V_{02}=R_C\beta_1\left(\frac{\beta_2 I_{b2}}{\beta_1 I_{b1}}-1\right)\frac{\left(V_{b1}-V_{b2}\right)}{\left(\frac{I_{b2}R_{g2}}{I_{b1}R_{g1}}-1\right)R_{g1}} \tag{2.24}$$

It is desired that

$$V_{01}=V_{02}$$

if the amplifier is at its *quiescent condition*, or zero signal input. Equation (2.24) shows that parameters which control the equality of $V_{01} = V_{02}$.

2.1.3. Current Mirror Circuit

Recall that Equation (2.11) was derived based on the assumption that R_E in Figure 2.4 is very large. Since we know that a current source has the property of high output resistance, the R_E in a differential amplifier is normally replaced by a constant current generator. Furthermore, one of the essential requirements for a linear amplifier is the stability of its biasing current at its quiescent or operating point. Figure 2.5 depicts a popular circuit known as the *current mirror*, which is usually used in the integrated circuit (IC) differential amplifier. This circuit can be used as a constant current generator, a biasing circuit, or an active load impedance for a transistor. The analysis follows.

Referring to Figure 2.5, we have

$$I_R=\alpha_1 I_{sat}\epsilon^{V_{BE1}/V_T}$$

$$I_L=\alpha_2 I_{sat}\epsilon^{V_{BE2}/V_T}$$

where α_1, α_2 is the Emitter-Collector current gain of Q_1 and Q_2, respectively, I_{sat} is the saturation current, V_{BE1}, V_{BE2} is the Base-Emitter voltage of Q_1 and Q_2, respectively, and V_T is the Temperature Equivalent voltage. Since we may assume that $\alpha_1 = \alpha_2$,

$$\frac{I_R}{I_L} = \epsilon^{(V_{BE1} - V_{BE2})V_T}$$

But $V_{BE1} = V_{BE2}$, so we have

$$I_R = I_L \tag{2.25}$$

Notice that although Q_1 is a transistor, it is connected as a diode, and

$$I_R = \frac{V_{EE} - V_{BE1}}{R}$$

If V_{EE}, V_{BE}, and R can be maintained with constant values, then I_R would be constant. Since $I_L = I_R$, Q_2 becomes a constant current generator which is not sensitive to R_L or V_{CC}. Thus, I_R is often called the *reference current*. Here, R_L need not be a resistor; it can be an equivalent load resistance of a network or any circuit connected to it. For example, R_L can be replaced by an NPN transistor collector-base terminal in such a way that the Q_2 circuitry becomes the active load (with very high dynamic resistance) of the NPN transistor. Another way of looking at this same circuit is that the Q_2 is biased at $I_C = I_R$ with high stability.

2.2. OPERATIONAL AMPLIFIER

The operational amplifier is a multiple-stage direct-coupled amplifier. Its first stage is generally a differential amplifier and the last stage is a

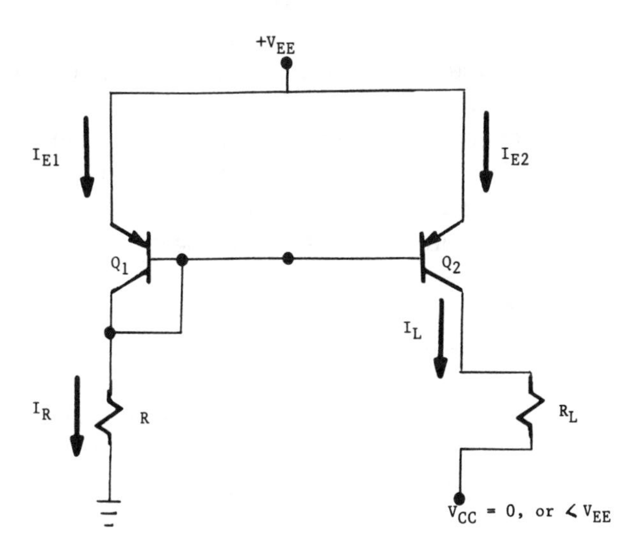

FIGURE 2.5 Current mirror circuit.

single-end buffer amplifier. One may consider an operational amplifier as one which is a differential amplifier followed by a high-gain dc amplifier. The operational amplifier has been the essential building block of an analog or linear computing system for decades. Since the introduction of large-scale integrated circuit (LSI) technology, the unit cost of some general purpose IC operational amplifiers has been reduced to almost a negligible level with respect to that of the system where operational amplifiers are being used as system components. A digital system designer can now consider an operational amplifier as a system component equivalent to a NAND gate. One may use it wherever necessary without being cost conscious. What follows are the basic principles of operation and practical considerations of the operational amplifier. Rigorous circuit analysis will not be covered; approximation techniques will be employed whenever feasible.

2.2.1. Simplified Schematic of a Typical Operational Amplifier

Figure 2.6 shows a simplified schematic diagram of the PM747 operational amplifier. In view of the diagram, one can readily identify that the current mirror circuits described in Section 2.1.3 have been employed many times here as current sources or active loads. Notice that the two input pins labeled as "–IN" and "+IN", respectively, constitute the inputs of the first-stage differential amplifier described in Section 2.1. In addition, as the name implies, the pin labeled "OUTPUT" delivers the amplified input signal. Although all IC operational amplifiers have a fairly complex internal structure, fortunately, as a digital system designer, one need not be concerned with the detailed internal structure. Mostly, we can take a "black box" approach by learning how to use the primary parameters or specifications provided by the manufacturer of the operational amplifiers in question. In other words, an operational amplifier can be considered as a system component as other IC chips, be it digital or analog, as long as we understand how it functions internally. In the following subsections 2.2.2 to 2.2.5, we will present information about operational amplifiers from the system designer's viewpoint.

2.2.2. Equivalent Circuit of an Operational Amplifier

Figure 2.7 depicts a simplified equivalent circuit of an operational amplifier which has three essential terminals for processing signals, i.e., two differential inputs labeled respectively with minus and plus signs, and one output labeled with V_o. Z_{in} is the input impedance, Z_o the output impedance, and A is the amplifier voltage gain. As the polar-

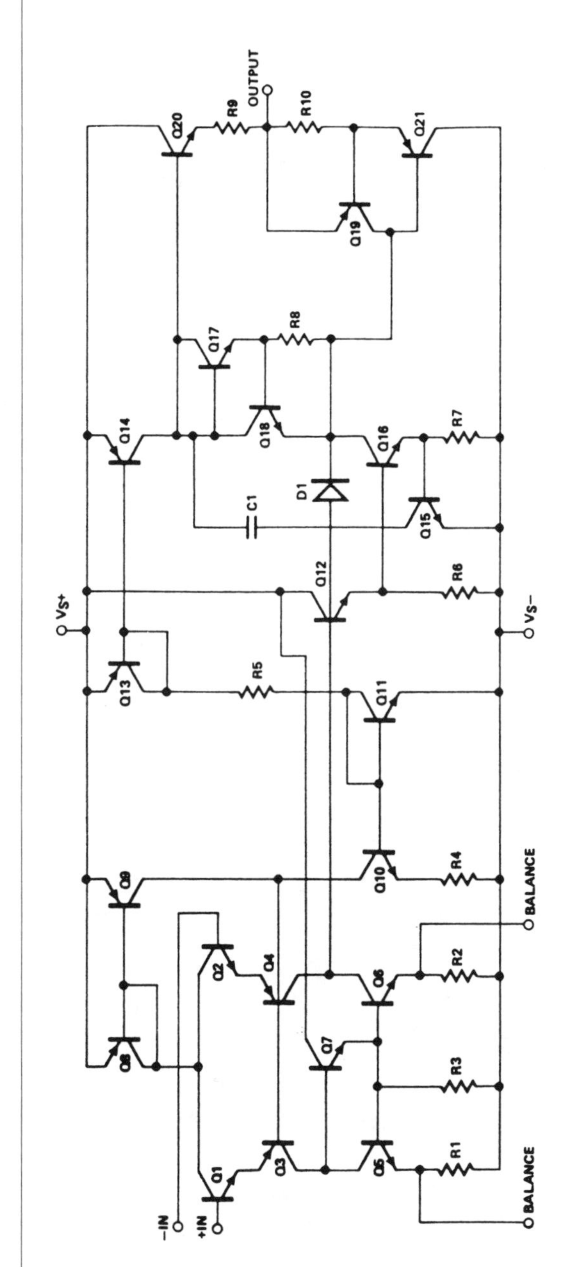

FIGURE 2.6 A simplified schematic of PM747 operational amplifier (courtesy of Precision Monolithics, Inc.).

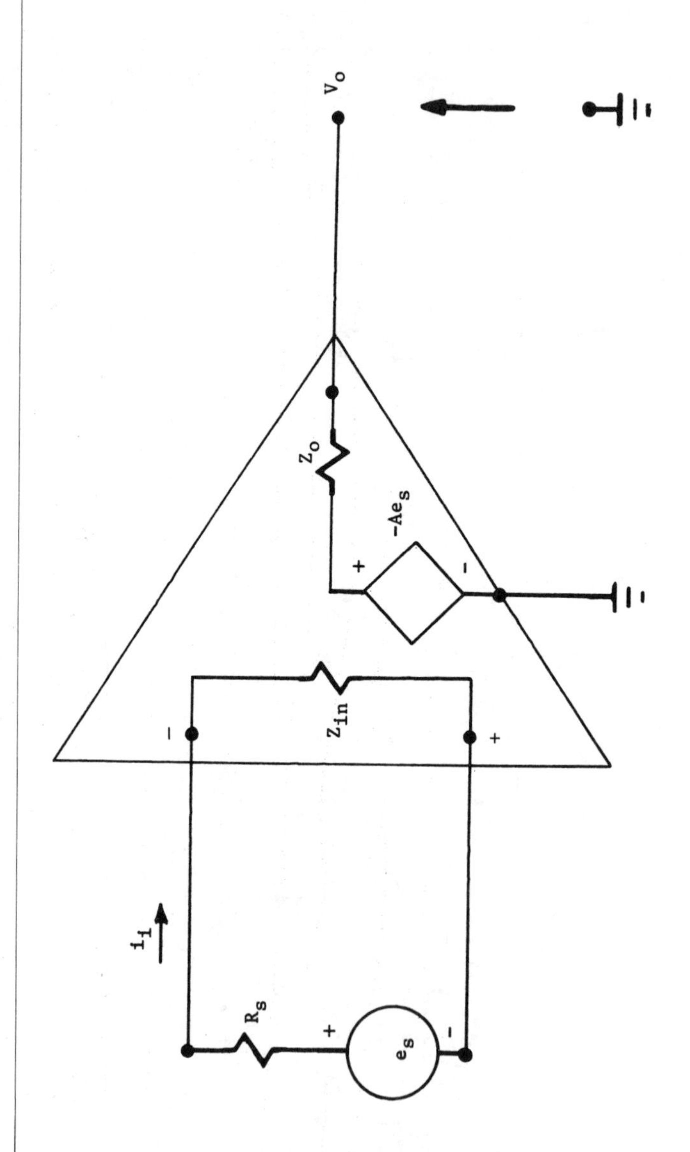

FIGURE 2.7 Equivalent circuit of an operational amplifier.

ity assigned for the input signal voltage e_s, the output voltage will be

$$V_o = -Ae_s$$

Note that all these parameters are those of the operational amplifier to which no external circuitry is added. Thus, they are called open-loop Z_{in}, Z_o, and A, respectively. The ideal model of an operational amplifier, however, is normally defined as follows:

$$Z_{in} = \infty, \quad Z_o = 0$$

$$A = \infty, \quad \text{but } V_0 = 0 \text{ when } e_s = 0$$

where A is called open-loop voltage gain,

$$\begin{aligned} \text{Bandwidth} &= \infty \\ \text{Offset current} &= 0 \\ \text{Offset voltage} &= 0 \\ \text{Common mode rejection} &= \infty \end{aligned}$$

In practice, the ideal model often is evidently oversimplified for some circuit analysis and design. Depending on the nature of the application, for instance, if the input, e_s, is a low-level dc signal, the offset current and offset voltage effect on the output voltage cannot be neglected. In light of Equation (2.24), we see that a mismatch of the base-emitter junction voltages V_{B1} and V_{B2}, as well as of $\beta_1 I_{B1}$ and β_{2IB2}, would cause $V_{01} - V_{02} \neq 0$. The former is known as *offset voltage* and the latter is known as *offset current.* Fortunately, in most applications, such as ac coupled circuit configuration, or large dc signal, these problems are not significant. For applications in low-level dc analog signal or analog integration operation, however, these problems can be severe and special purpose operational amplifiers, or low-drift or chopper-stabilized instrumentation amplifiers should be employed.

In general, as a system designer, one only needs to be concerned with the first level or primary parameters such as Z_{in}, Z_o, A, and frequency response bandwidth. As a rule of thumb, we may assume that Z_{in} is on the order of hundreds of thousand ohms or higher, Z_o is less than 50 Ω, and A is greater than 100,000 or 100 dB. The exact values for all these parameters are usually provided by the manufacturer in their data sheets, or we may determine them experimentally. Before we leave this topic, it is important to point out the meaning of the input pins that are labeled with "+" and "–". They do not indicate the input signal voltage polarity. Instead, they merely refer to the relative phase between the input and output signals. For example, if the "–" input pin is tied to ground, and the input signal is applied between the input terminal pair, then the output signal is *in phase* with the input; however, if the "+" input pin is grounded, then the output is 180° *out of phase* with the input. This type of configuration is called *single-end input* and *single-end output*. In many cases, neither one of the input terminals is tied to ground. In that case, if we assign the "–" pin as the positive polarity of the input signal, then the output signal is 180° out of phase with the input, otherwise, the output signal is in phase with the input. When none of the input terminals is tied to ground, as in this case, the circuit configuration is called *differential-end input* with *single-end output*.

2.2.3. The Four Useful and Essential Feedback Circuit Configurations

In most applications, the operational amplifier is used for processing analog or continuous signals. An operational amplifier is simply a "raw"component which needs to be externally modified or tailored to meet certain requirements. Descriptions of the four most popular and useful feedback circuit configurations and their unique features follow.

a. Shunt-Shunt Feedback (Inverted Voltage Amplifier)

The circuit configuration is shown in Figure 2.8(a). It is important to point out that the operational amplifier needs external resistor networks to serve as a bias network to supply bias currents regardless of the circuit configuration. Here, R_1, R_2, and R_b are used for biasing purposes, but R_1 and R_2, in addition, determine the voltage gain of the network. For maintaining the circuit symmetry of the differential inputs and minimizing the drifting effect, i.e., maintaining the condition that output voltage equals zero with zero input, the following conditions must be met. If the coupling capacitor C is not used and there is a direct current path from R_1 to ground through the input signal source, then

$$R_b = \frac{R_1 R_2}{R_1 + R_2}$$

otherwise

$$R_b = R_2$$

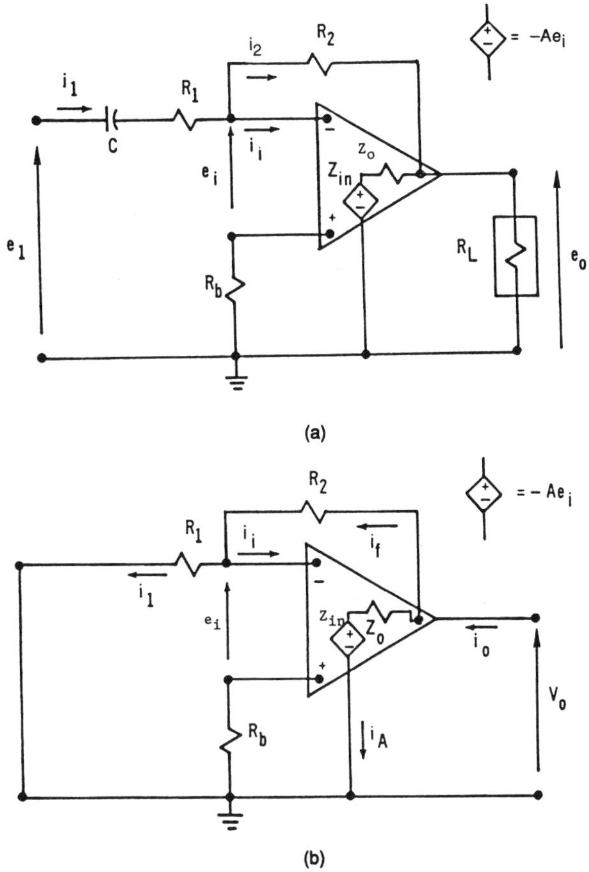

(a)

(b)

FIGURE 2.8 (a) ac coupled shunt-shunt feedback. (b) Output impedance of shunt-shunt feedback.

For convenience, but without sacrificing the generality, we shall consider the capacitor coupling as a special case, and therefore, the analysis from here will be based on the circuitry without the coupling capacitor.

Voltage Gain of the Feedback Network: A_{fb}

Referring to Figure 2.8(a), we have

$$i_1 = i_i + i_2$$

Since e_i and e_o are finite, and A, relatively speaking, is very high, then

$$e_i = \frac{e_o}{A} \simeq \frac{e_o}{\infty} = 0$$

where A is the open-loop voltage gain of the operational amplifier. Since

$$i_i = \frac{e_i}{z_{in}} \simeq \frac{0}{z_{in}} = 0$$

then

$$i_1 = i_2$$

But,

$$e_o = e_1 - i_1R_1 - i_2R_2$$

thus

$$e_o = e_1 - i_1(R_1 + R_2) \tag{2.26}$$

From the input loop,

$$e_1 = i_1R_1 + e_i + i_iR_b = i_1R_1 + e_i$$

$$i_1 = \frac{1}{R_1}(e_1 - e_i) = \frac{1}{R_1}\left(e_1 - \frac{e_o}{-A}\right) = \frac{1}{R_1}\left(e_1 + \frac{e_o}{A}\right) \tag{2.27}$$

Substituting Equation (2.27) into Equation (2.26),

$$e_o = e_1 - \frac{1}{R_1}\left(e_1 + \frac{e_o}{A}\right)(R_1 + R_2)$$

$$e_o\left(1 + \frac{R_1 + R_2}{R_1A}\right) = e_1\left(1 - \frac{R_1 + R_2}{R_1}\right) = e_1\left(-\frac{R_2}{R_1}\right)$$

$$\frac{e_o}{e_1} = \frac{-R_2/R_1}{1 + \frac{1 + R_2/R_1}{A}} = -\frac{R_2}{R_1} \quad \text{if } A >> \left(1 + \frac{R_2}{R_1}\right)$$

Finally,

$$A_{fb} = \frac{e_o}{e_1} = -\frac{R_2}{R_1}$$

where fb stands for feedback configuration.

Input Impedance of the Feedback Network: Z_{ifb}

Since

$$Z_{ifb} \overset{\Delta}{=} \frac{e_1}{i_1} = \frac{i_1 R_1 + e_i}{i_1}$$

but

$$e_i \simeq 0, \text{ or } e_i << i_1 R_1$$

we have

$$Z_{ifb} = R_1$$

Output Impedance of the Feedback Network: Z_{ofb}

Referring to Figure 2.8(b), and based on the definition of output impedance, that is, with the input terminals short-circuited and a voltage source V_o applied at the output terminals, Z_{ofb} is defined as V_o/i_o. We have

$$i_o = i_f + i_A \tag{2.28}$$

$$i_f = i_1 + i_i = i_1$$

$$e_i = i_f R_1 - i_i R_b \simeq i_f R_1 \quad (\text{Assume } i_i \simeq 0, \text{ or } i_i << i_f) \tag{2.29}$$

$$V_o = i_f R_2 + e_i = i_f R_2 + i_f R_1 = i_f (R_2 + R_1) \tag{2.30}$$

$$V_o = i_A Z_o + (-Ae_i) \tag{2.31}$$

Substituting Equations (2.28) and (2.29) into Equation (2.31),

$$V_o = (i_o - i_f) Z_o - A i_f R_1$$

$$= i_o Z_o - i_f (Z_o + AR_1) = i_o Z_o - i_f A(\frac{Z_o}{A} + R_1) \simeq i_o Z_o - i_f AR_1 \tag{2.32}$$

since $\frac{Z_o}{A} << R_1$. Substituting Equation (2.30) into Equation (2.32),

$$V_o = i_o Z_o - \left(\frac{V_o}{R_2 + R_1}\right) AR_1$$

so

$$V_o\left[1 + A\frac{R_1}{R_2 + R_1}\right] = i_o Z_o$$

Thus,

$$Z_{ofb} = \frac{V_o}{i_o} = \frac{Z_o}{1 + A\left(\dfrac{R_1}{R_1 + R_2}\right)}$$

b. Shunt-Series Feedback (Noninverted Voltage Amplifier)

Figure 2.9(a) shows the circuit configuration.

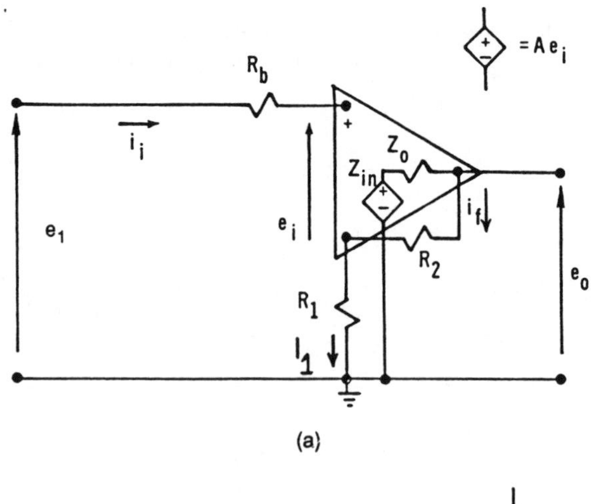

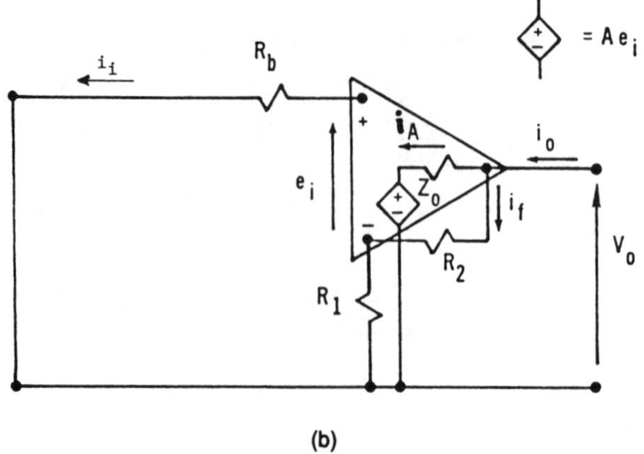

FIGURE 2.9 (a) Shunt-series feedback. (b) Output impedance of shunt-series feedback.

Voltage Gain of the Feedback Network: A_{fb}

$$e_1 = i_i R_b + e_i + i_1 R_1$$

where

$$R_b = R_1 R_2/(R_1 + R_2)$$

$$i_1 = (i_i + i_f)$$

For the same reason as described in Section 2.2.3.a,

$$i_i \simeq 0,\ e_i \simeq 0,\ \text{or } i_f >> i_i$$

we have

$$i_f = i_1$$

$$e_1 = i_f R_1 \tag{2.33}$$

$$e_o = i_f R_2 + (i_i + i_f) R_1 = i_f (R_1 + R_2) \tag{2.34}$$

Then, take Equation (2.34) divided by Equation (2.33),

$$\frac{e_o}{e_1} = \frac{R_1 + R_2}{R_1} = 1 + \frac{R_2}{R_1}$$

$$A_{fb} = \frac{e_o}{e_1} = 1 + \frac{R_2}{R_1}$$

Input Impedance of the Feedback Network: Z_{ifb}

$$e_1 = i_i(R_b + Z_{in}) + (i_i + i_f)R_1$$

$$= i_i(Z_{in} + R_b + R_1) + i_f R_1 \tag{2.35}$$

$$e_o = i_f(R_1 + R_2)$$

$$Ae_i = e_o + i_f Z_o = i_f(R_1 + R_2 + Z_o)$$

$$e_i = i_i Z_{in}$$

$$Ai_i Z_{in} = i_f(R_1 + R_2 + Z_o) \tag{2.36}$$

Substituting Equation (2.36) into Equation (2.35),

$$e_1 = i_i(Z_{in} + R_b + R_1) + R_1 \frac{Ai_iZ_{in}}{R_1 + R_2 + Z_o}$$

$$= i_i\left[Z_{in} + R_b + R_1 + R_1 \frac{AZ_{in}}{R_1 + R_2 + Z_o}\right]$$

$$= i_i\left[Z_{in}\left(1 + A \frac{1}{R_1 + R_2 + Z_o}\right) + R_b + R_1\right]$$

Since $Z_0 << (R_1 + R_2)$,

$$Z_{in}\left(1 + A \frac{R_1}{R_1 + R_2 + Z_o}\right) >> (R_b + R_1)$$

So we have

$$e_1 \simeq i_iZ_{in}\left(1 + A \frac{R_1}{R_1 + R_2}\right)$$

$$\frac{e_1}{i_i} = Z_{in}\left(1 + A \frac{R_1}{R_1 + R_2}\right)$$

and

$$Z_{ifb} = Z_{in}\left(1 + A \frac{R_1}{R_1 + R_2}\right)$$

Output Impedance of the Feedback Network: Z_{ofb}

Referring to Figure 2.9(b),

$$i_o = i_f + i_A \tag{2.37}$$

$$V_o = Ae_i + i_AZ_o \tag{2.38}$$

$$V_o = i_f(R_2 + R_1) \tag{2.39}$$

$$e_i = -i_fR_1, \text{ since } |i_i| << |i_f| \tag{2.40}$$

Substituting Equations (2.37) and (2.40) into Equation (2.38), we get

$$V_o = -Ai_f R_1 + (i_o - i_f)Z_o$$

$$= i_o Z_o - i_f(Z_o + AR_1) \quad (2.41)$$

Then, substituting Equation (2.39) into Equation (2.41),

$$V_o = i_o Z_o - \frac{V_o}{R_1 + R_2}(Z_o + AR_1)$$

$$= i_o Z_o - \frac{V_o}{R_1 + R_2}(AR_1), \text{ since } Z_o << AR_1$$

$$V_o\left(1 + A\frac{R_1}{R_1 + R_2}\right) = i_o Z_o$$

$$\frac{V_o}{i_o} = \frac{Z_o}{1 + A\dfrac{R_1}{R_1 + R_2}}$$

So,

$$Z_{ofb} \triangleq \frac{V_o}{i_o} = \frac{Z_o}{1 + A\dfrac{R_1}{R_1 + R_2}}$$

c. Series-Series Feedback (Voltage to Current Converter)

Figure 2.10(a) shows the series-series feedback circuitry. As the name implies, the feedback element R_1 is a series element in the input as well as output loop. Here, R_L represents the load resistance and R_b the bias resistor, whose value should be equal to the parallel resistance of R_L and R_1.

Voltage Gain of the Feedback Network: $A_{fb} \triangleq e_L/e_1$

In Figure 2.10(a), we have

$$e_1 = i_i R_b + e_i + i_1 R_1$$

$$i_1 = i_L + i_i \simeq i_L \text{ for } i_L >> i_i$$

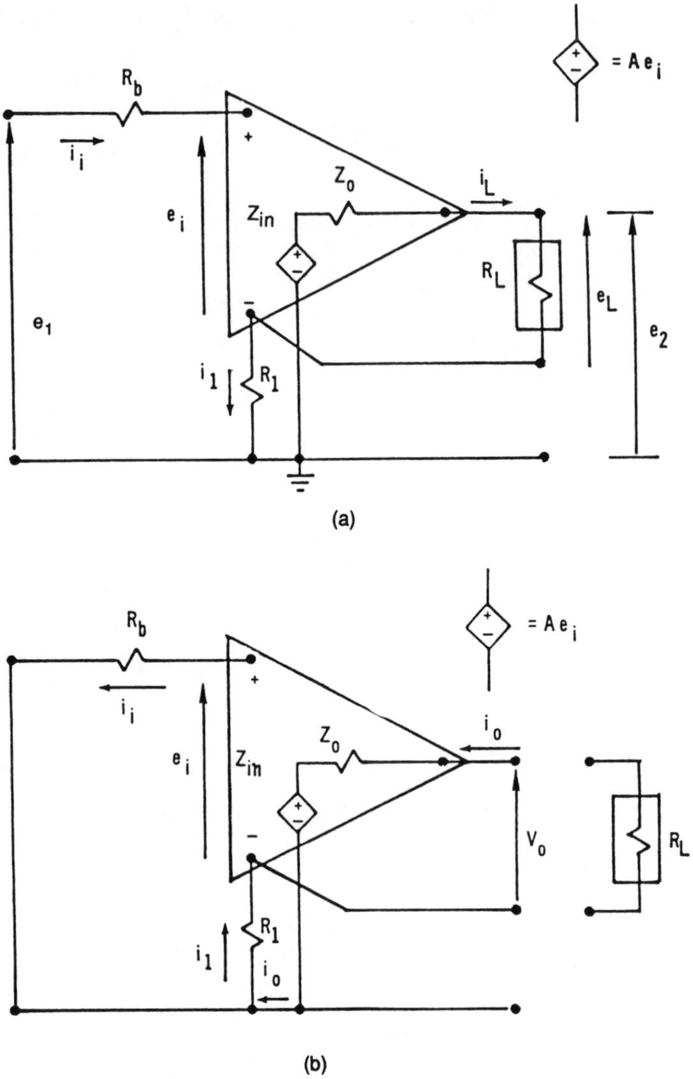

(a)

(b)

FIGURE 2.10 (a) Series-series feedback. (b) Output impedance of series-series feedback.

Thus,

$$e_1 \simeq i_L R_1$$

since

$$e_i \simeq 0$$

And,

$$\frac{e_2}{e_1} \simeq \frac{i_L(R_L + R_1)}{i_L R_1} = 1 + \frac{R_L}{R_1} \tag{2.42}$$

But,

$$e_L = i_L R_L \tag{2.43}$$

$$e_L = e_2 - i_L R_1$$

Thus,

$$e_L = e_2 - \frac{e_L}{R_L} R_1$$

and

$$e_2 = e_L\left(1 + \frac{R_1}{R_L}\right) \tag{2.44}$$

Substituting Equation (2.44) into Equation (2.42), then

$$A_{fb} = \frac{e_L}{e_1} = \frac{\left(1 + \frac{R_L}{R_1}\right)}{\left(1 + \frac{R_1}{R_L}\right)} = \frac{R_L}{R_1}$$

If we replace e_L according to Equation (2.43), we have

$$\frac{i_L}{e_1} = \frac{1}{R_1} \quad \text{or} \quad e_1 = R_1 i_L$$

The last equation constitutes the voltage (e_1) to current (i_L) conversion. It is important to point out that in this configuration, neither terminals of the load, R_L, can be tied directly to ground. That is, the load must be floated from ground.

Input Impedance of the Feedback Circuit: Z_{ifb}

Since the input circuit configuration is identical with respect to the input loop of shunt-series feedback configuration with $R_2 = R_L$, we have

$$Z_{ifb} = Z_{in}\left(1 + A\frac{R_1}{R_1 + R_L}\right)$$

Output Impedance of the Feedback Circuit: Z_{ofb}
In reference to Figure 2.10(b), since

$$R_b < R_1 << Z_{in}$$

and

$$i_1 = \frac{Z_{in} + R_b}{Z_{in} + R_b + R_1} i_o$$

we have

$$i_1 \simeq i_o \quad \text{or} \quad i_i << i_1$$

But,

$$V_o = Ae_i + i_oZ_o + i_1R_1 \tag{2.45}$$

and

$$e_i = i_iR_b + i_1R_1 \simeq i_1R_1 = i_oR_1 \tag{2.46}$$

Substituting Equation (2.46) into Equation (2.45),

$$\begin{aligned} V_o &= Ai_oR_1 + i_oZ_o + i_oR_1 \\ &= I_o[Z_o + R_1(1 + A)] \\ \frac{V_o}{I_o} &= Z_o + R_1(1 + A) \end{aligned}$$

So,

$$Z_{ofb} = Z_o + R_1(1 + A)$$

d. Series-Shunt Feedback (Current Amplifier)
Figure 2.11(a) shows the series-shunt feedback circuitry. Note that the voltage drop across R_3 is fed back in parallel with the input through R_2. Because the input is a current source which can be assumed to

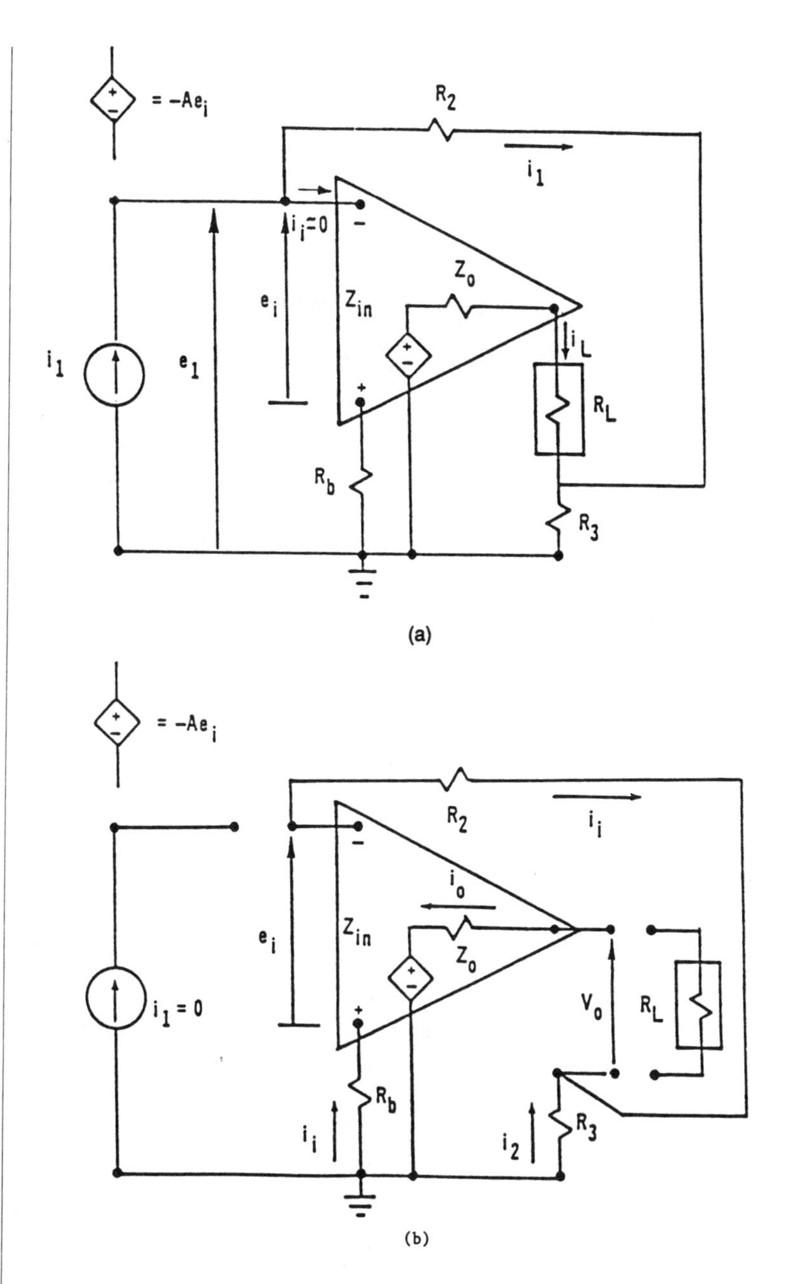

FIGURE 2.11. (a) Series-shunt feedback. (b) Output impedance of series-shunt feedback.

have very high source impedance, the bias-resistor R_b can be designed as such that

$$R_b \simeq R_2 + \frac{R_3 R_L}{R_3 + R_L}$$

Current Gain of the Feedback Network: $G_{fb} \triangleq i_L/i_i$

In Figure 2.11(a),

$$e_i = i_1(R_2 + R_3) + i_L R_3$$

$$-Ae_i = i_L(Z_o + R_L + R_3) + i_1 R_3$$

By eliminating e_i, we have

$$-A[i_1(R_2 + R_3) + i_L R_3] = i_L(Z_o + R_L + R_3) + i_1 R_3$$

$$i_L[-AR_3 - (Z_o + R_L + R_3)] = i_1[R_3 + A(R_2 + R_3)]$$

$$\frac{i_L}{i_1} = -\frac{R_3 + \dfrac{A}{1+A}R_2}{R_3 + \dfrac{Z_o + R_L}{1+A}} \simeq -\left(1 + \frac{R_2}{R_3}\right)$$

since

$$R_3 >> \frac{R_L + Z_o}{1 + A}$$

Thus,

$$G_{fb} \triangleq \frac{i_L}{i_1} = -\left(1 + \frac{R_2}{R_3}\right)$$

Input Impedance of the Feedback Loop: Z_{ifb}

The input loop equation of the circuit is

$$e_1 = i_1 R_2 + (i_L + i_1)R_3$$

$$= i_1(R_3 + R_2) + i_L R_3 \tag{2.47}$$

From the current gain equation, we assume

$$R_3 >> \frac{R_L + Z_o}{1 + A}$$

so we have

$$i_L = i_1 G_{fb} \simeq -\left(1 + \frac{R_2}{R_3}\right) i_1 \tag{2.48}$$

Substituting Equation (2.48) into Equation (2.47),

$$e_1 \simeq i_1(R_3 + R_2) - i_1\left(1 + \frac{R_2}{R_3}\right) R_3$$

$$\frac{e_1}{i_1} \simeq 0$$

so

$$Z_{ifb} \triangleq \frac{e_1}{i_1} \simeq 0$$

This is quite reasonable, since the input loop of this circuit is similar to shunt-shunt feedback circuit except that $R_1 = 0$ here. Therefore, Z_{ifb} for this circuit is approximately zero.

Output Impedance of the Feedback Circuit: $Z_{ofb} \triangleq V_0/i_o$

Since the input is expected to be an independent current source, it should be disconnected for deriving the output impedance, as shown in Figure 2.11(b). Since

$$i_o = i_i + i_2$$

$$v_o = i_o Z_o + (-A e_i) + i_2 R_3 \tag{2.49}$$

$$e_i = i_i R_2 - i_2 R_3 + i_i R_b$$

$$R_3 << (R_b + Z_{int} + R_2), \text{ or } i_2 >> i_i$$

we have

$$i_o \simeq i_2$$

$$e_i \simeq -i_2 R_3 \simeq -i_o R_3 \tag{2.50}$$

Substituting Equation (2.50) into Equation (2.49),

$$v_o = i_o Z_o + A i_o R_3 + i_o R_3$$

$$= i_o[Z_o + (1 + A)R_3)]$$

$$\frac{v_o}{i_o} = Z_o + (1 + A)R_3$$

Thus,

$$Z_{ofb} = Z_o + (1 + A)R_3$$

2.2.4. Consideration of Gain-Bandwidth Property

It was defined previously that an ideal operational amplifier has an infinite frequency bandwidth. That is, an ideal operational amplifier has a constant voltage or current gain for any signal of any frequency. In practice, however, an operational amplifier has a finite bandwidth which can usually be found in the specification sheets provided by the manufacturer. In a more detailed specification sheet, the gain-bandwidth characteristic of a typical operational amplifier is normally described by its Bode plot or mathematically.

$$\text{Voltage gain (in dB)} = 20 \log \left| \frac{E_o}{E_i} \right| = 20 \log | A |$$

which is a function of the signal's frequency as shown in Figure 2.12. Note that the 3 dB point is at a frequency equal to 100 Hz, and the Bode plot shown has a gain of 100 dB with 100 Hz bandwidth. Often the gain-bandwidth of an operational amplifier is specified at 0 dB or unity gain, thus the unity-gain-bandwidth of this amplifier is 10 MHz. The gain-bandwidth information for an amplifier is very important to the system designer. To process a signal with a given bandwidth, an amplifier should be selected with a gain-bandwidth at least equal to the desirable gain-bandwidth of the signal.

It is important to point out that in practice, operational amplifiers are mostly used in conjunction with feedback circuit configurations as described in Section 2.2.3. Note that the gain with feedback, A_{fb}, is normally considerably lower than the gain without feedback which is also known as open-loop gain, A. Thanks to the discovery of the linear negative feedback theory during World War II, the gain-loss in feedback configuration is used to pay for the desirable features. That is, the feedback network is relatively free from the variation of amplifier parameters due to environmental changes or aging effects of the circuit components. As for bandwidth, the feedback configuration will result in a bandwidth virtually wider than that of the open-loop configuration.

As shown in Figure 2.12, with an open-loop gain of 100 dB, the amplifier has a bandwidth of 100 Hz. With a feedback gain A_{fb} = 10 or 20 dB, however, the amplifier has a bandwidth of 1 MHz. In other words, without feedback the amplifier can only be used to process signals with a bandwidth of 100 Hz, while with feedback, the amplifier can process signals with a bandwidth of 1 MHz. But the latter has considerably lower gain. In some cases a designer may want to have both gain and bandwidth at the same time. Fortunately, there are operational amplifiers having pins accessible to the designer to add external frequency compensation circuitry known as lead/lag compensation networks in linear feedback theory. Figure 2.13(a) shows the test circuitry for the operational amplifier μA709 which provides the accessible pins for frequency compensation networks: R_1, C_1; R_2, C_2. The solid lines shown in Figure 2.13(b) are the Bode plots of the same amplifier with different values of the compensation networks, R_3 and R_4, respectively. The dotted lines are the Bode plots of the amplifier with different feedback networks. If the amplifier is the type with internal frequency compensation, it would have a Bode plot similar to the one shown with R_1 = 1.5 K, C_1 = 5000 pf, R_2 = 50, and C_2 = 200 pf; then for A_{fb} = 60 dB or R_3/R_4 = 1000, the amplifier would have a bandwidth of 1 kHz. Note that due to the availability of external compensation pins, with R_1 = 0, C_1 = 10 pf, R_2 = 50, C_2 = 3 pf, we can now have a bandwidth of 1 MHz in this case,

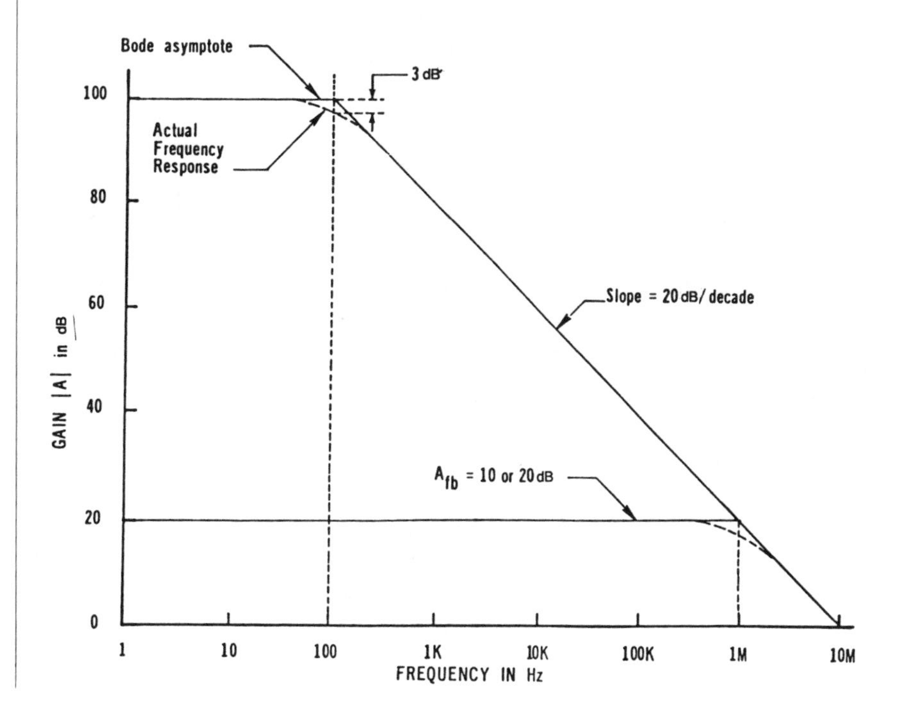

FIGURE 2.12 Bode diagram of a typical operational amplifier.

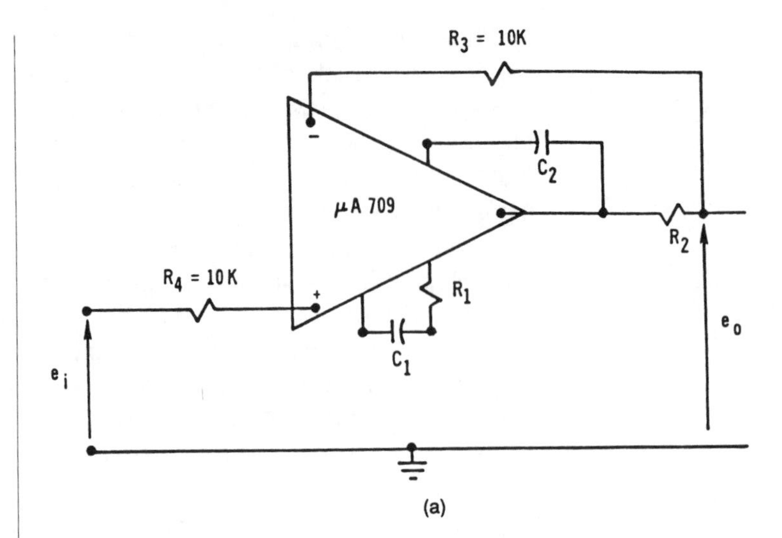

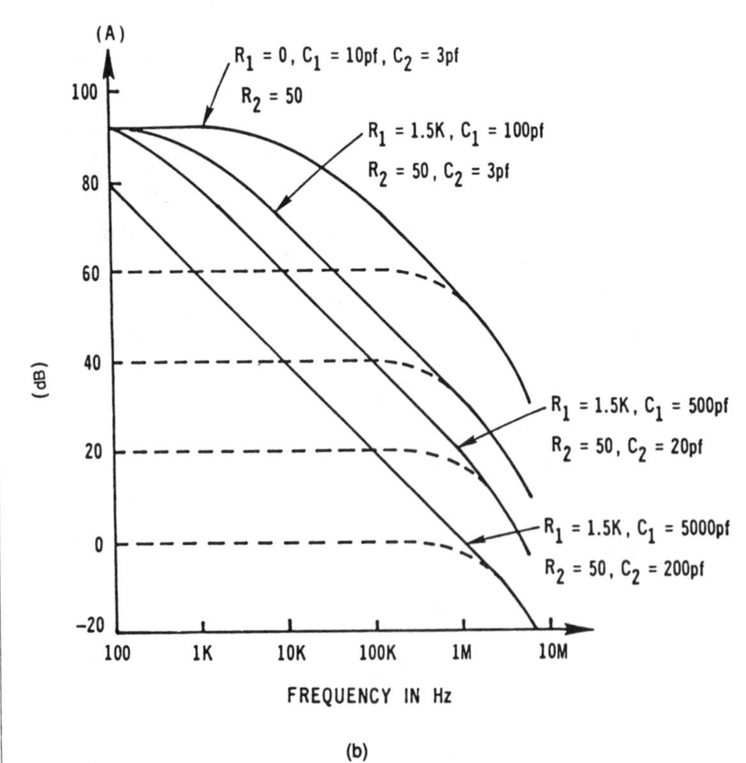

FIGURE 2.13 (a) Operational amplifier with external frequency compensation. (b) Bode diagram of the operational amplifier with external compensation network (courtesy of Texas Instruments, Inc.).

i.e., the 60 dB dotted line. At this point, one might ask what would happen if one uses the same values of compensation network components having lower A_{fb}, say A_{fb} = 20 dB. The result will undoubtedly be an unstable network. This is because of the fact that the Bode plot

TABLE 2.1

CIRCUIT CONFIGURATION	GAIN	INPUT IMPEDANCE	OUTPUT IMPEDANCE	COMMENTS
R_1, R_2, R_b, −, +, 0	Voltage Gain: $-\frac{R_2}{R_1}$	R_1	$\frac{z_o}{1 + A\left(\frac{R_1}{R_1 + R_2}\right)}$	Inverted Output. Controllable Input Impedance Very Low Output Impedance
R_b, R_2, R_1, +, −	Voltage Gain: $1 + \frac{R_2}{R_1}$	$Z_{in}\left[1 + A\,\frac{R_1}{R_1 + R_2}\right]$	$\frac{z_0}{1 + A\left(\frac{R_1}{R_1 + R_2}\right)}$	Non-Inverted Output. Very High Input Impedance Very Low Output Impedance Generally Used as Buffer Amplifier
R_b, e_1, R_1, R_L, i_L, +, −	$i_L = e_1/R_1$	$Z_{in}\left[1 + A\,\frac{R_1}{R_1 + R_L}\right]$	$z_o + R_1\,(1 + A)$	Voltage to Current Conversion High Input Impedance High Output Impedance R_L Must Be Floating From Ground i_L is Independent of RL
R_2, i_1, i_L, R_L, R_b, R_3, −, +	Current Gain: $\frac{i_L}{i_1} = -\left(1 + \frac{R_2}{R_3}\right)$	0	$z_o + (1 + A)\,R_3$	Current to Current Conversion Low Input Impedance Very High Output Impedance R_L Must Be Floating From Ground

with the first set of values, i.e., $R_1 = 0$, $C_1 = 10$ pf, $R_2 = 50$, $C_2 = 3$ pf, will have a phase-shift greater than 180° at the gain, $A_{fb} = 20$ dB. Thus, it is important that the designer should use the suggested values for the compensation network for a given closed-loop gain, A_{fb}, or he/she should select the ones having internal compensation networks, but be satisfied with a narrower bandwidth or whatever the bandwidth result in a given A_{fb}. In general, however, the bandwidth is virtually increased as the amount of feedback known as Feedback Factor, i.e., $(1 + A\beta)$, where A is the open-loop gain and β is the feedback ratio determined by the feedback network.

2.2.5. Summary: Comparison of the Four Feedback Configurations

It is important to point out that except for special applications, such as analog comparators, etc., an operational amplifier is normally being used with one of the four kinds of feedback circuit configurations described in Section 2.2.3. Therefore, it is desirable to have a comparison table for fast reference. Table 2.1 shows the major characteristics.

The main advantage of using feedback circuitry, as shown in this table, is that all the key parameters such as gain and input/output impedances are not primarily functions of the characteristics of the individual amplifier; rather, they are virtually determined by the external resistive network which can be selected depending on the requirement of the applications. In other words, it can be made almost as accurate or as stable as the resistors used.

2.3. EXPERIMENTAL EXAMPLES FOR TYPICAL APPLICATIONS

Based on the fundamentals developed in Section 2.2.3, we shall now show some typical applications with examples and experimental results. For simplicity, *the biasing resistors required at the input pins are not shown.*

2.3.1. Analog Adder-Subtractor

Figure 2.14(a) shows the circuit configuration of an adder/subtractor. From Kirchoff's current law,

$$\frac{e_1 - e_n}{R_1} + \frac{e_2 - e_n}{R_2} = \frac{e_n - e_o}{R_f} \tag{2.51}$$

$$\frac{e_3 - e_p}{R_3} + \frac{e_4 - e_p}{R_4} = \frac{e_p}{R_f} \tag{2.52}$$

$$\frac{R_f}{R_1} e_1 + \frac{R_f}{R_2} e_2 = e_n\left[1 + \frac{R_f}{R_1} + \frac{R_f}{R_2}\right] - e_o \tag{2.53}$$

$$\frac{R_f}{R_3} e_3 + \frac{R_f}{R_4} e_4 = e_p\left[1 + \frac{R_f}{R_3} + \frac{R_f}{R_4}\right] \tag{2.54}$$

Let

$$R_f\left(\frac{1}{R_1} + \frac{1}{R_2}\right) = R_f\left(\frac{1}{R_3} + \frac{1}{R_4}\right) = k \tag{2.55}$$

Substituting into Equations (2.53) and (2.54), and subtracting Equation (2.54) from Equation (2.53) yields

$$R_f\left[\frac{1}{R_1} e_1 + \frac{1}{R_2} e_2 - \frac{1}{R_3} e_3 - \frac{1}{R_4} e_4\right]$$

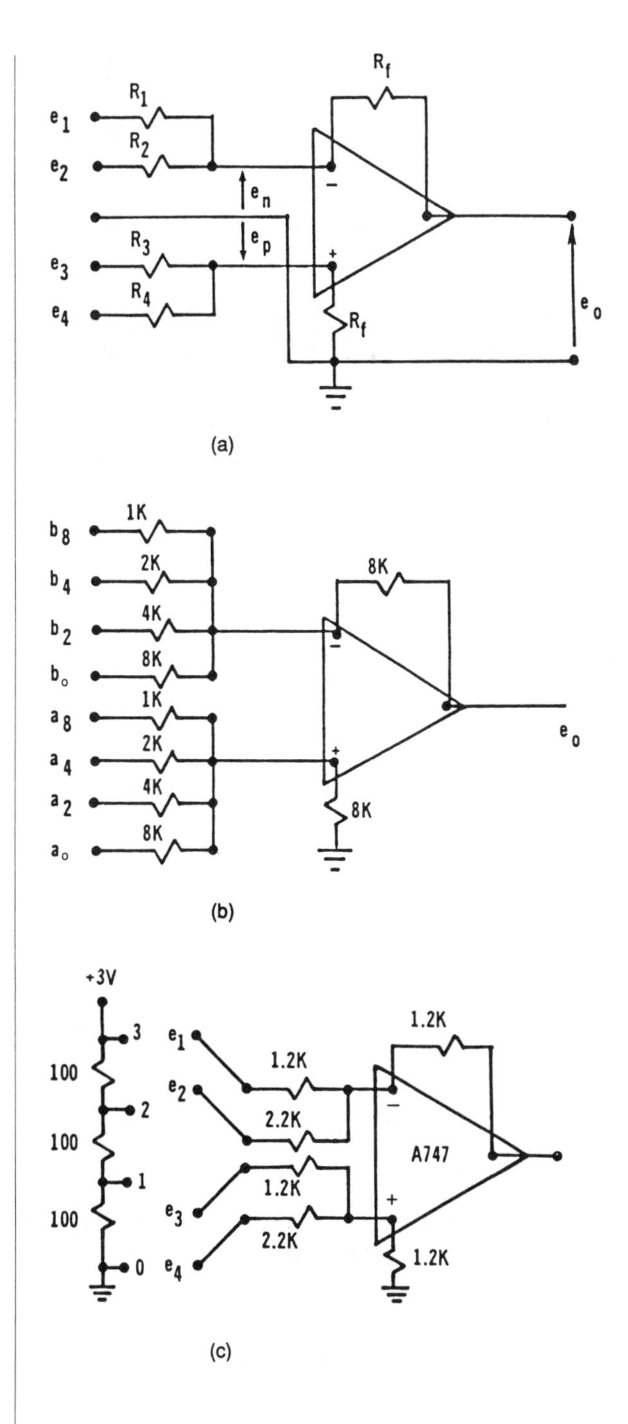

FIGURE 2.14 (a) Analog adder/subtractor. (b) 4-bit binary adder/subtractor. (c) Experimental circuit. (d) Experimental results.

e_1	e_2	e_3	e_4	e_0	
				Measured	Calculated
1	1	0	0	−1.40	−1.55
1	0	0	1	−0.45	−0.45
0	0	1	1	+1.50	+1.55
2	0	1	1	−0.50	−0.45
1	2	1	2	0	0
2	1	1	2	−0.5	−0.45
2	3	1	3	−1.0	−1.00
3	2	1	0	−3.0	−3.10

(d)

FIGURE 2.14 continued. (a) Analog adder/subtractor. (b) 4-bit binary adder/subtractor. (c) Experimental circuit. (d) Experimental results.

$$= (e_n - e_p)(1 + k) - e_o$$

$$= -\frac{e_o}{A}(1 + k) - e_o$$

$$= -e_o\left(1 + \frac{1 + k}{A}\right)$$

where open-loop gain is

$$-A = \frac{e_o}{e_n - e_p}$$

Let $A_o >> (1 + k)$. We then have

$$R_f\left[\frac{1}{R_3} e_3 + \frac{1}{R_4} e_4\right] - R_f\left[\frac{1}{R_1} e_1 - \frac{1}{R_2} e_2\right] = e_o \qquad (2.56)$$

Based on Equation (2.56), a 4-bit binary adder/subtractor with analog output can easily be implemented with the circuit configuration shown in Figure 2.14(b).

Let a_1, a_2, a_3, a_4 and b_1, b_2, b_3, b_4 be equal to a binary value, 0 or V volts, respectively. Then we have

$$\begin{aligned} e_o &= V[2^3 a_8 + 2^2 a_4 + 2^1 a_2 + 2^0 a_o] \\ &\quad - V[2^3 b_8 + 2^2 b_4 + 2^1 b_2 + 2^0 b_o] \end{aligned} \qquad (2.57)$$

where $a_o, \ldots, a_4$ and $b_o, \ldots, b_4$ are either 0 or 1. The output voltage e_o is the analog value of the binary adder/subtractor. This circuit therefore is a 4-bit binary adder/subtractor with a built-in digital-to-analog converter.

Experimental Example

To verify the theoretical derivation, the experimental circuit shown in Figure 2.14(c) was used, and the experimental results vs. the theoretical results are shown in Figure 2.14(d). The theoretical values are determined by

$$e_o = \frac{1.2}{1.2}e_3 + \frac{1.2}{2.2}e_4 - \left[\frac{1.2}{1.2}e_1 + \frac{1.2}{2.2}e_2\right]$$

$$= e_3 + 0.55e_4 - e_1 - 0.55e_2$$

The μA747 operational amplifier with internal frequency compensation network was used. In all cases, the theoretical values agree with the experimental results within ±10%. This is within the tolerances of the resistors and measuring equipment.

2.3.2. Square Wave and Triangular Wave Generator

Figure 2.15(a) shows a simple circuit which will generate both triangular and square waves at the same time and frequency. As shown, e_{tr} denotes triangular and e_{sq} square waves, respectively. The circuit operates in the following way.

Let us assume that the e_{sq} terminal is slightly positive at the beginning. It would then produce a positive voltage, $e_{sq}R_1/(R_1 + R_2)$, at the noninverted input terminal, Y. Since this is a positive feedback configuration, it will almost instantly drive the output more positive until the saturation voltage, V_s, of the operational amplifier is reached, which is limited by the voltage of the power supplies. At this point, the voltage at Y will be equal to $V_sR_1/(R_1 + R_2)$. As a result, capacitor C will be charged toward $+V_s$ by e_{sq} through resistor R_f. That is, the voltage at the inverted terminal X will rise exponentially with a time constant equal to R_fC toward V_s. However, as the voltage exceeds the voltage at the noninverted terminal, Y, i.e., $V_sR_1/(R_1 + R_2)$, the amplifier output flips to the other extreme, $-V_s$. This, in turn, would result in a negative voltage at Y. This process repeats itself, and the circuit becomes an oscillator which yields square waves at e_{sq} and triangular waves at e_{tr}. Figure 2.15(b) depicts the two superimposed waveforms. The triangular

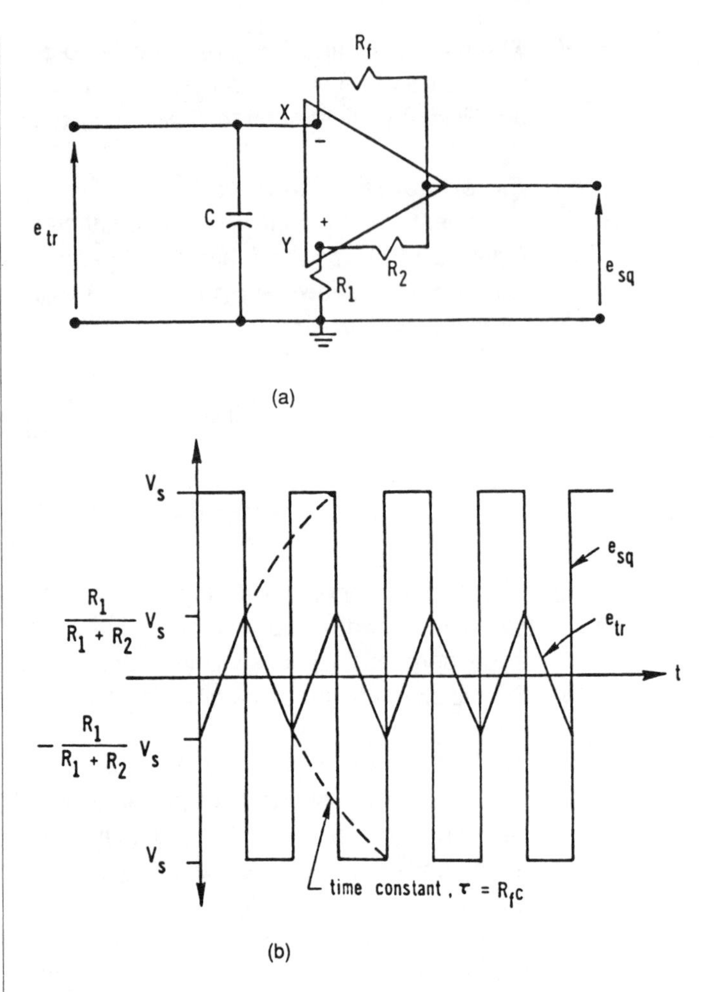

FIGURE 2.15 (a) Square and triangular wave generator. (b) Square and triangular waveforms.

waveform is the linear portion of the exponential charging (discharging) waveform,

$$v(t) = V(1 - e^{-t/\tau})$$

where

$$V = V_s + \frac{R_1}{R_1 + R_2} V_s = V_s \left(1 + \frac{R_1}{R_1 + R_2}\right)$$

$$\tau = R_f C \tag{2.58}$$

In view of Equation (2.58) and Figure 2.15(b), the peak-to-peak amplitude of the triangular signal is determined by

$$2\left(\frac{R_1}{R_1 + R_2} V_s\right)$$

and the time period of the generated waveforms is twice that of the ΔT, which is determined by Equation (2.58), as $v(t) = 2R_1/(R_1 + R_2)\, V_s$, i.e.,

$$\frac{2R_1}{R_1 + R_2} V_s = V(1 - e^{-\Delta T/\tau})$$

or

$$\Delta T = R_f C \ln\left(1 + 2\frac{R_1}{R_2}\right) \tag{2.59}$$

and the frequency is

$$f = \frac{1}{2\Delta T} \tag{2.60}$$

Experimental Example

To verify the derivation, the circuit shown in Figure 2.15(a) was implemented with the following circuit components:

C = 0.1 µf
R_f = 10 K
R_2 = 10 K
R_1 = 1.2 K
Operational amplifier = µA747
Power supplies = ±18 V
V_s = 16 V

A comparison between the calculated and experimental values is shown below:

Description	Calculated values	Measured values	Difference
Square wave amplitude	±16 V	±16 V	None
Triangular wave amplitude	±1.7 V	±2.0 V	0.3 V
Frequency	2344 Hz	2439 Hz	95 Hz

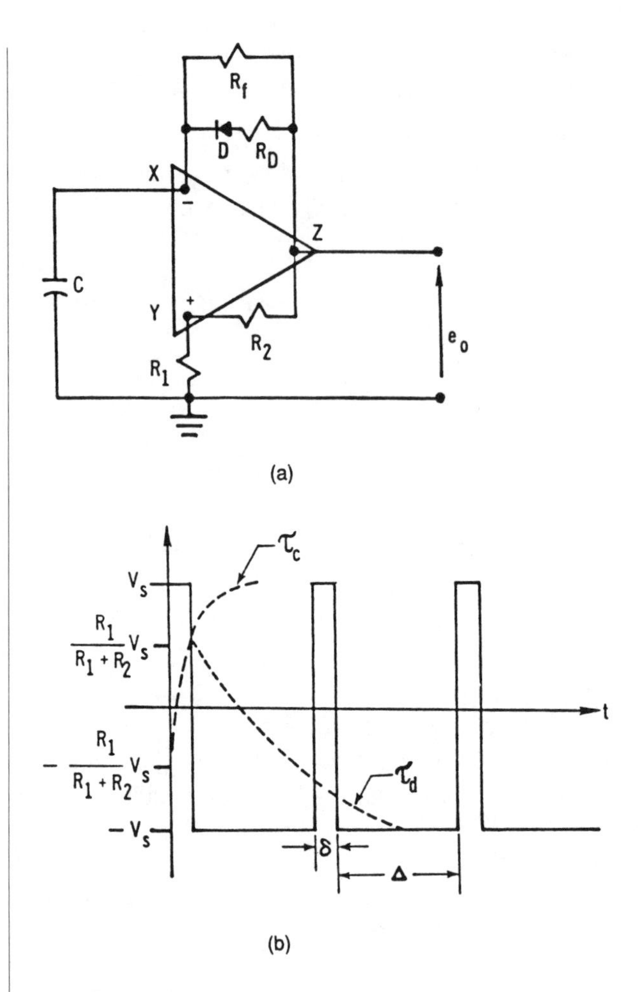

FIGURE 2.16 (a) Pulse generator. (b) Pulse generator waveform.

2.3.3. Pulse Generator

The circuit shown in Figure 2.16(a) is a pulse generator; the pulse width and frequency are controlled by the circuit components. This circuit is similar to that of triangular/square wave generators, even in principle of operation. The addition of a diode and a resistor R_D in parallel with R_f causes the capacitor to be charged and discharged with different time constants, τ_c and τ_d, whereas the triangular/square wave generator has the same time constant, R_fC. For the pulse generator circuitry, the charging time constant $\tau_c = CR_DR_f/(R_D + R_f)$, while the discharging time constant $\tau_d = R_fC$, or $\tau_c < \tau_d$. Like the triangular/square wave generator, as the output voltage e_o at Z is positive or V_s, the diode D is forward biased, and the capacitor C is being charged through resistors R_f and R_d with a time constant of τ_c. As the voltage at X increases to slightly greater than $V_sR_1/(R_1 + R_2)$, the output voltage flips from $+V_s$ to $-V_s$, which turns off the diode D and the capacitor is then discharged through R_f with a time constant τ_d. As shown in Figure 2.16(b), it can be seen that τ_c and $V_sR_1/(R_1 + R_2)$ control the pulse width δ, and τ_d and $-V_sR_1/(R_1 + R_2)$ control the pulse-to-pulse duration.

Experimental Example

The pulse generator circuit was tested with the following circuit components:

$$
\begin{aligned}
C &= 0.1\ \mu f \\
R_f &= 10\ K \\
R_d &= 1.2\ K \\
R_2 &= 10\ K \\
R_1 &= 1.2\ K \\
\text{Operational amplifier} &= \mu A747 \\
\text{Power supplies} &= \pm 18\ V\ V_s = 15\ V
\end{aligned}
$$

The experimental results are shown below:

Description	Calculated values	Measured values	Difference
Frequency (Hz)	4240	4545	305
δ (ms)	0.213	0.2	0.013
Δ (ms)	0.228	0.02	0.028

Comments: If only positive pulses are desired, a diode can be connected in series with the output terminal Z, pointing away from Z. This diode will eliminate the negative portion of the waveform. As the value of R_f is ten or more times greater than that of R_D, the pulse width can be controlled by varying R_D, while the frequency can be controlled by R_f.

2.3.4. Second Order Active Filters

a. Low-Pass Filter

Figure 2.17(a) shows the circuit of a second order low-pass active filter. Note that the operational amplifier is connected in the shunt-series feedback configuration, Figure 2.9(a), with the nominal feedback network $R_2 = 0$, $R_1 = \infty$. As a result, the voltage gain of this circuit is

$$\frac{e_o}{e_y} = 1 + \frac{R_2}{R_1} = 1$$

which is known as a voltage follower, i.e.,

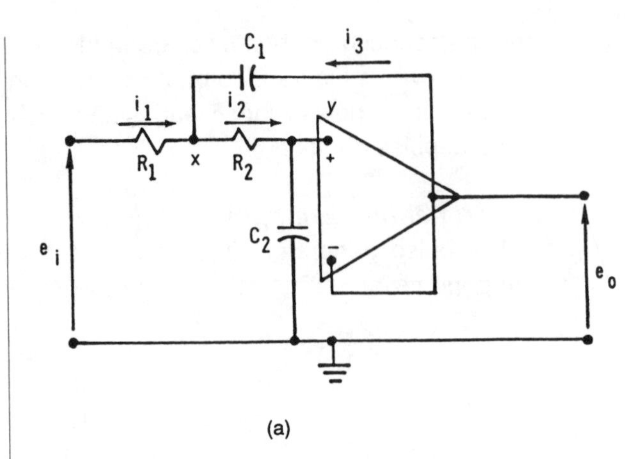

(a)

Frequency (Hz)	e_o* (volt)	Voltage Gain: Gain	Voltage Gain: dB
10	1.0	1.0	0
20	0.97	0.97	–0.26
50	0.96	0.96	–0.35
75	0.83	0.83	–1.62
100	0.76	0.76	–2.38
150	0.60	0.60	–4.44
200	0.47	0.47	–6.56
250	0.35	0.35	–9.12
300	0.28	0.28	–11.10
500	0.14	0.14	–17.10
1000	0.05	0.05	–26.00
2000	0.02	0.02	–33.90

*e_i = 1 volt (p–p) sine wave

(b)

FIGURE 2.17 (a) Low-pass filter. (b) Experimental data. (c) Frequency response (low-pass filter).

$$e_o = e_y \tag{2.61}$$

where e_y is the voltage at terminal y.

Referring to Figure 2.17(a) again, we have

$$i_2 = i_1 + i_3$$

where

$$i_1 = \frac{1}{R_1}(e_i - e_x)$$

$$i_2 = \frac{1}{R_2}(e_x - e_y)$$

$$i_3 = sc_1(e_o - e_x)$$

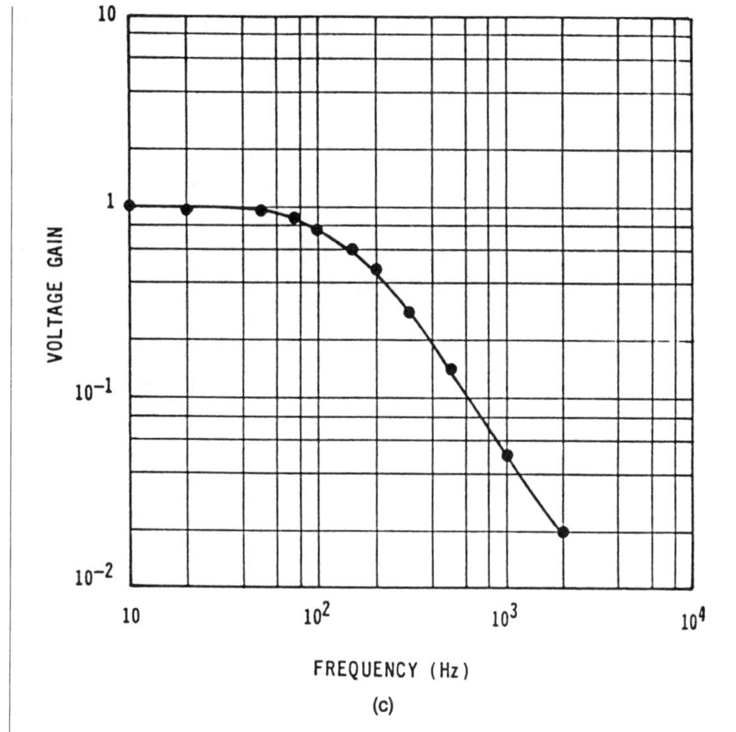

FIGURE 2.17 continued. (a) Low-pass filter. (b) Experimental data. (c) Frequency response (low-pass filter).

Thus,

$$\frac{e_x}{R_2} - \frac{e_y}{R_2} = \frac{e_i}{R_1} - \frac{e_x}{R_1} + sc_1e_o - sc_1e_x$$

$$e_x\left(\frac{1}{R_1} + \frac{1}{R_2} + sc_1\right) = \frac{e_i}{R_1} + sc_1e_o + \frac{e_y}{R_2} \tag{2.62}$$

But,

$$e_y = \frac{\frac{1}{sc_2}}{R_2 + \frac{1}{sc_2}} e_x \tag{2.63}$$

From Equations (2.61) to (2.63),

$$\frac{e_o}{e_i} = \frac{\frac{1}{c_1c_2R_1R_2}}{s^2 + s\left[\frac{1}{c_1R_1} + \frac{1}{c_1R_2}\right] + \frac{1}{c_1c_2R_1R_2}} \tag{2.64}$$

From the s-domain or pole-zero concept, the transfer function of a second order low-pass filter is

$$H(s) = \frac{w_o^2}{s^2 + 2\xi w_o s + w_0^2} \tag{2.65}$$

From Equations (2.64) and (2.65) we have the cut-off or roll-off frequency of the filter:

$$w_o = \frac{1}{\sqrt{R_1 R_2 c_1 c_2}}$$

damping factor:

$$\xi = \sqrt{\frac{c_2}{c_1}} \frac{1}{2} \frac{R_1 + R_2}{\sqrt{R_1 R_2}}$$

For experimental verification, the following circuit components were used:

$$C_1 = C_2 = C = 0.0094\ \mu f$$
$$R_1 = R_2 = R = 100\ K$$
Operational amplifier = μA747

Figure 2.17(b) shows the data of the frequency response for this circuit, and Figure 2.17(c) depicts the corresponding Bode diagram. The comparison between the calculated and measured values is listed below:

Description	Calculated values	Measured values
Cut-off frequency (Hz)	169	102
Gain slope	–12 dB/octave	–8.8 dB/octave
Low frequency gain	1	1

b. High-Pass Filter

Figure 2.18(a) shows the circuit of a second order high-pass filter. The derivation of the transfer function in terms of circuit components, which is similar to that of the low-pass filter presented in Section 2.3.4.a, is, however, being omitted from this and the succeeding section, except for the key formulations, i.e.,

Transfer function:

$$H(s) = \frac{s^2}{s^2 + 2\xi w_o s + w_o^2} \tag{2.66}$$

Cut-off frequency:

$$w_o = \frac{1}{\sqrt{R_1 R_2 C_1 C_2}}$$

Damping factor:

$$\xi = \sqrt{\frac{R_1}{R_2}}$$

For experimental verification,

$$\begin{aligned} C_1 &= C_2 = C = 0.0047\ \mu f \\ R_1 &= R_2 = R = 10\ K \\ \text{Operational amplifier} &= \mu A747 \end{aligned}$$

Figure 2.18(b) shows the data of the frequency response for this circuit, and Figure 2.18(c) depicts the corresponding Bode diagram. The comparison between the calculated and measured values is listed below:

Description	Calculated values	Measured values
Cut-off frequency (Hz)	3386	3300
Gain slope	12 dB/octave	9.6 dB/octave
High frequency gain	1	1

c. Band-Pass Filter

Figure 2.19(a) shows the circuit of a second order band-pass filter. Its key formulations are as follows:

Transfer function:

$$H(s) = \frac{A_o(w_o/Q)s}{s^2 + (w_o/Q)s + w_0^2} \tag{2.67}$$

where

A_o = Voltage gain at center frequency
Q = Quality factor
w_o = Center frequency

Center frequency:

$$w_o = \frac{\sqrt{2}}{CR}$$

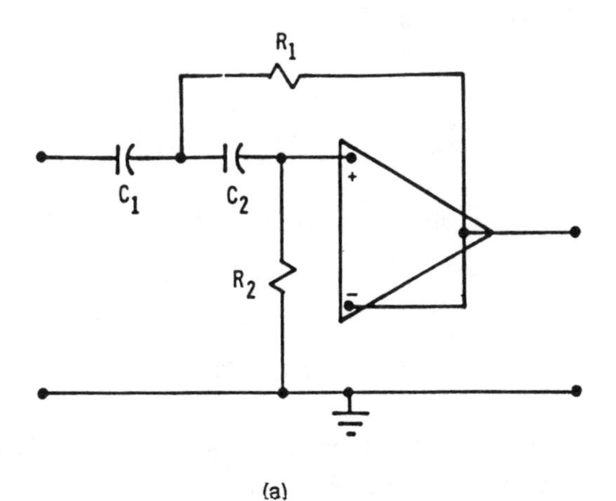

(a)

Frequency (Hz)	e^*_0 (volt)	Gain
100	0.015	0.015
200	0.025	0.025
300	0.030	0.030
400	0.050	0.050
500	0.065	0.065
750	0.125	0.125
1K	0.200	0.200
1.5K	0.350	0.350
2.0K	0.480	0.480
2.5K	0.600	0.600
3.0K	0.650	0.650
5.0K	0.800	0.800
7.5K	0.900	0.900
10.0K	0.920	0.920
15.0K	0.950	0.950
20.0K	0.970	0.970
25.0K	0.980	0.980
50.0K	1.00	1.00

$^*e_i$ = 1 volt (p−p) sine wave

(b)

FIGURE 2.18 (a) High-pass filter. (b) Experimental data. (c) Frequency response (high-pass filter).

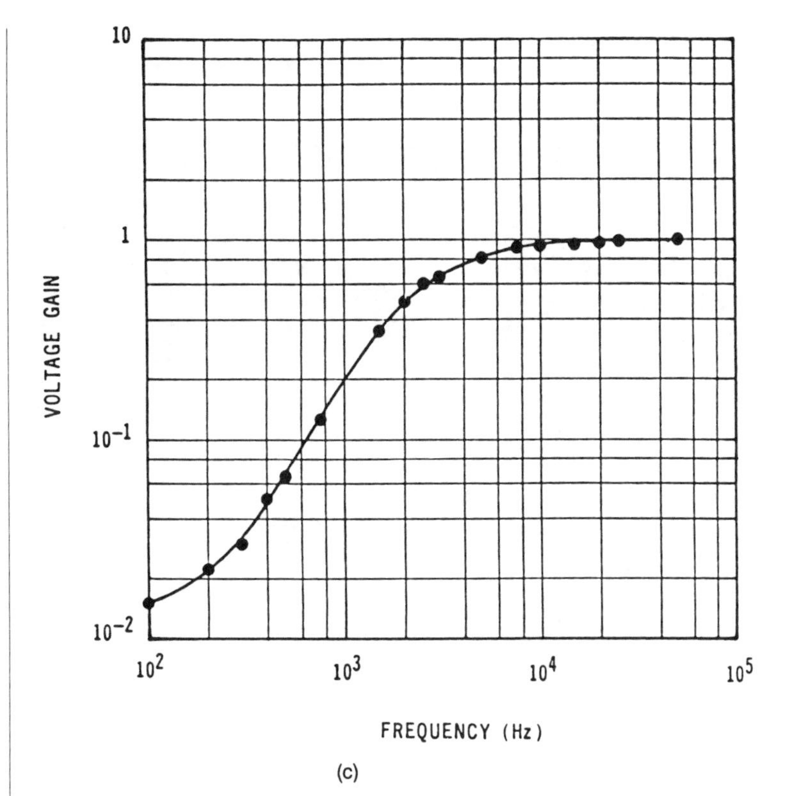

(c)

FIGURE 2.18 continued. (a) High-pass filter. (b) Experimental data. (c) Frequency response (high-pass filter).

Quality factor:

$$Q = \frac{\sqrt{2}}{5 - A_{fb}}$$

Center frequency gain:

$$A_o = \frac{A_{fb}}{5 - A_{fb}}$$

where

$$A_{fb} = \text{Operational Amplifier Gain with Feedback} = 1 + \frac{R_2}{R_1}$$

For experimental verification,

$R = 27$ K $\qquad C = 0.0047\ \mu$f
$R_1 = 1.2$ K $\qquad R_2 = 3.9$ K
Operational amplifier = μA747

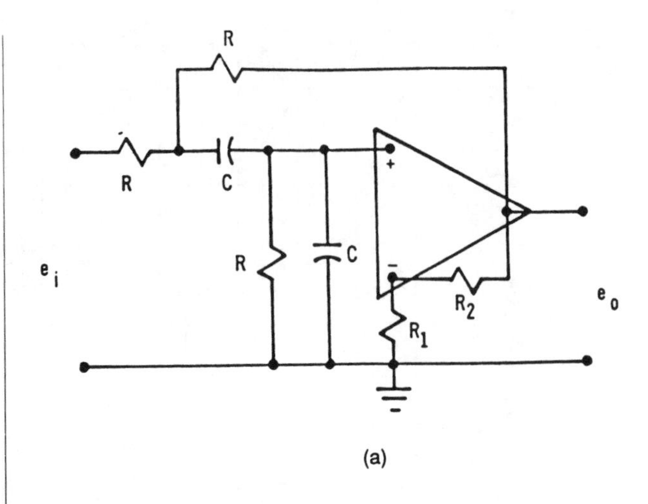

(a)

Frequency(Hz)	e^*_o(volt)	Gain
10	0.025	0.025
20	0.05	0.05
50	0.10	0.10
100	0.16	0.16
200	0.31	0.31
250	0.40	0.40
400	0.60	0.60
500	0.79	0.79
600	1.00	1.00
700	1.20	1.20
800	1.40	1.40
900	1.60	1.60
1.0K	1.85	1.85
1.5K	3.70	3.70
2.0K	6.00	6.00
2.5K	4.50	4.50
3.0K	3.00	3.00
3.5K	2.40	2.40
4.0K	1.75	1.75
5.0K	1.40	1.40
10.0K	0.62	0.62
20.0K	0.30	0.30
50.0K	0.12	0.12
100.0K	0.06	0.06

$^*e_i$ = 1 volt (p-p) Sine wave

(b)

FIGURE 2.19 (a) Band-pass filter. (b) Experimental data. (c) Frequency response (band-pass filter).

Figure 2.19(b) shows the data of the frequency response for the circuit, and Figure 2.19(c) depicts the corresponding Bode diagram. The comparison between the calculated and measured values is listed below:

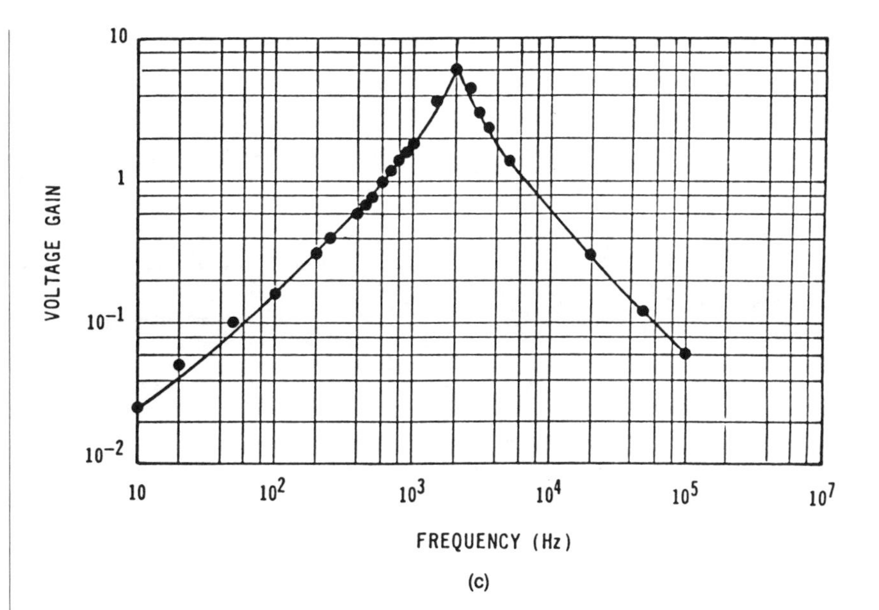

(c)

FIGURE 2.19 continued. (a) Band-pass filter. (b) Experimental data. (c) Frequency response (band-pass filter).

Descriptions	Calculated values	Measured values
Center frequency (Hz)	1774	2050
Q	1.89	2.28
Bandwidth (Hz)	936	900
Center frequency gain	5.66	6.0
Low frequency slope	12 dB/octave	10 dB/octave
High frequency slope	–12 dB/octave	–11.2 dB/octave

d. Band Reject Filter

Figure 2.20(a) shows the circuit of a band reject filter. Note that the network at the input end is known as a Twin-T network, which would pass through signals of all frequencies except the frequencies for which the network is tuned. Its key formulations are as follows.

Center of reject frequency:

$$w_o = \frac{1}{RC}$$

Quality factor:

$$Q = \frac{R_1}{4R}$$

and

$$C_1 = \left[\frac{1}{2Q}\right] C$$

For experimental verification the following data are used for the design:

Center of reject frequency $f_o = \dfrac{w_o}{2\pi} = 1$ kHz

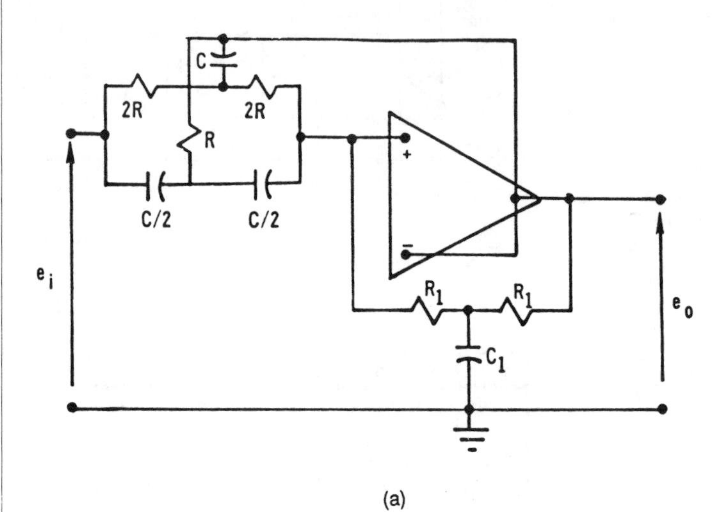

(a)

Frequency(Hz)	e_o^*(volt)	Gain
10	1	1
50	1	1
100	1	1
200	1	1
500	1	1
800	0.97	0.97
1K	0.92	0.92
1.1K	0.72	0.72
1.15K	0.40	0.40
1.20K	0.14	0.14
1.25K	0.45	0.45
1.30K	0.68	0.68
1.50K	0.90	0.90
2K	0.97	0.97
5K	1	1
1.0K	1	1
100K	1	1
300K	0.9	0.9
500K	0.6	0.6
1M	0.3	0.3
2M	0.14	0.14

*e_i(= 1 volt (p-p) sine wave

(b)

FIGURE 2.20 (a) Band reject filter. (b) Experimental data. (c) Frequency response (band reject filter).

FIGURE 2.20 continued. (a) Band reject filter. (b) Experimental data. (c) Frequency response (band reject filter).

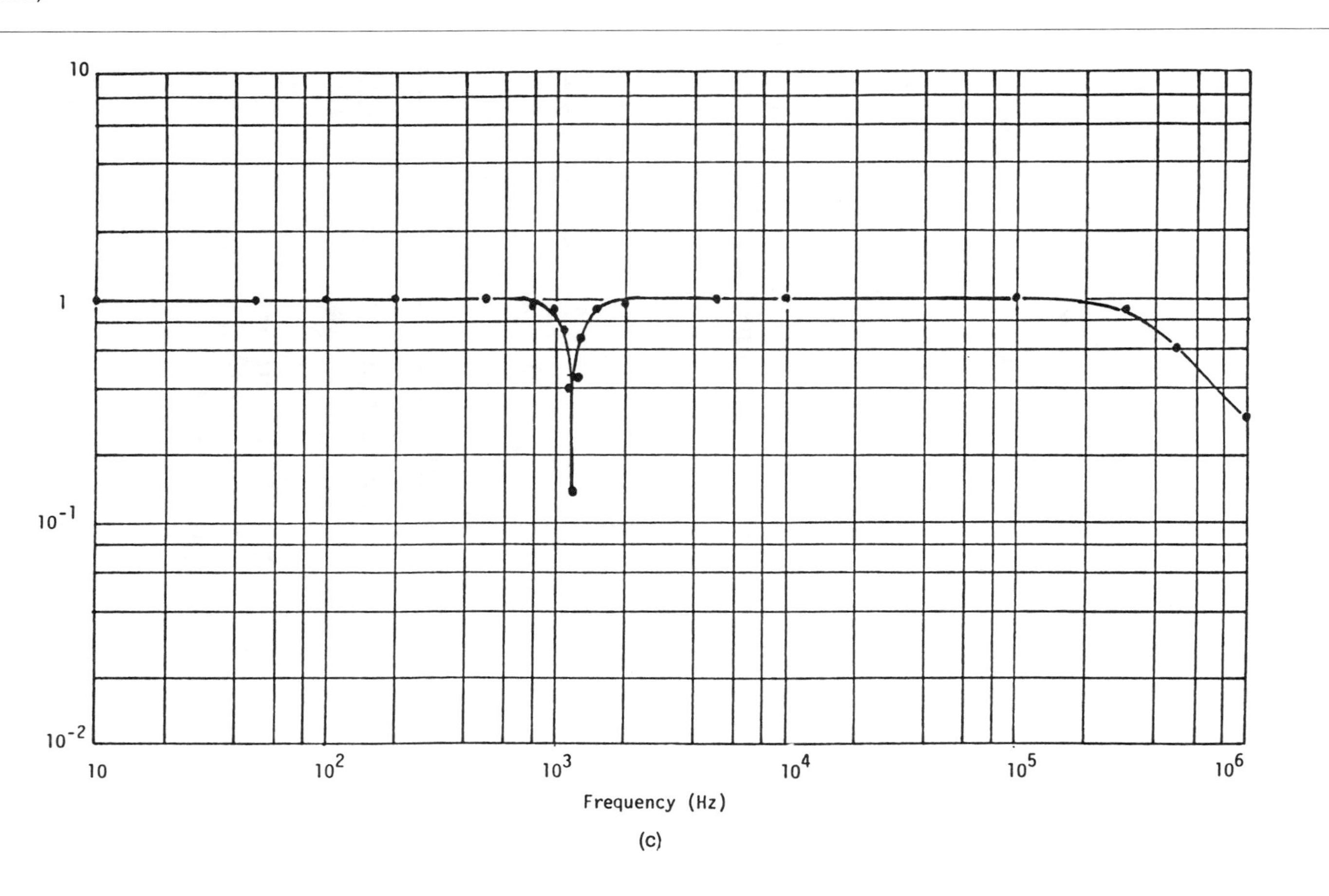

$$Q = 5$$

$$C = 0.0047\ \mu f$$

$$R = \frac{1}{w_o C} = 33.9\text{ K; 33 K was used}$$

$$R_1 = 4RQ = 660\text{ K; 680 K was used}$$

$$C_1 = \left[\frac{1}{2Q}\right] C = 470\text{ pf}$$

$$\frac{C}{2} = 0.00235\ \mu f\text{; 0.0022 was used}$$

$$2R = 68\text{ K}$$

Figure 2.20(b) shows the experimental data of the frequency response for the circuit and Figure 2.20(c) depicts the Bode diagram. Comparison between the calculated and measured values is listed below:

Descriptions	Calculated values	Measured values
Center of reject frequency (Hz)	1026	1200
Bandwidth (Hz)	200	200
Q	5.15	6
Gain (max)	1.0	1.0
Gain (min)	~0	0.14

2.4. A COLLECTION OF CIRCUITS FOR GENERAL APPLICATIONS

2.4.1. Integrator

Figure 2.21 shows a simple and useful analog integrator circuit. Its derivation follows.

$$e_1 = i_1 R + e_i$$

$$e_o = e_i - \frac{1}{C}\int i_C\, dt$$

$$e_o = -\frac{e_o}{A} - \frac{1}{C}\int i_C\, dt$$

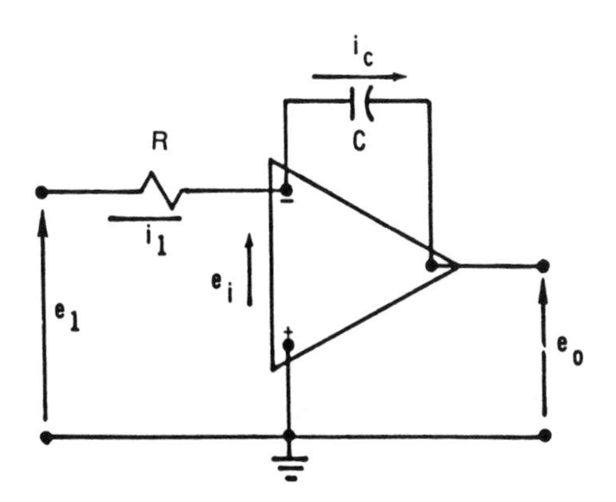

FIGURE 2.21 Integrator.

But,

$$1 >> \frac{1}{A}, \; i_1 \simeq i_c$$

We have

$$e_o = -\frac{1}{C}\int i_1 \, dt \simeq -\frac{1}{C}\int \frac{e_1}{R} \, dt$$

$$= -\frac{1}{RC}\int e_1 \, dt \qquad (2.68)$$

Note that this circuit can also be viewed as a shunted-shunted feedback circuit configuration, referring to Table 2.1, replacing R_1, R_2 with impedance notations as a function of the s-operator:

$$\frac{E_o(s)}{E_i(s)} = -\frac{Z_2(s)}{Z_1(s)} = -\frac{1/SC}{1/R} = -\frac{1}{RCS}$$

$$\frac{e_o}{e_i} = -\frac{1}{jwR_1C} \qquad (2.69)$$

Equation 2.69 shows that this circuit has an extremely high gain at low frequency.

For practice, see the differentiator-integrator circuit described in Section 2.4.3.

2.4.2. Differentiator

Figure 2.22 shows the circuit of an analog differentiator. Following a procedure similar to the one described in Section 2.4.1, we have

$$e_1 = \frac{1}{C}\int i_1\,dt + e_i \simeq \frac{1}{C}\int i_1\,dt$$

$$i_1 = C\frac{de_1}{dt}$$

$$e_o = -i_2R_2 + e_i \simeq -i_1R_2$$

$$e_o \simeq -R_2C\frac{de_1}{dt}$$

$$\frac{E_o(s)}{E_i(s)} = -\frac{R_2}{\frac{1}{SC}} = -R_2CS$$

$$\frac{e_o}{e_i} = -jR_2Cw \tag{2.70}$$

Equation (2.70) shows that this circuit has an extremely high gain at high frequency. For practice, see the differentiator-integrator circuit described in Section 2.4.3.

2.4.3. Differentiator-Integrator

Figure 2.23(a) shows a useful and practical circuit configuration. By properly selecting R_1, C_1 and R_2, C_2, the circuit can be used as a differentiator, an integrator, and even a first order band-pass filter. Its derivation follows. Let

$$Z_1(s) = R_1 + \frac{1}{SC_1} = \frac{1}{SC_1}(R_1C_1S + 1)$$

$$Z_2(s) = \frac{R_2\frac{1}{SC_2}}{R_2 + \frac{1}{SC_2}} = \frac{R_2}{R_2C_2S + 1}$$

then

$$\frac{E_o(s)}{E_1(s)} = -\frac{Z_2(s)}{Z_1(s)} = -\frac{R_2}{R_2C_2S + 1}\cdot\frac{SC_1}{R_1C_1S + 1}$$

$$= \frac{-S}{R_1C_2\left(S + \frac{1}{R_2C_2}\right)\left(S + \frac{1}{R_1C_1}\right)}$$

Let

$$w_1 = \frac{1}{R_1C_1}, \; w_2 = \frac{1}{R_2C_2}$$

then

$$\frac{E_o(s)}{E_1(s)} = -\frac{s}{R_1C_2\left(s + \frac{1}{w_1}\right)\left(s + \frac{1}{w_2}\right)}$$

$$\frac{e_o}{e_1} = -\frac{jw}{R_1C_2\left(jw + \frac{1}{w_1}\right)\left(jw + \frac{1}{w_2}\right)} \qquad (2.71)$$

Figure 2.23(b) shows the Bode asymptotic diagram. Note that the circuit is a differentiator for a signal with bandwidth less than w_1, and an integrator for a signal with bandwidth greater than w_2. Furthermore, it is a first order band-pass filter with a bandwidth equal to $w_2 - w_1$.

It was pointed out in Sections 2.4.2. and 2.4.3. that the differentiator has extremely high gain for high frequency so that it is sensitive to high frequency noise, and on the other hand, the integrator is sensitive to low frequency noise. Therefore, this circuit can be used as a differentiator, providing that the highest frequency component of the signal to be differentiated is less than w_1, and meanwhile R_2C_2 is used to attenuate the noise with frequency higher than w_1. Similarly, the network R_1C_1 serves as a noise filter when the circuit is used as an integrator for the signals having lowest frequency component greater

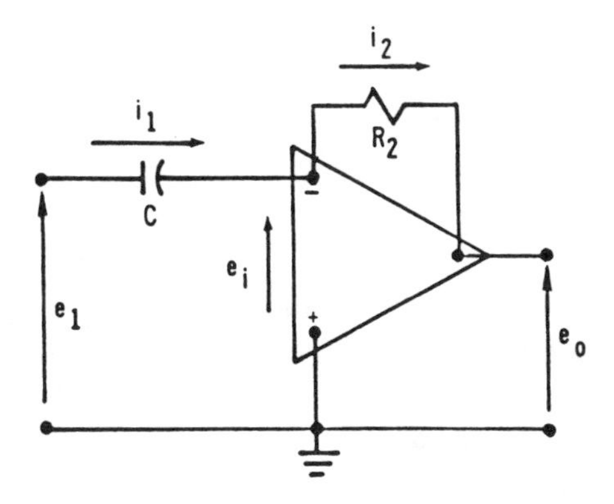

FIGURE 2.22 Differentiator.

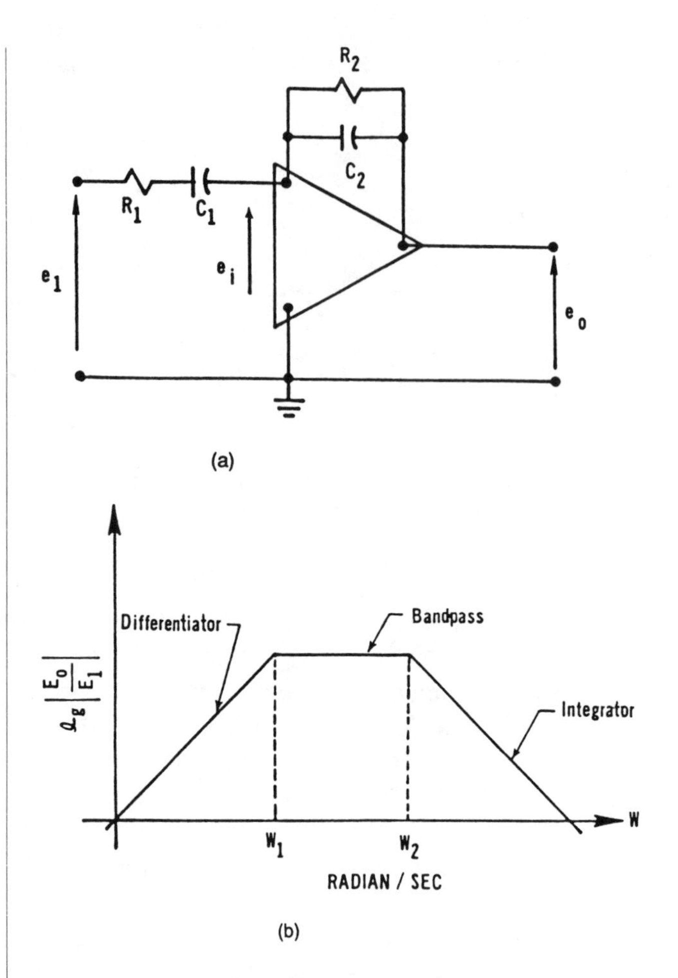

FIGURE 2.23 (a) Differentiator-integrator. (b) Bode diagram.

than w_2. For integrator or differentiator only, we may simply eliminate the bandportion shown in Figure 2.23(b) by designing

$$w_1 = w_2$$

or

$$R_1C_1 = R_2C_2 \tag{2.72}$$

then, the circuit is a differentiator for the signals with $w < w_1$ and integrator for those with $w > w_2$.

2.4.4. Comparator

A comparator is a special purpose operational amplifier which is normally used without any feedback circuit configuration. It is therefore designed to have wide bandwidth, high speed, or very fast response. Typically, the response time for low-to-high-level output is a few nanoseconds. Typical applications are sense amplifiers for memory systems, voltage comparators in analog-digital-analog conversion systems, zero-crossing detector, etc. A number of comparators provide strobe-input, offset balancing terminals, and they operate with single or dual power supply. The provision of strobe-input is extremely useful in noisy environments such as magnetic core memory systems.

2.4.5. Power Operational Amplifier Circuits

Thus far we have limited our discussion to operational amplifiers that process only small signals. There is a large area of applications, such as motor controllers, relay drivers, audio loud speaker drivers, etc., where power amplifiers that not only can amplify voltage, but also provide sufficient driving current. In this section we examine power operational amplifiers and their applications. Figure 2.24(a) depicts the equivalent circuit schematic of a typical power operational amplifier, SG1173, where the symbol of two overlapping circles stands for a constant current generator. Although the schematic looks similar to that of the small signal operational amplifier shown in Figure 2.6, the designer here must consider the power required in all applications. For example, the SG1173 is able to deliver 3.5 A with supply voltage range from ±5 to ±24 V. That is, the SG1173 can only drive or control a load between 17.5 and 84 W. In addition, the designer must be concerned with the heat dissipation of the amplifier. In general, a sufficient heat sink must be installed for the device. Another important point is that when the amplifier is driving an inductive load, such as a motor or solenoid-driven power relay, which normally produces "back-emf" or "kick-back" voltage that can damage the amplifier itself, two switching diodes must be added for protection purposes, as shown in Figure 2.24(b). Figure 2.24(c) illustrates a practical circuit for driving an audio loud speaker. Note that the power operational amplifier is configured with shunt-series feedback. The voltage gain of the circuit is determined by the R_F and R_S, and the R_S, C_S determine the low-end cut-off frequency. Capacitors at the power supply pins are for noise reduction purposes. In general, the power operational amplifier is designed with low voltage gain, but high current output capability. Thus, its input is usually connected to the output of a small signal operational amplifier which has high voltage gain. Note that by using symmetric plus/minus power supplies in this circuit, the output of the amplifier would stay at 0 V with zero input signal so that the speaker can be directly connected to the output of the amplifier. As a result, the

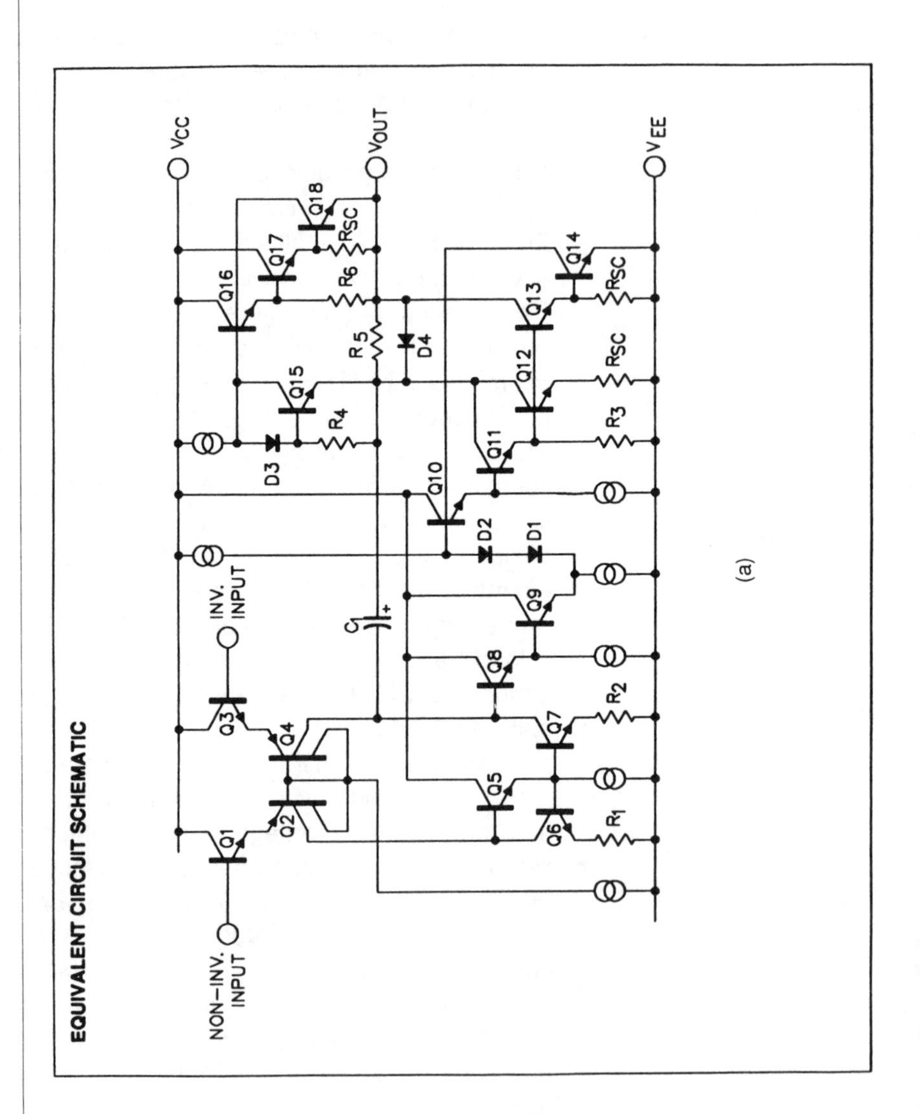

FIGURE 2.24 (a) Power operational amplifier SG1173 (courtesy of Silicon General, Inc.). (b) Inductive load. (c) Audio amplifier.

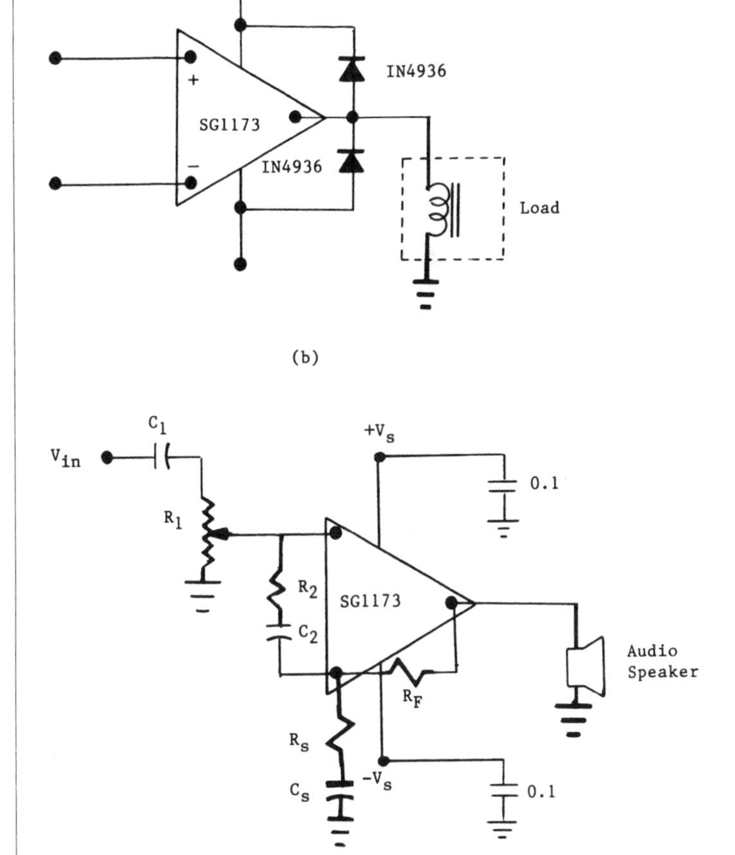

FIGURE 2.24 continued. (a) Power operational amplifier SG1173 (courtesy of Silicon General, Inc.). (b) Inductive load. (c) Audio amplifier.

huge coupling capacitor required between the amplifier and the speaker for blocking the sustained direct current flowing into the speaker is eliminated.

2.5. PHASE-LOCKED LOOP

2.5.1. Introduction

Phase-locked loops, known as PLL, like operational amplifiers, are an important building block in communication, control, and digital systems. They are basically feedback systems in themselves. As shown in Figure 2.25, $V_s(t)$ denotes the incoming signal with noise; $V_o(t)$, the output signal of the voltage or current controlled oscillator; and $v_e(t)$, the output of the phase comparator which is a function of the phase/frequency difference between $v_s(t)$ and $v_o(t)$. As the system is in

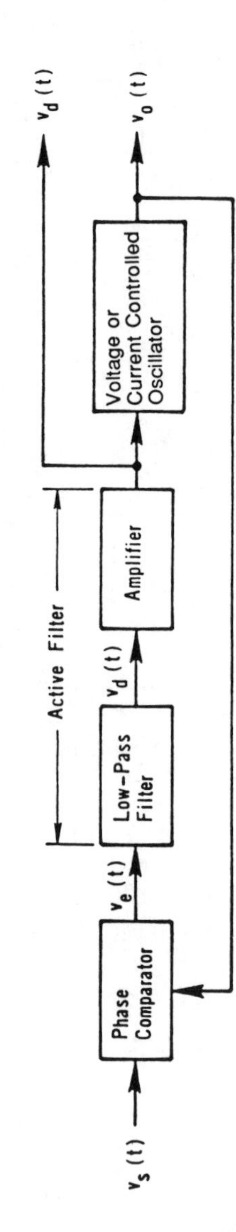

FIGURE 2.25 Block diagram of a phase-locked loop.

steady state, or locked-in state, $v_o(t)$ will be a strong signal which has the same frequency as $v_s(t)$, but with a phase shift about 90°; thus, the output virtually "represents" or "replaces" the incoming signal with a desirable signal-to-noise ratio. It is evident that this device is a good signal conditioning device which can recover signals being buried in noise apparently unidentifiable by the method of amplitude demodulation or discrimination.

2.5.2. Major Elements of the System

a. Phase Comparator and Its Mathematical Function

Figure 2.26(a) shows the simplified phase comparator circuitry. Its equivalent functional circuit is shown in Figure 2.26(b). It is basically an analog multiplier which operates as follows. Let

$$v_s(t) = V_s \sin w_s t$$

$$V_o(t) = \text{Square wave with } f_o \text{ frequency and amplitude} = 1$$

then by Fourier analysis,

$$v_o(t) = \sum_{n=0}^{\infty} \frac{4}{\pi(2n+1)} \sin[(2n+1)w_o t]$$

and

$$\begin{aligned} v_e(t) &= v_o(t)\, v_s(t) \\ &= V_s \sin w_s t \left[\sum_{n=0}^{\infty} \frac{4}{\pi(2n+1)} \sin(2n+1) w_o t \right] \\ &= V_s \sin w_s t \ \sin w_o t + V_s \sin w_s t \ \sin 3w_o t + \dots \\ &= \frac{V_s}{2}[\cos(w_o - w_s)t + \cos(w_o + w_s)t] \\ &\quad + \frac{V_s}{2}[\cos(3w_o - w_s)t + \cos(3w_o + w_s)t] + \dots \end{aligned}$$

If the cut-off frequency of the low-pass filter w_1 has a value of

$$w_o - w_s < w_1 < w_o + w_s$$

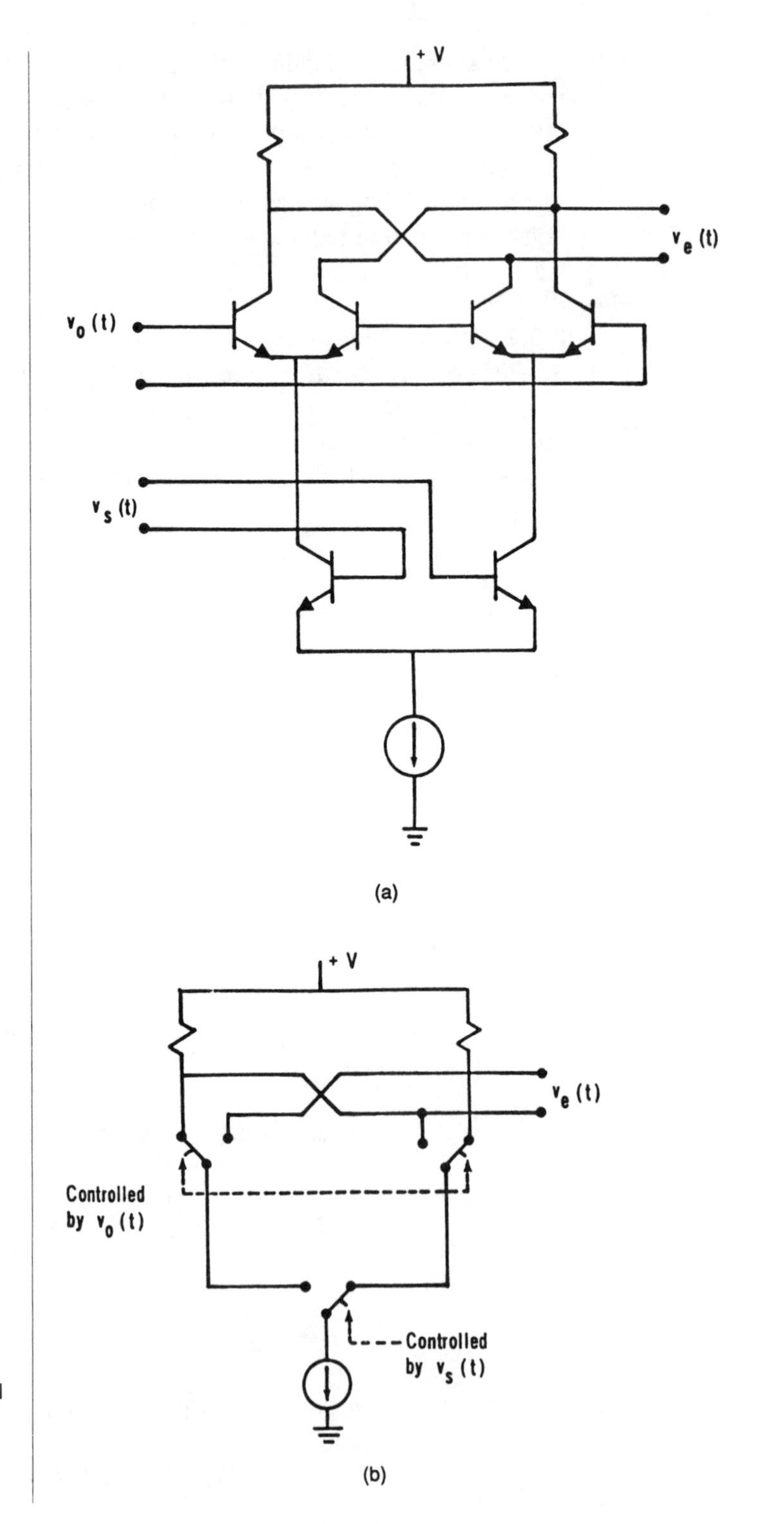

FIGURE 2.26 (a) Phase comparator. (b) Functional circuit of phase comparator. (c) Voltage-controlled oscillator.

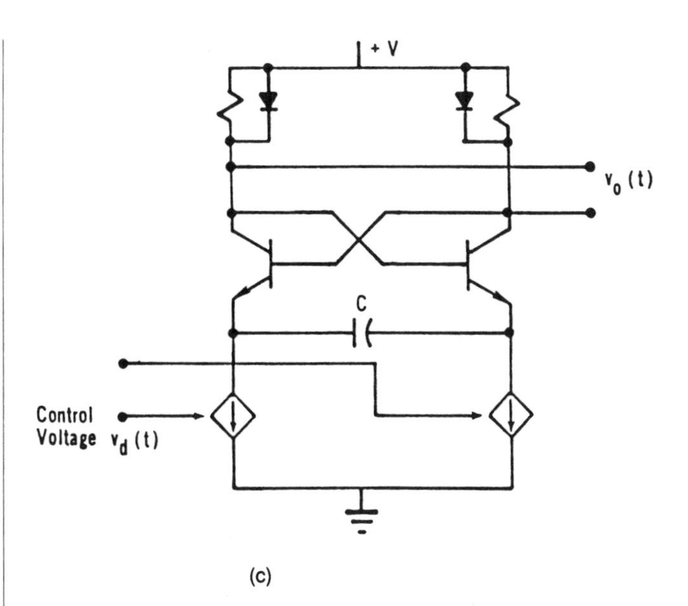

(c)

FIGURE 2.26 continued. (a) Phase comparator. (b) Functional circuit of phase comparator. (c) Voltage-controlled oscillator.

then the output voltage of the low-pass filter is

$$v_d(t) = \frac{V_s}{2}\cos(w_o - w_s)t \tag{2.72}$$

Consider the following two possible cases:

1. *Case I*: Assume $w_o \neq w_s$. Then,

$$v_d(t) = \frac{V_s}{2}\cos(w_o - w_S)t$$

   In this case we may consider v_d a time-varying dc signal. Its amplitude changes from zero to + $V_s/2$ and back to zero, then – $V_s/2$, etc. $v_d(t)$, however, is the input of the voltage controlled oscillator. That is, as $v_d = 0$ V, the oscillator will output a square wave at frequency w_o. When v_d increases, the frequency of the oscillator increases accordingly. In other words, as the v_d changes from + $V_s/2$ to zero to – $V_s/2$, the oscillator frequency will change from $(w_o + \Delta w)$ to $(w_o - \Delta w)$. If $w_o - w_s$ is within the range of Δw, the oscillator will eventually "catch up" with the signal frequency and "lock in" with it.

2. *Case II*: Assume $w_o = w_s$ and there is a phase difference between v_s and v_o, say ϕ. Then,

$$v_e(t) = [V_s \sin w_s t]\,[\sin(w_s t + \phi)]$$

$$= \frac{V_s}{2}\,[\cos\phi - \cos(2w_s t + \phi)]$$

At the output of the low-pass filter,

$$v_d = \frac{V_s}{2}\cos\phi$$

Therefore, as the oscillator locked in with the input signal, the value of v_d represents their phase difference: $v_d = 0$ as $\phi = 90°$.

b. Low-Pass Filter-Amplifier or Low-Pass Active Filter

This section can be a passive low-pass filter followed by an amplifier or an active filter with gain. As described in the preceding section, the cut-off frequency, w_1, of this filter should cover the range of $w_o - w_s$.

c. Voltage-Controlled Oscillator (VCO)

Figure 2.26(c) shows a simplified VCO circuit. It is a single capacitor free-running oscillator whose frequency is

$$f_o = \frac{V_d G_m}{2CV_{BE}}$$

where G_m and V_{BE} are the parameters of the controlled current generators. The diodes in the collector circuits are used to assure nonsaturation operation and limited voltage swing of the oscillator output. There are, of course, other circuitries that can serve the same purpose.

2.5.3. Linearized Negative Feedback System Model

During "lock-in" state, the system can be viewed and analyzed as a linear negative feedback system. Figure 2.27(a) shows a linearized system model where

ϕ_i = Phase angle of the input signal
ϕ_o = Phase angle of the output signal of the VCO
K_d = Conversion gain of the phase comparator (volt/rad)
$F(s)$ = Transfer function of the low-pass filter
A = Amplifier gain
K_o = VCO conversion gain. Since it converts voltage into frequency (f_o) and frequency into ϕ_o, while $f_o = \frac{d\phi_o}{dt}$, thus ϕ_o is an integration of f_o, so there is a block represented by 1/S.

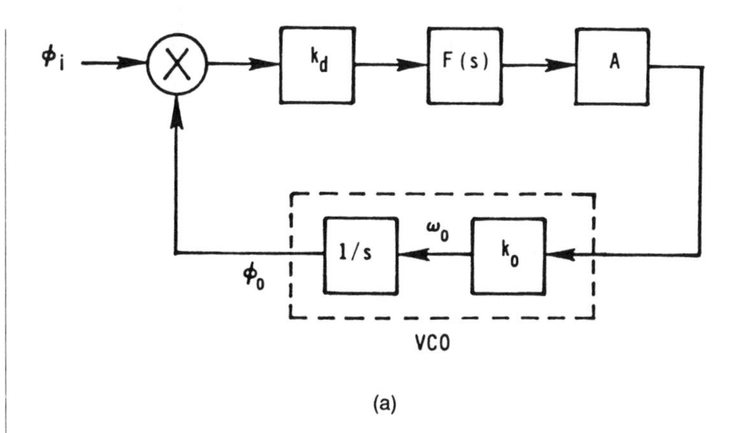

(a)

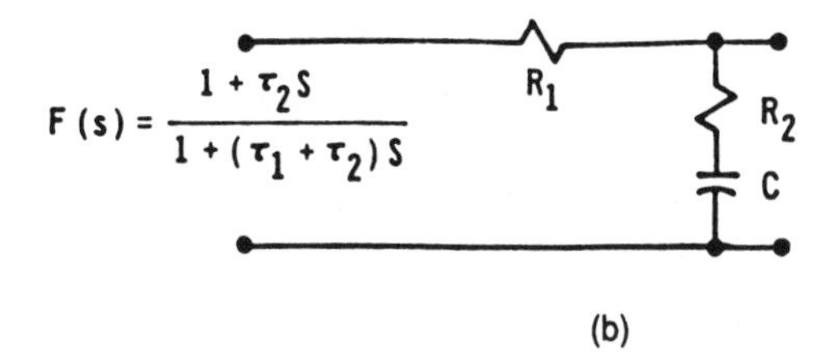

(b)

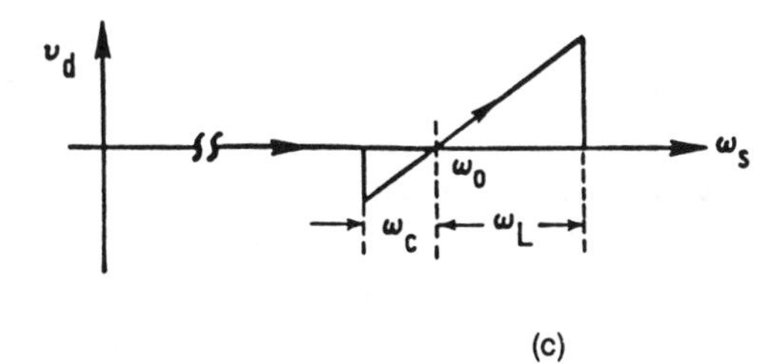

(c)

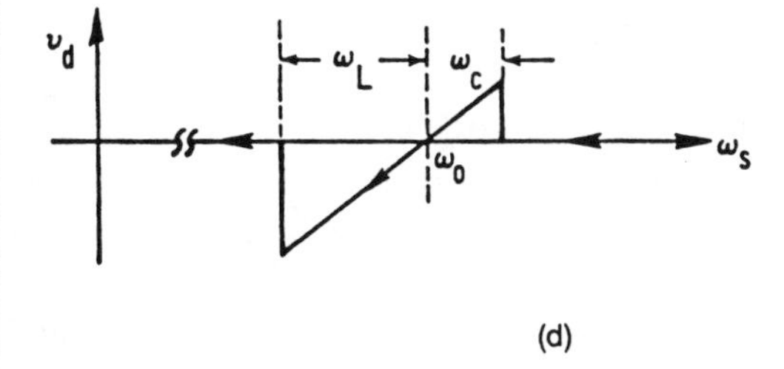

(d)

FIGURE 2.27 (a) Linearized negative feedback system model of PLL. (b) Lag-lead network. (c) w_S increasing locking process. (d) w_S decreasing locking process.

The system transfer function is

$$H(S) = \frac{AK_dF(s)}{1 + AK_d \frac{K_o}{S} F(S)}$$

To assure system stability, a lag-lead network can be used for the low-pass filter as shown in Fig. 2.27(b), which has the transfer function

$$F(s) = \frac{1 + \tau_2 S}{1 + (\tau_1 + \tau_2) S}$$

where

$$\tau_1 = R_1C, \tau_2 = R_2C$$

2.5.4. Capture Range and Lock-In Range

Figure 2.27(c) and (d) depict a different way to view the operation of the "locking" process. The horizontal axis represents the signal frequency, while the vertical axis represents the average or dc voltage of $v_d(t)$. Figure 2.27(c) shows the process of w_S increases from zero toward and past the VCO signal with a frequency equal to w_0, while Figure 2.27(d) shows the process of decreasing w_S.

In view of Equation (2.72), we see that $v_d(t)$ will be zero if $(w_0 - w_S) > w_1$, the cut-off frequency of the low-pass filter. If the network shown in Figure 2.27(b) is used, consider $w_1 = 1/(\tau_1 + \tau_2)$. As the w_S approaches w_0 and $w_0 - w_S$ entering the range between zero and w_1, the output of the low-pass filter becomes non-zero; as a result, the VCO responds to v_d adopting its frequency to w_S and the system is now "locked-in". As w_S increases continuously, the VCO will lock with w_S and the $w_0 - w_S$ becomes greater than w_1 and v_d drops to zero. Similar processes follow with the decreasing w_S. As shown in Figure 2.27(c) and (d), w_C is defined as the capture frequency and w_L, the locked-in frequency. Therefore, $2w_C$ is defined as the capture or acquisition range, and $2w_L$, the lock or tracking range.

2.5.5. Comments

It is interesting to point out that the low-pass filter actually serves two major functions, i.e., (a) control of the capture range of the system and (b) recapture of the signal in case the signal momentarily jumps out of the lock-range. In comparison with RC active filters, PLL has the following advantages and disadvantages.

Advantages:

a. High frequency capability
b. Independent controlling of selectivity and center frequency
c. Fewer external components
d. Ease of tuning

Disadvantages:

a. Lack of amplitude information of the input signal
b. Response to harmonics of the input signal
c. Difficult to provide automatic gain control

In view of the system diagram shown in Figure 2.25, PLL provides two outputs, namely, $v_d(t)$ and $v_o(t)$, depending on the nature of the application. A brief list of applications is given here:

a. FM demodulation:
   i. Broadcast FM detection
   ii. AM/FM telemetering decoding
   iii. FSK (Frequency Shift Keyed) demodulation
b. Frequency synchronization
c. Signal conditioning

REFERENCES

1. Millman, J. and Grabel, A., *Microelectronics*, 2nd ed., McGraw-Hill, New York, 1987.
2. Horowitz, P. and Hill, W., *The Art of Electronics*, 2nd ed., Cambridge University Press, New York, 1990.
3. Eimbinder, J., Ed., *Application Considerations for Linear Integrated Circuits*, John Wiley & Sons, New York, 1970.
4. Eimbinder, J., Ed., *Design with Linear Integrated Circuits*, John Wiley & Sons, New York, 1969.
5. Eimbinder, J., Ed., *Design with Linear Integrated Circuits: Theory and Applications*, John Wiley & Sons, New York, 1968.
6. Giles, J. N., Ed., Fairchild Semiconductor Linear Integrated Circuits Applications Handbook, Fairchild Semiconductor, 313 Fairchild Drive, Mountain View, CA, 1967.
7. Mochytz, G. S., The operational amplifier in linear active networks, *IEEE Spectrum*, 7(1), 42—50, January 1970.
8. Naylor, J. R., Digital and analog signal applications of operational amplifiers, Part I, *IEEE Spectrum*, 8(5), 79—87, May 1971.
9. Naylor, J. R., Digital and analog signal applications of operational amplifiers, Part II, *IEEE Spectrum*, 8(6), 38—46, June 1971.
10. Johnson, D. E. and Hilburn, J. L., *Rapid Practical Design of Active Filters*, John Wiley & Sons, New York, 1975.
11. Signetics Analog Data Manual, 1977.
12. Handbook of Operational Amplifier Applications, Burr-Brown, Tucson, Arizona, 85706.
13. Grebene, A. B., The monolithic phase-locked loop — a versatile building block, *IEEE Spectrum*, 8(3), 38—49, March 1971.
14. Signetic Linear Phase-Locked Loops Applications Book.
15. Grebene, A. B., The monolithic phase-locked loop — a versatile building block, *EDN*, 17(19), 26—33, October 1, 1972.
16. Kesner, D., Take the guesswork out of phase-locked design, *EDN*, 18(1), 54—60, January 5, 1973.

17. Gardner, F. M., *Phaselock Techniques*, 2nd ed., John Wiley & Sons, New York, 1979.
18. Grebene, A. B. and Camezind, H. R., Frequency selective I.C. using phase-lock techniques, *IEEE Journal of Solid-State Circuits*, SC-4, 216—225, August 1969.
19. Moschytz, G. S., Miniaturized RC filters using phase-locked loop, *Bell System Tech. J.*, 44, 823—870, May 1965.
20. Kalpper, J. and Frankle, J. T., *Phase-Locked and Frequency Feed-back Systems: Principle and Techniques*, Academic Press, New York, 1972.
21. Viterbi, A. J., *Principles of Coherent Communication*, McGraw-Hill, New York, 1971.
22. Van Trees, H. L., *Detection, Estimation and Modulation Theory Part II*, John Wiley & Sons, New York, 1971.
23. Lindsey, W. C., *Synchronization Systems on Communication and Control*, Prentice-Hall, Englewood Cliffs, NJ, 1972.
24. Holmes, J. K. and Tegnelia, C. R., A second-order all-digital phase-locked loop, *IEEE Trans. Comm.*, COM-22(1), 62—68, January 1974.
25. Reed, L. J. and Treadway, R. J., Test your PLL IQ, *EDN*, 14(24), 27—30, December 20, 1974.
26. Motorola Manual on Phase-Locked Loops.
27. Gupta, S. C., Phase-locked loops, *Proc. IEEE*, 63(2), 291—306, February 1975.

3 A REVIEW AND CLARIFICATION OF SWITCHING AND PULSE CIRCUIT FUNDAMENTALS

3.1. CLARIFICATIONS OF BASIC CONCEPTS

A review of the electrical engineering curriculum reveals that the training sequence for circuit or system designers has been direct current (dc), alternate current (ac), linear circuits, and steady state analysis of circuits and systems. Having thus been "brainwashed", designers often have difficulties adapting themselves to dealing with digital circuits and system analysis. In digital design, most of the circuit elements used are nonlinear, and techniques for transient state and nonlinear circuit analysis are vital tools. An attempt is made in this chapter to clarify some of the topics where confusion is commonly experienced. Basically, a digital designer should be concerned with time variables and nonlinear characteristics in addition to the four traditional parameters, i.e., current, voltage, impedance, and power. Fortunately, in digital system design, one needs only to deal with two logic states and the transient change from one state to the other; once this basic concept is clarified, there is no need for difficulty. As for analysis of nonlinear devices, there are three popular techniques used in digital design: mathematical, graphical and piece-wise linear. For linear analysis, such as the analysis of the operational amplifier described in Chapter 2, the signals considered are mostly periodic and small signals, so that linear models and Fourier series analysis can be used. In digital systems and circuits, one deals mostly with large and nonperiodic signals. The traditional general expression of impedance for a R-L-C circuit such as

$$Z(jw) = R + jwL - j\frac{1}{WC}$$

and linear models of voltage and current sources are seldom used. Instead, one would consider mostly the electric properties during the switching transition period and the binary steady states. Although the

Laplace Transform technique can be used for transient analysis, a simple time domain concept appears to be sufficient. What follows will be the details of these techniques applied to digital circuit elements or system analysis.

3.2. LINEAR ELEMENTS

3.2.1. Resistors

A resistor can be considered as a linear device. In other words, regardless of the magnitude of the voltage or current, it satisfies Ohm's Law:

For dc signal,

$$V = IR$$

where

V = Direct voltage, invariant with time
I = Direct current, invariant with time
R = Resistance, a constant.

For ac signal,

$$v(t) = i(t)\,R$$

where

$v(t)$ = Time variant voltage
$i(t)$ = Time variant current.

There are devices or circuit components that do not always satisfy Ohm's Law for all values of voltage and current. In such cases, graphical presentation of the voltage-current characteristics (known as the V-I curve) of a device would be most useful and appropriate. Although trivial, for completeness of discussion and comparison with other devices in later sections, the V-I curves of resistors with different values are shown in Figure 3.1.

3.2.2. Capacitors

Those who are familiar with the problems involved only with steady state sinusoidal signals would still use Ohm's Law and replace resistance by a capacitive reactance X_C, i.e.,

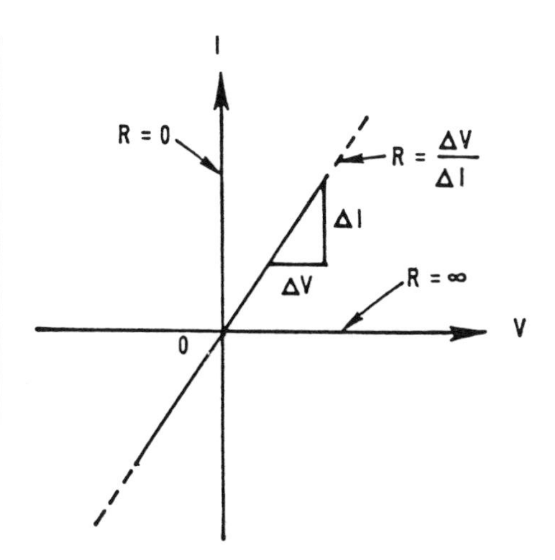

FIGURE 3.1 V-I curve for resistors.

$$v(wt) = X_c i(t) = \frac{1}{jwc} i(t)$$

where

$$\begin{aligned} w &= 2\pi f, \;\; f = \text{frequency in hertz} \\ c &= \text{Capacitance of the capacitor in farads.} \end{aligned}$$

However, in digital systems we deal mostly with pulse signals instead of sinusoidal signals. Therefore, we need a thorough conceptual clarification. Here the volt-current relationship for a capacitor is different from that for a resistor. For a direct or time-invariant voltage (steady state), the current through a capacitor is always zero, while for a time-variant voltage, the current is proportional to the change of the voltage with respect to time. Mathematically,

$$Q = CV$$

or

$$\frac{dv(t)}{dt} = \frac{1}{C}\frac{d}{dt}q(t) = \frac{1}{C}i(t) \tag{3.1}$$

where

$$\begin{aligned} Q &= \text{Electrical charge in coulombs} \\ C &= \text{Capacitance in farads} \end{aligned}$$

V = Constant voltage in volts
v(t) = Time-varying voltage.

Also, we have

$$v(t) = \frac{1}{C}\int_{t_1}^{t} i(t)\, dt + v(t_1) \tag{3.2}$$

If i(t) is finite, then

$$\lim_{t \to t_1} v(t) = 0 + v(t_1) \tag{3.3}$$

Physical interpretation of Equation (3.3) is interesting and important. It means that

THE VOLTAGE ACROSS A CAPACITOR CANNOT CHANGE INSTANTANEOUSLY IF THE CURRENT IS FINITE.

The following examples will clarify the concept.

Example 1

In Figure 3.2(a) the switch is closed at $t = t_1$. Since the maximum possible current of the circuit,

$$[i(t)]_{max} = \frac{E}{R}$$

is finite, Equation (3.3) is valid, and the voltage across the capacitor cannot change instantaneously. So,

$$v(t_1^-) = v(t_1^+) = 0$$

where

$$t_1^- \triangleq \text{an infinitesimal time before } t_1$$

$$t_1^+ \triangleq \text{an infinitesimal time after } t_1$$

and the symbol "$\triangleq$" means "defined as".

For the steady state condition, the voltage of the capacitor must be constant, so

$$\frac{dv(\infty)}{dt} = 0$$

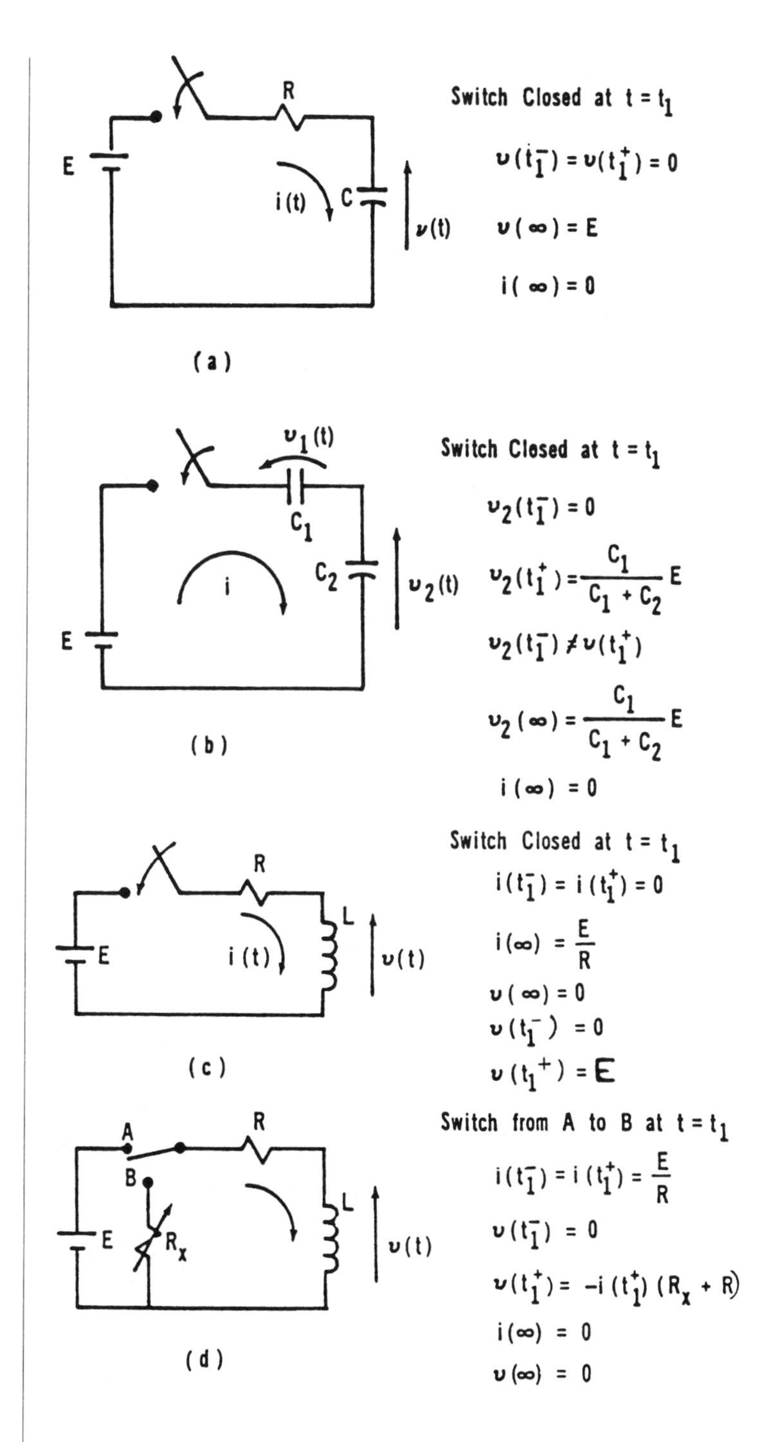

FIGURE 3.2 (a) R-C circuit. (b) Pure capacitive circuit. (c) R-L circuit. (d) R-L circuit.

then from Equation (3.1), the current is zero. The steady state condition is usually denoted by $t = \infty$ for convenience, hence we have

$$i(\infty) = 0$$

and the voltage across the capacitor in this example,

$$v(\infty) = e - Ri(\infty) = E$$

Example 2

In Figure 3.2(b) the switch is closed at $t = t_1$. However, in this example, the resistance is zero; therefore, the current would not be finite and the voltage $v_2(t)$ can jump from zero to

$$\frac{C_1}{C_1 + C_2} E$$

instantaneously, and we have

$$v_2(\infty) = \frac{C_1}{C_1 + C_2} E, \; i(\infty) = 0$$

One may wonder how we got this result. Based on the basic equation,

$$Q = CV$$

one can write the loop equation:

$$\begin{aligned} E &= v_1(t) + v_2(t) \\ &= \frac{1}{C_1} q_1(t) + \frac{1}{C_2} q_2(t) \end{aligned}$$

Since $q_1(t) = q_2(t)$, we have

$$E = \frac{C_1 + C_2}{C_1 C_2} q_1(t)$$

$$q_1(t) = \frac{C_1 C_2}{C_1 + C_2} E = q_2(t)$$

Hence,

$$v_2(t) = \frac{1}{C_2} q_2(t) = \frac{C_1}{C_1 + C_2} E$$

Similarly,

$$v_1(t) = \frac{1}{C_1} q_1(t) = \frac{C_2}{C_1 + C_2} E$$

Most of us are familiar with the characteristics shown in Example 1, but not those shown in Example 2. However, the latter are as important as the former. Readers are urged to bear these concepts in mind while analyzing the switching circuits.

3.2.3. Inductors

In Section 3.2.2, the electric properties of a capacitor were described in detail. Since the inductor is a "dual element" of a capacitor, we simply replace capacitance, C, by inductance, L, and current by voltage or vice versa, and all of the equations shown in the preceding section will be valid. We have

$$\frac{di(t)}{dt} = \frac{1}{L} v(t) \tag{3.4}$$

$$i(t) = \frac{1}{L} \int_{t_1}^{t} v(t)\, dt + i(t_1)$$

If v(t) is finite,

$$\lim_{t \to t_1} i(t) = 0 + i(t_1) \tag{3.5}$$

which means,

THE CURRENT OF AN INDUCTOR CANNOT CHANGE INSTANTANEOUSLY IF THE VOLTAGE ACROSS THE INDUCTOR IS FINITE.

Example 1

In Figure 3.2(c) the switch is closed at $t = t_1$. We have

$$E = i(t)R + v(t)$$

where i(t) is finite, hence v(t) is finite and

$$i(t_1^+) = i(t_1^-) = 0$$

$$v(t_1^-) = 0$$

$$v(t_1^+) = E$$

When $t \to \infty$, the circuit is in steady state and $\frac{di}{dt} = 0$, thus

$$v(\infty) = 0 \text{ and } i(\infty) = \frac{E}{R}$$

Example 2

In Figure 3.2(d), the switch is on "A" for a *long, long* time, and being switched to "B" at $t = t_1$, we have

$$v(t_1^-) = L\frac{di(t^-)}{dt} = 0 \tag{3.6}$$

$$i(t_1^-) = i(t_1^+) = \frac{E}{R}$$

The loop equation with the switch on "B" is

$$v(t^+) + i((t^+)(R_x + R) = 0$$

$$v(t^+) = -i(t^+)(R_x + R)$$

$$v(t_1^+) = -i(t_1^+)(R_x + R) \tag{3.7}$$

Since there is no generator or source in the loop with the switch on "B", we know that

$$i(\infty) = 0$$

$$v(\infty) = 0$$

Equations (3.6) and (3.7) show a very interesting property, namely, the voltage across the inductor drops from zero to negative abruptly, i.e., if $R_x = 0$,

$$v(t_1^+) = -i(t_1^+)R = \frac{E}{R} = -E$$

If $R_x = \infty$, or open-circuit,

$$v(t_1^+) = -\infty \text{ !!}$$

One should note that this example explains why the switching circuit sometimes yields a "negative spike" even though only a positive power supply is used. This negative spike is usually an unwelcome troublemaker. It generates noise which may cause error in the digital system and sometimes may even damage transistors in the circuit.

3.3. NONLINEAR ELEMENTS

3.3.1. Diodes

A diode is a nonlinear device. Unlike the resistor, its V-I characteristic cannot be simply described by Ohm's Law. Instead, it can be quite accurately expressed by

$$I = K_1[(e^{V/K_2}) - 1]$$

where K_1, K_2 are constants and V is the voltage across the P-N junction as shown in Figure 3.3(a). However, for functional analysis or for understanding how the device functions as a digital or a binary state device, the V-I curve can be approximated by two segments of straight line as shown in Figure 3.3(b). This technique is known as *piece-wise linear approximation.* The curve can be interpreted as

$$I = 0 \quad \text{if } V \leq V_\theta$$

$$I = \frac{V - V_\theta}{r_D} \quad \text{if } V > V_\theta$$

where

V_θ = Threshold voltage $\simeq 0.7$ V

r_D = Diode equivalent forward resistance $\simeq 50\ \Omega$.

Physically, this means that when the voltage across the diode is equal to or less than V_θ, the diode is practically open-circuit, otherwise it is

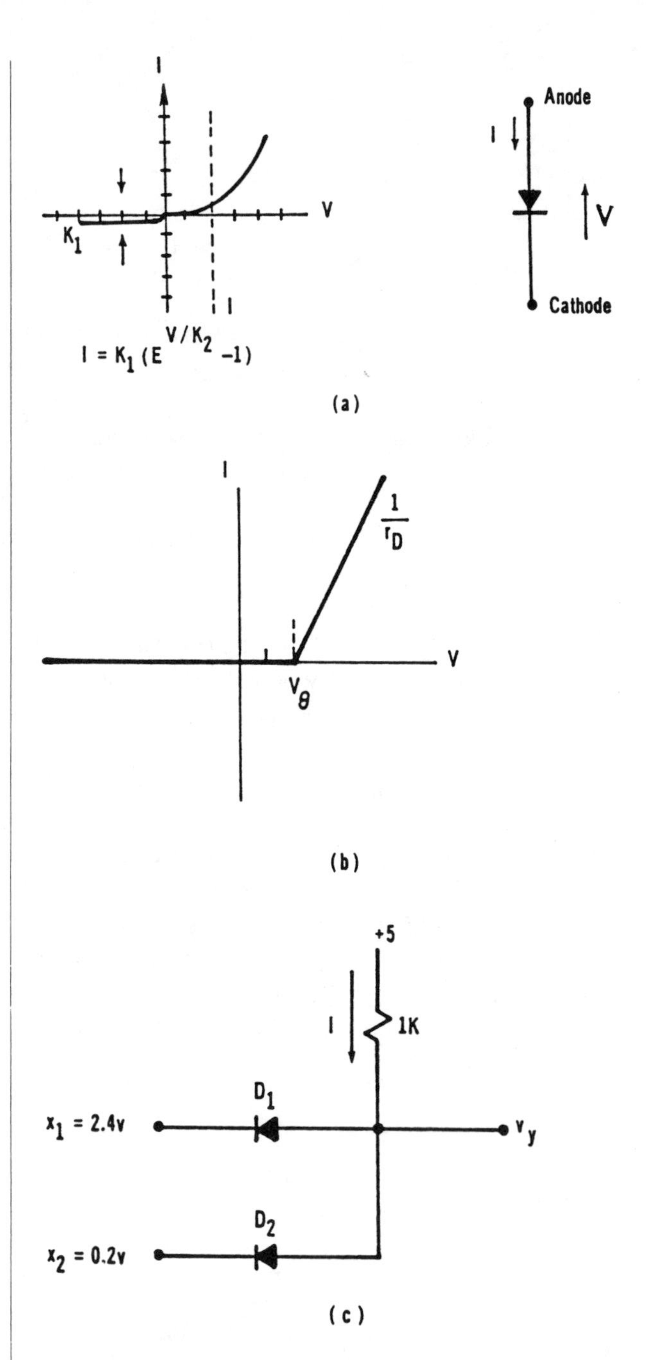

FIGURE 3.3 (a) Diode. (b) Piece-wise linear diode V-I curve. (c) Resistor-diode circuit. (d) Resistor-diode circuit.

closed with a resistance of 50 Ω in series. In a diode-resistor logic network, one would mostly want to know which states the diodes are in. Sometimes the problems are not so trivial, as we can illustrate here. The following two examples demonstrate some of the problems.

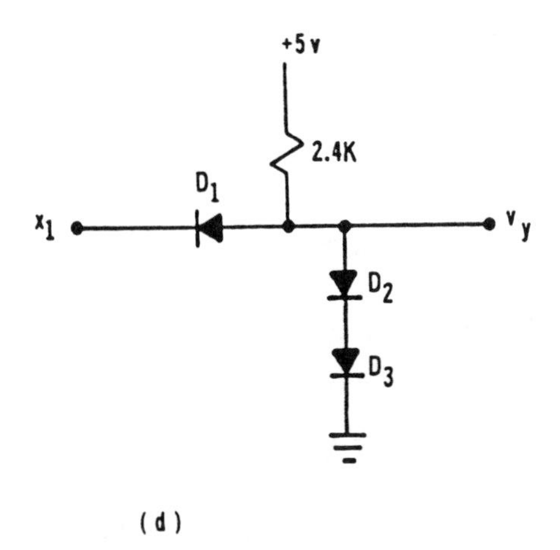

FIGURE 3.3 continued. (a) Diode. (b) Piece-wise linear diode V-I curve. (c) Resistor-diode circuit. (d) Resistor-diode circuit.

Example 1

For the diagram in Figure 3.3(c), let x_1 = logical 1 = 2.4 V, and x_2 = logical 0 = 0.2 V. It appears that diodes D_1 and D_2 will both be conducting since their anodes are connected to 5 V, while the cathode is at 2.4 and 0.2 V, respectively; the voltage difference between the anode and cathode for both diodes exceeds the threshold voltage $V_\theta \simeq 0.7$ V Actually, however, D_2 is "ON" and D_1 is "OFF". Although an experienced person can see this immediately, let us pretend to be a novice and set up a truth table:

Case	D_1	D_2
1	OFF	OFF
2	ON	OFF
3	ON	ON
4	OFF	ON

Case 1 is impossible because the voltage difference between terminals of both diodes exceeds V_θ.

Case 2 will result in

$$V_y = 2.4 + V_\theta + Ir_D = 2.4 + V_\theta + \frac{5 - V_y}{1\ K} \cdot r_D$$

Since $r_D << 1$ K,

$$V_y \simeq 2.4 + 0.7 = 3.1 \text{ V}$$

and $(3.1 - x_2) > V_\theta$, thus D_2 could not be "OFF".

In Case 3, we have

$$V_y = x_2 + V_\theta = 0.2 + 0.7 = 0.9 \text{ V and also}$$
$$V_y = x_1 + V_\theta = 2.4 + 0.7 = 3.1 \text{ V}$$

but V_y cannot have two values at the same time, thus, this is impossible.

Case 4 will result in $V_y = 0.9$ V, which will keep D_1 from conducting, since

$$V_y - x_1 = 0.9 - 2.4 = -1.5 < V_\theta$$

Hence, Case 4 is the only possible state for this circuit.

Example 2

In Figure 3.3(d), let x_1 = logical 1 = 2.4 V if D_1 is "ON", then $V_y \simeq 2.4 + 0.7 = 3.1$. Since D_2 and D_3 are in series, the minimum required voltage for D_2 and D_3 to be conducting is

$$V_y = 2V_\theta \simeq 1.4 \text{ V}$$

because $V_y = 3.1$ if D_1 is "ON". Thus, D_2 and D_3 have to be conducting. But if D_2 and D_3 are conducting, then

$$V_y \simeq 0.7 + 0.7 = 1.4 \text{ V}$$

which will require V_y to have two values at the same time, thus this will not be the case. But if D_2 and D_3 are conducting, then $V_y \simeq 1.4$, which will keep D_1 from conducting if $x_1 = 2.4$. Thus if $x_1 = 2.4$ V, D_1 will be "OFF" and D_2 and D_3 will be "ON". Now let x_1 = logical 0 = 0.2 V, then

$$V_y \simeq x_1 + V_\theta = 0.2 + 0.7 = 0.9 \text{ V}$$

which is less than $2V_\theta$. Thus, D_2 and D_3 will be "OFF", and D_1 will be "ON".

Comments

These two examples show the basic concept which will be used over and over again when the logic designer has to know how integrated circuit logic elements such as DTL, TTL, etc., function. For convenience, let us define this method of analysis as *binary state analysis*, or BSA, which is basically derived from the piece-wise linear V-I curve model.

3.3.2. Transistors (Bipolar): The Current Control Device

The transistor is also a nonlinear device which can be used as an electrically controllable switch or as an amplifier. However, in contrast to diodes, which are two-terminal devices, transistors are three-terminal devices. Since we are dealing with digital circuits, use of a transistor as a switch or a binary state device will be emphasized here.

In a switching circuit, a transistor appears analogous to a push-button switch as shown in Figure 3.4(a). For a push-button switch, one has to apply a force on the button to complete the electric circuit; the force required will depend on the strength of the spring inside the switch. For a transistor, the base terminal can be thought of as being the button of the switch. But instead of applying a mechanical force at the base, it requires a proper amount of current flowing through the base in order to turn on the transistor switch. In what follows, we will describe the analysis techniques and design examples for circuitry consisting of transistors and diodes.

a. Binary State Analysis (BSA) for a Transistor Switch

For simplicity, a transistor can be considered as two diodes, i.e., Base-Emitter and Base-Collector, being connected back to back with current gain property. That is, in the circuit diagram, I_C/I_B = current gain = β, which is normally much greater than 1. Let us neglect the leakage current of the transistor. As shown in Figure 3.4(a), since

$$V_y = 5 - I_C \cdot 1\text{ K} = 5 - \beta I_B \cdot 1\text{ K} \tag{3.8}$$

we have

$$V_y = 5\text{ V} \quad \text{if } I_B = 0$$

That is, if the transistor is considered as a switch, then the switch is open when the current $I_B = 0$. Now let $\beta = 20$. If $I_B = 0.25$ mA, then $I_C = \beta I_B = 5$ mA. Substituting into Equation (3.8), we have $V_y = 0$ and the transistor switch is now closed. Consider $I_B = 0.1$ mA, from Equation (3.8),

$$V_y = 5 - 20 \cdot 0.1 \cdot 10^{-3} \cdot 1\text{ K} = 3\text{ V}$$

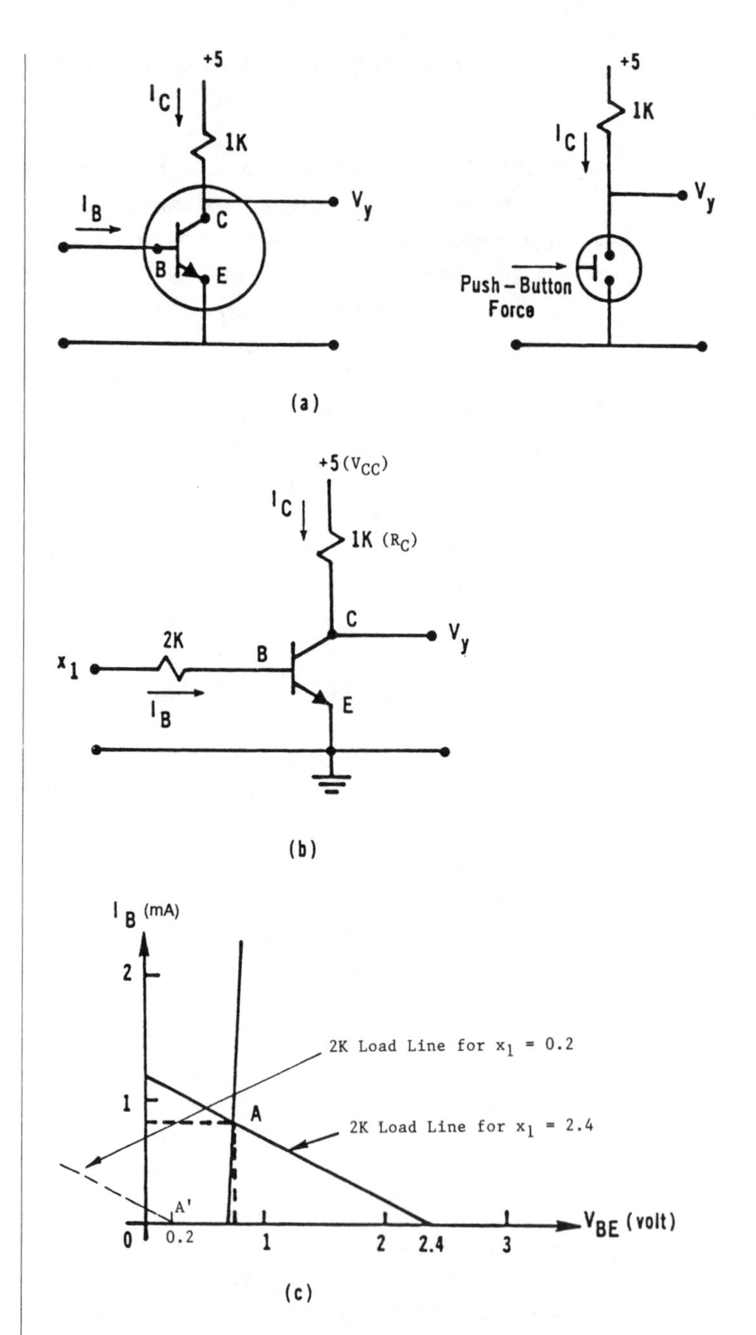

FIGURE 3.4 (a) Resistor-transistor circuit. (b) Resistor-transistor circuit. (c) Transistor input V-I curve. (d) Transistor output V-I curves. (e) Resistor-transistor circuit.

which means that the transistor switch is neither completely "ON" or "OFF". It behaves like a conventional switch having a poor contact because here we have a 2-V voltage-drop across the transistor C-E junction. Of course, this is not desirable, since for an ideal switch, the voltage-drop across it must be zero when closed. However,

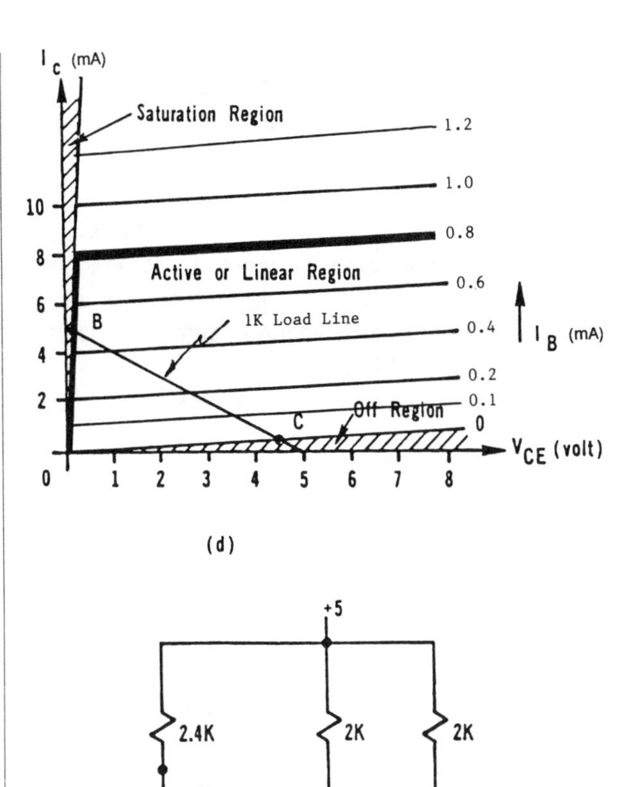

FIGURE 3.4 continued. (a) Resistor-transistor circuit. (b) Resistor-transistor circuit. (c) Transistor input V-I curve. (d) Transistor output V-I curves. (e) Resistor-transistor circuit.

$$(V_y)_{min} = 0, \text{ which cannot be negative, so from Equation (3.8)}$$

$$(IC)_{max} = 5 \text{ mA}$$

That is, the collector current of the transistor is limited by the power supply and the collector resistance.

Now consider $I_B = 1$ mA, with the transistor in its "ON" state. Although the control current I_B demands $I_C = \beta I_B = 20$ mA, the circuit can supply no more than 5 ma. This is equivalent to saying that one may push the button harder, but the switch is still only in the "ON" state. One can never make a lamp brighter by pushing the switch harder. A transistor being overdriven like this is referred to as being in *saturation state*. Most of the logic circuit is designed so that the transistor is either in the "OFF" or "saturation" state except extremely high speed logic circuits such as emitter-coupled logic (ECL) or current mode switching logic circuits. It is noted that when a push-button switch is "ON", $V_y = 0$, but for a transistor switch, $V_y \simeq 0.1 - 0.2$ V when it is driven to saturation. This voltage is called *saturation voltage*, which is not desirable in processing analog signals but is acceptable for logic network realization. In practice, the control current I_B is usually generated with a voltage source as shown in Figure 3.4(b).

Let x_1 = logical 1 = 2.4 V. Since Base-Emitter is a diode junction, the BSA technique used in Section 3.3.2.a can be applied here. We have

$$I_B = \frac{2.4 - V_\theta}{2\ K} \simeq \frac{2.4 - 0.7}{2\ K} = 0.85 \text{ mA}$$

and

$$\beta I_B = 20 \cdot 0.85 \text{ mA} = 17 \text{ mA} > 5 \text{ mA}$$

thus the transistor is saturated and the switch is turned "ON". Now let x_1 = logical 0 = 0.2 V, which is less than the threshold voltage V_θ, thus $I_B \simeq 0$, and the transistor switch is turned "OFF", $V_y = 5$ V.

b. Graphical Analysis Method

Graphical analysis is a convenient method for analyzing a circuit that consists of nonlinear devices or components such as diodes or transistors. As electrical engineers, we can describe any electric device, be it newly invented or old, by the relationship between the voltage across it and the current flowing through it. In other words, we can describe it either by a mathematical equation with V, voltage, and I, current, as the two variables, or graphically by its V-I curve. For example, to characterize a resistor, we can use the equation V = IR, or the V-I curve shown in Figure 3.1. Since the former is simpler, we never bother using the graphical expression for a resistor. Similarly, for a diode, we may use the equation shown in Section 3.3.1, or the V-I curve shown in Figure 3.3(a), or the simplified piece-wise linear V-I curve shown in Figure 3.3(b). Of course, one can still express the curve Figure 3.3(b) with equations:

$$I = 0 \quad \text{if } V \leq V_\theta$$

$$I = \frac{V}{r_D} \quad \text{if } V > V_\theta$$

However, when we are interested in analysis of a circuit consisting of both linear and nonlinear devices, the graphical method is a convenient one. A numerical example may clarify the concept. Let us analyze the circuit shown in Figure 3.4(b) by the graphical method as follows.

We may consider the transistor as a black box which is characterized by its input V-I curve (a piece-wise linear diode V-I curve) and output V-I curves (a family of I_C-V_{CE} curves) as depicted in Figure 3.4(c) and (d), respectively. Referring to Figure 3.4(b), for the input loop we have a 2 K resistor (linear device) mixed with the B-E diode junction (nonlinear device) of the transistor, while for the output loop the 1 K resistor is mixed with the N-P-N junctions of the transistor. Since the transistor is described by its input/output V-I curves, we can start with the given V-I curve without bothering with the internal detail. Actually, we can use graphical technique to analyze any device as long as its input/output V-I curves are given. Basically, what we are interested in finding out is, given an input of $X_1 = 2.4$ V, what would be the values of I_B, I_C, and V_y or V_{CE}?

For Input Loop:

$$2.4 = 2\,\text{K} \cdot I_B + V_{BE} \quad \text{and}$$

$$I_B = 0 \ \text{ for } V_{BE} < 0.7$$

$$I_B = \frac{V_{BE} - 0.7}{100} \ \text{ for } V_{BE} > 0.7$$

where we assume the forward resistance of the diode junction to be

$r_D \simeq 100\ \Omega$, as the V-I curve shown.

Here, we are to solve for the unknowns or variables, V_{BE} and I_B. Since we linearized the input V-I curve, we could still solve the equations shown algebraically. However, the graphical method could easily solve the problem even if the input V-I curve is not linearized. Notice that on the $V_{BE} - I_B$ space, we may plot two curves, one for the input V-I curve of the nonlinear (but here we use the piece-wise linear method for convenience) diode junction, and the other for the resistor. However, since we must plot the curves on the same space defined by the two variables we are interested in, the V-I curve for the resistor in this space is now

$$I_B = \frac{2.4 - V_{BE}}{2\ \text{K}}$$

Since it is a linear curve, we need only calculate two points for plotting the curve. Fortunately, there are two trivial points, namely $V_{BE} = 0$ and $I_B = 0$, that we can use to determine the two points from the equation shown by inspection. That is, for $V_{BE} = 0$, we have $I_B = (2.4/2\ K)$, which is 1.2 mA; for $I_B = 0$, we have $V_{BE} = 2.4$. With the two points (0 V, 1.2 mA) and (2.4 V, 0 mA) determined, we can plot the V-I curve for the resistor on the same space where the V-I curve of the diode junction is given. The two curves are shown in Figure 3.4(c), and the intersection of the two curves, A, at (0.75 V, 0.8 mA) is found, which is the solution for the input loop.

For Output Loop:

Since for the input loop we found that $I_B = 0.8$ mA, we can now pick the V-I curve (boldfaced) with $I_B = 0.8$ mA from the V-I curve family shown in Figure 3.4(d) as the V-I curve of the output loop for this circuit. Notice that we are now in the $V_{CE} - I_C$ space, and the V-I curve for the 1 K resistor in the output loop (Figure 3.4(b)) can be expressed in the $V_{CE} - I_C$ space as follows:

$$V_{CE} = 5 - 1\ K \cdot I_C$$

By the same token, for the 1 K resistor we determine the two points. For $V_{CE} = 0$, we have $I_C = 5$; for $I_C = 0$, $V_{CE} = 5$. As a result, we have the two curves shown in Figure 3.4(d): the $V_{CE} - I_C$ curve for $I_B = 0.8$ and that for the 1 K resistor. They intersect at point B (0.2 V, 4.9 mA). That is, the solution for this output loop is at point B and thus we have $V_y = 0.2$ V for $I_B = 0.8$. It is worth noting that the V-I curve of the resistor is also known as the load line of the transistor.

Now, let us consider the solution for $x_1 = 0.2$ V. Using the BSA technique, we find the B-E diode junction is OFF, thus $I_B = 0$. By the graphical technique, the V-I curve for the 2 K resistor in $V_{BE} - I_B$ space shown in Figure 3.4(c) will be shifted parallel to the left until $V_{BE} = 0.2$ as shown by the dotted line. The solution then is the intersection of the two curves, or Point A′ at (0.2 V, 0 mA). Accordingly, in the output space shown in Figure 3.4(d), we would in this case pick the V-I curve of the transistor with $I_B = 0$, and point C is then the solution. In practice, the V-I curve for $I_B = 0$ is nearly coincident with the V_{CE} axis, since there should be only the leakage current, which can only be a fraction of a microampere flowing through the 1 K resistor and should not be observable in this diagram. For convenience of demonstrating the concept, however, we make the curve visible. That is, we must point out the significance of the two points, B and C, shown in the figure. They are known as the *saturation* and *off* points, which respectively define these two logical states in digital systems. The two shaded areas are called saturation and off regions. Notice that point B will not move as I_B exceeds 0.5 mA. The advantage to using saturation state and off state for a transistor to implement logical 1 and logical 0 is now apparent. It provides the assurance of two distinct states for binary or digital operation. For example, if we design the circuit such that I_B is much greater than 0.5 mA in one state and 0 mA in the other, then the circuit would yield 0.2 V at one state and 5 V at the other. Thus the circuit would not be sensitive to variations of device properties and operation environment. However, logic elements operating at saturation mode have one disadvantage; they have a slower switching speed in comparison with nonsaturation logic, such as ECL and Schottky TTL, which will be described in Chapter 4, Sections 4.2.6 and 4.2.7.

c. Identification of Saturation State

There are two ways to check if a transistor is saturated or not. From the designer's viewpoint, one need only be sure that the circuit is designed in such a way that

$$\beta I_B > \frac{V_{CC}}{R_C}$$

where

VCC = Power supply in volts
R_C = The resistor in the collector circuit as shown in Figure 3.4(b)
β = Current gain of the transistor

As a rule of thumb, one should design βI_B to be ten times greater than V_{CC}/R_C. Or, one can determine graphically if point B is in the transistor's saturation region as shown in Figure 3.4(d). In the laboratory, however, one can tell that a transistor is in its saturation state if by measurement $V_{CE} \leq 0.2$ v. Or, $V_{CE} < V_{BE}$.

d. Comparison Between BSA and Graphical Techniques

It is apparent that the graphical techique yields more complete and detailed information than the BSA. Examine Figure 3.4(d) again, and note that we can find the solutions for any given values of I_B, and determine the criterion for the design of saturation/nonsaturation logic. On the other hand, the BSA technique yields quick and adequate information for ON/OFF or logic 1/logic 0 analysis. The next example illustrates this point.

Example 1

We want to determine the states of transistors Q_1, Q_2, and Q_3 shown in Figure 3.4(e) with x_1 = logical 1 = 2.4 V, and x_1 = logical 0 = 0.2 V. Basically, the circuit has the diode junctions of the Base-Collector of Q_1 and the Base-Emitter junctions of Q_2 and Q_3 in series, which is shunting the diode junction of the Base-Emitter of Q_1. The BSA technique of Example 2 in Section 3.3.1 can be used here. We have for x_1 = 2.4 V the three "diode junctions", i.e., Base-Collector of Q_1 and Base-Emitter of Q_2 and Q_3 are conducting, which only requires $3 \times 0.7 = 2.1$ to turn on, thus Q_2 and Q_3 are saturated and

$$V_y \simeq 0.2 \text{ V}$$

Similarly, for x_2 = 0.2 V, the Base-Emitter junction of Q_1 is "ON", and Q_2 and Q_3 are "OFF", thus V_y = +5 V.

e. Design Examples

Although in digital system design transistors and diodes are mostly employed as switching devices, we may find in many cases that devices being designed and operated in linear mode are buried in the jungle of MSI (medium-scale integration) or LSA (large-scale integration) switching circuitry. For example, the single-stage differential amplifier described in Chapter 2 is being used in ECL, transmission line drivers, and comparator circuits. In the following we present three design examples which are frequently used as "building blocks" in switching and logic circuitry, but are operated in linear modes. As a by-product, they will illustrate how the graphical analysis concept introduced in Section 3.3.2.b is used for the design.

Voltage Reference

Design — Figure 3.5(a) shows a simple circuit to produce a reference voltage, V_r. V_x is the input voltage source which varies randomly. Let us arbitrarily pick V_x = 10 V (± 25%), V_r = 5 V, and a ¼ W zener diode of 5 V as the design example. We are to determine the value of the resistor, R_z, for the circuit. Figure 3.5(b) shows the design with graphical analysis technique by showing the circuit for R_z = 250 and 1 K Ω, respectively, and how they respond to the variations in the input voltage, V_x. Here we see the V-I curves for the zener diode, and the 250 resistor in solid lines, and the 1 K resistor in dotted lines, responding to the V_x which changes from 7.5 to 12.5 V. Notice that the V-I curves intersect with the zener diode's V-I curve at points A and B for the 250 resistor and at points A′ and B′ for the 1 K resistor. They

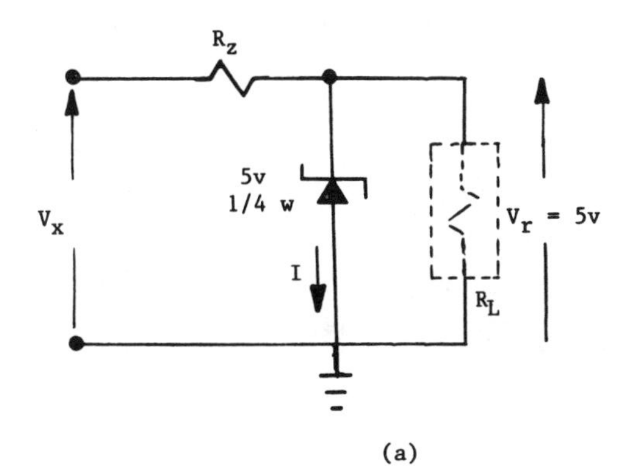

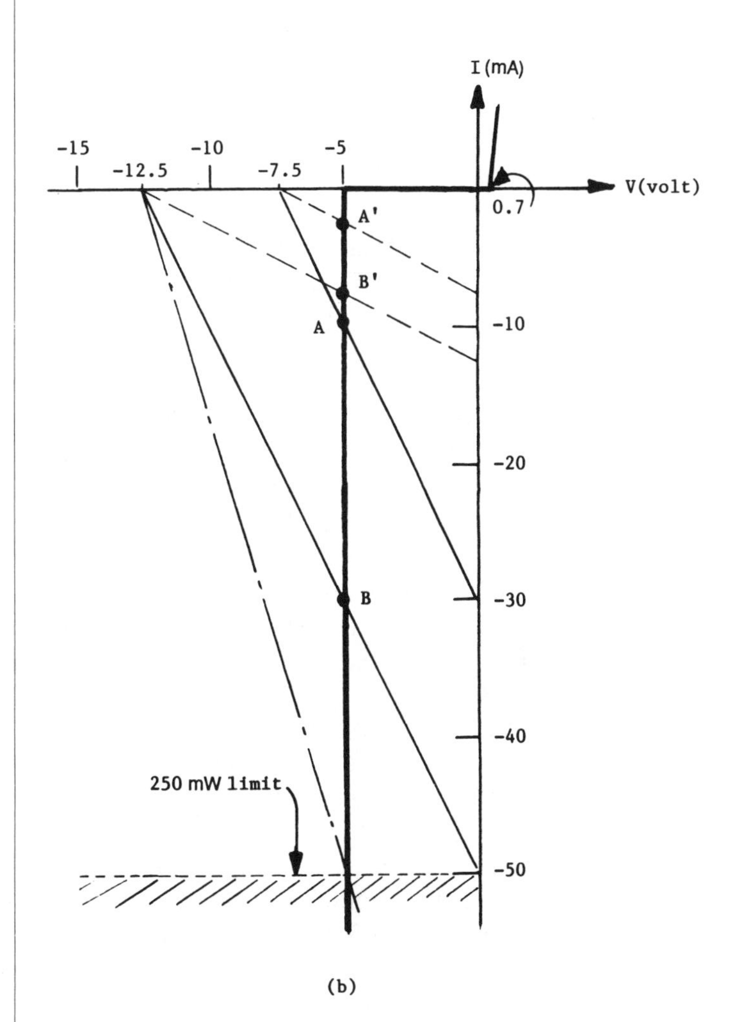

FIGURE 3.5 Reference voltage.

reveal that for the former, I_z varies from negative 10 to 30 mA, while for the latter, approximately from negative 2 to 8 mA. However, the voltage across the zener diode, the reference voltage V_r, stays at 5 V. Thus, one may conclude that either one of the two resistor values would satisfy the circuit specifications.

Discussion — In the design presented here, we have neglected the effect of the load-resistance, R_L. To consider the effect of R_L in this circuit, we must first determine the Thevenin voltage,

$$\frac{R_L}{R_z + R_L} V_x$$

and the Thevenin resistance,

$$\frac{R_z \cdot R_L}{R_z + R_L}$$

The Thevenin voltage must be greater than V_r and the V-I curves for the resistors must be constructed with the Thevenin voltage and resistance, not V_x or R_z alone. In this example, we assume $R_L >> R_z$.

Although either the 250 resistor or the 1 K resistor would satisfy the specification for $R_L = \infty$, the circuit with the 1 K resistor would dissipate less heat or waste less power. The maximum power wasted for the two circuits can be determined as follows:

$$\text{For } 250\ \Omega : (30\text{ mA})^2 \cdot 250 + (30\text{ mA}) \cdot 5 = 375\text{ mW}$$
$$\text{For } 1\text{K}\ \Omega : (8\text{ mA})^2 \cdot 1\text{ K} + (8\text{ mA}) \cdot 5 = 104\text{ mW}$$

It can readily be seen that since the zener diode is a 250 mW device, the minimum value of the resistor $R_z = 150\ \Omega$ as shown by the dashed line in the figure.

Emitter-Follower/Darlington Circuit

Design — The circuit shown in Figure 3.6(a) is known as the *emitter-follower.* It is a current amplifier and must be designed in such a way that it is always operated in linear or nonsaturated mode. It is very popular and useful in both analog and digital circuitry. As the name implies, the output of the circuit is at the emitter terminal of the transistor, while the input is at the base. The output voltage "follows" the input with a voltage gain almost or close to one. Since the circuit is normally used in direct-coupled configuration in digital system design,

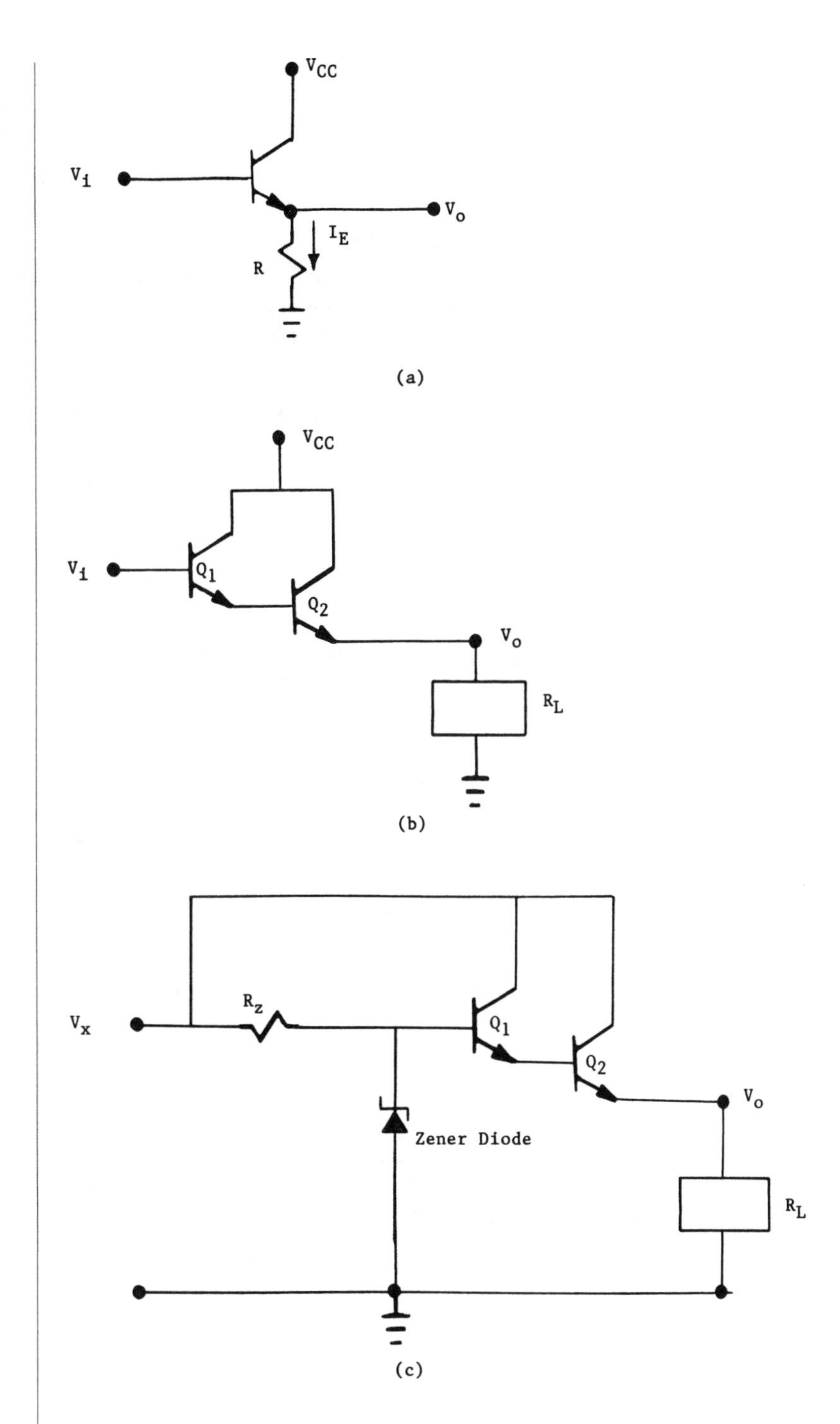

FIGURE 3.6 Regulated power supply.

we will limit our discussion to this area. First, let us examine the circuit by the BSA technique and write the input loop equation:

$$V_i = V_{BE} + R \cdot I_E = V_{BE} + V_o$$

By the BSA method, $V_{BE} = 0.7$ V. Thus, $V_o = V_i - 0.7$. That is, the output voltage always follows the input voltage, but is 0.7 V less. One may ask what is the use of this circuit if there is no voltage gain. As mentioned previously, it is intended to be a current amplifier, not a voltage amplifier. Therefore, it is normally used as a buffer circuit which isolates (buffers) the load that demands driving power from a voltage amplifier or logic circuit, because in many cases voltage amplifier or logic circuit cannot supply enough power for that load. One can get a clearer picture if a more rigorous analysis technique is used to examine the circuit. This method is almost identical to the one derived in Chapter 2, Section 2.1.1 for a single-stage differential amplifier. Recall that for analytical purposes, we split the differential amplifier into two identical single transistor circuits with R_C in the collector circuit and $2R_E$ in the emitter circuit. By setting the collector resistor R_C and $2R_E$ of the split single transistor circuit to zero and R, respectively, the derivation result developed in that section can be used here. Therefore we have for input impedance:

$$\begin{aligned} V_i &= i_b \cdot h_{ie} + i_e \cdot R \\ &= i_b \cdot h_{ie} + i_b(1 + h_{fe}) \cdot R \\ &= i_b(h_{ie} + (1 + h_{fe}) \cdot R) \\ Z_{in} &= \frac{V_i}{i_b} = (h_{ie} + (1 + h_{fe}) \cdot R) \end{aligned}$$

where h_{ie} is normally about 1 K Ω and $h_{fe} = 100$. If R = 1 K, we would have $Z_{in} = 102$ K Ω. As a result, the circuit that precedes the emitter-follower would see 101 times the resistance of R instead of just R.

The output impedance before (or excluding) R, is

$$Z_o = \frac{R_g + h_{ie}}{1 + h_{fe}}$$

where R_g is the output impedance of the circuit preceding this emitter-follower. As a result, the output impedance of the preceding stage is lower by a factor of $1 + h_{fe}$. that is why the emitter-follower has more power to drive its succeeding circuit, and it is thus also known as an impedance converter (it has high input impedance, but low output impedance) in addition to being called a current gain amplifier. It is important to point out that this circuit has an error of V_{BE} which is about 0.7 V between the input and output voltage. If this error is significant to the designer, one should then consider using an operational amplifier with shunt-series configuration as described in Chapter 2, which has very high input impedance and very low output impedance, and yet the voltage gain error can be designed to have a negligible value.

Figure 3.6(b) shows a circuit known as a Darlington circuit which is basically an emitter-follower with very high current gain. That is, the transistors Q_1 and Q_2 replace the single transistor in the emitter-follower circuit. Thus, we now have an equivalent current gain of

$$h_{fe} = h_{fe1} \cdot h_{fe2}$$

With this circuit configuration, we would have higher input impedance and lower output impedance. However, we must realize that this circuit would have a 1.4 V voltage difference between the input and output voltages.

A Simple Regulated Power Supply

Figure 3.6(c) illustrates the application of the circuits shown in Figure 3.6(a) and (b). It is a simple regulated power supply. Notice that the output voltage V_o follows the voltage of the zener diode. Thus, by applying the design procedure described in the examples in Section 3.3.2.e, the voltage across the load R_L will be practically constant regardless of the variation of the source voltage, V_x.

3.3.3. Field-Effect Transistors (FET) — Voltage Control Device

The FET is basically a voltage control device. It draws practically no dc current at the input. Tables 3.1.1 and 3.1.2 show a summary of the classification, symbol, and major electrical properties of the FET.

FETs have the following major advantages when used in digital circuitry: (1) high fan-out due to high input impedance, (2) low noise, and (3) extremely low power dissipation. The following examples show how the FET can be used in a digital system.

Example 1

The diagram shown in Figure 3.7(a) is a simple switching circuit us-

Table 3.1.1

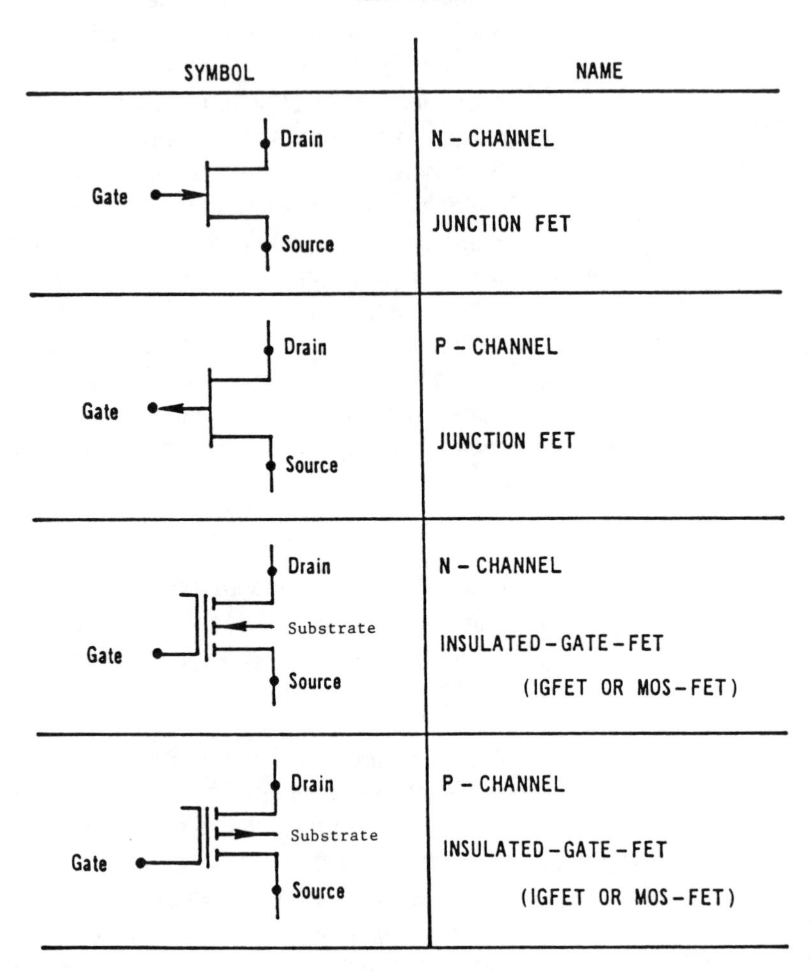

SYMBOL	NAME
Drain Gate Source	N – CHANNEL JUNCTION FET
Drain Gate Source	P – CHANNEL JUNCTION FET
Drain Substrate Gate Source	N – CHANNEL INSULATED – GATE – FET (IGFET OR MOS – FET)
Drain Substrate Gate Source	P – CHANNEL INSULATED – GATE – FET (IGFET OR MOS – FET)

ing a junction N-channel depletion type FET. From Table 3.1.2, the FET switch will be OFF if $V_{GS} \leq -V_p$, the pinch-off voltage, and will be ON if $V_{GS} = 0$. Let the V_p of the FET be –7; then one may use

$$V_{DD} = +15$$

$$V_{GG} = -15$$

$$R_S = R_G = 1\ \text{M}\Omega$$

$$R_D = 15\ \text{K}$$

and define

$$x_1 = \text{logical } 1 = +15 \text{ V}$$

then

$$V_y = 0 \text{ V if } x_1 = +15; \; V_y = 15 \text{ V if } x_1 = 0$$

Example 2

The circuit shown in Figure 3.7(b) is a simple complementary IGFET, also known as CMOS. It is called "complementary" because both P-channel (Q_1) and N-channel (Q_2) are used at the same time. The devices are enhancement types. The electrical characteristics for both P-

Table 3.1.2

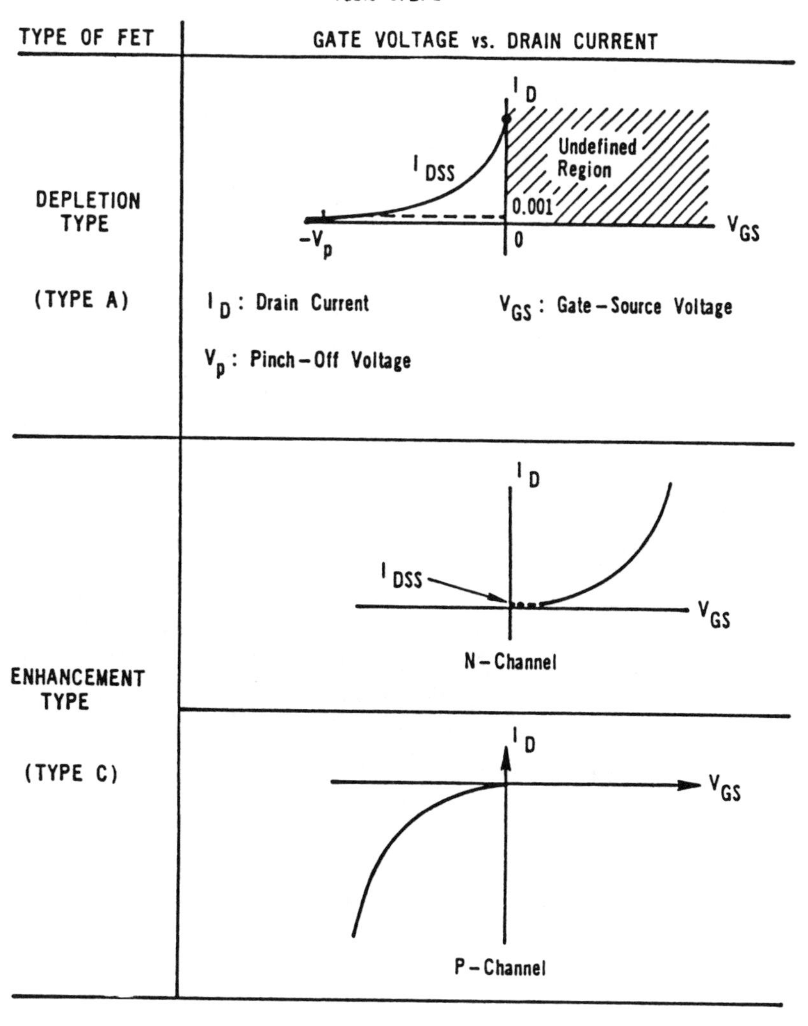

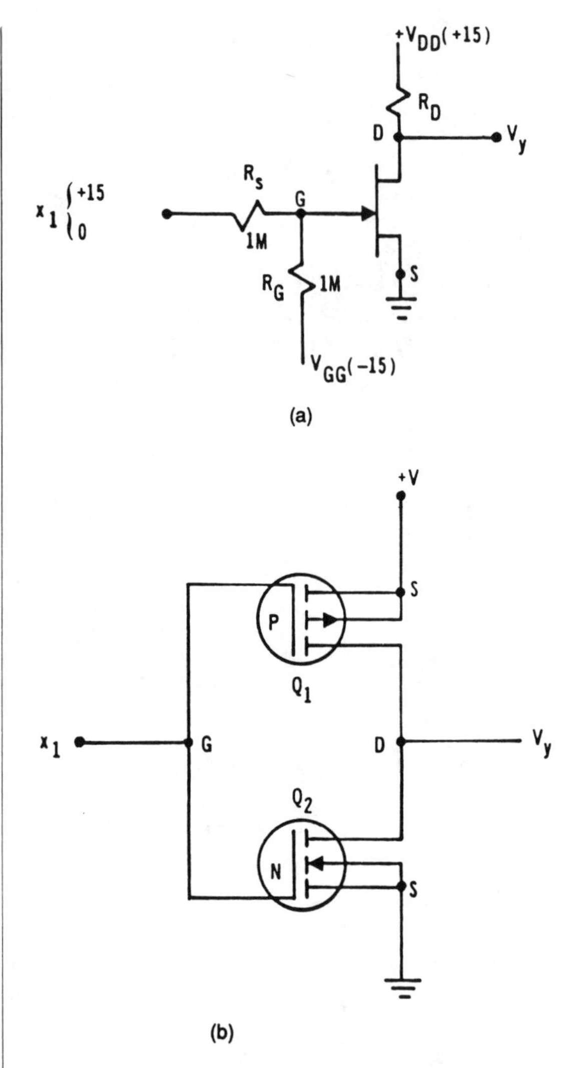

FIGURE 3.7 (a) Junction FET. (b) CMOS inverter.

channel and N-channel IGFET are shown in Table 3.1.2. For P-channel IGFET just change the sign of I_D and V_{GS} from positive to negative, i.e., the device will be ON if V_{GS} is negative and OFF if $V_{GS} = 0$. Now, let x_1 = logical 1 = +V; then for Q_1, $V_{GS} = 0$ and for Q_2, $V_{GS} = +V$. Thus, Q_1 will be OFF and Q_2 will be ON, and

$$V_y = 0$$

If x_1 = logical 0 = 0 V, then for Q_1, $V_{GS} = -V$ and for Q_2, $V_{GS} = 0$. Thus, Q_1 will be ON and Q_2 will be OFF, and

$$V_y = +V$$

The circuit is analogously operating as a single pole double throw switch with its arm connected to V_y. It is interesting to note that if the load of this circuit is composed of another IGFET gate, the circuit has practically no resistive load (the input resistance of an IGFET is about 10^{19} Ω) and dissipates zero power at the steady state. However, the circuit will dissipate power during the state transition time. Thus the circuit is extremely attractive when low power dissipation is primarily important.

Example 3

Although in any digital system the major circuit elements are binary in nature, the system usually is interfaced with the outside physical world which is mostly analog in nature. Therefore, a digital system usually contains analog-digital and digital-analog converters. This example is to show how an FET can be used as an analog switch for switching a low-level analog signal.

As described in the previous section, when a conventional transistor (bipolar transistor) is turned on, a saturation voltage of 0.1 ~ 0.2 V would exist between collector and emitter. The magnitude of this voltage is sensitive to temperature and the individual device. This would introduce considerable error if used as a switch for switching low-level analog signals, on the order of 1 mV or lower. However, since the path between drain and source of an FET in the neighborhood of $V_{DS} = 0$ behaves just like a linear resistor, the FET becomes a good analog switch. Figure 3.8(a) shows a typical low level output characteristic of an FET. Notice that for $V_{GS} = 0$, the V-I curve is horizontal, which means the device has very high resistance. For $V_{GS} = -10$, the device behaves like a resistor of 400 Ω, as long as $-0.2 \le V_{DS} \le +0.2$. As the voltage V_{DS} increases, the V-I curve in Figure 3.8(a) will not be linear like a resistor. Thus, in Figure 3.8(b), as long as the signal voltage E_S is within ±0.2 V, and $R_L >> 400$ Ω, but much less than the off resistance of the FET, the device is a good analog switch.

3.4. PULSE CIRCUIT FUNDAMENTALS

In this section, an attempt is made to clarify the time response of a network to a pulse. Specifically, how an R-C network responds to a pulse or a switching signal will be described and a review on R-C networks mixed with diode and transistor follows. A thorough understanding of their electrical properties with respect to time is essential and it would make the study of multivibrators in later sections painless and pleasant.

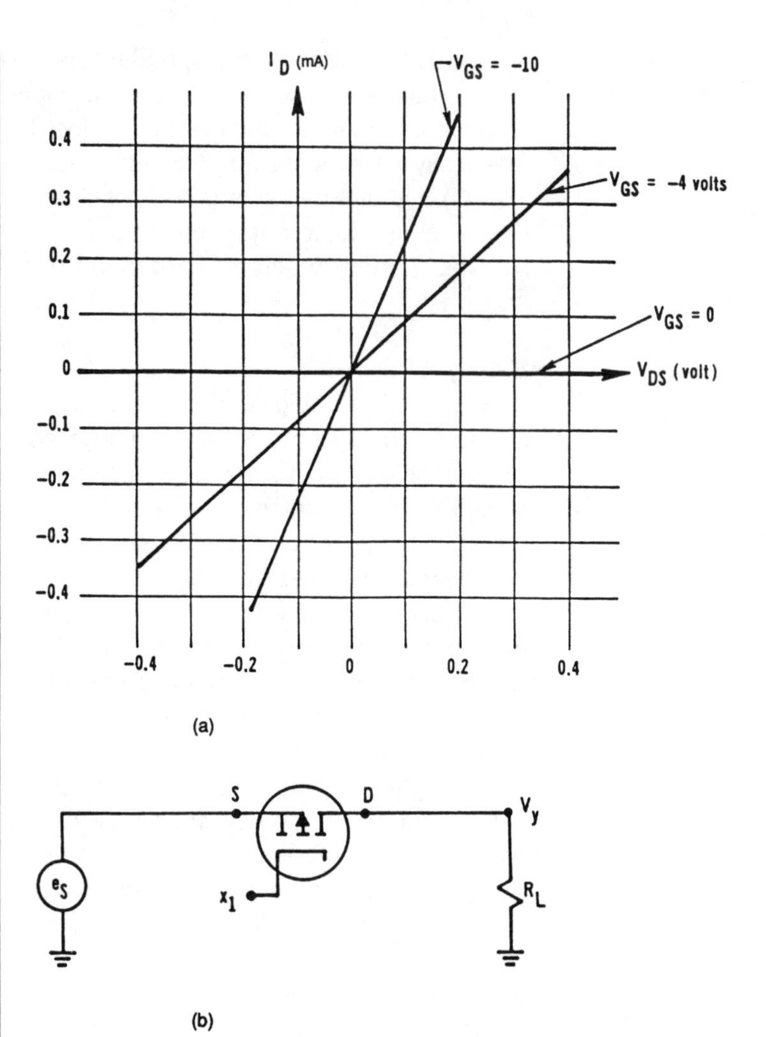

FIGURE 3.8 (a) FET low-level output characteristic. (b) FET as an analog switch.

3.4.1. Complete Time Response of a Basic Network

The circuit shown in Figure 3.9 is an old friend of ours. If the switch S is moved from A to B at t = 0, we have the following conditions:

$$V_C(0^-) = V_C(0^+) = E_a$$

$$V_C(\infty) = E_b$$

$$i(0^-) = 0$$

$$i(0^+) = \frac{E_b - V_C(0^+)}{R}$$

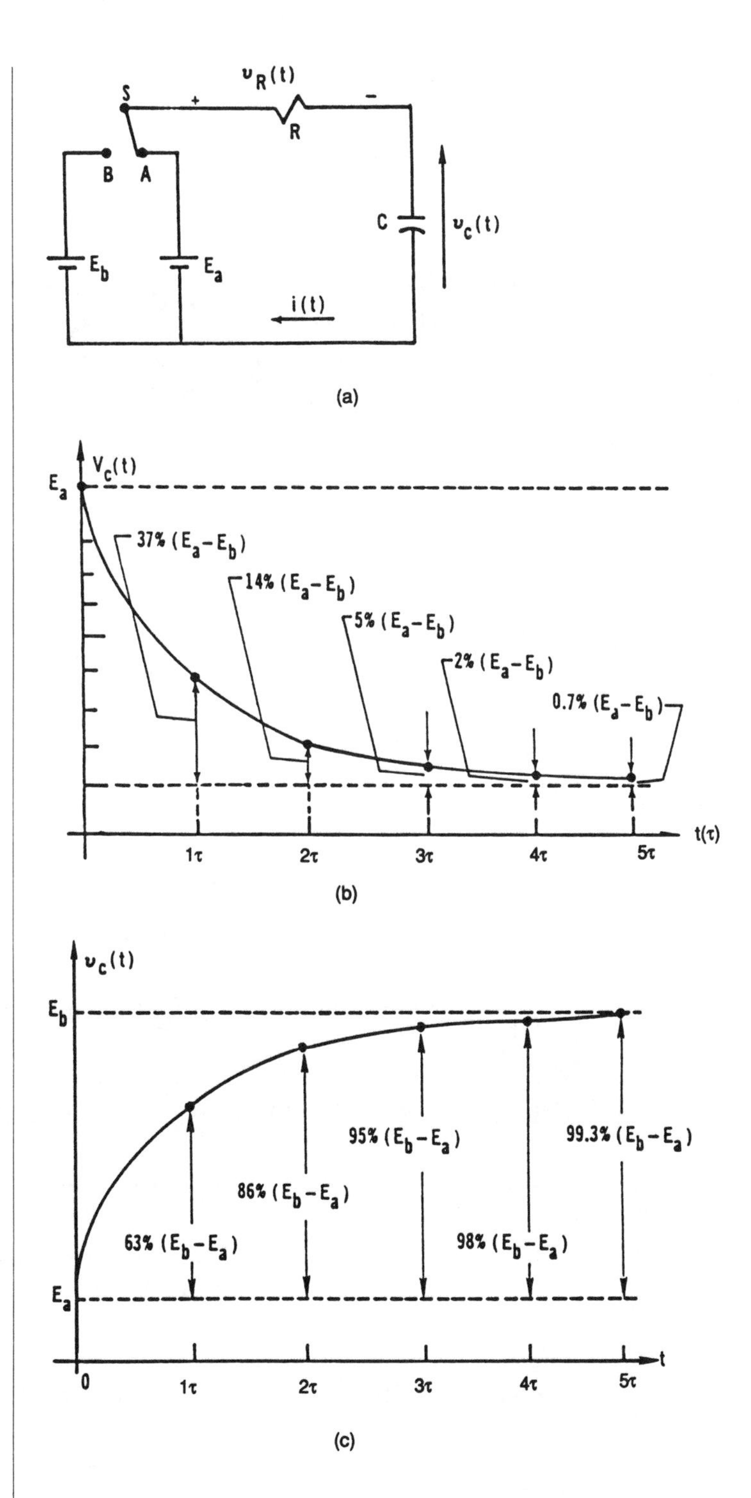

FIGURE 3.9 (a) R-C circuit. (b) $V_c(t)$ for $E_b < E_a$. (c) $V_c(t)$ for $E_b > E_a$. (d) i(t) for $E_b > E_a$.

$$V_R(0^-) = 0$$

$$V_R(0^+) = E_b - V_C(0^+) = E_b - V_C(0^-) = E_b - E_b$$

$$V_R(\infty) = E_b - V_C(\infty) = 0$$

However, we are interested in knowing how the circuit behaves for $0^+ < t < \infty$. In this circuit, for $0^+ < t < \infty$ the loop equation is

$$E_b = V_R(t) + V_C(t)$$

$$= Ri(t) + \frac{1}{C}\int i(t)\,dt$$

which is simply a first order differential equation, whose general solution is

$$i(t) = K_1 + K_2 e^{-t/RC} \tag{3.9}$$

for $t = 0^+$

$$i(0^+) = K_1 + K_2 \tag{3.10}$$

for $t = \infty$

$$i(\infty) = K_1 + 0 = K_1 \tag{3.11}$$

Substituting Equation (3.11) into Equation (3.10), $K2 = i(0^+) - K_1 = i(0^+) - i(\infty)$. Hence,

$$i(t) = i(\infty) + \{i(0^+) - i(\infty)\}e^{-t/RC} \tag{3.12}$$

where

$$RC \triangleq \text{circuit time constant}$$
$$i(0^+) \triangleq \text{initial value}$$
$$i(\infty) \triangleq \text{final value}$$

Equation (3.12) can be generalized and made applicable to any circuit which can be lumped into *one resistor* and *one energy storage element* (memory element) which can be either a capacitor or an inductor. The generalized equation can be written as

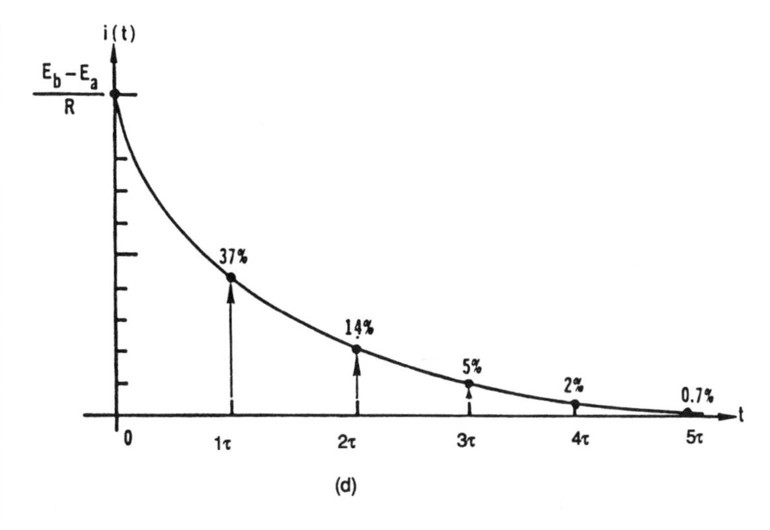

FIGURE 3.9 continued. (a) R-C circuit. (b) $V_c(t)$ for $E_b < E_a$. (c) $V_c(t)$ for $E_b > E_a$. (d) i(t) for $E_b > E_a$.

$$x(t) = x(\infty) + \{x(0^+) - x(\infty)\}e^{-t/\tau} \tag{3.13}$$

where x(t) can be either voltage or current, and

$$\tau = \text{Time constant} = RC \text{ for resistor-capacitor network}$$

$$\tau = \frac{L}{R} \text{ for resistor-inductor network}$$

Example 1

Applying Equation (3.13) to the circuit shown in Figure 3.9, one may easily determine the complete time response of $V_c(t)$ and i(t) in the following way:

$$\begin{aligned} V_C(t) &= V_C(\infty) + \{V_C(0^+) - V_C(\infty)\}e^{-t/RC} \\ &= E_b + (E_a - E_b)e^{-t/RC} \end{aligned} \tag{3.14}$$

and

$$\begin{aligned} i(t) &= i(\infty) + \{i(0^+) - i(\infty)\}e^{-t/RC} \\ &= 0 + \left\{\frac{E_b - E_a}{R} - 0\right\} e^{-t/RC} \\ &= \frac{E_b - E_a}{R} e^{-t/RC} \end{aligned} \tag{3.15}$$

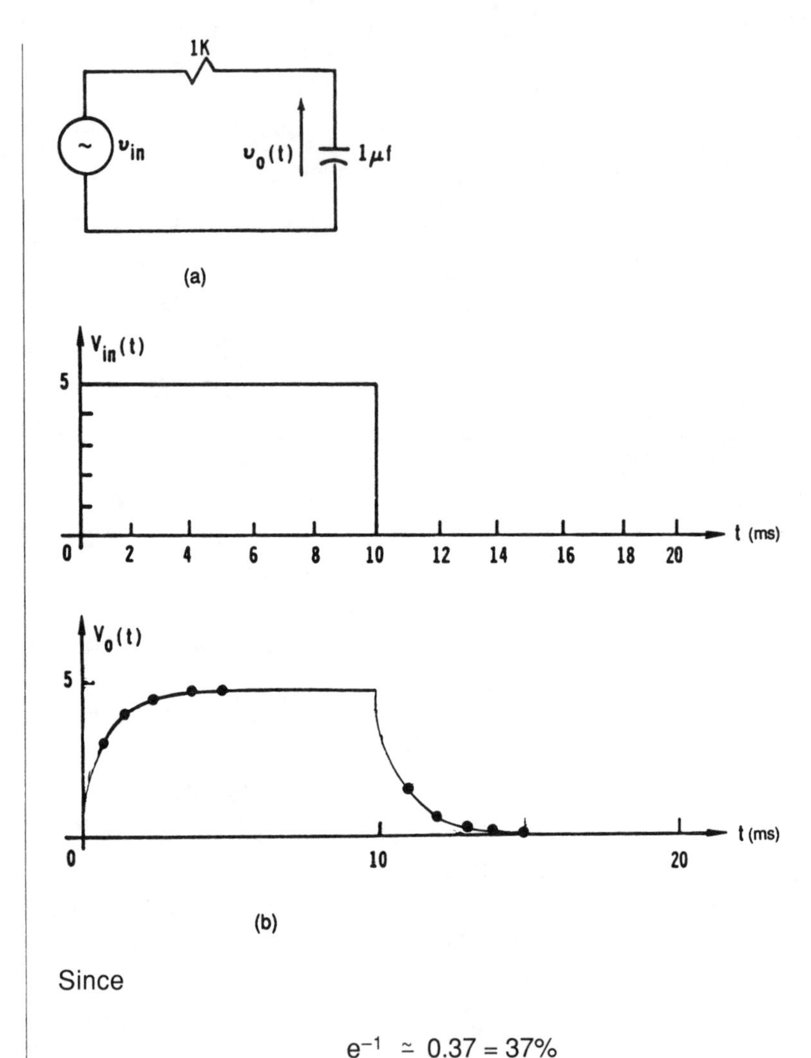

FIGURE 3.10 (a) R-C circuit. (b) Signal pulse width greater than 5τ. (c) Signal pulse width less than 5τ.

Since

$$
\begin{aligned}
e^{-1} &\simeq 0.37 = 37\% \\
e^{-2} &= 0.14 = 14\% \\
e^{-3} &\simeq 0.05 = 5\% \\
e^{-4} &\simeq 0.02 = 2\% \\
e^{-5} &= 0.007 = 0.7\%
\end{aligned}
$$

the time waveform of Equations (3.14) and (3.15) can easily be sketched as shown in Figure 3.9(b), (c) and (d). Notice that the key points for determination of the complete time response of the circuit are (a) initial value, (b) final value, and (c) time constant.

Example 2

In Figure 3.10(a), determine and sketch the output waveform for

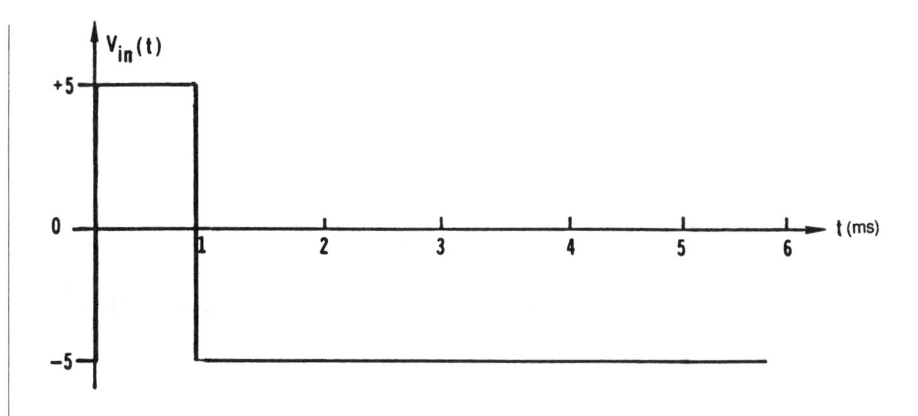

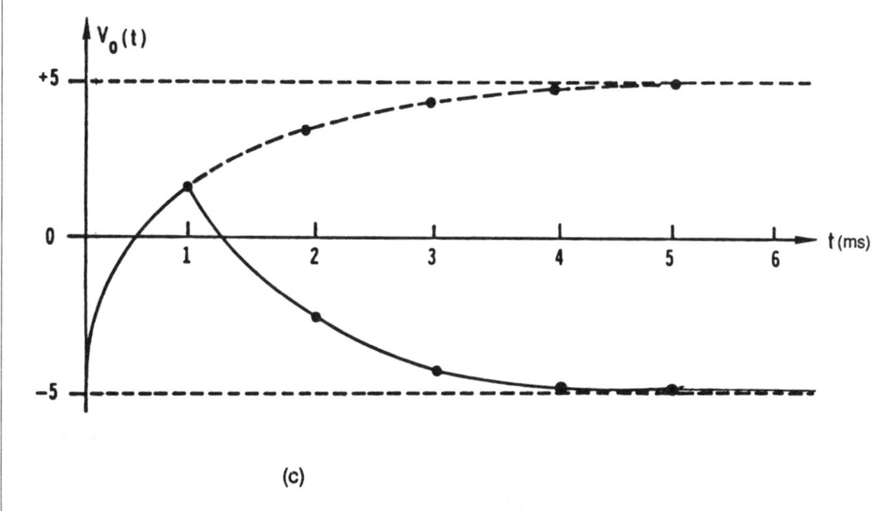

(c)

FIGURE 3.10 continued. (a) R-C circuit. (b) Signal pulse width greater than 5τ. (c) Signal pulse width less than 5τ.

1. $V_{in}(t)$, a pulse with amplitude equal to 5 V and pulse width equal to 10 ms, and
2. $V_{in}(t)$, a pulse with amplitude changes from –5 to +5 V and pulse width 1 ms.

Solution:

1.
$$\tau = RC = 1\text{ ms. For } 0^+ < t < 10\text{ ms}$$

$$[V_0]_{initial} = 0\text{ V}$$

$$[V_0]_{final} = 5\text{ V}$$

$$V_0(t) = 5 + (0 - 5)e^{-t/10^{-3}}$$

$$V_0(t = 10\text{ ms}) = 5 - 5e^{10\times10^{-3}/10^{-3}} = 5 - 5e^{-10} = 5$$

For $t > 10$ ms,

$$[V_0]_{initial} = 5 \text{ V}$$

$$[V_0]_{final} = 0 \text{ V}$$

Thus, $V_0(t \geq 10 \text{ ms}) = 5e^{-(t-10 \text{ ms})/10^{-3}}$

2. $t = 1$ ms. For $0^+ < t < 1$ ms

$$[V_0]_{initial} = -5 \text{ V}$$

$$[V_0]_{final} = +5 \text{ V}$$

$$V_0(t) = 5 + (-5 - 5)e^{-t/10^{-3}}$$

$$V_0(t) = 5 - 10e^{-t/10^{-3}}$$

For $t > 1$ ms,

$$[V_0]_{initial} = V_0(t = 1 \text{ ms}) = 5 - 10e^{-1} = 5 - 3.7 = 1.3 \text{ V}$$

The waveforms are shown in Figures 3.10(b) and (c). It is now clear that if the time constant of the circuit is much, much greater than the pulse width, the input pulse will not show at the output.

Example 3

For the circuit shown in Figure 3.11(a), determine and sketch the waveforms of $V_C(t)$, $z(t)$, and $y(t)$ if $x_1 = 5$ V and x_2 is a pulse with $\delta = 1$ ms as shown.

Solution: Let the pulse of x_2 occur at $t = 0$, then we have

$$y(0^-) = \left(\frac{5 - 0.7}{4 \text{ K}}\right) \cdot 2 \text{ K} = 2.15 \text{ V}$$

where 0.7 = diode voltage-drop.

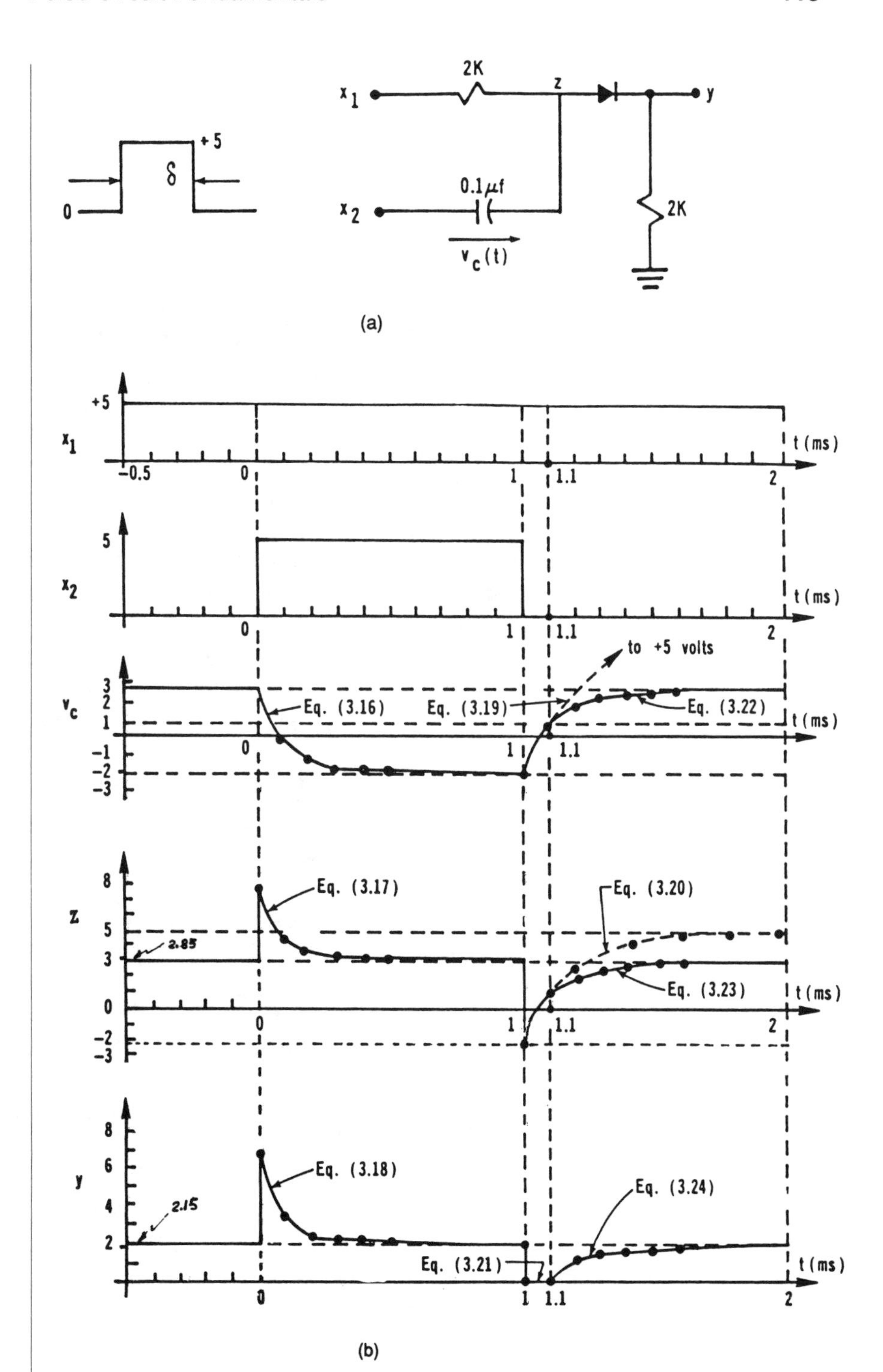

FIGURE 3.11 (a) R-C diode circuit. (b) Waveforms of Example 3.

$$V_C(0^-) = z(0^-) - x_2(0^-) = 2.15 + 0.7 - 0 = 2.85\ V$$

$$V_C(0^+) = V_C(0^-) = 2.85\ V$$

$$z(0^-) = 2.85\ V$$

$$z(0^+) = V_C(0^+) + x_2(0^+) = 2.85 + 5 = 7.85\ V$$

Since the circuit would not know when the pulse would be terminated, it assumes $\delta = \infty$, thus

$$z(\infty) = 2.85$$

$$y(\infty) = z(\infty) - 0.7 = 2.15$$

Now we shall determine the time constant of the circuit. To do that, we simply short circuit all the independent voltage sources, i.e., x_1 and x_2, then we have (1) time constant $\tau = (0.1\ \mu f)\ 2\ K = 0.2$ ms if the diode is OFF, and (2) $\tau = (0.1\ \mu f)\ 2\ K//2\ K = 0.1$ ms if the diode is ON.

Since during the period of $0 < t < 1$ ms the diode is conducting, the time constant $\tau = 0.1$ ms should be used. Then from Equation (3.13),

$$V_C(t) = V_C(\infty) + \{V_C(0^+) - V_C(\infty)\}e^{-t/\tau}$$

where

$$V_C(\infty) \triangleq \text{the final value of } V_C(t) \text{ if } \delta = \infty$$

and

$$V_C(\infty) = z(\infty) - x_2(\infty)$$

$$= 2.85 - 5 = -2.15$$

Thus,

$$V_C(0^+ < t < 1) = -2.15 + \{2.85 + 2.15\}e^{-t/0.1}$$

$$= -2.15 + 5e^{-t/0.1} \quad (3.16)$$

$$z(0^+ < t < 1) = z(\infty) + \{z(0^+) - z(\infty)\}e^{-t/0.1}$$

$$= 2.85 + (7.85 - 2.85)e^{-t/0.1}$$

$$= 2.85 + 5e^{-t/0.1} \quad (3.17)$$

$$y(0^+ < t < 1) = 2.15 + 5e^{-t/0.1} \quad (3.18)$$

Now let us analyze the period of $t > \delta = 1$ ms. From Equations (3.16) to (3.18),

$$V_C(1^-) = V_C(1^+) = -2.15$$

$$z(1^-) = 2.85$$

$$y(1^-) = 2.15$$

and

$$z(1^+) = x_2(1^+) + V_C(1^+) = 0 - 2.15 = -2.15$$

$$y(1^+) = 0 \text{ since the diode is OFF at } t = 1^+$$

But for $1^+ < t < \infty$,

$$V_C(\infty) = x_1(\infty) = 5 \text{ if the diode is not conducting}$$

$$V_C(\infty) = z(\infty) = 2.85 \text{ if the diode is conducting}$$

That is, V_C will change from –2.15 to either 5 or 2.85 V depending on whether the diode is conducting or not. According to Binary State Analysis (BSA), the diode will be turned on when $V_C \geq 0.7$. Therefore, the problem becomes even more interesting. V_C is heading toward $z = x_1 = 5$ before V_C reaches 0.7 and changes its course, heading to $z = 2.85$ immediately after $V_C = 0.7$. The problem is now to determine at what time $V_C = 0.7$. Let $V_C(t_1) = 0.7$, since for $1^+ < t < t_1$ the diode is OFF; then

$$V_C(1\text{ ms} < t < t_1) = V_C(\infty) + \{V_C(0^+) - V_C(\infty)\}e^{-t/\tau}$$

$$= 5 + \{-2.15 - 5\}e^{-(t-1)/0.2}$$

$$= 5 - 7.15e^{-(t-1)/0.2} \tag{3.19}$$

or

$$0.7 = 5 - 7.15e^{-(t_1-1)/0.2}$$

$$t_1 = 1.1\text{ ms}$$

and

$$z(1 < t < t_1) = x_2(1 < t < t_1) + V_C(1 < t < t_1)$$

$$= 0 + V_C(1 < t < t_1)$$

$$= 5 - 7.15e^{-(t-1)/0.2} \tag{3.20}$$

$$y(1 < t < t_1) = 0$$

Now, since $t_1 < t < \infty$ the diode is conducting, we have

$$V_C(t > t_1) = V_C(\infty) + \{V_C(0) - V_C(\infty)\}e^{-(t-t_1)/\tau}$$

$$= 2.85 + \{0.7 - 2.85)e^{-(t-t_1)/0.1}$$

$$= 2.85 - 2.15e^{-(t-t_1)/0.1} \tag{3.22}$$

and

$$z(t > t_1) = x_2(t > t_1) + V_C(t > t_1)$$

$$= 0 + V_C(t > t_1)$$

$$= 2.85 - 2.15e^{-(t-t_1)/0.1} \tag{3.23}$$

$$y(t > t_1) = z(t > t_1) - 0.7$$

$$= 2.15 - 2.15e^{-(t-t_1)/0.1} \tag{3.24}$$

The waveforms of x_1, x_2, V_C, z and y are plotted in Figure 3.11(b).

Important Remark: The reader may feel that the first two examples are too simple and the last example is unnecessarily complex. However, the first two examples are designed to provide the reader with a good review of the fundamentals of timing circuits, and the third example is specially designed to cover the analysis techniques that one needs to understand the principles of operation of the sequential circuit elements described in later chapters. The reader is therefore urged to re-read these examples and to understand each step described in them thoroughly.

3.5. LOGIC FUNDAMENTALS IN BRIEF

In this section, a brief review of logic fundamentals is presented for completeness of this handbook and for the convenience of persons who may need to refresh themselves on the fundamentals from time to time. FIrst, a summary of the definitions and theorems for logical operations is presented.

3.5.1. Definitions

Switching variable $\triangleq$ any letter which assumes only two possible values: "1" and "0".

Logical AND operation $\triangleq$ "·"
Logical OR operation $\triangleq$ "+"
Logic inversion $\triangleq$ "′"

3.5.2. Theorems

1. $1 \cdot 1 = 1$
2. $0 \cdot 0 = 0$
3. $1 + 1 = 1$
4. $0 + 0 = 0$
5. $1 \cdot 0 = 0 \cdot 1 = 0$
6. $1 + 0 = 0 + 1 = 1$
7. $0' = 1,\ 1' = 0$
8. $x \cdot 1 = x,\ x + 1 = 1$
9. $x \cdot 0 = 0,\ x + 0 = x$
10. $x + x = x,\ x \cdot x = x$
11. $(x)' = x',\ (x')' = x$
12. $x + x' = 1,\ x \cdot x' = 0$
13. $x + y = y + x,\ x \cdot y \triangleq xy = yx \triangleq y \cdot x$

14. $x + xy = x, x(x + y) = x \cdot x + x \cdot y = x$
15. $xy' + y = x + y$
16. $(w + x)(y + z) = wy + xy + wz + xz$
17. $(x + y)(y + z)(z + x') = (x + y)(z + x')$
18. $xy + yz + zx' = xy + zx'$
19. $(x + y + z + \ldots)' = x'y'z'\ldots$
20. $(xyz\ldots)' = x' + y' + z' + \ldots$
21. $f(x_1, x_2, \ldots, x_n, +, \cdot)' = f(x_1', x_2', \ldots, x_n', \cdot, +)$
22. $f(x_1, x_2, \ldots, x_n) = x_1 f(1, x_2, \ldots, x_n) + x_1' f(0, x_2, \ldots, x_n)$
23. $f(x_1, x_2, \ldots, x_n) = [x_1 + f(0, x_2, \ldots, x_n)][x_1' + f(1, x_2, \ldots, x_n)]$

3.5.3. Minimization of Switching Function by Karnaugh Map

a. Three-Variable Karnaugh Map

Switching function:

$$F = xyz' + x'y'z + xyz + xy'z$$

Karnaugh Map:

z \ xy	00	01	11	10
0	0	0	1	0
1	1	0	1	1

F

Minimized switching function:

$$F = xy + y'z$$

b. Four-Variable Karnaugh Map

Switching Function:

$$F = w'xy'z + w'x'yz + w'xy'z + wxy'z + w'xyz + wxyz + wx'yz$$

Karnaugh Map:

yz \ wx	00	01	11	10
00	0	1	0	0
01	0	1	1	0
11	1	1	1	1
10	0	0	0	0

F

Minimized switching function:

$$F = w'xy' + xz + yz$$

REFERENCES

1. Millman, J. and Taub, H., *Pulse, DIgital, and Switching Waveforms*, McGraw-Hill, New York, 1965.
2. Millman, J., *Microelectronics: Digital and Analog, Circuits and Systems,* McGraw-Hill, New York, 1979.
3. Krieger, M., *Basic Switching Circuit Theory*, MacMillan, New York, 1967.
4. McCluskey, E. J., *Introduction to the Theory of Switching Circuits*, McGraw-Hill, New York, 1965.
5. Marcus, M. P., *Switching Circuits for Engineers*, 3rd ed., Prentice-Hall, Englewood Cliffs, NJ, 1975.
6. Miller, R. E., *Switching Circuit Theory*, Vols. I and 2, John Wiley & Sons, New York, 1965.
7. Peatman, J. B., *The Design of Digital Systems*, McGraw-Hill, New York, 1972.
8. Blakeslee, T. R., *Digital Design with Standard MSI and LSI*, 2nd ed., John Wiley & Sons, New York, 1979.
9. Horowitz, P. and Hill, W., *The Art of Electronics*, 2nd ed., Cambridge University Press, New York, 1990.
10. Millman, J. and Grabel, A., *Microelectronics*, 2nd ed., McGraw-Hill, New York, 1987.
11. Johnson, E. L. and Karim, M. A., *Digital Design*, PWS Publishers, Boston, 1987.
12. Bell, D. A., *Solid State Pulse Circuits*, 3rd ed., Prentice-Hall, Englewood Cliffs, NJ, 1988.

4 BASIC ELECTRONIC LOGIC ELEMENTS

4.1. BACKGROUND

In general, a digital system designer is given, either verbally or in writing, a list of specifications for a system to be designed and implemented with electric hardware. The designer normally starts with a system block diagram according to the specifications. For each block, the desired inputs and outputs are defined, and its internal characteristic or transfer function is described by switching functions. Eventually, the mathematical descriptions are implemented by electronic elements. As shown by the definitions and switching functions described in Chapter 3, Section 3.5, there are only three basic operations, i.e., "AND", "OR", and "INVERT" in any switching function. Those basic elements are called gates. By adding some basic functional blocks such as Flip-Flop, clock, one shot, etc., to the switching network, the whole system will then be realized. Those electronic elements will be described in Sections 4.2 and 4.3.

4.2. LOGIC CIRCUITS

4.2.1. Introduction

Having studied the properties of the basic circuit elements in Chapter 3, we shall now describe how the different kinds of logic circuits or gates function. The memory logic circuits or multivibrators will be studied in later chapters. In this section, we shall limit ourselves to the study of the gates for logic networks. From Section 4.1, we know that the basic logic elements for realizing any switching function are Inverter, OR and AND gates, or just simply AND-INVERT (NAND) or OR-INVERT (NOR) gate only. Before studying the function of the logic circuits, it is necessary to define a few terms which are important for describing the properties of any logic circuits.

Since the values of the variables in the switching functions that we are dealing with are limited to binary states, one may use any device which has two distinct states to realize the switching variables. One

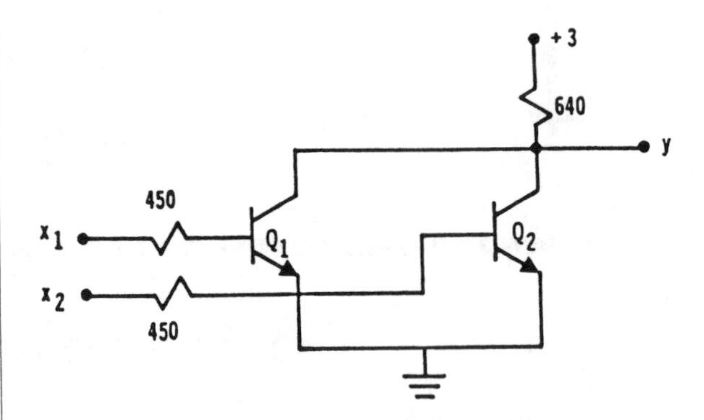

FIGURE 4.1 R-T logic circuit.

may assign logical 1 to the state in which a device is "ON" and logical 0 when the device is "OFF", or conversely. But generally, the voltage of a logic circuit is used to represent the logic levels. For example,

$$\left.\begin{array}{c} \text{high voltage, say } v \geq 2.4 \text{ V}, \triangleq \text{logical } 1 \\ \text{low voltage, say } v \leq 0.5 \text{ V}, \triangleq \text{logical } 0 \\ 0.5 < v < 2.4 \triangleq \text{undefined} \end{array}\right\} \quad (4.1)$$

or

$$\left.\begin{array}{c} \text{low voltage, say } v \leq 0.5 \text{ V}, \triangleq \text{logical } 1 \\ \text{high voltage, say } v \geq 2.4 \text{ V}, \triangleq \text{logical } 0 \\ 0.5 < v < 2.4 \triangleq \text{undefined} \end{array}\right\} \quad (4.2)$$

where v is a switching variable and the symbol "$\triangleq$" means "is defined as". The system using the definition of Equation (4.1) is called a *positive logic* system; the system using that of Equation (4.2) is called a *negative logic* system.

4.2.2. Resistor-Transistor Logic (RTL)

The circuit shown in Figure 4.1 is a typical RTL circuit. Let the input signals x_1 and x_2 be either 3 V or 0.2 V. Then, by applying the Binary State Analysis technique discussed in Section 3.3.2(a), one may construct Table 4.1 as shown. In this table, let 3 V $\triangleq$ H $\triangleq$ logical 1, and 0.2 V $\triangleq$ L $\triangleq$ logical 0. Then Table 4.1 can be rewritten as shown in Table 4.2.

TABLE 4.1

x_1 (volt)	Q_1	x_2 (volt)	Q_2	y (volt)
0.2	OFF	0.2	OFF	3
3	ON	0.2	OFF	0.2
0.2	OFF	3	ON	0.2
3	ON	3	ON	0.2

TABLE 4.2

x_1	x_2	y
0(L)	0(L)	1(H)
1(H)	0(L)	0(L)
0(L)	1(H)	0(L)
1(H)	1(H)	0(L)

The information in Table 4.2 can easily be translated into a switching function:

$$y = \overline{x_1 + x_2} \tag{4.3}$$

For a *positive* logic system, Equation (4.3) is a NOR switching function, and the circuit is a NOR gate. Now consider the case where 3 V $\triangleq$ H $\triangleq$ logical 0, and 0.2 V $\triangleq$ L $\triangleq$ logical 1. Table 4.2 then becomes

TABLE 4.3

x_1	x_2	y
1	1	0
0	1	1
1	0	1
0	0	1

from which we have the switching function

$$y = \overline{x_1 \cdot x_2} \tag{4.4}$$

But Equation (4.4) is a NAND function. This is because we are in a *negative* logic system, thus the same circuit becomes a NAND gate! It is important to point out that the logic designer should bear in mind that one cannot define the circuit as an OR gate or an AND gate, nor as a NOR gate or a NAND gate unless the positive or negative logic system is first defined. Sometimes a logic designer may find it advantageous to think in a mixed positive and negative logic system, known as

Kintner's system. In that case, one may label the logic circuit with symbols such as H for high and L for low to carry out the logic design. For example, from Table 4.2 we have

TABLE 4.4

x_1	x_2	y
L	L	H
H	L	L
L	H	L
H	H	L

Now, if one used positive logic at the input and negative logic for the output, then this table in logical form would be as shown in Table 4.5,

TABLE 4.5

x_1	x_2	y
0	0	0
1	0	1
0	1	1
1	1	1

which realizes an OR function.

The author does not intend to confuse the reader with this concept, but rather to show the flexibility of logic assignment and to pave an easy way for the reader to convert a switching function into an integrated logic circuit. It is important to point out that this concept is not limited to RTL, but is also applicable to TTL, DTL, and ECL, which will be described in Sections 4.2.3, 4.2.4, and 4.2.6.

In addition, let us introduce another term, the *transfer curve*. It is very useful in studying the electrical properties of a logic circuit. The transfer curve is best explained by an example.

The circuit shown in Figure 4.2(a) is an RTL inverter, because if x = HIGH, then y = LOW and vice versa. Figure 4.2(b) is a sketch of the input vs. output transfer curve. Let the current gain of the transistor be 4, and the threshold voltage, v_θ, be 0.7. Then

$$\text{(I)}\quad I_c = 0,\ y = 3 \text{ for } 0 < x \leq 0.7$$

$$\text{(II)}\quad 0 < I_c < 4.3 \text{ mA},\ y < 3 \text{ for } 0.7 < x < 1.2$$

$$\text{(III)}\quad I_c = 4.3 \text{ mA},\ y = 0.2 \text{ for } x > 1.2$$

There are three distinct regions:

in Region (I), y = H, x = L

in Region (II), y = L, x = H and

in Region (III), the circuit is undefined

In positive logic, logical 1 is normally defined as the voltage is greater than 1.2 V, and logical 0 as it is less than 0.5 V, otherwise the circuit is undefined. Evidently, there is no digital system that can tolerate any of the logic circuits remaining in the undefined state. With the help of the transfer curve concept, a few more terms can now be easily defined.

a. Fan Out

Figure 4.3(a) shows an RTL inverter driving another three identical inverters, and Figure 4.3(b) shows the transfer curves of the driver

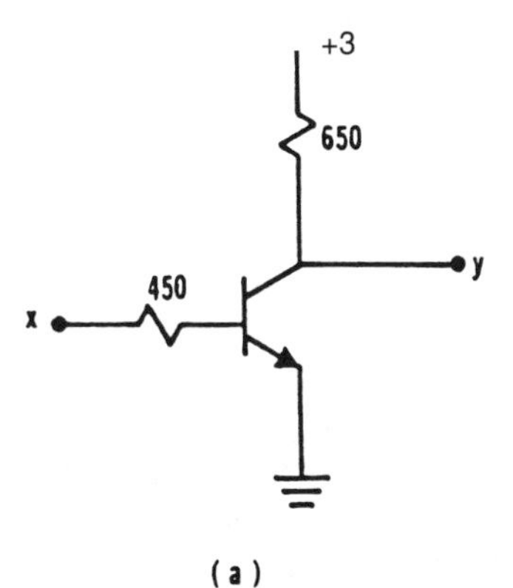

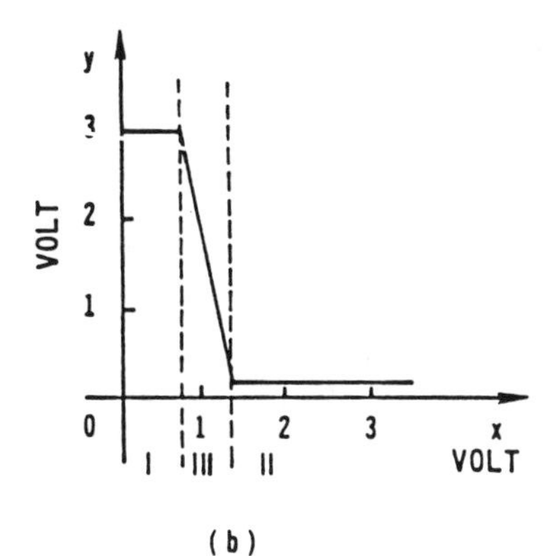

FIGURE 4.2 (a) R-T-L circuit. (b) R-T-L circuit output V-I curve.

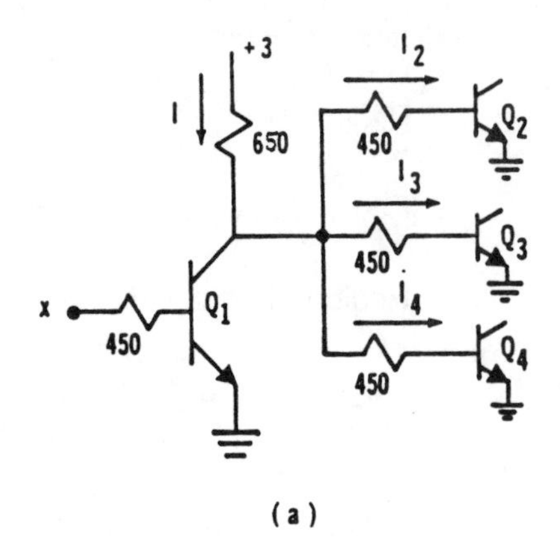

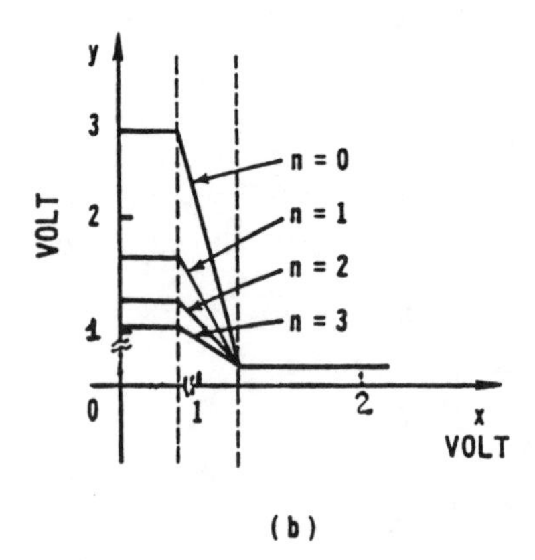

FIGURE 4.3 (a) R-T-L circuit. (b) R-T-L circuit output V-I curve.

which is driving one, two and three gates, respectively. Notice that when $x \geq 1.2$, Q_1 is saturated and $y = 0.2$ so that Q_2, Q_3, and Q_4 will all be OFF, the driver behaves exactly the same as in Figure 4.2(b). However, as $x \leq 0.5$, Q_1 will be OFF and Q_2, Q_3, and Q_4 will be ON and y will no longer be 3 V, since

$$y = 3 - I \cdot 650 = 3 - (I_2 + I_3 + I_4) \cdot 650$$

and

$$I = \frac{3 - V_\theta}{650 + \dfrac{450}{n}} = \frac{3 - 0.7}{650 + \dfrac{450}{n}}$$

where n = 1, 2, 3, or the number of the driven circuits. Thus, for $x \leq 0.5$,

$$y = 1.65 \text{ V} \qquad \text{for } n = 1$$

$$= 1.3 \text{ V} \qquad \text{for } n = 2 \text{ and}$$

$$= 1.1 \text{ V} \qquad \text{for } n = 3$$

Notice that when $n = 1$, $y > 1.2$ and the output is logical 1 and the inverter is working properly. But for $n \geq 3$, $y \leq 1.2$ and the circuit is in the undefined state and the network will not function properly. We now define the maximum number of gates a circuit can drive properly as the *fan-out* of that circuit. Hence, this RTL inverter has a fan-out = 2. This information is extremely important to the system designer; if the designer assigns more gates than the specified fan-out of a driving circuit, the system will not function properly.

b. Noise Immunity

In Figure 4.2, let us assume that logical 1 $\triangleq$ 3 V and logical 0 $\triangleq$ 0.2 V. That is, when

$$x = L \rightarrow 0.2 \text{ V}, y = H \rightarrow 3 \text{ V}$$

$$\text{or} \qquad x = H \rightarrow 3 \text{ V}, y = L \rightarrow 0.2 \text{ V}$$

where "$\rightarrow$" means "implies". Consider $x = 0.2$ V; if there is a noise of +0.8 V, either a dc level drifting or an induced noise spike, at the input terminal, then $x' =$ signal + noise = 1.0 V, which will result in $y = 1.5$ V according to the transfer curve. That is, the output is driven into the undefined region and the circuit will not function properly. However, the system will be all right if the noise is ≤ 0.5 V. In this case, 0.5 is defined as the *noise immunity* of the circuit. A similar concept can be applied to the "high" state except that the negative dc drift or negative spike will drive the circuit into the undefined region. There is a family of logic circuits called Zener-Transistor Logic or High Threshold Logic (HTL) available which may have noise immunity of 5 to 10 V. It is designed for use in noisy environments.

Because of the simplicity of the RTL, the author has used it as a vehicle to define these terminologies. However, the definitions hold for all other types of logic circuit as well.

4.2.3. Diode-Transistor Logic (DTL)

The circuit shown in Figure 4.4 is a typical DTL circuit. By applying

BSA techniques as described in Chapter 3, Section 3.3.2(a), one can easily verify the following table (positive logic is assumed).

TABLE 4.6

x_1	x_2	D_1	D_2	Q_1	D_3	Q_2	y
[0]	[0]	ON	ON	OFF	OFF	OFF	[1]
[1]	[0]	OFF	ON	OFF	OFF	OFF	[1]
[0]	[1]	ON	OFF	OFF	OFF	OFF	[1]
[1]	[1]	OFF	OFF	ON	ON	ON	[0]

where [0] $\triangleq$ logical 0 and [1] $\triangleq$ logical 1. Thus, $y = \overline{x_1 \cdot x_2}$ and the circuit is a NAND gate for positive logic. Note that as x_1 = H (high), D_1 is OFF and practically draws no current. Recall that in RTL, the driven circuit draws considerable current from the driving circuit, causing a fan-out problem; the DTL circuit seems to eliminate this problem. Unfortunately, there is a problem when y = [0] and it is driving many other gates. This is because when y = [0], all the gates connected to it will be conducting and the conducting current of each diode of the driven circuits will be flowing into the y terminal which becomes a current sink. This will cause the voltage at y to rise and eventually exceed the noise immunity margin, resulting in a malfunction if the number of driven circuits exceeds the fan-out of the circuit.

At this point, one may ask what is the function of the 5 K resistor shunting the base-emitter junction of Q_2. Actually the resistor is for the purpose of reducing the leakage current and speeding up the switching time of Q_2. For BSA, this resistor can be neglected.

4.2.4. Transistor-Transistor Logic (T^2L or TTL)

The circuit shown in Figure 4.5 is a basic TTL circuit. The circuit is almost the same as the one shown in Figure 3.4(e), except that here

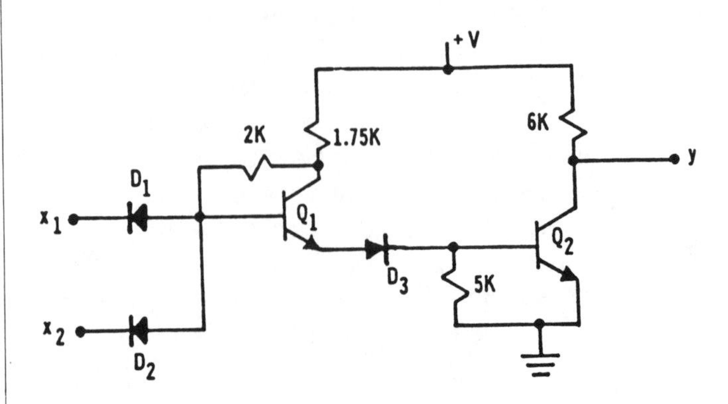

FIGURE 4.4 DTL logic circuit.

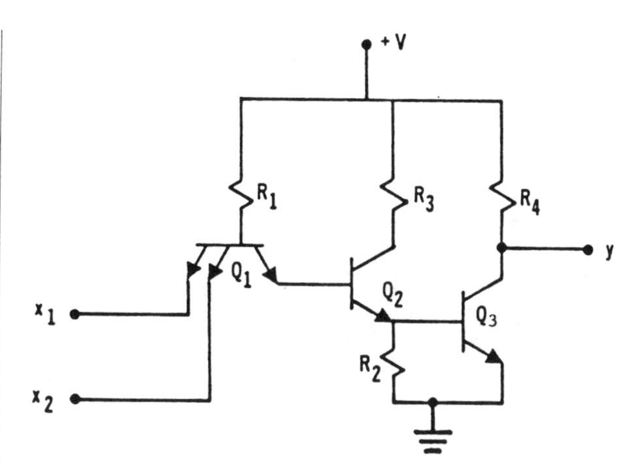

FIGURE 4.5 Transistor-transistor logic circuit.

Q_1 has two emitters. Applying BSA, Q_1 can be viewed as three fictitious diodes D_1, D_2, and D_3. The anodes of all three diodes are connected together to resistor R_1, and their cathodes are connected to x_1, x_2, and the base of Q_2, respectively. One can then verify the states listed in the following table.

TABLE 4.7

x_1	x_2	D_1	D_2	D_3	Q_2	Q_3	y
[0]	[0]	ON	ON	OFF	OFF	OFF	[1]
[1]	[0]	OFF	ON	OFF	OFF	OFF	[1]
[0]	[1]	ON	OFF	OFF	OFF	OFF	[1]
[1]	[1]	OFF	OFF	ON	ON	ON	[0]

where $y = \overline{x_1 \cdot x_2}$, a NAND gate for positive logic. Since TTL is widely used, the following examples will help familiarize the reader with it.

Example 1

Joe was a "Puritan" of logic design. He knew very little about circuits. One day, he used the circuit shown in Figure 4.6(a) to implement the switching function

$$y = \overline{x_1 \cdot x_2} = \bar{x}_1 + \bar{x}_2$$

and positive logic was chosen. According to his calculations, when the photocell was being exposed to light,

$$x_2 = \frac{\lambda}{100K + \lambda} \cdot 5 = \frac{7}{107} \cdot 5 = 0.33 \text{ V} \rightarrow [0]$$

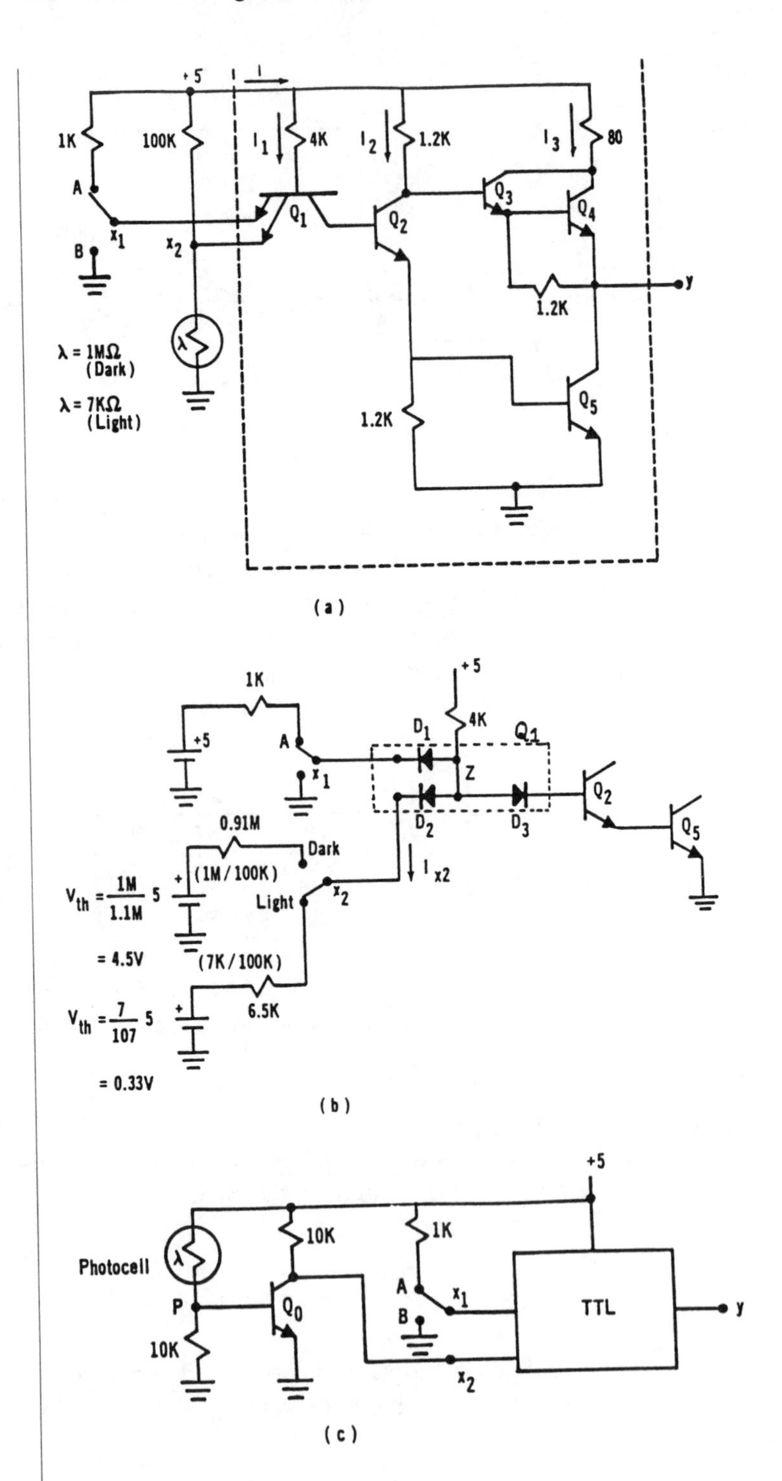

FIGURE 4.6 (a—d) TTL circuit. (e) TTL dynamic analysis.

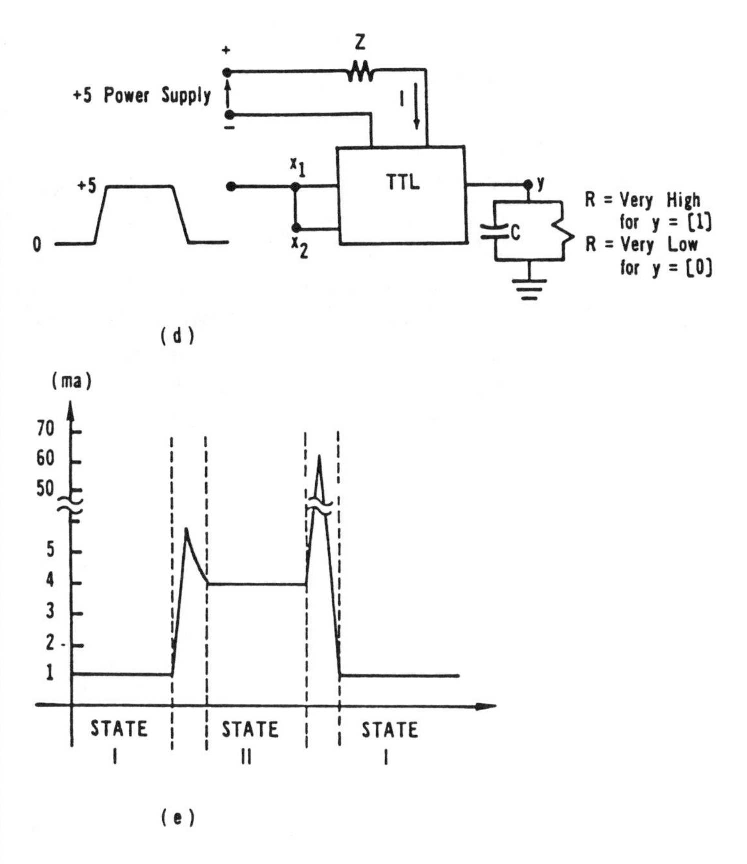

FIGURE 4.6 continued. (a—d) TTL circuit. (e) TTL dynamic analysis.

where the arrow "→" means "imply" or "implies".

When it's dark,

$$x_2 = \frac{\lambda}{100K + \lambda} \cdot 5 = \frac{1M}{1.1M} \cdot 5 = 4.5\,V \rightarrow [1]$$

However, he found that the circuit failed to realize the switching function, and he had the following responses:

TABLE 4.8

x_1		x_2			
State	Volt	State	Volt	y	Remarks
B	0	Light	0.33	[1]	Expected
B	0	Dark	4.5	[1]	Expected
A	5	Light	1.4	[0]	Unexpected
A	5	Dark	4.5	[0]	Expected

where [1] $\triangleq$ logical 1 and [0] $\triangleq$ logical 0.

At first, Joe thought he had a defective gate. But after he had tried a few brand new ones, and even tried different brands, he was convinced that something must be terribly wrong. Then he called his friend John for help, since John was a circuit expert although he did not know much about logic design. After studying the circuit for awhile, John said, "Of course it won't work, since the photocell has high 'output resistance'!". John then drew the circuit shown in Figure 4.6(b), to explain Case 3 from Table 4.8, which has x_1 at A and x_2 at the state of "light". He used three fictitious diodes to replace Q_1 and then applied Thevenin theory to get the equivalent circuit of the photocell as shown in the figure. Then he explained that if the circuit is functioning properly, D_2 should be ON and D_1, D_3, Q_2, and Q_5 should be OFF. If this were the case, then

$$Z = 5 - I_{x2} \cdot 4\,K \tag{4.5}$$

with

$$I_{x2} = \frac{5 - V_\theta - 0.33}{4\ K + 6.5\ K} = 0.38\ mA$$

Thus,

$$Z = 5 - 0.38\ mA \cdot 4\ K = 3.48\ V$$

However, the threshold voltage for D_3, Q_2, and Q_5 is only $3V_\theta \simeq 2.1$, and the potential of Z will keep D_1 OFF but turn D_3, Q_2, and Q_5 ON, which will result in Z = 2.1 V and still keep D_1 OFF. As Z = 2.1 V, D_2 will still be ON, but now $x_2 = 2.1 - V_\theta$ of $D_2 = 2.1 - 0.7 = 1.4$ V! Thus, we have the condition shown in the third line of the table.

Later, John suggested a solution, the circuit shown in Figure 4.6(c). He expected the circuit would respond as shown in Table 4.9.

TABLE 4.9

				Q_1							
x_1	Photo	Q_0	x_2	D_1	D_2	D_3	Q_2	Q_3	Q_4	Q_5	y
0	Light	ON	0.2	ON	OFF	OFF	OFF	ON	ON	OFF	[1]
0	Dark	OFF	5	ON	OFF	OFF	OFF	ON	ON	OFF	[1]
5	Light	ON	0.2	OFF	ON	OFF	OFF	ON	ON	OFF	[1]
5	Dark	OFF	5	OFF	OFF	ON	ON	OFF	OFF	ON	[0]

Joe took John's suggestion and found that John was right. Then Joe said to himself that he wished that simple circuit analysis techniques were taught in the logic design course.

Example 2

Joe designed a simple digital system with a few NAND gates and MSI binary counters. After it was fabricated, he found that the counters did not function properly. John came to help him but they could not find the problem. Later, Joe's high school classmate Carl came to visit him. Carl did not go to college, but he had been an electronic technician for years. He took a careful look at the system and said, "My boss, who has been an engineer for many years, always makes me put a 0.1 μf capacitor across $V_{CC}(+5)$ pin, 14, and ground pin, 7. Why don't you try that?". Joe took Carl's advice, and the system worked. Then John became curious. He investigated the basic circuit of TTL, the inverter, which is equivalent to connecting x_1 and x_2 of Figure 4.6(a) together as the input terminal. He assumed that the inverter was driving a load of a capacitor in parallel with a resistor, which simulates the equivalent load of a TTL being driven. He knew that this is a good model to investigate how a circuit behaves in the system. He also knew that the voltage across a capacitor could not change instantaneously. As shown in Figure 4.6(d), he assumed an input signal changed from 0 to 5 V and back to 0 again. Since the circuit response for steady state, such as when the input is [0] or [1], is the same as shown in Row 1 or Row 4 of Table 4.9, it is not repeated here. However, John analyzed the circuit during the transition period in the following manner. With reference to Figure 4.6(a), for input signal changing from low to high (Row 1 to Row 4 in Table 4.9), we have

Q_1: $\begin{cases}\text{Base-Emitter junction changes from ON to OFF}\\ \text{Base-Collector junction changes from OFF to ON}\end{cases}$

Q_2, Q_5: Changes from OFF to ON
Q_3, Q_4: Changes from ON to OFF

and the loading capacitor C then discharges through Q_5. However, all transistors have to pass through their active region regardless of whether they are switching from ON to OFF or from OFF to ON. "Active Region" is defined as

$$0 < \frac{I_C}{I_B} = \beta < \frac{(I_C)_{max}}{I_B} \quad \text{with}$$

$$(I_C)_{max} = \frac{V_{CC}}{R_C}$$

where

I_C = Collector current
I_B = Base current
β = Current gain
V_{CC} = Voltage of the power supply
R_C = Resistance of the collector resistor
$(I_C)_{max}$ = Maximum possible collector current while the transistor is saturated

Therefore, there is a short period in which all transistors are in the active region. Some of them are switching from ON to OFF, others from OFF to ON, and they all draw current from the power supply. Application of BSA follows. Referring to Figure 4.6(a), but remembering that the inputs x_1 and x_2 have been tied together in this example, consider steady state I, i.e., input = [0]. Then

$$I = I_1 + I_2 + I_3$$

with

$$I_1 = \frac{5 - 0.7}{4\ K} = 1.08\ mA$$

Since y = [1] and equivalent resistance are very high, I_2 and I_3 can be neglected. Now consider steady state II, with input = [1].

$$I_1 = \frac{5 - 2.1}{4\ K} = 0.73\ \text{mA}$$

$$I_2 = \frac{5 - \text{saturation voltage of } Q_2 - V_\theta \text{ of } Q_5}{1.2\ K}$$

$$= \frac{5 - 0.2 - 0.7}{1.2\ K} = 3.4\ \text{mA}$$

$$I_3 \simeq 0$$

$$I = I_1 + I_2 + I_3 = 0.73 + 3.4 = 4.13\ \text{mA}$$

For the transition period (y = [0] to y = [1]), since the loading capacitor C momentarily behaves as a short circuit,

$$I_3 = \frac{5 - \text{saturation voltage of } Q_4}{80} = \frac{5 - 0.2}{80} = 60\ \text{mA}$$

Figure 4.6(e) shows the approximate current waveform for the power supply. Note that there is a "spike current" during each transition. During the first transition period, all transistors are conducting, but the capacitor is discharging and demanding no current from the supply, and the maximum possible current will be the sum of the current required for both steady states, i.e., 1.08 + 4.13 = 5.21 mA. During the second transition period, the circuit has to charge the capacitor to high voltage. As calculated, it requires 60 mA charging current from the power supply, plus the transition current for all transistors. Hence, a much larger current spike would be generated.

John suddenly realized that because of the output impedance of the power supply and wires, $Z \simeq 10\ \Omega$ (Figure 4.6(d)), there would be an 0.6 V voltage-drop which would cause a problem. A capacitor of 0.1 μf connected directly across Pins 14 and 7 can filter out the spiking noise.

4.2.5. Some Special and Useful TTL Logic Elements

a. Open-Collector Gate

The open-collector logic element is one of the most popular and useful devices in logic or digital system design. Figure 4.7(a) shows the circuit diagram. Its BSA is shown in the following table.

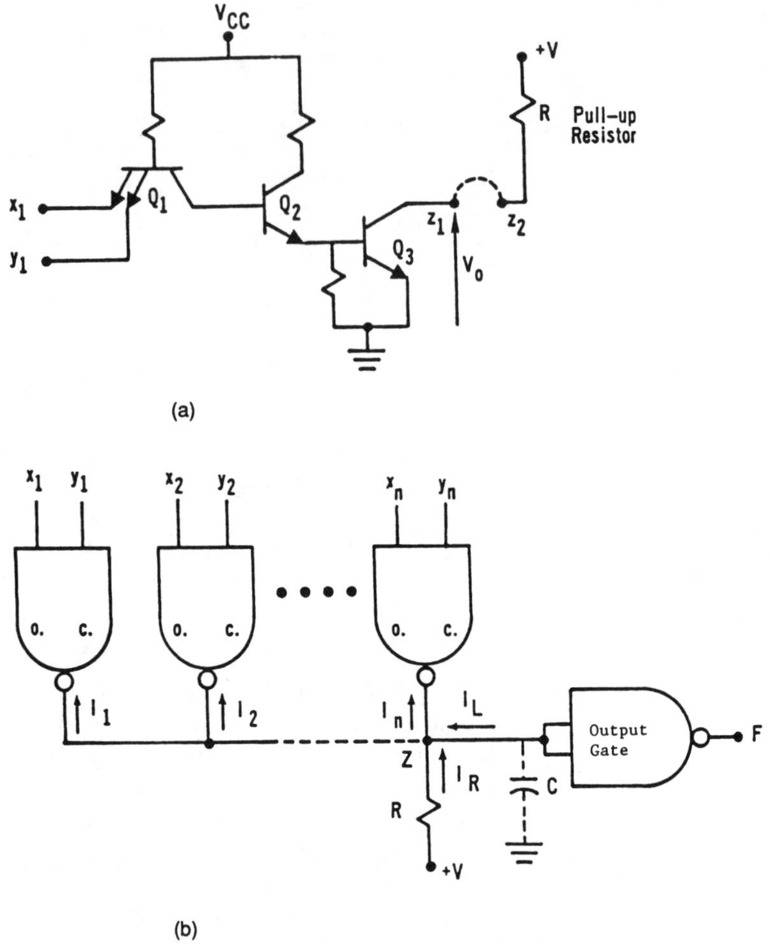

(a)

(b)

FIGURE 4.7 (a) Open collector logic element. (b) Switching function realization with open collectors.

				V_o		Output Impedance	
				z_1, z_2 open	z_1, z_2 closed	z_1, z_2 open	z_1, z_2 closed
x_1	y_1	Q_2	Q_3				
0	0	OFF	OFF	0 v	+V	HI	R
0	1	OFF	OFF	0 v	+V	HI	R
1	0	OFF	OFF	0 v	+V	HI	R
1	1	ON	ON	0.7 v	0.2 v	HI	< 1 ohm

Note that the circuit will not function properly without the external resistor R, which is usually called the "pull-up" resistor. Based on this feature, the outputs of a number of open-collector gates can be directly tied together as shown in Figure 4.7(b) to realize the switching function

$$F = x_1y_1 + x_2y_2 + \ldots + x_ny_n$$

Otherwise, an n-Fan-In OR gate would be required. The value of the pull-up resistor, however, cannot be arbitrary. In order to assure that the voltage at z satisfies the logic levels of 0.5 and 2.4 V, the value of the resistor cannot be out of the range of R_{max} and R_{min}. Calculation of these values follows.

Let us define

R_{max} = Maximum allowable value of R
R_{min} = Minimum allowable value of R
I_{OFF} = Reverse of leakage current of each gate
I_{SAT} = Saturated circuit current of each gate
I_L = Sink current from the output gate.

Then, to assure $V_z \geq 2.4$, we have

$$R_{max} = \frac{V - 2.4}{I_R} = \frac{V - 2.4}{(n + 1)I_{OFF}}$$

where the leakage current of the output gate is included.

Similarly, to assure $V_z \leq 0.5$,

$$V - I_R R_{min} = 0.5 \text{ V}$$

Here, the worst case occurs when the output of one and only one of the open-collector gates is LOW, because under that condition the equivalent output impedance of the n open-collector gates tied together is the highest. Therefore, we have

$$I_R + I_L = I_{SAT} + (n - 1)\, I_{OFF}$$

or

$$I_R = I_{SAT} + (n - 1)\, I_{OFF} - I_L$$

Thus,

$$V - [I_{SAT} + (n - 1)\, I_{OFF} - I_L]\, R_{min} = 0.5 \text{ V}$$

and

$$R_{min} = \frac{V - 0.5}{I_{SAT} + (n - 1)\, I_{OFF} - I_L}$$

It is important to point out that by increasing the voltage V that supplies R, the voltage swing of V_z can be increased accordingly. A few words are needed on the rise time of V_z. If C in Figure 4.7(b) is the input capacitance of the output gate, then the time constant RC will determine the rise time of V_z. For logic circuits where speed is critical, the value of R should then be as low as possible. For circuits where power dissipation is important, however, the value of R should be as high as possible.

b. Three-State or Tri-State Gate

The three-state or tri-state gate is another useful and popular TTL gate. Figure 4.8(a) shows a typical circuit diagram of a tri-state gate. Its BSA is shown below.

$\bar{E}$	x	V_{BC} of Q_1	Q_3	Q_4	Q_5	D	Q_6	Q_7	Q_8	Z_o	V_{out}
1	0	OFF	ON	ON	OFF	ON	ON	OFF	OFF	High impedance	Float
1	1	OFF	ON	ON	OFF	ON	ON	OFF	OFF	High impedance	Float
0	0	OFF	OFF	OFF	OFF	OFF	ON*	ON*	OFF	1	$\simeq V_{CC}$
0	1	ON	OFF	OFF	ON	OFF	OFF	OFF	ON	0	0.2

* Cascaded Emitter-Follower, always in linear mode

Note that the output has three states, namely, high impedance, logic zero, and logic one. When the NOT ENABLE is low, the gate functions as a conventional inverter. When the NOT ENABLE is at the logic 1 state, the output enters a high impedance state, i.e., the gate is basically disconnected from its output or load. A typical application is shown in Figure 4.8(b). By using the tri-state gate, *m* devices of n-bit data can be connected to the same data bus of *n* lines. By setting $\bar{E}_i$ of the ith device low and all others high, the ith device will be connected to the bus, and the *n* lines of data from the ith device will enter into the bus; other devices will be "floating" from the bus. This is the most desirable feature for multiplexing many devices to one bus.

4.2.6. Current Mode Logic (Emitter-Coupled Logic — ECL) — Nonsaturation Type Logic

a. Introduction

All of the logic circuits discussed so far are saturation types which have two distinct states and normally are not sensitive to the variation of parameters, such as current gain, of switching devices. However, because of the excess charges in the diode junctions of the devices, the saturation type logic gates are inherently slow. For extremely high speed applications, the saturation logic gates will not work satisfactorily. In this section, the emitter-coupled logic circuit, which is nonsaturation type logic, will be analyzed. First, an emitter-coupled-pair circuit, the heart of the ECL, should be studied. In Figure 4.9(a), Q_1 and Q_2 operate like a balance (It is basically a single-stage differential amplifier described in Chapter 2, Section 2.1.). That is, while in an equilibrium state, both sides will be at the same height from the "ground". If one side, say the left side, is heavier, then the left side will be low and the

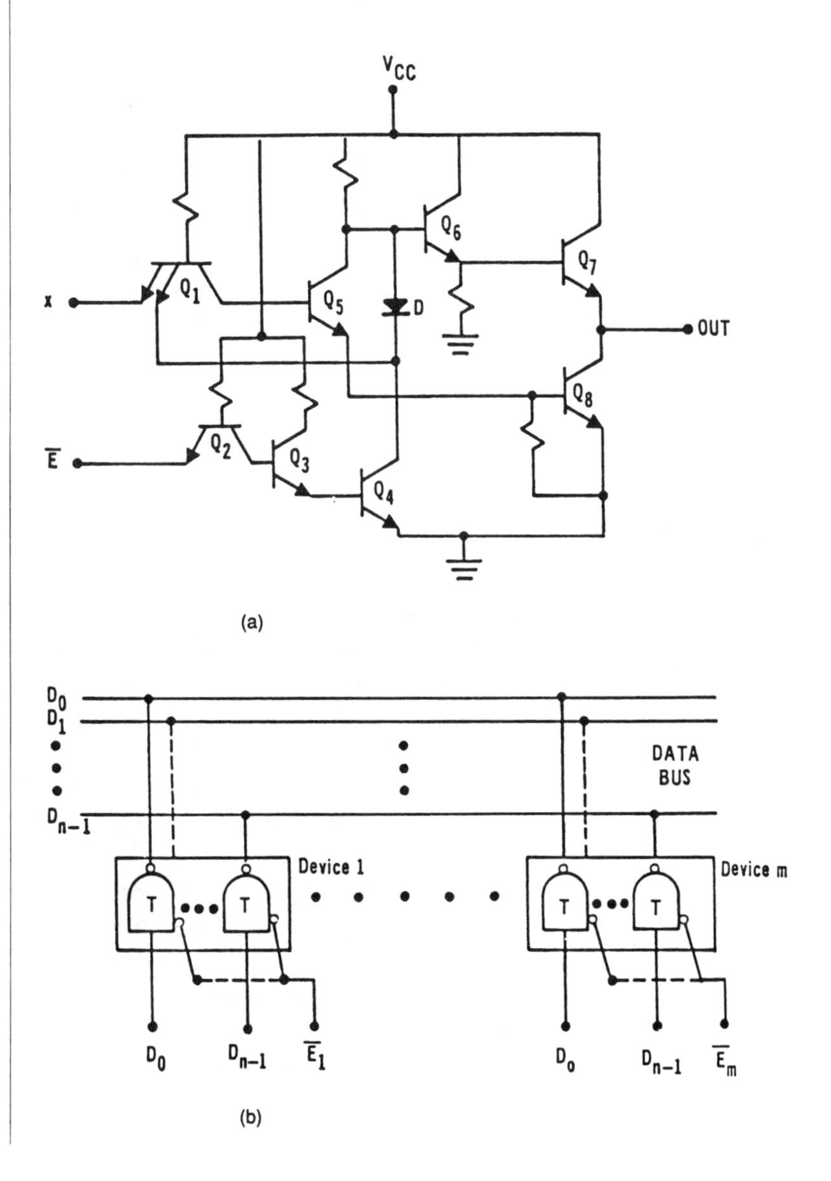

FIGURE 4.8 (a) Circuit diagram of a tri-state gate. (b) A typical application of a tri-state gate.

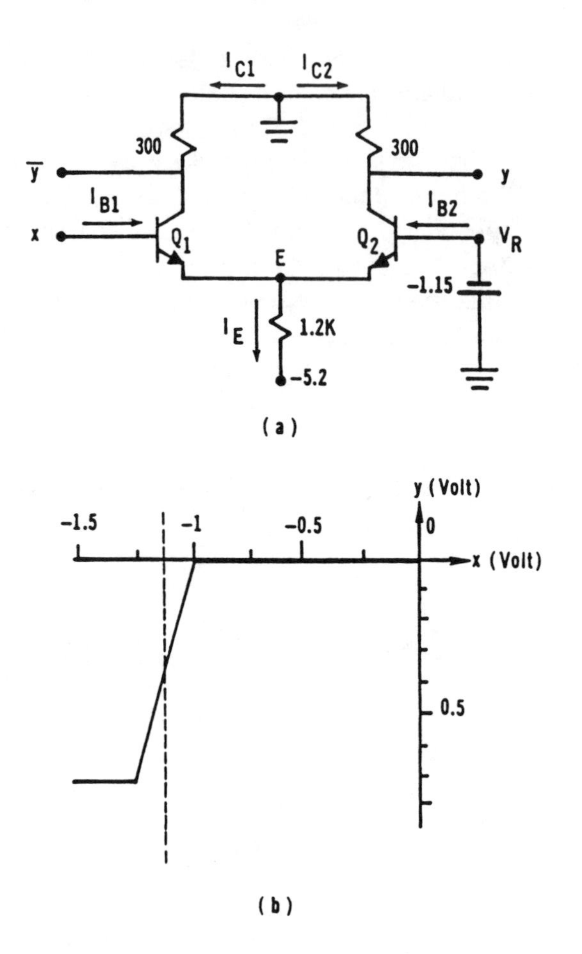

FIGURE 4.9 Analysis of current mode logic circuit.

right side will be high in the air. Analogously, if x is more positive than V_R, then Q_1 will pull more "weight" than Q_2 and $\overline{y}$ will be relatively low and y will be high in voltage. In other words, the diode junctions of the base-emitter of Q_1 and Q_2 are either both conducting or just one is conducting, depending on the voltage difference between x and V_R. For example, if x = –1.5, V_R = –1.15, then Q_2 is conducting and E = –1.15 – 0.7 = –1.85. Since the difference between x and E is less than the threshold voltage, V_θ, Q_1 is OFF. In this case $I_E = \frac{E - (-5.2)}{1.2\text{ K}} = \frac{3.35}{1.2\text{ K}}$ = 2.8 mA, which is approximately equal to I_{C2}. Since Q_1 is OFF, $I_{C1} = 0$. Thus, y = 0 – I_{C2} 300 = –0.84 V. Now let x = –0.7. Q_1 will be conducting and E = –0.7 – 0.7 = –1.4. The voltage difference between E and V_R is less than V_θ. Hence, Q_2 is OFF. In this case, y = 0 V. Figure 4.9(b) shows the transfer curve of x vs. y. Notice that the maximum possible collector current occurs when Q_2 or Q_1 is saturated, i.e.,

$$(I_{C2})_{max} = \frac{5.2 - \text{saturation voltage of } Q_2}{1.2\ K + 300} = \frac{5}{1.5\ K} = 3.3\ mA$$

However, as Q_1 is OFF and Q_2 is ON, the $I_{C2} = 2.8$ mA as we have just calculated. Thus $I_{C2} < (I_{C2})_{max}$, so the transistor Q_2 can never be in the saturation state. Similarly, one can show that Q_1 can never be saturated if $x \geq -1.15$ V. Therefore, the circuit could only be working either in OFF or in the active region, which assures high speed switching.

b. *Emitter-Coupled Logic (ECL)*

Figure 4.10(a) shows a typical ECL two-input NOR/OR gate circuit.

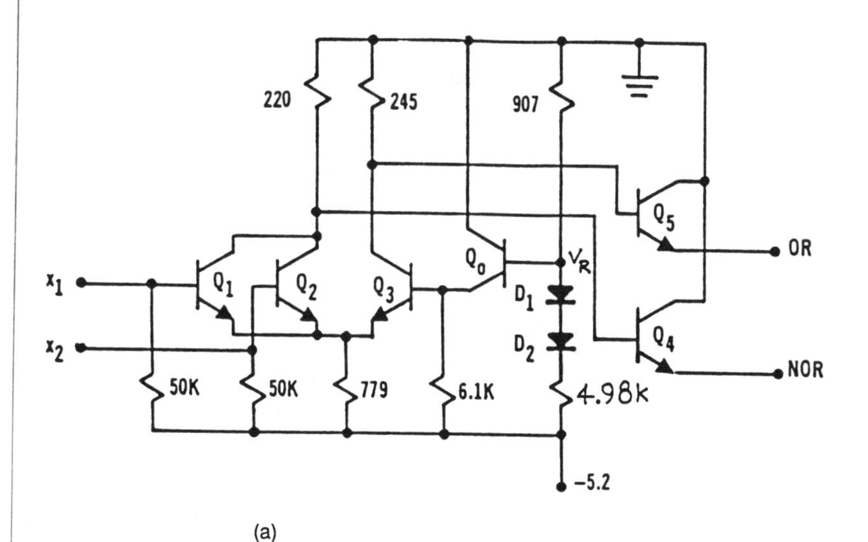

(a)

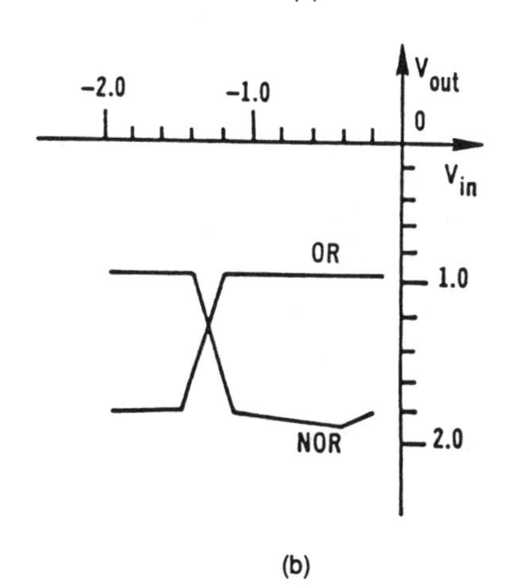

(b)

FIGURE 4.10 (a) A typical ECL circuit. (b) Input/output transfer curves.

Q_2 and Q_3 form the heart of the circuit. Diodes D_1, D_2, and the 4.98 K resistor provide the temperature compensation voltage reference, V_R. Q_4 and Q_5 are emitter-followers which are always active and serve as output buffers. It is interesting to point out that this device has two high fan-out logic outputs, i.e., NOR and OR. Therefore, the circuit shown can be used as either an OR gate or a NOR gate, or both. Figure 4.10(b) depicts the input/output transfer curves. Accordingly, for positive logic, we have logical 1 = –0.9 V and logical 0 = –1.8 V. Since the device is operated in the active region, it is by far the fastest logic available to the designer. Note that the voltage swing, however, for this logic is only from negative 0.9 to 1.8 V. As a result, this logic has poor noise immunity in comparison with TTL. The following table shows the steady states for each transistor with different inputs.

x_1	x_2	Q_0	Q_1	Q_2	Q_3	OR	NOR
–0.9	–0.9	I_r	I_H	I_H	I_L	V_H	V_L
–1.8	–0.9	I_r	OFF	I_H	I_L	V_H	V_L
–0.9	–1.8	I_r	I_H	OFF	I_L	V_H	V_L
–1.8	–1.8	I_r	OFF	OFF	I_H	V_L	V_H

where

$$I_r \triangleq \text{reference current}$$
$$I_H \triangleq \text{high current level}, \; I_L \triangleq \text{low current level}$$
$$V_H \triangleq \text{high voltage level}, \; V_L \triangleq \text{low voltage level}$$

4.2.7. Schottky TTL — Nonsaturation Type Logic

Schottky Transistor-Transistor Logic is an improved TTL circuitry. The Schottky diode-clamp technique prevents the TTL from being driven into the saturation region. As shown in Figure 4.11(a), a Schottky diode which has short recovery time and very low forward voltage drop, 0.1 ~ 0.2 V, is connected between the base and collector terminals. Figure 4.11(b) shows the graphical analysis of the basic circuit. Consider the circuit without the clamping diode. Assume that the transistor is overdriven and thus operates as a saturation logic. That is, it would be in either "saturated" state or OFF state. In saturation state, $V_{BE} \simeq 0.7$ V, $V_{CE} \simeq 0.2$ V and $V_{BC} = 0.7 - 0.2 = 0.5$ V. If the Schottky diode is connected as shown, it would clamp the collector voltage to the base voltage. As a result, the collector voltage would be 0.1 to 0.2 V lower than the base voltage due to the voltage-drop of the Schottky diode. Thus the voltage between collector and emitter would be 0.6 to 0.5 V, as shown in the output V-I curve of the transistor,

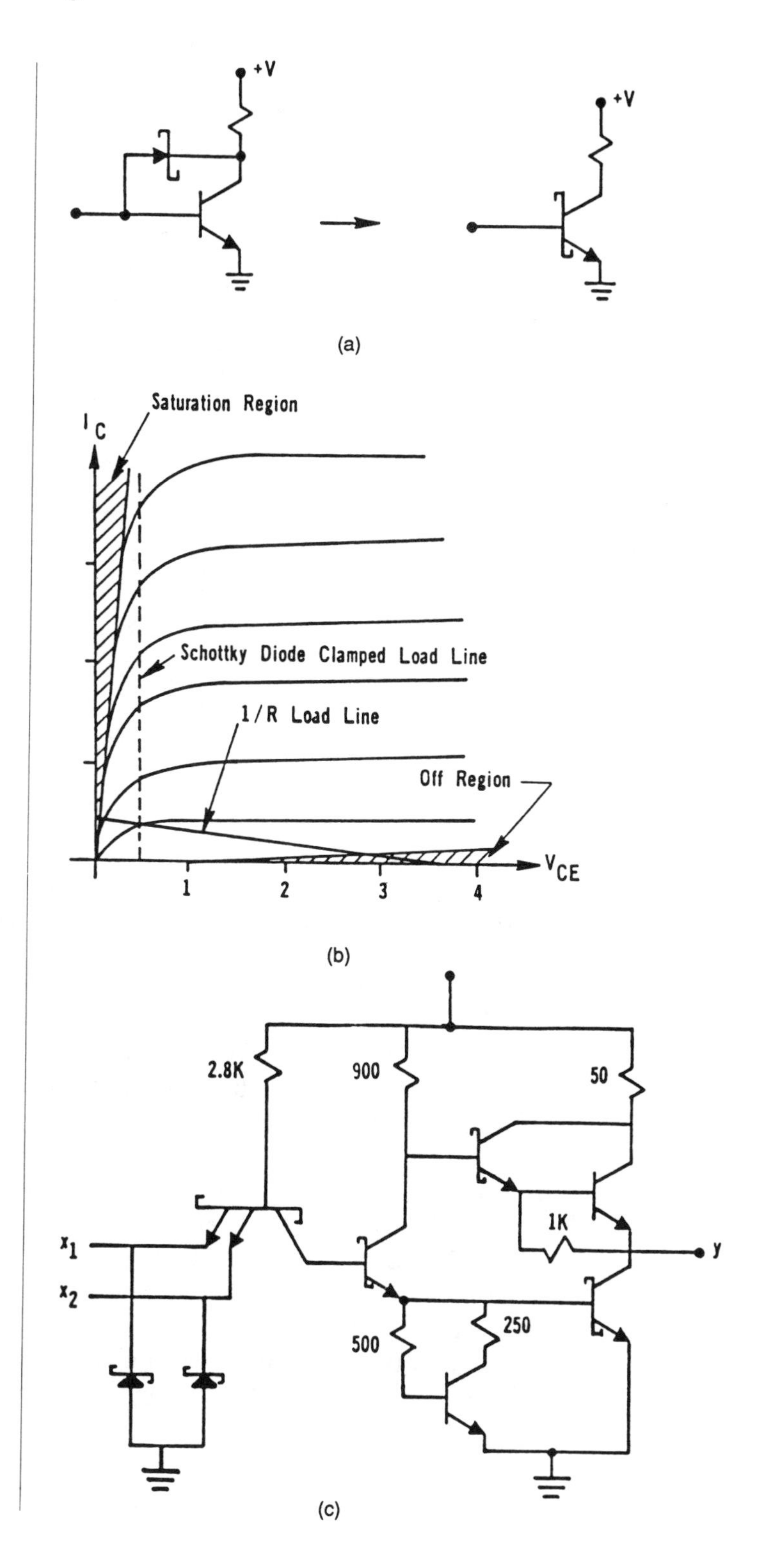

FIGURE 4.11 (a) Schottky transistor circuit and symbol. (b) Schottky diode clamped operation. (c) Regular Schottky TTL. (d) Low power Schottky TTL.

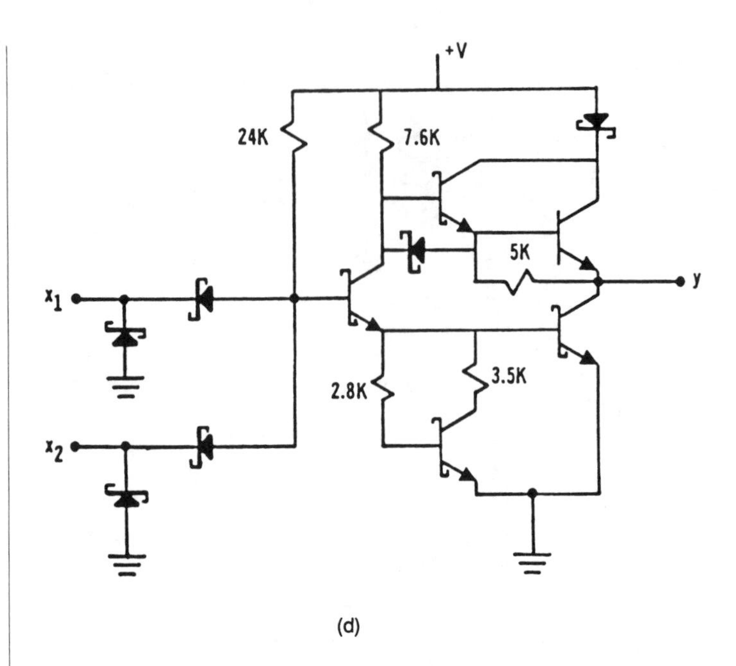

FIGURE 4.11 continued. (a) Schottky transistor circuit and symbol. (b) Schottky diode clamped operation. (c) Regular Schottky TTL. (d) Low power Schottky TTL.

which will keep the device away from its saturation region. In contrast with TTL, the "S" shape "base-bar" is used to denote the Schottky TTL. There are two classes of Schottky TTL families, namely, regular and low power. Figure 4.11(c) and (d) show typical circuits for regular and low power Schottky TTL, respectively. The former typically consumes 20 mw power with a propagation delay time about 3 ~ 5 ns and the latter, 2 mW with operation speed about 5 ns.

4.2.8. CMOS (Complementary Symmetric MOSFET)

In Section 3.3.3, the basic characteristics of field-effect transistors (FETs) were described. We now present the analysis of CMOS logic elements. There are two basic elements in the CMOS logic family, namely, the inverter and the transmission gate.

a. Inverter

Figure 4.12(a) shows a CMOS inverter circuit which consists of two MOSFETs, i.e., P-channel (Q_1) and N-channel (Q_2), and both of them are enhancement types. For clarification, the electric characteristics of each type are shown in Figure 4.12(b) and (c), respectively. From Figure 4.12(a), we see that unlike the conventional circuit, the inverter contains no resistor. In other words, the load resistor of the N-channel MOSFET (the lower one) is now replaced by the P-channel MOSFET. Since these are all nonlinear devices, the load line is no longer linear. We shall now apply first BSA and then graphical analysis. Consider the binary input levels of the inverter to be 0 and +5 V, and it is driving another CMOS logic circuit. Based on the electrical characteristics shown in Figure 4.12(b) and (c) and recalling that the nonconducting static resistance of CMOS is extremely high, one can easily verify the response of the circuit shown below.

x	V_{GS1}	Q_1	V_{GS2}	Q_2	y	I_D	Power dissipation
0	–5	ON	0	OFF	5	0	0
5	0	OFF	5	ON	0	0	0

It is interesting to point out that the circuit appears to be dissipating 0 power. For static analysis, this conclusion is very true; for dynamic operation, however, it is an entirely different story. By means of graphical analysis, the dynamic operation of the circuit can clearly be shown. Figure 4.12(d) shows the graphical analysis. Here, the V-I curve of $I_D - V_{DS}$ of P-channel (Q_1) is shown by the dotted line, superimposed over the $I_D - V_{DS}$ curve of N-channel (Q_2), so that the load line or the operating point locus for variable input voltage can be graphically determined. Typically, points A, B, and C are operating points corresponding to V_{in} of 5, 2.5, and 0 V, respectively. That is, at point A we have $V_{in} = 5$ V, Q_2 ON and Q_1 OFF. Point A is the intersection of the V-I curve of Q_2 with $V_{GS2} = 5$, and the V-I curve of Q_1 with $V_{GS1} = 0$. Accordingly, the I_D flows and the power dissipation of the circuit occurs during the input voltage transitions between logical 0 and logical 1, as shown in Figure 4.12(e) and (f). It is now apparent that the CMOS logic circuit consumes virtually no power during steady state; it does, however, consume power during the transition state. Therefore, we may conclude that the CMOS logic element consumes extremely low stand-by power, and its power dissipation is proportional to the switching frequency of the signals as well as other factors, such as power supply voltage.

b. Transmission Gate

The transmission gate is a unique basic element in the CMOS logic family. It is a bilateral switch which can be used as both a digital and analog switch when an inverter is used as a control device in conjunction with it. Figure 4.12(g) shows the circuit diagram, equivalent circuit, and a BSA table. An important limitatin that the designer should bear in mind is that the voltage level of the data or signal to be transmitted should never exceed the range of V_{SS} and V_{DD}, i.e.,

$$V_{SS} < \text{signal voltage swing} < V_{DD}$$

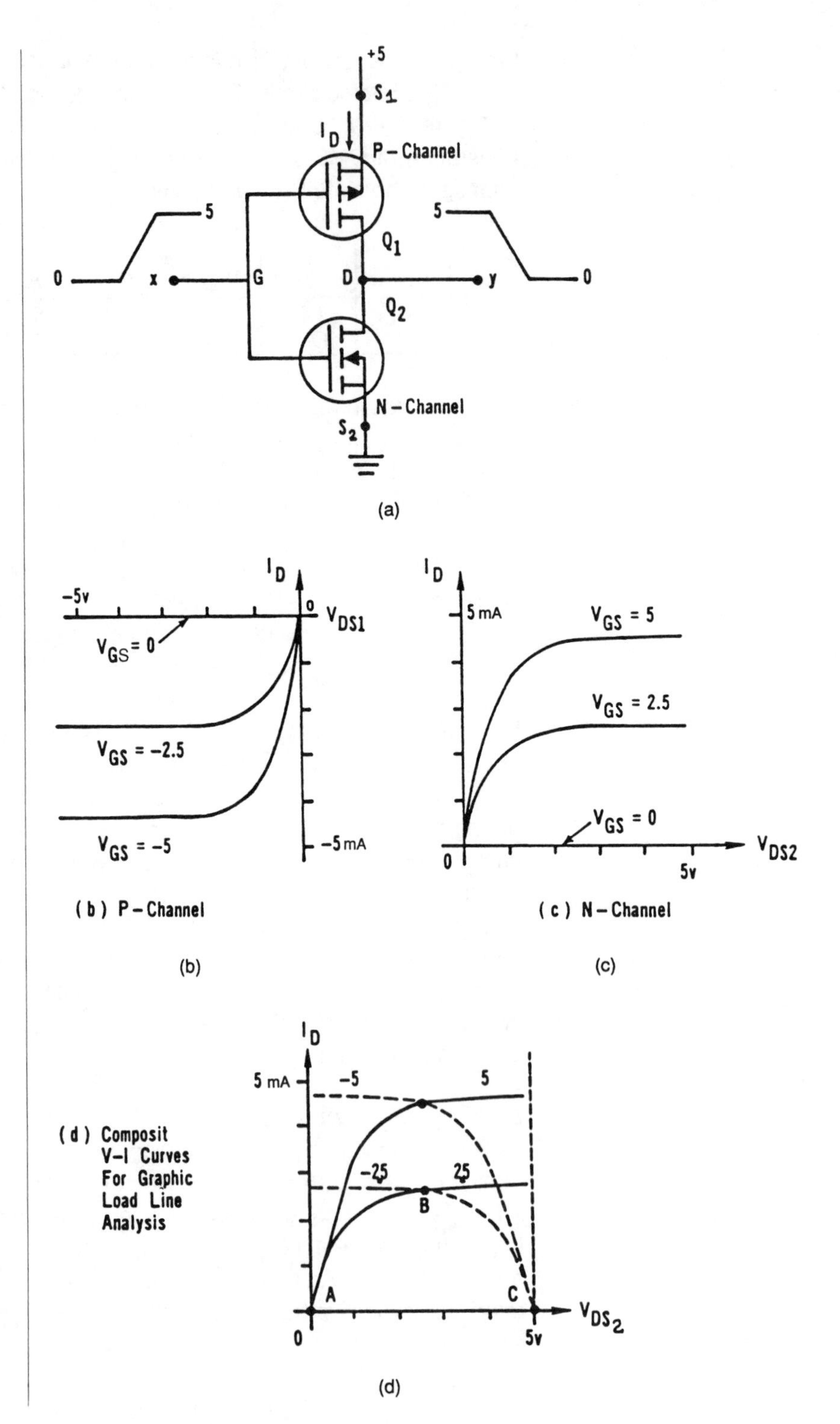

FIGURE 4.12 (a) CMOS inverter. (b, c) CMOS V-I curves. (d, e, and f) CMOS inverter graphical analysis. (g) CMOS transmission gate.

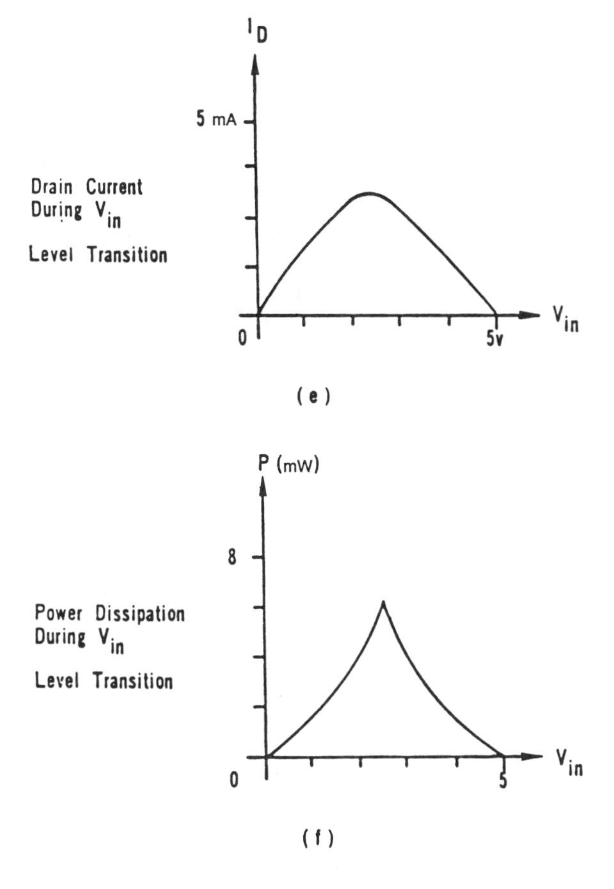

FIGURE 4.12 continued. (a) CMOS inverter. (b, c) CMOS V-I curves. (d, e, and f) CMOS inverter graphical analysis. (g) CMOS transmission gate.

4.2.9. Integrated Injection Logic (I^2L)

Integrated Injection Logic (I^2L) is another kind of bipolar logic. In comparison with TTL, it consumes less electric power and has higher packing density. Figure 4.13(a) and (b) show a single I^2L element and its equivalent. Figure 4.13(c) shows a typical NOR gate and a table showing its BSA result. It is important to point out that the logic level or voltage swing of I^2L is within the range of 0.1 to 0.7 V, which is considerably lower than that of TTL. Its operation speed is faster than TTL but faster than CMOS.

4.2.10. Comparison of the Different Logic Families

The following table shows the quantitative electrical properties of the typical logic families described in the preceding sections. It offers the designer some guidelines in selecting logic elements for his or her applications. In the future, the values shown may improve.

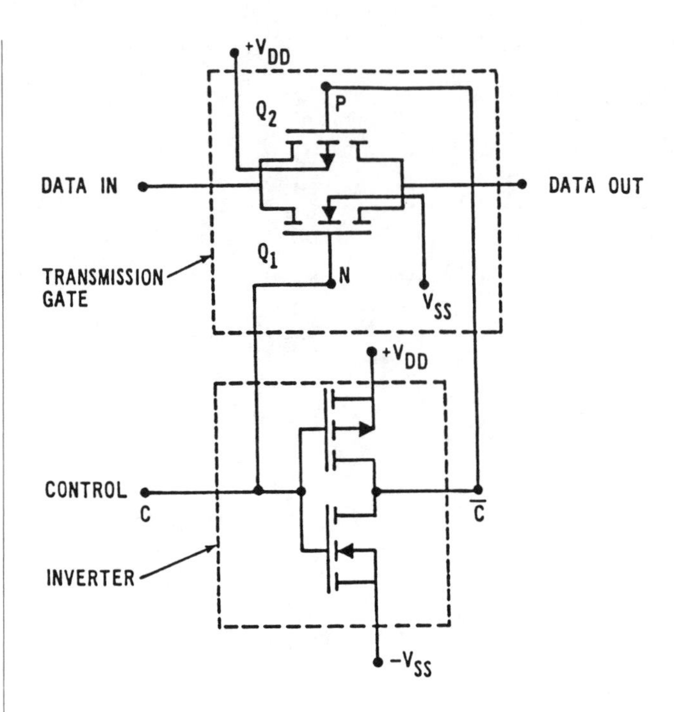

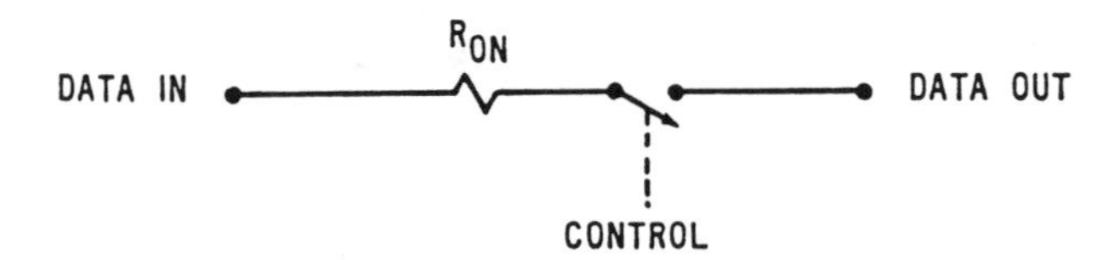

IN	C	$\bar{C}$	Q_1	Q_2	OUTPUT
1	0	1	OFF	OFF	DISCONNECTED FROM INPUT
0			OFF	OFF	
1	1	0	ON	ON	1
0			ON	ON	0

(g)

FIGURE 4.12 continued. (a) CMOS inverter. (b, c) CMOS V-I curves. (d, e, and f) CMOS inverter graphical analysis. (g) CMOS transmission gate.

	Standard TTL	74LS Low power Schottky	CMOS	ECL	I^2L
Gate propagation delay	10 ns	9 ns	25 ns	1 ns	5 ns
Quiescent power per gate	10 mW	2 mW	0.6 μW	30 mW	1 mW
Noise immunity	1 V	0.8 V	2 V	0.4 V	N/A
Fan-out	10	20	50	>50	N/A

N/A = Data not available

4.2.11. Interfacing Logic Elements of Different Families

In the preceding sections we introduced logic elements of different families. We found that each family has its own unique properties;

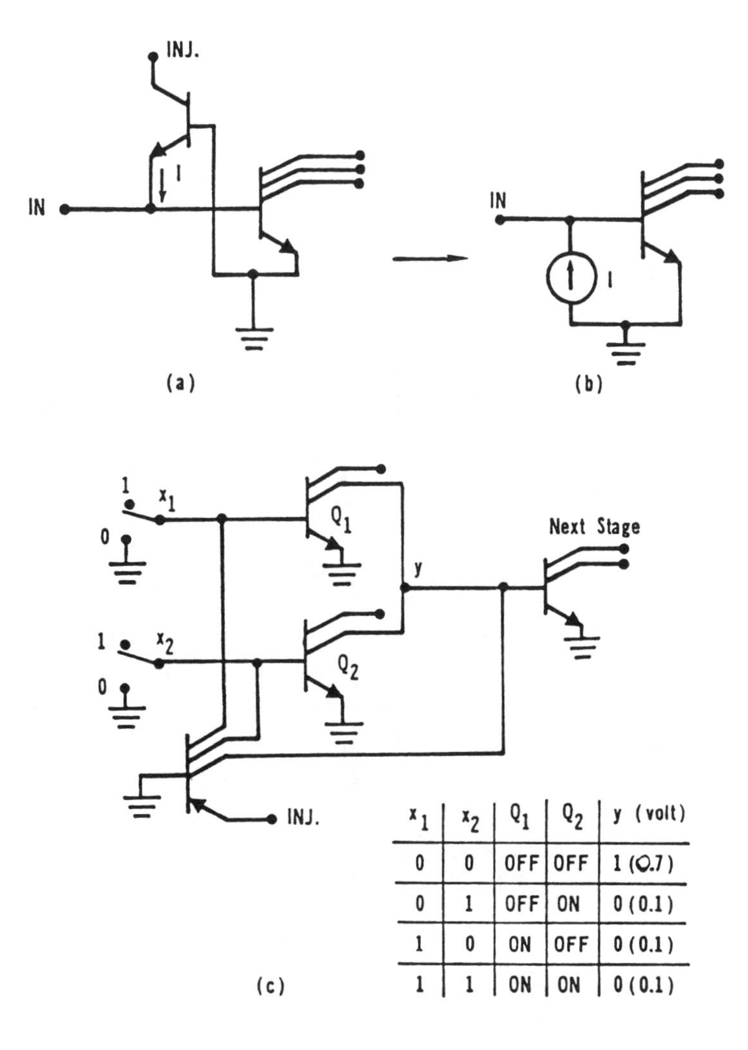

x_1	x_2	Q_1	Q_2	y (volt)
0	0	OFF	OFF	1 (0.7)
0	1	OFF	ON	0 (0.1)
1	0	ON	OFF	0 (0.1)
1	1	ON	ON	0 (0.1)

FIGURE 4.13 Integrated injection logic.

therefore, in some applications, the designer will have to deal with the problems of interfacing logic elements of different families. Although there are now available devices specifically designed for these applications, an understanding of the problems involved and their solutions is essential for a system designer.

a. Interfacing Between CMOS and TTL

There are two problems for interfacing between CMOS and TTL logic elements. First, for TTL the binary logic levels are logical 1 $\geq$ 2.4 V, logical 0 $\leq$ 0.5 V, while for CMOS, logical 1 $\geq$ 70% of ΔV, logical 0 $\leq$ 30% of ΔV, where ΔV is the voltage difference between V_{DD} and V_{SS}. For instance, if V_{DD} = 5 and V_{SS} = 0, then

$$\text{logical } 1 \geq 3.5 \text{ V and}$$

$$\text{logical } 0 \leq 1.5 \text{ V}$$

Therefore, there is an incompatibility of logic levels between the two logic families. Second, CMOS must be able to sink sufficient current from TTL in the logic 0 state and still maintain a voltage level $\leq$0.5 V to assure the logical 0 level for TTL. Figure 4.14(a) and (b) show the level conversion circuits for interfacing TTL to CMOS and CMOS to TTL, respectively. In Figure 4.14(a), the open collector TTL will provide V_{DD} volts in HIGH state and 0.2 V in LOW state, while Figure 4.14(b) shows that the sink current should be sufficiently low to assure a LOW state input to TTL. As a result, it is recommended that one should use either a high-current CMOS buffer or low-power TTL which has sufficiently high input impedance at LOW state input. As shown in Figure 4.14(b), from graphical analysis, it is important that the resistor R shown for TTL be sufficiently large to assure a proper value of I_{sink} so that V_{LOW} of the CMOS output is less than 0.5 V.

b. Interfacing Between CMOS and ECL

The problem in interfacing CMOS and ECL is again incompatibility of logic levels. Figure 4.15(a) and (b) show the level conversion circuits for ECL to CMOS and CMOS to ECL, respectively. Referring to Figure 4.15(a), when the voltage at the emitter Q_1 is higher than that of Q_2, Q_3, and Q_4 , the output of Q_4 will be LOW; however, as the voltage at the emitter of Q_2 is higher than that of Q_1, Q_3, and Q_4 will be in the OFF state and the output of Q_4 will be HIGH. By properly selecting the values of R_1, R_2, R_3, and R_4, the output of Q_4 will swing between zero and V_{CC}. Now consider Figure 4.15(b). Since the output of CMOS swings

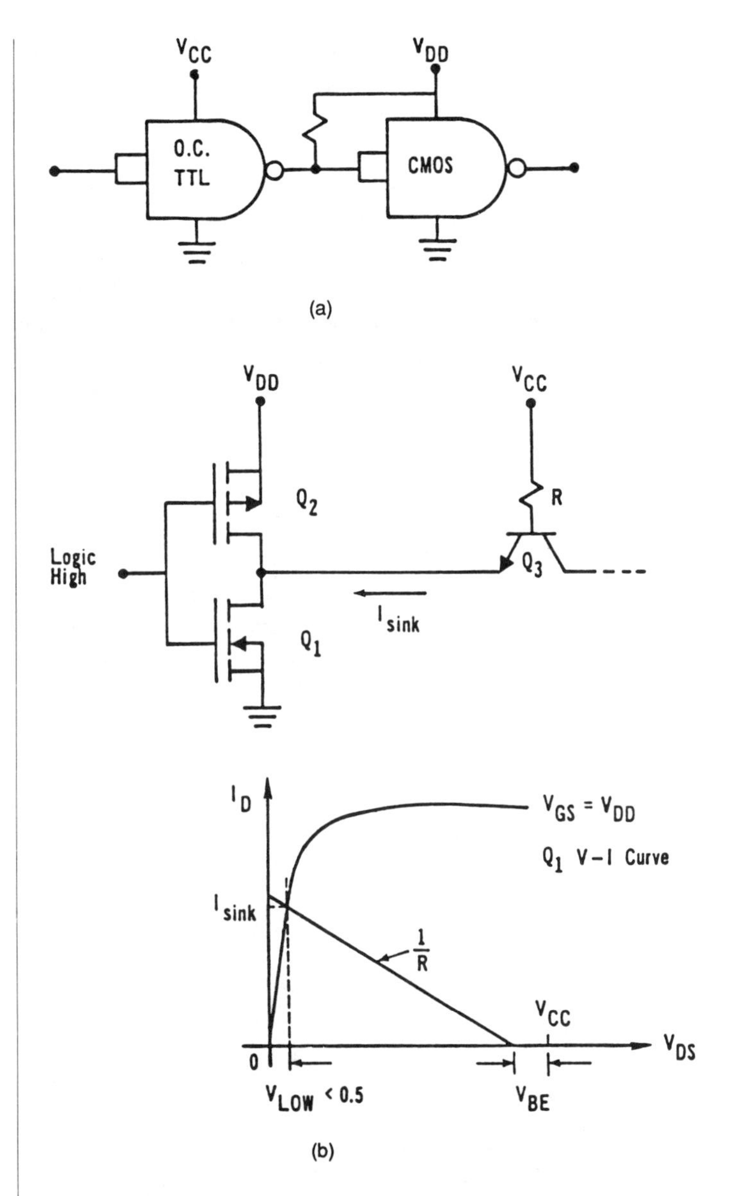

FIGURE 4.14 (a) TTL to CMOS interface. (b) CMOS to TTL.

between ground and V_{DD}, R_1, and R_3 can be selected so that the base voltage of Q_1 will swing between logic high, say –0.9 V and logic low, say some voltage lower than –1.8 V (recall that the ECL normally swings between –0.9 and –1.8 V). Nevertheless, it is important for the designer to be aware of interfacing problems. One could either design the necessary interface networks or use some commercially available devices such as the RCA 10125 and 10124, or other devices, for logic level conversion.

4.2.12. Other Important Properties of Logic Circuits

a. Time Response of the Logic Circuit

All the descriptions of logic circuits given so far have been limited to a steady state response. The time characteristic has not yet been explored. We shall now get acquainted with some important terminologies and leave the more challenging timing problems for later chapters where sequential logic design is described.

Figure 4.16(a) shows a simple RTL inverter. Figure 4.16(b) shows the typical time response of the inverter. However, the terminology introduced here is valid for other logic circuits as well. In Figure 4.16(b), we define

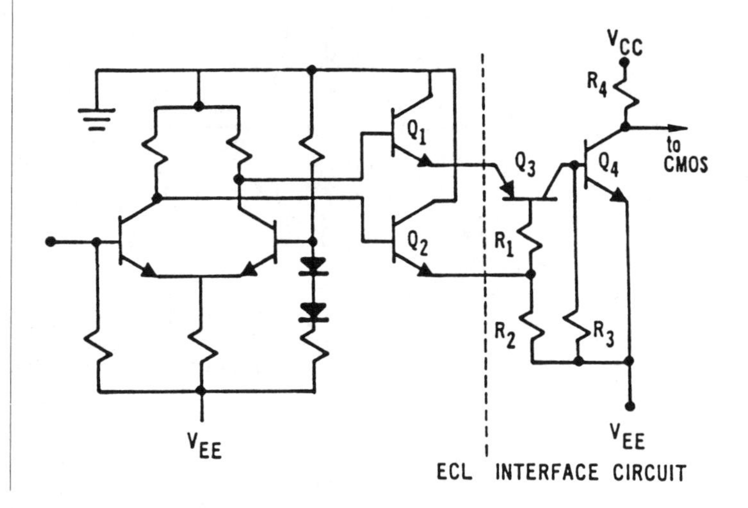

(a)

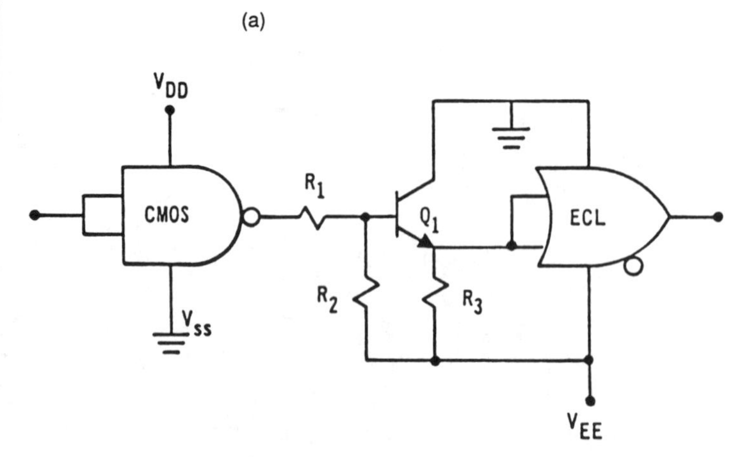

(b)

FIGURE 4.15 (a) ECL to CMOS. (b) CMOS to ECL.

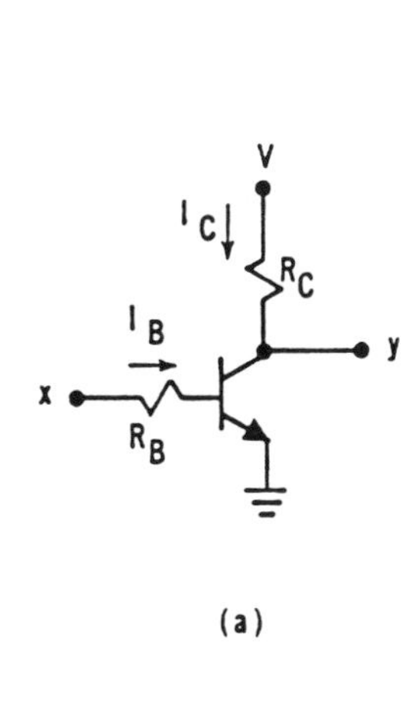

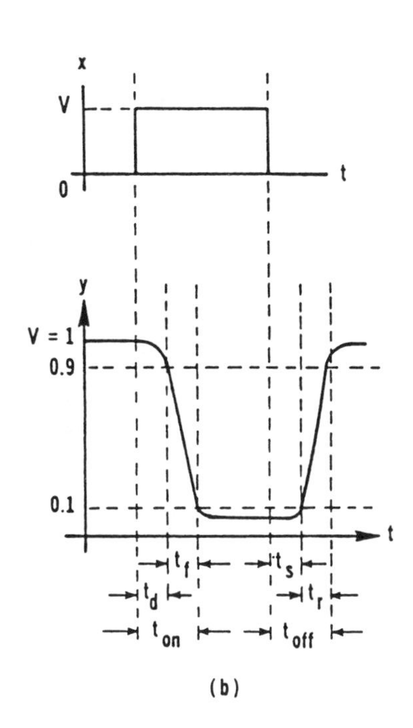

FIGURE 4.16 Time response characteristics.

t_d (time delay)	$\triangleq$	the time between the input ON-SET time and the time when the output drops to 10%
t_f (fall time)	$\triangleq$	the time required for the output to drop from 90 to 10%
t_{on} (turn-on time)	$\triangleq$	$t_d + t_f$
t_s (storage time)	$\triangleq$	the interval between the time when the input turns off and the output rises to 10%
t_r (rise time)	$\triangleq$	the time required for the output to rise from 10% to 90%
t_{off} (off time)	$\triangleq$	$t_s + t_r$
propagation time	$\triangleq$	the time required for a unit to transmit binary information from input to output, $\frac{1}{2}(t_{on} + t_{off})$

b. Power Consumption

The power consumption of a logic circuit is the electrical power being converted to heat within the unit. This information is needed by the system designer so he or she can predict the total heat that may be generated within the system. To understand the problem, consider the circuit shown in Figure 4.16(a). When x = [1], the transistor is turned ON, and the total power consumption is

$$(P_t)_{on} \simeq \frac{V^2}{R_B} + \frac{V^2}{R_C} + \left(\frac{V}{R_C}\right)V_{sat}$$

where $V_{sat} \triangleq$ the saturation voltage. When x = [0], the transistor is turned OFF, and the power consumption in this case is

$$(P_t)_{off} \simeq 0$$

However, in a circuit such as Figure 4.8(a), $(P_t)_{off}$ will not be 0. This is because in either state, there are some transistors in the ON state and others in the OFF state.

4.2.13. Conclusion

In this chapter, the BSA techniques have been introduced and their application to understanding the principles of operation of different logic circuits has been described. The main theme of this chapter is the contrast between the modeling of linear and nonlinear circuit elements. For this we need only know about the steady state and transition state responses of logic circuits. The BSA techniques introduced are general and can be applied to other logic circuit families. The table below provides a qualitative comparison of the major characteristics of the logic circuits described in this chapter. For more detailed information, one can refer to the manufacturers' bulletins.

Properties	RTL	DTL	TTL	ECL	CMOS
Speed	Low	Low	High	Ultra-high	Medium
Noise immunization	Poor	Good	Good	Fair	Good
Fan-out	Poor	Fair	Fair	Good	Excellent
Power consumption	Medium	Low	Fair	High	Very low

4.3. SWITCHING INDUCTIVE LOAD BY A TRANSISTOR

In digital system design, a designer often must design drivers or controllers, such as motors or relays, or inductive loads. An understanding of the basic principle of switching inductive load is therefore important. Consider a transistor used to drive an inductive load. From the transient characteristics described in Section 3.2.3, we know that the current flowing in an inductor cannot change instantaneously. Therefore, the circuit shown in Figure 4.17(a) requires special consideration. Assume the switch is at position B for a long, long time, and thus diode D, being backward biased, has no effect on the circuit. The circuit is then at its steady state or the transistor is in its saturation state. As a result, a steady current I_s is flowing from V_{CC} through the inductor and transistor to ground. When the switch S is turned to position A at t = 0, since the current in an inductor cannot change instantaneously, the operational path on the V-I curve of the transistor will no longer follow the dc load line. Instead, it follows the dotted line horizontally at time $t = 0^+$ and eventually, along the avalanche line of the transistor reaches point A as shown in Figure 4.17(b), with a time constant L/R_L. For the same reason, when the switch is suddenly turned to position B again, the turn-on path shown in Figure 4.17(b) is followed. Although the turn-on path would not cause trouble, the turn-off path which follows the avalanche path is evidently not desirable. Therefore, a switching diode should be connected across the inductor as shown in Figure 4.17(a). The diode is mainly used to pass the saturation current I_S as the transistor is turned off instantly. As a result, the operational path follows the perpendicular dotted line instead of the avalanche line shown in Figure 4.17(b). In conclusion, it is important to point out that a diode with proper current capacity and reverse voltage rating should always be used as shown for any inductive load, such as a relay or mo-

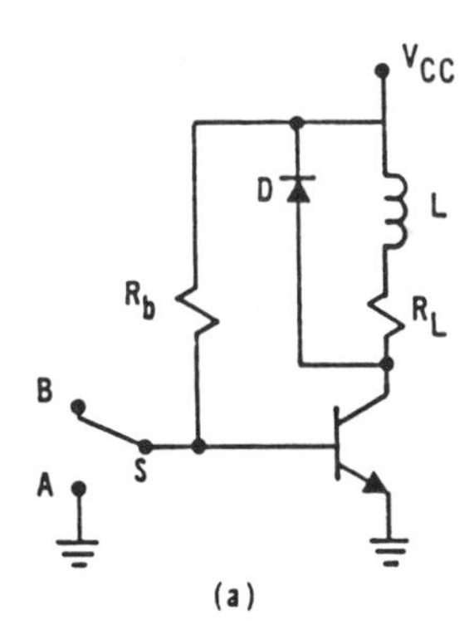

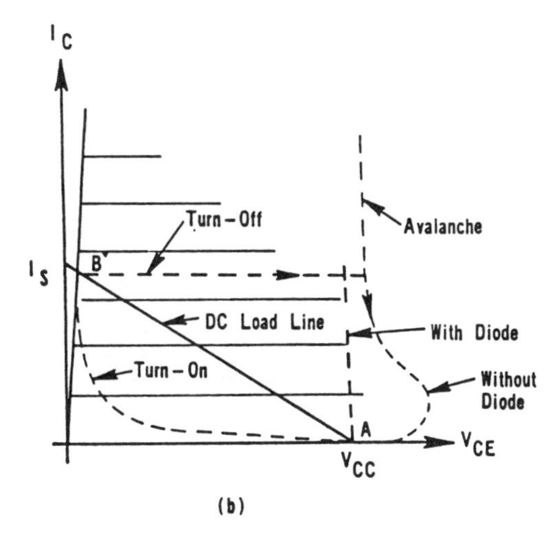

FIGURE 4.17 Switching an inductive load.

tor, in any digital system design. For example, for the circuit shown, the reverse voltage of the diode must be > V_{CC}, and its forward current rating must be > (V_{CC}/R_L).

REFERENCES

1. Altman, L., Logic's leap ahead creates new design tools for old and new applications, *Electronics*, 47(4), 81—90, February 21, 1974.
2. Altman, L., C-Mos enlarges its territory, *Electronics,* 48(10), 77—88, May 15, 1975.
3. Carr, W. N. and Mize, J. P., *MOS/LSI Design and Application*, McGraw-Hill, New York, 1972.
4. Hart, C. M., Slob, A., and Wulms, H. E. J., Bipolar LSI takes a new direction with integrated injection logic, *Electronics,* 47(20), 111—118, October 3, 1974.
5. Horton, R. L., Englade, J., and McGee, G., I^2L takes bipolar integration a significant step forward, *Electronics,* pp. 83—90, February 6, 1975.
6. Pederson, R. D., Integrated injection logic: a bipolar LSI technique, *IEEE Computer,* 9(2), 24—29, February 1976.
7. Torrero, E. A., Special Report on I^2L Applications, *IEEE Spectrum*, 14(6), 28—36, June 1977.
8. Capece, R. P., Faster, lower-power TTL looks for work, *Electronics,* 52(2), 88—89, February 1, 1979.
9. *ECL Handbook,* Fairchild Semiconductor, Mountain View, CA, July 1974.
10. *ECL System Design Handbook,* Motorola Semiconductor, Phoenix, AZ.
11. Horowitz, P. and Hill, W., *The Art of Electronics,* 2nd ed., Cambridge University Press, New York, 1990.
12. Millman, J. and Grabel, A., *Microelectronics,* 2nd ed., McGraw-Hill, New York, 1987.
13. Bell, D. A., *Solid State Pulse Circuits,* 3rd ed., Prentice-Hall, Englewood Cliffs, NJ, 1988.
14. Johnson, E. L. and Karim, M. A., *Digital Design,* PWS Publishers, Boston, 1987.

5 BASIC FUNCTIONAL MODULES

In addition to logic gates, a digital system uses many other functional modules built on gates or energy storage devices or resistor/capacitor timing elements. These include the bistable multivibrator or flip-flop, the monostable multivibrator or one-shot, and the astable multivibrator or free running or clock. Among them the flip-flop is used most frequently, and thus it is presented first. A flip-flop is simply a digital memory device which can "memorize" one and only one thing at a time. Hence, in a flip-flop, one can store either "YES" or "NO", "HIGH" or "LOW", or "ONE" or "ZERO". In the terminology of the logic designer, it can memorize or store only one "bit" at a time. Actually, in our daily life we use flip-flops quite often. For example, the wall switch in our home is one type of flip-flop device. When a switch is turned on, it means that "ON" information is stored and the switch memorizes it and keeps the light burning; when the switch is turned off, the "OFF" information is then stored. In fact, the switch called "Rockette" type on the market is actually a set-reset or R-S flip-flop, since if one presses one end of the switch arm to turn it ON, then he/she can turn it OFF by pressing the other end. Since most beginners have trouble understanding how a flip-flop works, we shall start with a basic circuit, translate it from a circuit designer's language into a logic designer's language, and then progress to more complicated ones.

5.1. FLIP-FLOPS BISTABLE MULTIVIBRATORS

5.1.1. The Basic Flip-Flop Circuit

The circuit shown in Figure 5.1 is a simple two-stage RTL circuit connected in a special way which yields a very interesting result. The circuit has two inputs, S and R, and two outputs, y_1 and y_2. They are related in a nontrivial way. A flip-flop to a logic designer is like an automobile to a driver; it can be either a useful vehicle or a number one killer depending on how well the driver handles the machine! The reader is urged to study this section carefully so that he/she will be able to understand the sections that follow.

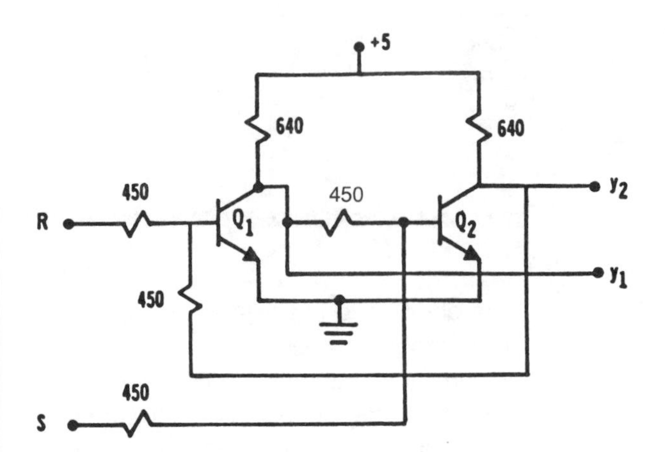

FIGURE 5.1 Basic flip-flop circuit.

Let us analyze the circuit by the BSA technique. There are two kinds of input signals: (a) dc level inputs, such as "0" or "+5" V, and (b) positive going pulse inputs, such as a pulse from 0 to 5 and then to 0 again. We shall analyze both cases.

a. dc Level Inputs

The following table shows the circuit responses for dc level inputs.

TABLE 5.1.1

S	R	Q_1	y_1	Q_2	y_2
+5	+5	ON	0.2	ON	0.2
0	+5	ON	0.2	OFF	2.47
+5	0	OFF	2.47	ON	0.2
0	0	?	?	?	?

The reader should have no trouble verifying the entries of the first three rows. However, one should think twice before making any decisions on the fourth row. Notice that Q_1 is "ON" or "OFF" depending not only on the voltage level at R, but also the voltage level at y_2. That is, Q_1 is on if R or y_2 or both are high. Similarly, Q_2 is ON depending on y_1 and S. If y_2 is high, which implies Q_2 is OFF, then Q_1 must be ON. If y_1 is high, which implies Q_1 is OFF, then Q_2 must be ON. In other words, if R = S = 0, then Q_1 can be ON if Q_2 is OFF and vice versa. So if R = S = 0, then there are only two possible cases: either Q_1 = ON, Q_2 = OFF or Q_1 = OFF, Q_2 = ON. They cannot be both ON or both OFF at the same time. One might then ask which of these two possible cases will hold if R = S = 0. The answer depends on the past history of Q_1 and Q_2. In other words, Q_1 and Q_2 will be the same as they were just before R and S both became zero. Consider the analogy of flipping a coin on a table and walking away. If nobody touches it, what will the state of the coin be, "head" or "tail"? The answer is that there will be "No Change". If it was head, it will still be head! What would happen if Q_1 and Q_2 were both ON or both OFF? In this circuit configuration, it is impossible to have the "Both OFF" condition, but the "Both ON" condition occurs if R = S = 5. However, if R = S = 0 follows R = S = 5, the response can be either y_1 = H, y_2 = L or y_1 = L, y_2 = H, depending on which one of the two inputs reaches low first. In order to design a reliable system, this condition should not be allowed. So now we can modify Table 5.1.1 as follows:

TABLE 5.1.2

S	R	y_1	y_2
H	H	Not allowed	Not allowed
L	H	L	H
H	L	H	L
L	L	H	L
		L	H

Notice that y_2 is always a negation of y_1 if the first row is not allowed. Thus, Table 5.1.2 can be further simplified if we define y_1 = Q.

TABLE 5.1.3

S	R	Q	$\bar{Q}$
H	H	Not allowed	Not allowed
L	H	L	H
H	L	H	L
L	L	No change	No change

Let us define Q_n as the present (old) state of y_1, and Q_{n+1} as the next (new) state of y_1. Then we can rewrite Table 5.1.3 as follows:

TABLE 5.1.4

S_n	R_n	Q_{n+1} (new)	$\bar{Q}_{n+1}$ (new)
H	H	Not allowed	Not allowed
L	H	L	H
H	L	H	L
L	L	Q_n (old)	$\bar{Q}_n$ (old)

where the entries in columns S_n and R_n are the cause and those in columns Q_{n+1} and $\bar{Q}_{n+1}$ are the results, or the next state of the flip-flop.

b. Positive Going Pulse Inputs

If

$Q_{n+1} \triangleq$ the circuit response shortly after the pulse rising edge
$Q_n \triangleq$ the circuit response shortly before the pulse rising edge
$p \triangleq$ pulse
$\bar{p} \triangleq$ no pulse,

then we have

TABLE 5.1.5

S_n	R_n	Q_{n+1} (new)	$\bar{Q}_{n+1}$ (new)
p	p	Not allowed	Not allowed
$\bar{p}$	p	L	H
p	$\bar{p}$	H	L
$\bar{p}$	$\bar{p}$	Q_n	$\bar{Q}_n$

It is important to point out that the circuit actually "remembers" whether a pulse was applied at the input terminals R or S. Since normally there is no pulse at R or S, during this period if y_1 is "HIGH" it means that there was a pulse at S, and if y_1 is "LOW" there was a pulse applied at R. This property gives the circuit the right to claim the glorious name of "Memory Device". *In reference to Q*, the designer can store a logical 1 in the circuit by applying a pulse at S and a logical 0 by applying a pulse at R. The "S" stands for "Set" and the "R" for "Reset". Some designers or manufacturers prefer to use C instead of R, where "C" stands for "Clear", because if a pulse is applied at R, the [1] at $y_1(Q)$ will be wiped or cleared off. Since there are only two possible steady states, the circuit is also called a bistable circuit or bistable multivibrator. Specifically, this one is known as an *R-S flip-flop*. There are many kinds of flip-flops, but this circuit is the heart of all of them.

c. Logic Presentation

Notice that the circuit actually contains two cross-connected resistor-transistor NOR gates, where one NOR gate has y_1 as output, and R and y_2 as inputs, while the other NOR gate has S and y_1 as inputs, and y_2 as output. By using MIL-STD-806B symbols, the circuit can be redrawn as shown in Figure 5.2. Due to its logical configuration, this circuit is also called a cross-coupled flip-flop.

We have now moved from the electronic circuit presentation into logical symbol presentation. Next, we should try to link the circuit with Boolean presentation, which is usually done by direct translation of a truth table. Since the fourth columns in Tables 5.1.4 and 5.1.5 are a negation of the third columns, they can be ignored and one can write the switching functions from Tables 5.1.4 and 5.1.5 as follows.

In Table 5.1.4, since S = R = H is not allowed, we may describe it by the equation, RS = 0. In addition, if the subscript n+1 is defined as the next state, and n implies the present state, then by translating Table 5.1.4 into Boolean presentation, we have

$$Q_{n+1} = S_n\overline{R}_n + \overline{S}_n\overline{R}_nQ_n = S_n\overline{R}_n + 0 + \overline{S}_n\overline{R}_nQ_n$$

But we set RS = 0 as the restriction, thus

$$Q_{n+1} = S_n\overline{R}_n + R_nS_n + \overline{S}_n\overline{R}_nQ_n$$

$$= S_n(\overline{R}_n + R_n) + \overline{S}_n\overline{R}_nQ_n$$

$$= S_n + \overline{S}_n\overline{R}_nQ_n$$

$$= (S_n + \overline{S}_n)(S_n + \overline{R}_nQ_n)$$

$$= S_n + \overline{R}_nQ_n \quad \text{(for positive logic using NOR gates)} \tag{5.1}$$

Similarly, from Table 5.1.5, we have

$$Q_{n+1} = p_S + \overline{p}_R \cdot Q_n \tag{5.2}$$

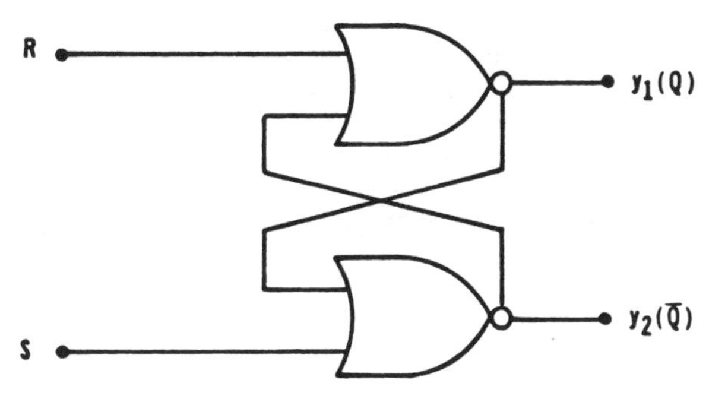

FIGURE 5.2 Using two NOR gates cross-coupled flip-flop.

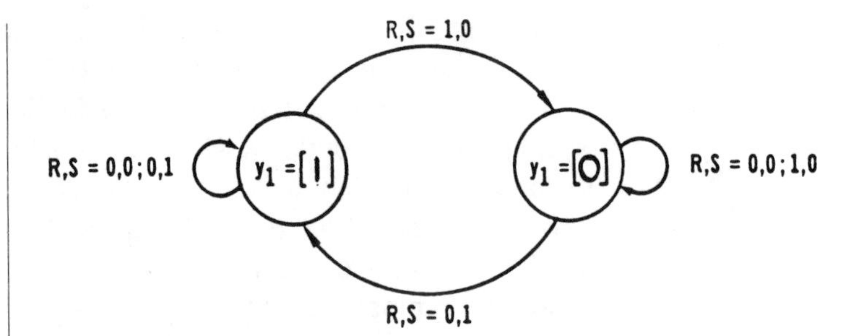

FIGURE 5.3 State diagram of NOR gate cross-coupled flip-flop.

where p_S denotes the pulse at terminal S and p_R denotes the pulse at terminal R. Notice that in the above equations, Q_{n+1} and Q_n refer to the same terminal y_1, where Q_{n+1} denotes the logical state or value at the time t = n+1, and Q_n denotes the state at time t = n. For example, in Equation (5.2), Q_{n+1} = [1] if there is a pulse occurring at the S input. If there is no pulse at R or S, then Q_{n+1} takes the old value Q_n; there will thus be no change in the logical value of Q. Equations (5.1) and (5.2) are called the *input equation of the R-S flip-flop.*

d. State Presentation

The logic diagram presentation, although clear in its logical sense, lacks time sequence information. Although switching functions do contain time sequence information they are somewhat abstract. Since it is found that the *state diagram* gives clearer time sequence information, it is considered here.

The state diagram shown in Figure 5.3 gives a clear presentation of the time sequence response. For convenience but without losing generality, positive logic is used here. The two big circles indicate the two possible steady states of the output terminal y_1. At any steady state instant, y_1 will be either [1] or [0]. Consider y_1 = [1] at this moment, i.e., the circuit stays in the circle on the left. Now, if RS = 00, or RS = 01, then the logic value of y_1 will remain unchanged. Therefore, a "self-circling" line is shown. If the input R becomes [1] and S = [0], y_1 will change from state [1] to state [0], thus the diagram shows an arrow for y_1 = [1] which "jumps" over to the state of y_1 = [0]. Since RS = 11 is not allowed in this circuit configuration, it is not shown in this diagram. The reader is urged to verify the rest of the operation shown in the diagram.

e. Conclusion

In this section, a basic flip-flop circuit composed of two RTL NOR gates was used as a vehicle to show how one can derive switching functions, logic diagrams, and state diagrams for a given electronic circuit. The reader is urged to practice this technique on some other electronic circuit so he/she will develop the ability to interpret any new electronic circuit which logic designers normally use. It is worthwhile to point out that for a low number of inputs or switching variables, the state diagram presentation is a good tool to describe how a sequential circuit "behaves" dynamically under different input conditions.

5.1.2. Basic Integrated Circuit Flip-Flops

In Section 5.1.1, the basic flip-flop circuit made of two discrete NOR gates was investigated. In this section, the basic flip-flop that contains two IC NAND gates will be described. Figure 5.4 shows two 54/74

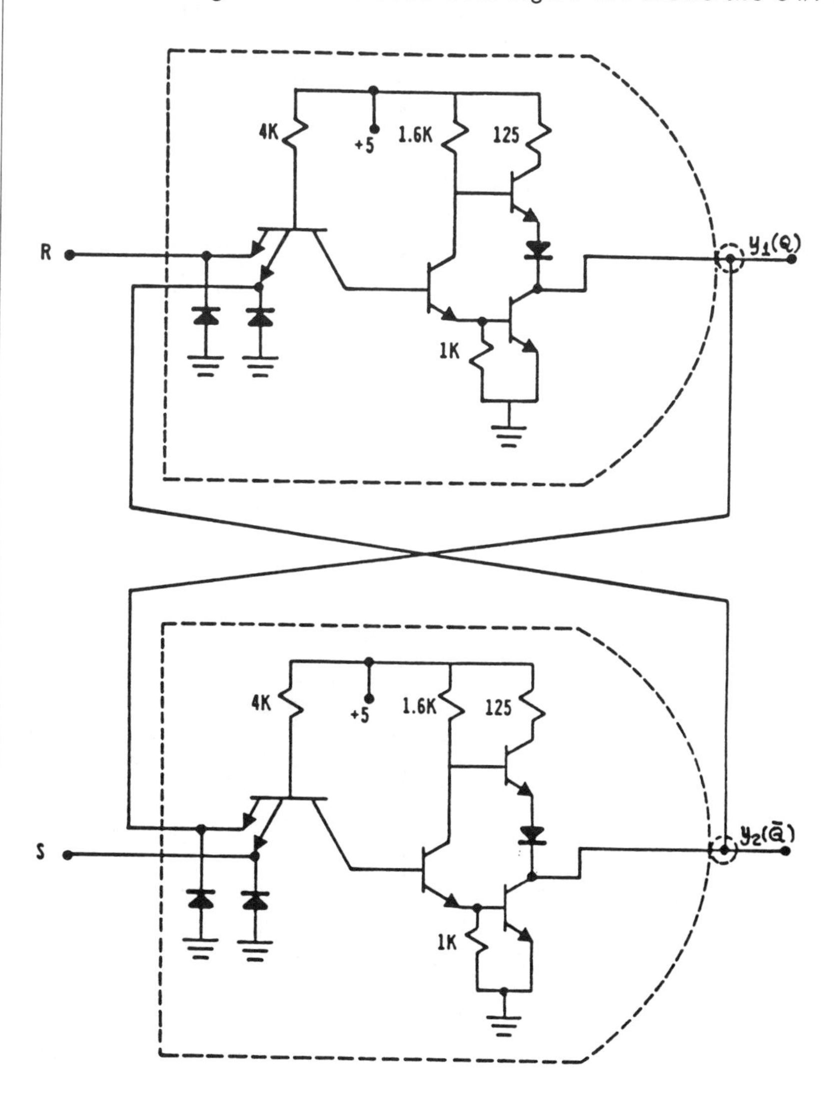

FIGURE 5.4 Cross-coupled flip-flop using two NAND gates.

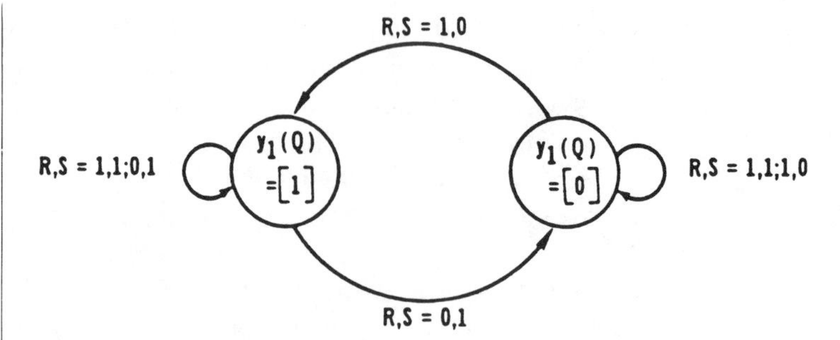

FIGURE 5.5 State diagram of NAND gate cross-coupled flip-flop.

series IC NAND gates with cross-coupled connection. The dotted lines show the MIL-STD-806B NAND gate symbols, and the integrated circuits are shown inside the symbol. Although this circuit appears more complicated than the discrete circuit described previously, in principle, it is the same except for one minor detail which will be discussed here.

By applying the BSA technique, one can easily obtain the truth table shown in Table 5.1.6. Notice that with this circuit it is impossible to have $y_1 = y_2 = L$. With R = S = H, we have the state of unchange. That is, there are two possible states for this circuit, namely $y_1 = H$, $y_2 = L$ or $y_1 = L$, $y_2 = H$. Following the same argument presented in the previous section, in this circuit R = S = L should not be allowed. Thus the switching function of this circuit in positive logic is

$$\overline{RS} = 0$$

$$Q_{n+1} = \overline{R} + SQ_n \quad \text{(positive logic using NAND gates)} \tag{5.3}$$

and its *state diagram* is shown in Figure 5.5.

TABLE 5.1.6

R	S	y_1	y_2
L	L	H	H
L	H	H	L
H	L	L	H
H	H	H	L
		L	H

5.1.3. Integrated Circuit Flip-Flops

The reader may find that there are too many kinds of IC flip-flops available on the market. Sometimes a logic designer may have a problem choosing the proper one for his/her system. The best way to overcome this kind of difficulty is to analyze a few representative ones. In Sections 5.1.1 and 5.1.2, two basic flip-flop circuits, cross-coupled NOR gates and NAND gates, were analyzed and their corresponding logic symbols were introduced. We shall eventually leave the detailed electronic circuit analysis technique behind and instead decompose the IC flip-flops down to the basic symbolic flip-flops and gate level for functional analysis. Readers who do not feel confident at the "logic presentation level" can always go back to the preceding sections for help.

5.1.4. Clocked R-S Flip-Flops

For the basic cross-coupled NAND gate flip-flop described in Section 5.1.2, it is clear that as soon as the set-line has become LOW, the output Q terminal becomes HIGH and a "ONE" is stored in the flip-flop. However, in some applications the designer wishes to store the information at a specific time. This can be done with the circuit shown in Figure 5.6.

Note that there are two NAND gates and one cross-coupled flip-flop whose electronic circuits and principle of operation were described in the preceding sections. Let us say a logical 1 is to be stored in this device. The S is set to HIGH and the R is set to LOW. The clock line is sitting at LOW. During this time, x_1 and x_2 are both HIGH, therefore there will be no change in the cross-coupled flip-flop. However, as soon as the clock pulse has arrived, the clock line becomes HIGH and $x_1 = [0]$ and $x_2 = [1]$, which results in $Q = [1]$. When the clock pulse disappears, the clock line returns to LOW and $x_1 = x_2 = [1]$ and Q remains unchanged until the next clock pulse arrived. This circuit is basically the same as the basic R-S flip-flop except that the information is stored at a specific time. Here R = S = HIGH is not allowed, since that condition

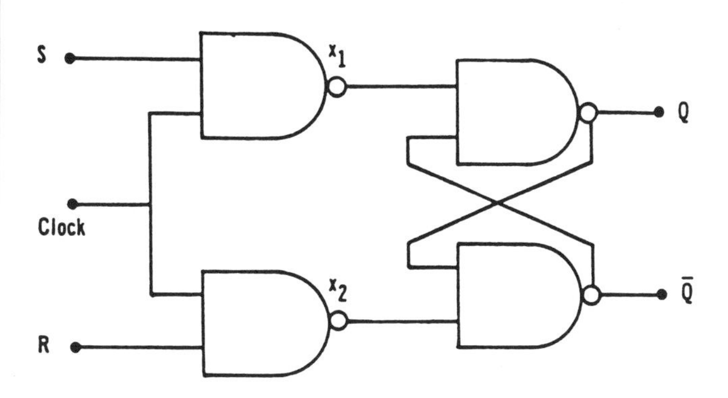

FIGURE 5.6 Clocked R-S flip-flop.

would result in x_1 and x_2 both being LOW, while the clock line stays HIGH; recall that $x_1 = x_2 =$ LOW is not allowed for cross-coupled NAND gate flip-flops. Notice that while the clock line is HIGH, the states of S and R have direct influence on Q, therefore it is important for the states of S and R to be maintained at the proper levels while the clock line is HIGH.

5.1.5. J-K Master-Slave Flip-Flops

a. Introduction

Since there is always one state which is not allowed in R-S type flip-flops (e.g., R = S = [0] in cross-coupled NAND flip-flops), the designer should always make sure that the "not allowed" state can never occur in his/her system. However, in the J-K flip-flop, all possible states are allowed. Figure 5.7 shows the functional diagram of a J-K *master-slave* flip-flop. In Figure 5.7(a), there are two flip-flops, master and slave, and two switching blocks, SW1 and SW2. The control clock pulses close and open SW1 and SW2 alternately. That is, when the clock line is LOW, SW1 is open and SW2 is closed. The information in the "master" flip-flop is then transferred to the "slave" flip-flop. When the clock line is HIGH, SW1 is closed and SW2 is open. Thus the master flip-flop is isolated from the slave flip-flop, and the input information is stored in the master flip-flop. Figure 5.7(b) is the logic diagram of the J-K master-slave flip-flop; it is a direct translation of the block diagram shown in FIgure 5.7(a). Gates 1, 2, 5, and 6 are the switches. Gates 3 and 4 are the master flip-flop, which is nothing but a simple cross-coupled NOR gate flip-flop. Gates 7 and 8 make up the cross-coupled NAND gate flip-flop which acts as the "slave" in this network. There are two "feedback" lines connecting the output $\overline{Q}_S$ to the input gate No. 1 and Q_S to the input gate No. 2, respectively. These two feedback lines are used for eliminating the "not allowed" state existing in an R-S flip-flop. From Figure 5.7(b) one may write the switching functions:

$$S_M = C \cdot J \cdot \overline{Q}_S \tag{5.4}$$

$$R_M = C \cdot K \cdot Q_S \tag{5.5}$$

$$S_S = Q_M + C \tag{5.6}$$

$$R_S = \overline{Q}_M + C \tag{5.7}$$

For the master flip-flop,

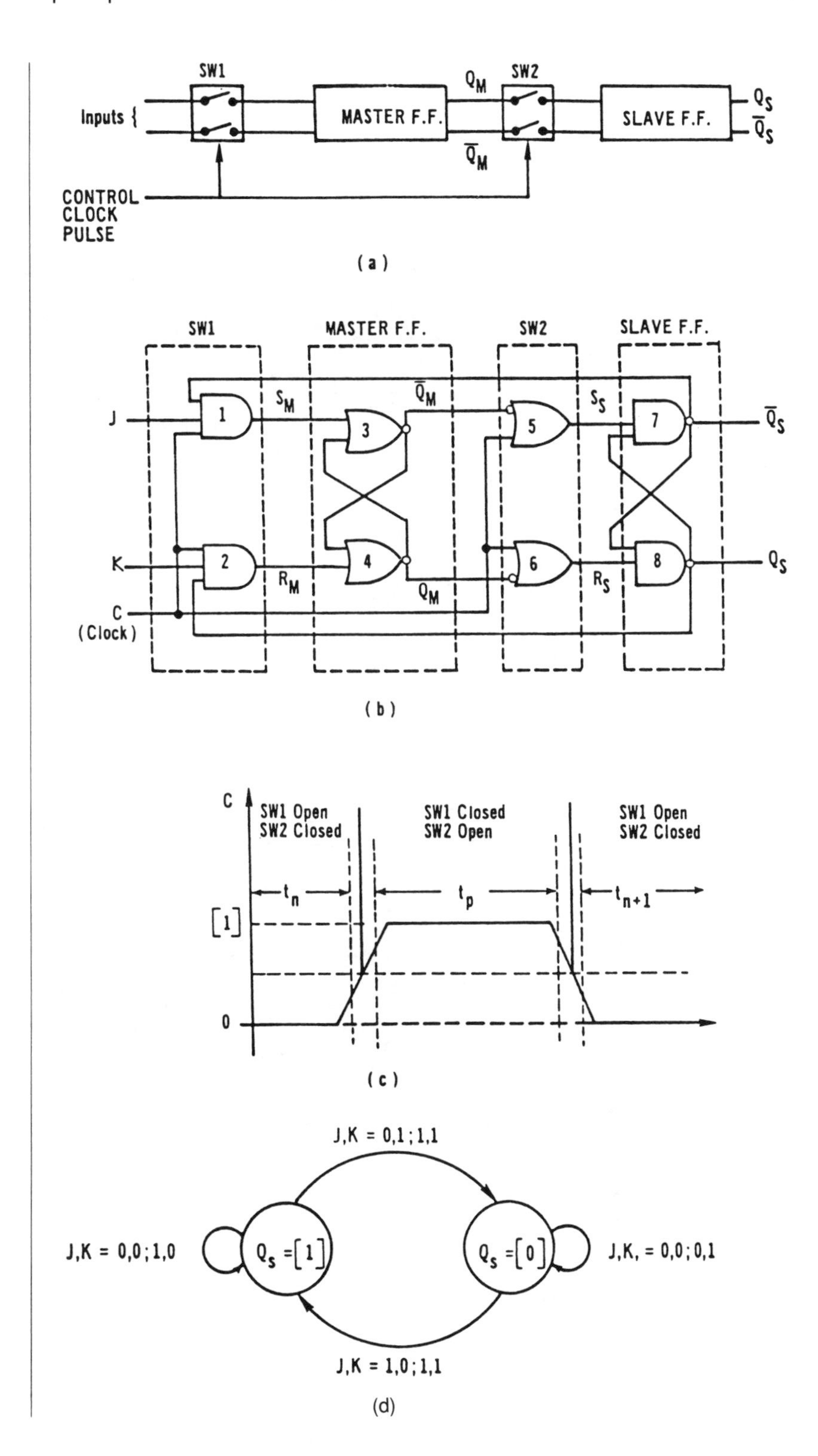

FIGURE 5.7 (a, b, c) J-K master-slave flip-flop. (d) State diagram of J-K master-slave flip-flop. (e) Circuit diagram of J-K master-slave flip-flop with preset and clear. (f) Logic diagram of J-K master-slave flip-flop with preset and clear.

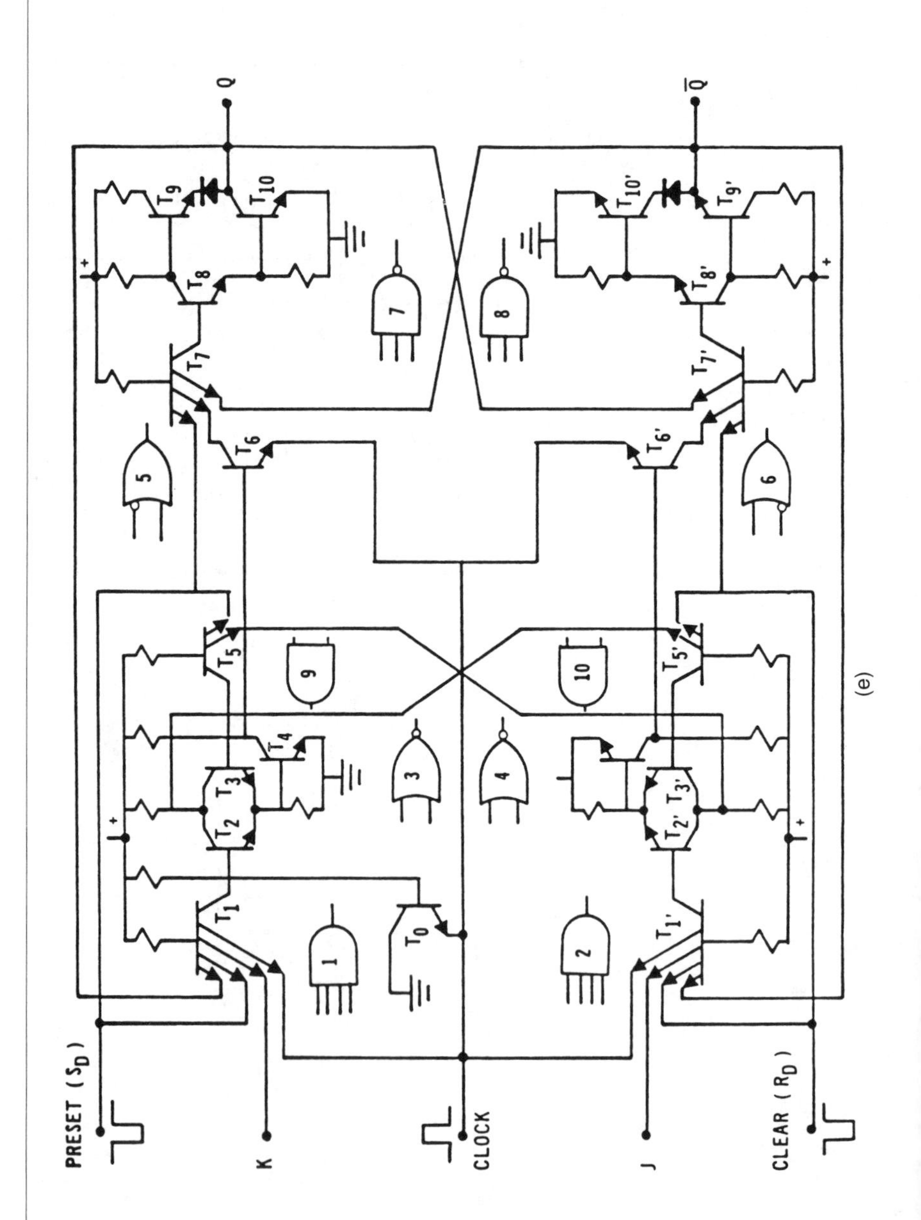

FIGURE 5.7 continued. (a, b, c) J-K master-slave flip-flop. (d) State diagram of J-K master-slave flip-flop. (e) Circuit diagram of J-K master-slave flip-flop with preset and clear. (f) Logic diagram of J-K master-slave flip-flop with preset and clear.

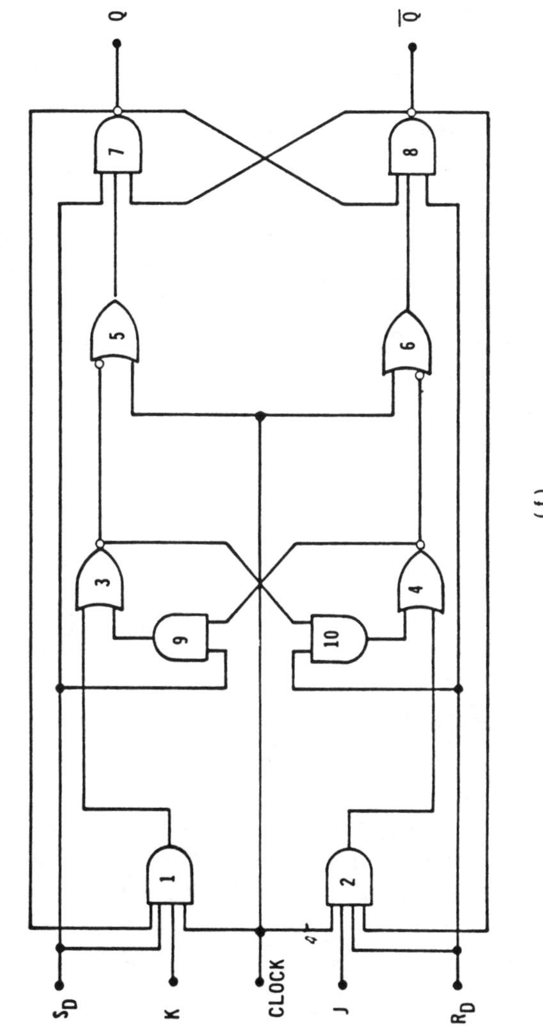

FIGURE 5.7 continued. (a, b, c) J-K master-slave flip-flop. (d) State diagram of J-K master-slave flip-flop. (e) Circuit diagram of J-K master-slave flip-flop with preset and clear. (f) Logic diagram of J-K master-slave flip-flop with preset and clear.

$$S_M R_M = 0 \tag{5.8}$$

$$(Q_M)_{n+1} = (S_M)_n + (\bar{R}_M)_n (Q_M)_n \tag{5.9}$$

For the slave flip-flop,

$$\bar{R}_S \cdot \bar{S}_S = 0 \tag{5.10}$$

$$(Q_S)_{n+1} = (\bar{R}_S)_n + (S_S)_n (Q_S)_n \tag{5.11}$$

Suppose that one wishes to store a [1] in this circuit and Q_S is initially [0]. First, one should set J = [1] and K = [0]. But nothing will happen as long as the clock line is low. Because $S_M = R_M = [0]$ from Equations (5.4) and (5.5), thus Q_M and $\bar{Q}_M$ remain unchanged, according to Table 5.1.4. From Equations (5.6) and (5.7), $S_S = Q_M$, $R_S = \bar{Q}_M$ or $\bar{R}_S = S_S$. Since it was assumed that $Q_S = [0]$, then we must have $\bar{Q}_S = [1]$ and $R_S = 1$, hence $S_S = [0]$. Thus, $Q_M = [0]$, $\bar{Q}_M = [1]$. Now, as the clock line becomes [1], from Equations (5.4) through (5.11) we have

$$S_M = [1]\,J\,[1] = [1] \tag{5.12}$$

$$R_M = [1]\,K\,[0] = [0] \tag{5.13}$$

$$S_S = R_S = [1] \tag{5.14}$$

$$Q_M = [1] \tag{5.15}$$

$$(Q_S)_{n+1} = (Q_S)_n \tag{5.16}$$

Effectively, gates 1 and 2 connect S_M to $J \cdot \bar{Q}_S$ and R_M to $K \cdot Q_S$, respectively. This stores the input information in the master flip-flop. However, because C = [1], $S_S = R_S = [1]$ regardless of the current states of Q_M and $\bar{Q}_M$, which effectively isolates the "slave" from the "master". Since at this time $S_S = R_S = [1]$, the slave flip-flop remains unchanged, i.e., $Q_S = [0]$.

When the clock line becomes LOW again, $S_M = R_M = [0]$ and gates 1 and 2 effectively disconnect the master flip-flop from $J \cdot \bar{Q}_S$ and $K \cdot Q_S$. Gates 5 and 6 cause $S_S = Q_M = [1]$ and $R_S = \bar{Q}_M = [0]$, which effectively connects the master flip-flop to the slave flip-flop. Therefore, Q_S becomes [1], and the operation of storing [1] is complete. Following the same steps, one can show how a [0] can be stored in the circuit by setting J = [0], K = [1]. Figure 5.7(c) summarizes the sequence of this operation as follows.

At time t_n: the master flip-flop is connected to the slave flip-flop

t_p: the period during which the clock line is high, the master flip-flop is connected to the inputs and is isolated from the slave flip-flop; the input information is stored in the master flip-flop

t_{n+1}: the master flip-flop is isolated from the inputs and is connected to the slave flip-flop, so that the new information in the master flip-flop is stored in the slave flip-flop

Let us now consider the cases when both inputs are equal. If J = K = [0], then $S_M = R_M = [0]$, and nothing will change. If J = K = [1] and the circuit is in the t_n period, there is still no change. However, in the t_p period when the clock line is high, or C = [1], from Equations (5.4) to (5.9) we have

$$
\begin{aligned}
S_M &= (\bar{Q}_S)_n \\
R_M &= (Q_S)_n \\
S_S &= R_S = [1] \\
(Q_M)_{n+1} &= (\bar{Q}_S)_n + (\bar{Q}_S)_n (Q_M)_n \\
&= (\bar{Q}_S)_n
\end{aligned}
$$

Then, in the t_{n+1} period where C = [0],

$$
\begin{aligned}
R_S &= (\bar{Q}_M)_{n+1} = (Q_S)_n \\
S_S &= (Q_M)_{n+1}) = (\bar{Q}_S)_n
\end{aligned}
$$

Thus, from Equation (5.11)

$$
\begin{aligned}
(Q_S)_{n+1} &= (\bar{Q}_S)_n + (\bar{Q}_S)_n (Q_S)_n \\
&= (\bar{Q}_S)_n
\end{aligned}
$$

Equation (5.12) shows that if J = K = [1] at $\bar{t}_p$ and through the t_p period, the negation of the old information in the slave flip-flop is stored in the master flip-flop. Thus, in the t_{n+1} period, the "new" information in the master flip-flop, which is nothing more than the negation of the old information in the slave flip-flop, is stored in the slave flip-flop. In other words, J = K = [1] in the t_p period will cause the slave flip-flop to negate its own old information after the clock pulse.

Physically, if one connects a light bulb at Q_S and sets J = K = [1], the clock pulse is analogous to someone pressing a push-button light switch. One push will turn off the light if the light was on, and will turn on the light if it was off.

Table 5.1.7 shows the operation of the J-K master-slave flip-flop. Note that there is no forbidden state.

TABLE 5.1.7

J_n	K_n	$(Q_S)_{n+1}$
0	0	$(Q_S)_n$
0	1	[0]
1	0	[1]
1	1	$(\bar{Q}_S)_n$

From this table one can easily find the switching function of this flip-flop, i.e.,

$$\begin{aligned}(Q_S)_{n+1} &= \bar{J}_n\bar{K}_n(Q_S)_n + J_n\bar{K}_n + J_nK_n(\bar{Q}_S)_n \\ &= \bar{J}_n\bar{K}_n(Q_S)_n + J_n\bar{K}_n\,[(Q_S)_n + (\bar{Q}_S)_n] + J_nK_n(\bar{Q}_S)_n \\ &= \bar{K}_n(Q_S)_n + J_n(\bar{Q}_S)_n\end{aligned}$$

and the state diagram is shown in Figure 5.7.

b. A Practical IC Circuit

Figure 5.7(e) shows a typical circuit diagram of an IC J-K master-slave flip-flop with preset and clear. At first glance, the circuit appears to be too complicated for a logic designer to comprehend. However, if the circuit is partitioned in subsections according to their logical functions, it will become quite clear. Thus, in this figure the circuit components are clustered in several subgroups, and each group is labeled with a logic symbol according to its logic operation. Since the Q-side and $\bar{Q}$-side circuits are identical, we shall discuss only the Q-side. Transistor T_1 is a 4-input AND-gate. T_2, T_3, and T_4 constitute a NOR gate having two outputs, one at the collectors of T_2 and T_3 and the other at the collector of T_4, so that the two outputs are electronically separated but logically the same. T_5 is a 2-input AND gate. T_7, T_8, T_9, and T_{10} constitute the familiar NAND gate. It is interesting to note the function of T_6. This single transistor functions as an OR gate with one output (the collector) and two inputs (the base and the emitter). The "base" input terminal automatically inverts the switching variable. Transistor T_0 is used as a clamping device which assures that the clock line drops to negative no more than a few tenths of a volt if the clock pulse is ringing.

Figure 5.7(f) shows the logic diagram of the circuit shown in Figure 5.7(e), which is basically the same as that shown in Figure 5.7(b) except that the AND gates, No. 9 and No. 10, were inserted in the cross-coupled NOR flip-flop for the additional inputs of Preset (S_D) and Clear (R_D). Because of these additions, the flip-flop becomes more versatile. It can be used as a set-reset flip-flop. a triggered or toggle flip-flop, or a J-K flip-flop. It is important to note that the J, K, and clock terminals are normally at LOW, while the S_D and R_D terminals are normally at HIGH. From the electronic circuit in Figure 5.7(e), it is obvious that a negative pulse going from HIGH to LOW at R_D or S_D, but not both, will cause the action, which will overrule the J-K and clock inputs and force the master and slave flip-flops to stay in the desired state set by R_D and S_D.

5.1.6. D Flip-Flops

The D flip-flop is sometimes called the delay flip-flop or data-latched flip-flop. Basically, it has two outputs, Q and $\bar{Q}$, and two inputs, D, the data input and C, the clock input. The data bit at D will be stored in the flip-flop when the clock line is high. Thus, the switching function of this type of flip-flop is simply

$$Q_{n+1} = C \cdot D_n$$

The simplest data-latch flip-flop is shown in Figure 5.8. In this diagram, the cross-coupled NAND gate flip-flop is the memory device. Gates 1 and 2 are connected in such a way that the condition $\bar{R} \cdot \bar{S} = 0$ is assured, and the following truth table can easily be verified.

D_n	C	S_n	R_n	Q_{n+1}
0	0	1	1	Q_n
1	0	1	1	Q_n
0	1	0	1	[0]
1	1	1	0	[1]

The table simply means that the flip-flop is isolated from the D input before and after the clock, but that the information at D is stored during the clock pulse. Although it is a simple matter to convert an R-S flip-flop

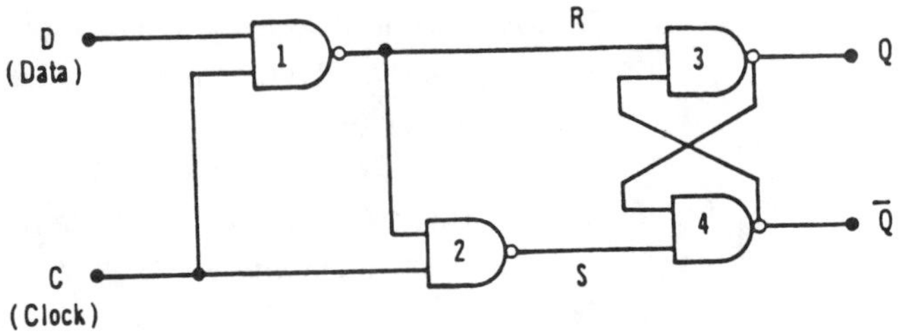

FIGURE 5.8 Logic diagram of D flip-flop.

or J-K flip-flop into a D flip-flop, we shall now analyze a different type of flip-flop called the Edge-Triggered D flip-flop which accepts information only at the rising or falling edge of the clock pulse. This feature is desirable for a system where timing is critical. Since the device does not respond to the input during the clock pulse, the width of the clock pulse is no longer critical.

Figure 5.9 shows a typical commercially available edge-triggered D flip-flop. The circuit diagram is shown in Figure 5.9(a) and its corresponding logic diagram is shown in Figure 5.9(b). Notice that the transistors T_1, T_2, T_3, T_4; and T_1', T_2', T_3', T_4', respectively, constitute the NAND gates 1, 2, 3 and 4 shown in Figure 5.9(b). They share the same diode D as shown in Figure 5.9(a). NAND gates 5 and 6 are made of standard circuitry which we need not analyze here. However, the circuitry for gates 1 to 4 is deceptively simple. Actually, if one imagines that there is a resistor connection from the collector of T_2 to the power supply +V, then T_1 and T_2 constitute a straightforward AND-INVERTED or NAND gate since T_1 is a simple AND gate and T_2 is a single transistor inverter. In this circuit, T_1, T_2, T_3, and T_4 with the cross-connections between the collector of T_2 and emitter of T_3, form the cross-coupled NAND gate flip-flop and similarly between the collector of T_4 and emitter of T_1. Thus, one may imagine a resistor in the collector circuit of T_2 which would be in parallel with the resistor in the base of T_3, and there is no reason why T_2 cannot share the resistor in the base of T_3. This is why the imagined resistor is a redundant one and is omitted from the circuit. With this understanding of how the simple circuit constitutes the NAND gate, we shall now investigate how the logic diagram in Figure 5.9(b) functions as a D flip-flop. The solid line portion of the diagram shows the basic D flip-flop, and the dotted lines show the additional features of "set" and "reset" which are normally at HIGH. To understand just the D flip-flop, let us pretend that the dotted lines do not exist (since HIGH is equivalent to open circuit); then, all gates except No. 3 become 2-input NAND gates. The clock line is normally at LOW, which causes R = S = [1] and isolates the 5-6 cross-coupled flip-flop from the inputs; the old information remains undisturbed. Now let $\theta \triangleq$

the threshold of the gate for changing state, and $y_i \triangleq$ the output of the ith gate. Then, let $t = t_n$ as shown in Figure 5.9(b) and D = [1]. Since $R_n = S_n$ = [1], we have $y_4 = \overline{D}$ = [0], y_1 = [1]. As C rises to $\geq \theta$, R = [0], S = [1], thus Q = [1] = D. Consider $t = t_p$, or C = [1]. But D changes to [0], then $y_4 = \overline{D}$ = [1]. However, R = [0], which locks y_1 = [1], thus S = [1]. Therefore, the output flip-flop remains unchanged during $t = t_p$ even

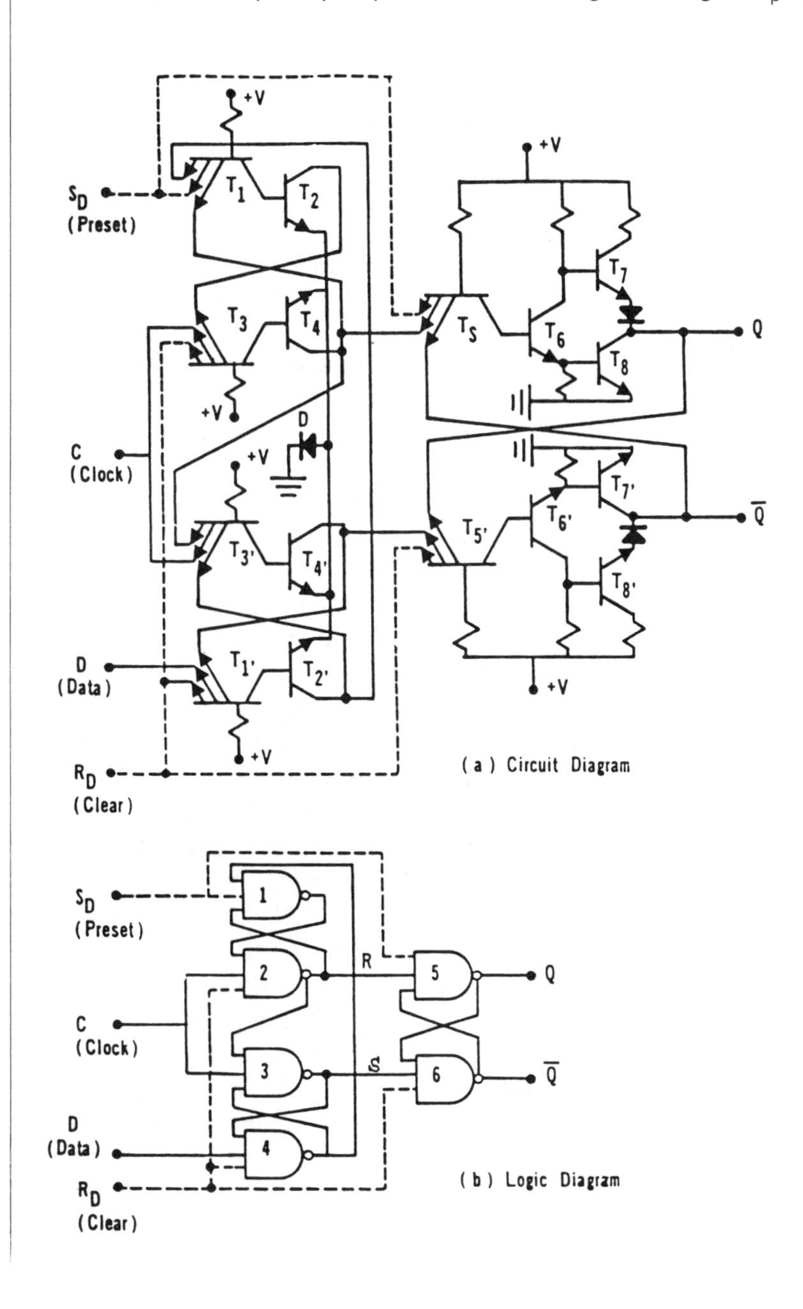

FIGURE 5.9 Edge-triggered D flip-flop.

though D changed to [0]. As C returns to [0], or $t = t_{n+1}$, both R and S return to [1]. The output flip-flop is then isolated from the input.

Similarly, consider D = [0] while $t = t_n$. Since $R_n = S_n$ = [1], y_4 = [1] and y_1 = [0]. As C rises to $\geq \theta$, S = [0]. But R will still be [1]. Therefore, Q = [0] = D and the information is transferred to the output flip-flop. Now let D change to [1] while $t = t_p$; since S = [0], y_4 = [1] regardless of what state D is in. Hence, it is equivalent to shutting off gate 4, which will not respond to the data line. Now, when C returns to zero or $t = t_{n+1}$, R = S = [1], which isolates the output flip-flop from the input and the stored information remains unchanged until the next clock pulse arrives. Therefore, this D flip-flop only responds to the rising edge of the clock pulse.

5.1.7. Conclusion

In this section, the BSA technique was used to analyze the basic cross-coupled flip-flops made of NOR gates and NAND gates. It was an attempt to link the electronic circuit, logic diagram, and switching functions of the basic flip-flops together in a coherent manner. Then, some practical IC flip-flops were analyzed by first showing the actual circuit diagram, then partitioning the diagram into cross-coupled flip-flops and gates, and finally translating it into a logic diagram. The basic repeating logic element was analyzed from the electronic circuit point of view, the whole flip-flop was treated as a small digital system or a logic network, and its logical characteristic equation or switching function was derived. From now on, the flip-flop can be considered as a black box characterized by its switching function. However, a novice is usually fascinated and confused by the many varieties of flip-flops available on the market. Actually, it is just like the many models of automobiles on the market. Although there are many different varieties, their basic principle of operation is the same; they are different only in optional features. According to their characteristic switching functions, one may classify flip-flops into four main classes, i.e., R-S, J-K, D, and T (toggle) flip-flops. Table 5.1.8 shows how they are related to each other. However, according to their triggering methods, one may classify them into *direct, clocked, master-slave,* and *edge-triggered* flip-flops. Although we have shown only the J-K master-slave flip-flop, there is also a D type master-slave flip-flop. But the interesting point here is the distinctive characteristics of each triggering method which are as follows. (1) For *direct triggering,* the output changes as the input does. (2) *Clocked* flip-flops respond only when the clock line is enabled, otherwise the device is isolated from the input. (3) In the *master-slave* flip-flop, the master flip-flop is connected to the input in the rising pe-

TABLE 5.1.8 Flip-Flop Conversion Table

FROM \ TO	J-K FLIP-FLOP	D FLIP-FLOP	T (TOGGLE) FLIP-FLOP
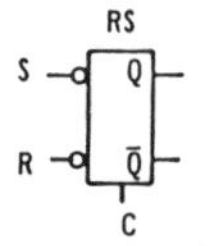 RS	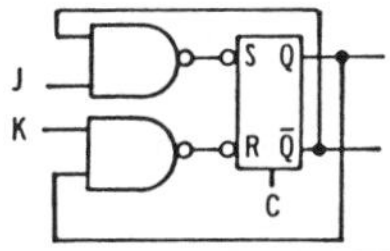	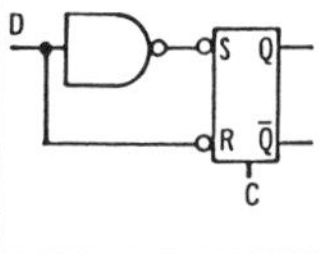	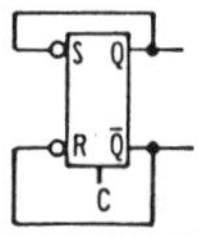
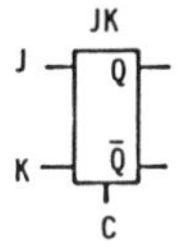 JK		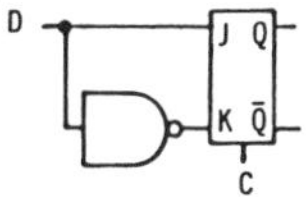	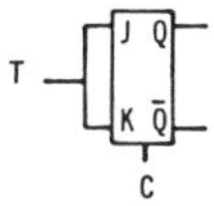
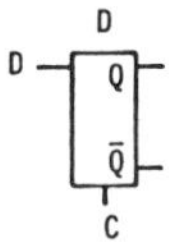 D			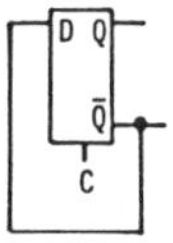

riod of the clock pulse, while it is isolated from the inputs and connected to the slave flip-flop during the falling of the clock pulse and the non-clocked periods. Thus, the input and output are mostly separated, which is desirable for closed loop systems that do not want feedback to be active during the nonclocked period. (4) *Edge-triggered* flip-flops only respond to the input at the rising edge (or falling edge) of the clock pulse; they keep the information faithfully unchanged at all other times. For one thing, the pulse width of the clock is not critical for this device, and for a high-speed synchronous system, this device would cause less noise problem than others.

5.2. MONOSTABLE MULTIVIBRATORS (ONE-SHOTS)

Since we know from the preceding section that a flip-flop has two stable states, i.e., Q = [1] or Q = [0], it may stay in any one of the two states for an indefinite length of time. However, the one-shot normally stays in one state when its input is not "excited" and enters another state if excited. But it can only stay in the other state for a finite period of time determined by its designer. The major applications of one-shots are in waveshaping, pulse width modification, and time delay. The discrete circuit one-shot is very sensitive to noise. The designer would not use it unless there were no other choice. However, the integrated circuit one-shots now on the market do possess desirable noise immunity.

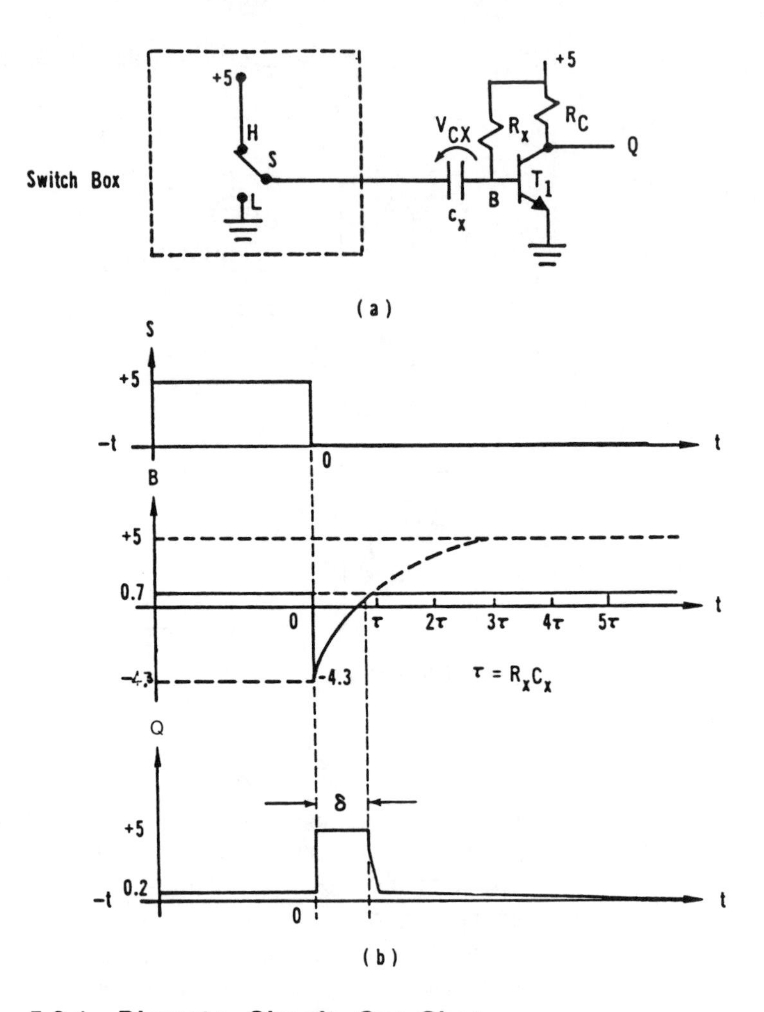

FIGURE 5.10 Simplified one-shot circuit.

5.2.1. Discrete Circuit One-Shot

Since the discrete one-shot is easier to understand and has the same principle of operation as any IC one-shots, we shall start with this circuit. Figure 5.10(a) shows a simplified one-shot circuit which we can use to demonstrate the basic principle of operation. Consider that the switch has been on position "H" for a long, long time. The capacitor C_x has been fully charged, and it can be treated as an open circuit, thus the voltage V_{cx} across it is +4.3 V in reference to the polarity assigned by the arrow. Let

$$\frac{5 - 0.2}{R_C} << \beta \frac{5 - 0.7}{R_x}$$

where β is the current gain of the transistor. Then T_1 is saturated, Q = 0.2 and B = 0.7. Now, at t = 0, the switch is closed to "L", the B becomes –4.3, which turns off the transistor T_1, and Q jumps to +5 V. Effectively, the point B is disconnected from the base, and R_x and C_x become a simple R-C network connecting from +5 through L to ground. The capacitor is thus being charged toward +5 with time constant $\tau = R_xC_x$. Thus we can write

$$B(t) = B(\infty) + \{B(0^+) - B(\infty)\}e^{-t/\tau}$$

$$= 5 + (-4.3 - 5)\, e^{-t/\tau}$$

But the transistor will be turned on again as soon as B(t) has reached 0.7 V. To determine the time δ at which the transistor T_1 conducts, we can solve the equation

$$0.7 = 5 - 9.3e^{-\delta/\tau}$$

to get

$$\delta = \tau \ln\left(\frac{9.3}{4.3}\right) = R_xC_x \ln(2.16)$$

$$= 0.73\, R_xC$$

Since

$$\beta >> \frac{R_x}{R_C}$$

the transistor enters the saturation region almost immediately after t = 0.73 R_xC_x. The waveforms of S, B, and Q are shown in Figure 5.10(b). It is important to point out that the output pulse starts at t = 0 with a pulse width equal to 0.73 R_xC_x which is a function of R_x and C_x and independent of the input waveform. However, in order to repeat this process, the switch has to be switched back to H for a period of time to allow the capacitor to be charged to 4.3 V again.

Figure 5.11(a) shows a simple discrete one-shot. By comparing this circuit with Figure 5.10(a), one can see that the switch S is replaced by the transistor T_2 and its control circuit R_1 and R_2. The circuit functions as follows. Let $V_i = 0$, as the circuit is at rest; C_x is fully charged and no current flows in or out of the capacitor. The capacitor is effectively an open circuit branch. T_1 is saturated, Q = 0.2. For convenience, let $R_1 = R_2$, then the voltage at the base of T_2 is 1/2 Q = 0.1, which will keep T_2

OFF, and C_x is connected to +5 through R_{C2}. Now, a positive going pulse of +5 V is applied to V_i, which will drive T_2 into saturation. Thus, S drops from +5 to 0.2 V, which is equivalent to turning the switch S (Figure 5.10(a)) from H to L. The responses of the right-hand side of the circuit shown in Figure 5.11(a) closely follow what has been described in Figure 5.10(b), except that the switch in Figure 5.10(a) does not automatically go back to H, but the transistor switch T_2 in Figure 5.11(a) does. Because the resistor R_1 senses the state of Q, and resistor R_1, R_2, and T_2 constitute an RTL NOR gate, the collector S of T_2 stays LOW as long as Q or V_i or both are HIGH. As soon as both V_i and Q have become LOW, T_2 turns OFF, which connects the capacitor C_X

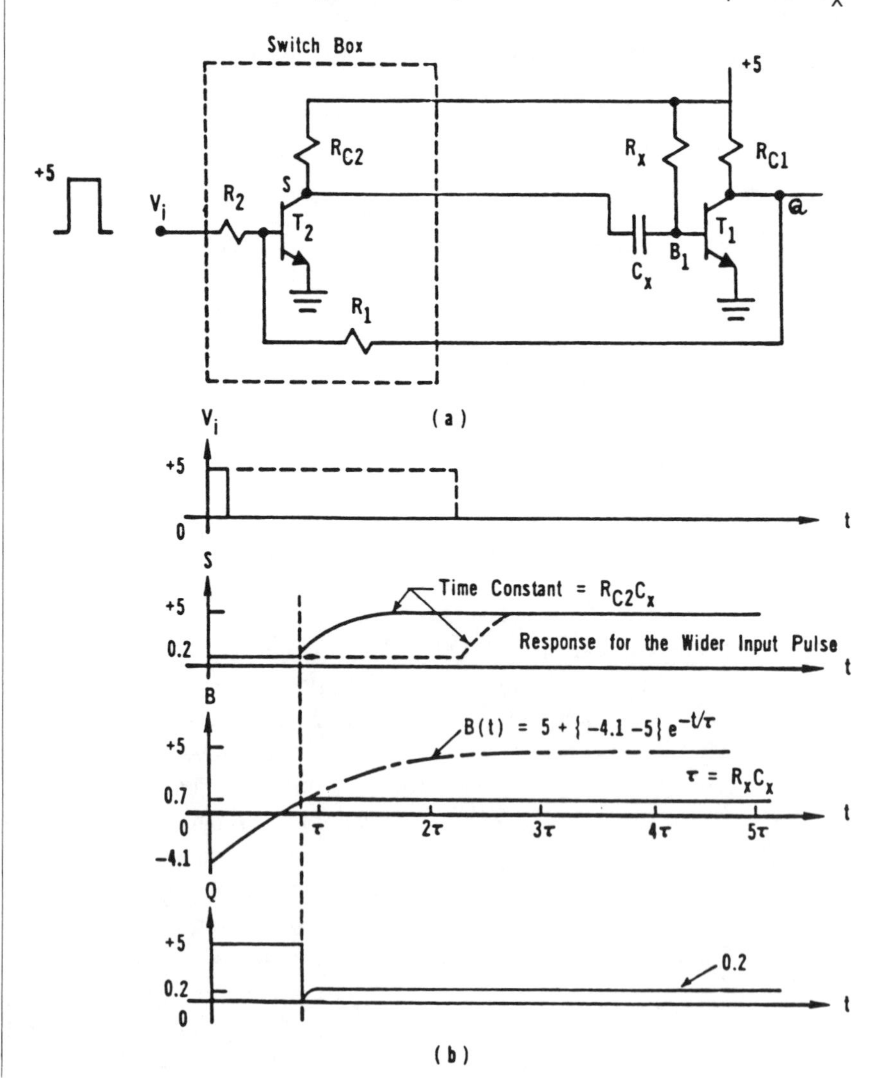

FIGURE 5.11 Discrete one-shot circuit.

back to +5 through R_{C2}. The capacitor C_x is then being charged to 5 – 0.7 with time constant $R_{C2}C_x$. After approximately $5R_{C2}C_x$, the C_x is fully charged and the circuit is ready to receive another excitation from V_i to generate another one-shot. The circuit responses to input pulses with different pulse widths are shown in Figure 5.11(b); the dotted lines show the response of the input with wider pulse width. Notice that the output pulse width is independent of the input pulse width, but is, however, dependent on R_xC_x. The reader is urged to verify the waveforms shown in Figure 5.11(b). Why does it have initial voltage of B = –4.1 instead of –4.3 V? Why does S have a slow rising-edge, and why is the falling-edge of Q much steeper? What would be the width of the output pulse, Q?

5.2.2. IC One-Shots

Figure 5.12 shows a typical IC one-shot. Although the circuit appears quite complicated, for functional analysis it need not be difficult. The key components are (1) R_x and C_x, which are provided externally for determination of the output pulse width. The manufacturer usually provides an equation which gives the pulse width in terms of C_x and R_x. (2) Transistors T_1 and T_4, which are equivalent to T_1 and T_2, respectively, in the discrete one-shot shown in Figure 5.11(a), are functioning as the two switching devices. (3) The trigger input V_{in} and the gate terminal allows or prohibits the trigger signal to enter the one-shot. (4) Transistors T_3 and T_7 are the output buffer amplifiers for $\bar{Q}$ and Q; T_2 and T_6, respectively interface T_1, T_3 and T_4, T_7; T_8 is for recharging C_x after the one-shot; T_5 interfaces the collector of T_1 and the base of T_4, which is equivalent to R_1 in Figure 5.11(a). T_2, T_5, and T_6 function exactly the same way as the multiemitter transistor at the input of a NAND gate that we have studied, in which either the emitter-base diode junction is ON or the collector-base junction is ON, but not both. Notice that the triggering scheme is slightly different from the discrete one-shot. In the IC one-shot, the triggering input V_{in} is applied to T_1 for turning T_1 OFF, while in the discrete circuit the trigger is applied to T_2 in Figure 5.11(a) for turning T_2 ON and in turn switching T_1 OFF. They serve the same purpose, but via different routes. Now we are ready to see how the IC one-shot functions. At rest, T_1 is ON which results in (1) base-emitter junction of T_2 will be ON which will turn T_3 OFF, and (2) base-emitter junction of T_5 will be ON which turns T_4 OFF, and the base-collector of T_6 will be ON so that T_7 will be saturated. Thus we have $\bar{Q}$ = [1], Q = [0]. Consider now, the gate = [0] and a pulse is applied at V_{in}. The rising edge of V_{in} causes a positive spike at S, but the diode D_1 blocks the spike and nothing will happen to the one-shot.

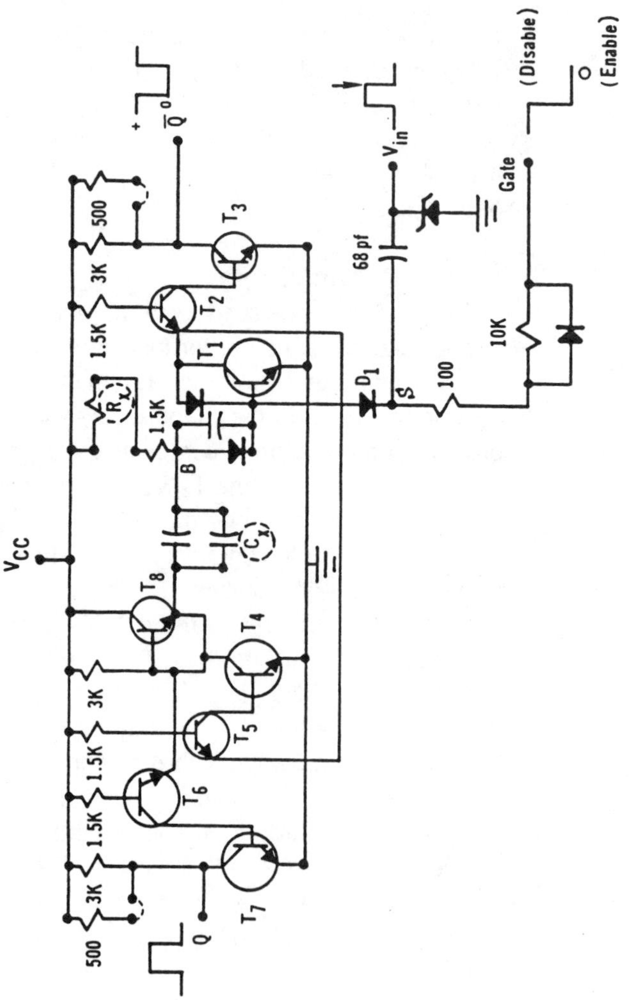

FIGURE 5.12 Integrated circuit one-shot.

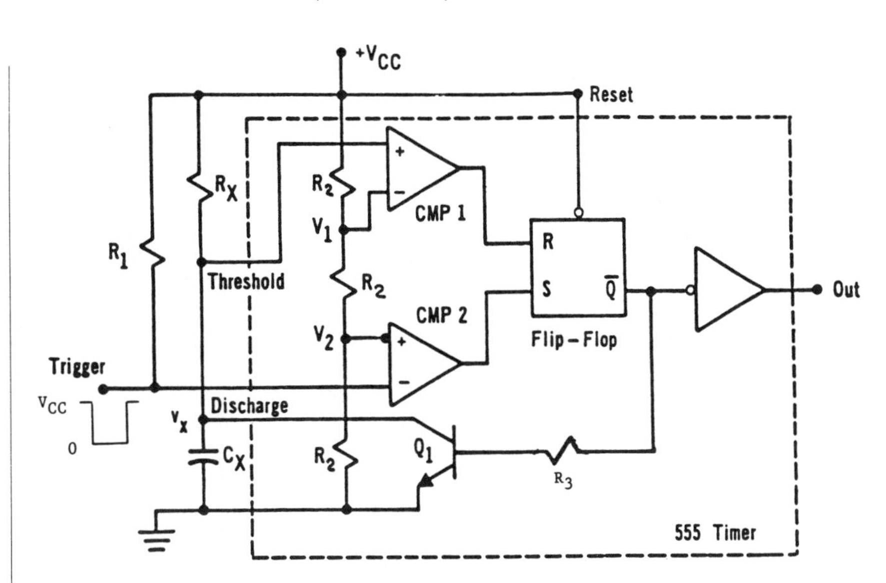

FIGURE 5.13 Using 555 timer as a one-shot.

However, the falling edge of V_{in} causes the S to go negative which turns D_1 ON and pulls the base of T_1 to negative, thus T_1 will be OFF and the following chain-reaction occurs: (1) base-collector of T_2 will be ON and T_3 becomes saturated; (2) base-collector of T_5 will be ON, which would saturate T_4 so that the base-emitter of T_6 will be ON and T_7 is OFF; and (3) since T_4 is saturated, it effectively ties the left-hand side of C_x to ground; therefore, B becomes negative which keeps T_1 OFF, although by this time the V_{in} pulse is gone. The circuit remains in this state until C_x is charged through (R_x + 1.5 K) and the voltage at B becomes positive enough to turn on T_1. The sequence then is basically the same as that which was described for the discrete one-shot. Notice that the provision of the control gate is very desirable since the circuit can be shut off while idling so that the noise will not trigger the one-shot falsely.

5.2.3. One-Shot Using IC 555 Chips

Figure 5.13 shows the functional block diagram of a one-shot configuration using an IC 555 timer. CMP1 and CMP2 are two analog comparators with $V_1 = \frac{2}{3} V_{CC}$ and $V_2 = \frac{1}{3} V_{CC}$ as one of the two inputs for CMP1 and CMP2, respectively. R_1, R_x, and C_x are the external circuit components whose values are determined by the system designer. Just as in the simple discrete one-shot, R_x and C_x control the pulse width of the one-shot output. R_1, however, assures that the trigger input pin is at V_{CC} volts in quiescent state, therefore its value can be 10 K or 100 K Ω. From the circuit diagram of the 555 provided by the manufacturer, one can derive the special characteristic of the R-S flip-flop as follows (Table 5.2.1).

TABLE 5.2.1

S	R	$\bar{Q}_{n+1}$
0	0	$\bar{Q}_n$
1	0	0
0	1	1
1	1	0

Consider that the circuit is presently in the following state:

$$\left.\begin{aligned} \text{Trigger} &= \text{HIGH} \rightarrow S = [0] \\ v_x(t) &= \text{HIGH} \rightarrow R = [1] \end{aligned}\right\} \rightarrow \bar{Q} = [1] \rightarrow \text{OUT} = [0]$$

where "$\rightarrow$" = "imply" or "result in". Then Q_1 conducts and discharges C_x with a time constant $\tau_d = C_xR_s$, where R_S is the saturation resistance of Q_1, which normally is on the order of a fraction of 1 Ω. As a result, the output of CMP1 goes LOW, and the circuit enters the following state:

$$\left.\begin{aligned} \text{Trigger} &= \text{HIGH} \rightarrow S = [0] \\ v_x(t) &= \text{LOW} \rightarrow R = [0] \end{aligned}\right\} \rightarrow \bar{Q} = [1] \text{ (no change)} \rightarrow \text{OUT} = [0]$$

The circuit will stay in this state indefinitely until a negative going pulse sets the trigger terminal to LOW at time t = 0, or any voltage lower than v_2, which is 1/3 V_{CC}, then

$$\left.\begin{aligned} \text{Trigger} &= \text{LOW} \rightarrow S = [1] \\ v_x(0^-) = v_x(0^+) &= \text{LOW} \rightarrow R = [0] \end{aligned}\right\} \rightarrow \bar{Q} = [0] \rightarrow \text{OUT} = [1]$$

As a result, Q_1 is OFF, capacitor C_x is then being charged by V_{CC} through resistor R_x with a time constant $\tau_c = R_xC_x$, and

$$v_x(t > 0) = v_x(\infty) + [v_x(0) - v(\infty)]e^{-t/\tau_c}$$

where

$$v_x(0) = 0 \text{ and } v_x(\infty) = V_{CC}$$

Thus,

$$v_x(t) = V_{CC}[1 - e^{-t/\tau_c}] \tag{5.13}$$

However, when

$$v_x(t) \geq V_1 = \frac{2}{3} V_{CC}$$

then R = 1. At this point, if the Trigger is still low, then we have

$$\left.\begin{array}{l} S = 1 \\ R = 1 \end{array}\right\} \rightarrow \bar{Q} = 0, \text{OUT} = 1$$

However, if the trigger pulse is gone, then

$$\left.\begin{array}{l} S = 0 \\ R = 1 \end{array}\right\} \rightarrow \bar{Q} = 1, \text{OUT} = 0$$

Here, $\bar{Q}$ causes the transistor Q_1 to be ON and discharging C_x. Let t = T when $v_x(t)$ reaches $\frac{2}{3} V_{CC}$, then Equation (5.13) becomes

$$\begin{aligned} v_x(T) &= \frac{2}{3} V_{CC} = V_{CC}[1 - e^{-T/\tau_c}] \\ T &= \tau_c \ln 3 \\ &= 1.1\ R_x C_x \end{aligned} \tag{5.14}$$

In conclusion, the output will remain at HIGH state for a T = 1.1 $R_x C_x$ period if the trigger pulse width is ≤T, otherwise, the output pulse width will be equal to that of the trigger pulse. Since C_x requires a finite time to be discharged, this one-shot is not a retriggering type. Equation (5.14) reveals that the output pulse width T can be virtually designed as precisely as the components R_x and C_x. That is, T is independent of V_{CC} and other circuit variables.

5.3. ASTABLE MULTIVIBRATORS (FREE-RUNNING)

The astable multivibrator is simply a square wave oscillator whose frequency is determined by two identical pairs of R-C networks. By controlling the R-C network individually, the time period for the HIGH state and LOW state of the circuit can be asymmetrical. The essential application of this circuit in a digital system is being used as a CLOCK. There are many kinds of astable multivibrators available on the market. For applications which require accurate frequency reference, one may select crystal-controlled and temperature-compensated astable multivibrators; however, if the designer is only concerned with the proper sequencing in a system, he/she may use the basic circuit described in the following section.

5.3.1. Basic Astable Multivibrators

The circuit shown in Figure 5.14 is a basic astable circuit with two pairs of R-C networks, R_xC_x and $R_x'C_x'$ for frequency control. The circuit can be reduced to the one-shot circuit shown in Figure 5.11(a) by (1) eliminating all diodes, R_2 and R_2'; (2) shortening $\overline{Q}'$ to $\overline{Q}$, Q' to Q; (3) replacing C_x' by a resistor; and (4) disconnecting R_x' from +5 and using it as the input resistor as in the one-shot circuit. Thus T_2

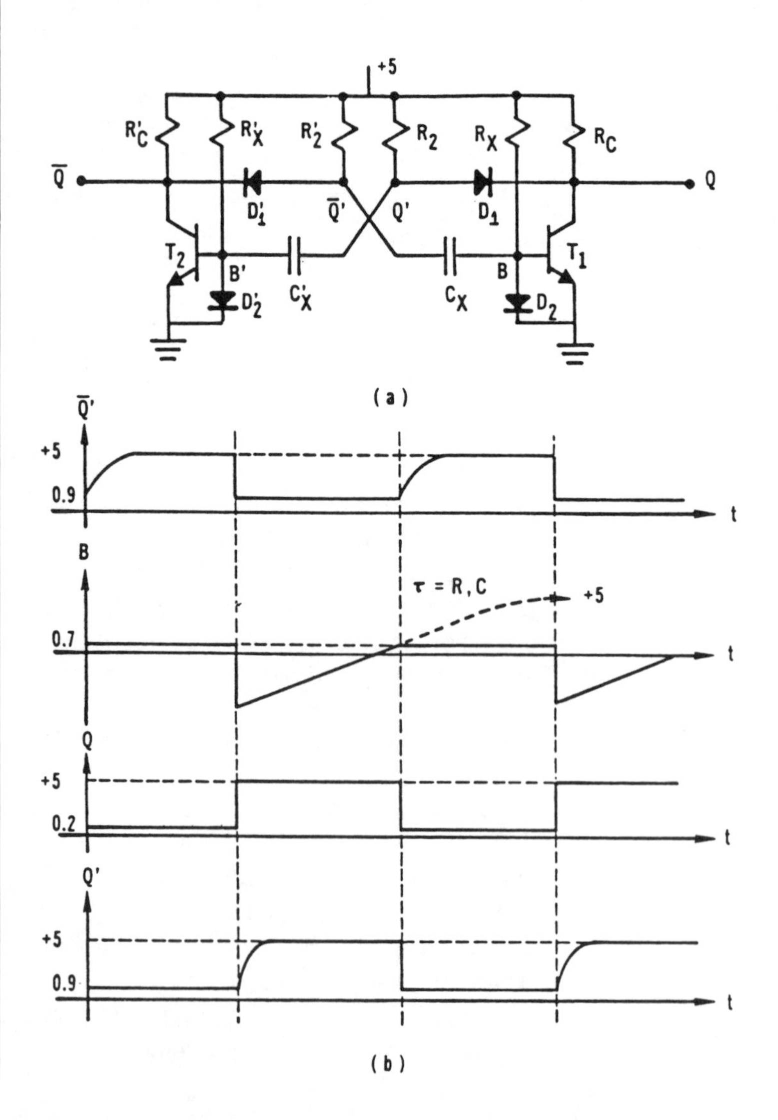

FIGURE 5.14 Basic astable multivibrator.

becomes the "switch box" of the one-shot. Similarly, one may use T_1 as the "switch box", and R_x as the input resistor; then one has another one-shot with $\bar{Q}$ as its input. Therefore, the astable can be treated as two one-shot circuits which are cross-connected such that they are "self-exciting" to maintain the free-running requirement for the astable circuit. The reader may recall that in Figure 5.11, the capacitor C_x should be fully charged to 4.3 V before the next "action" is initiated, which causes the slow rise voltage waveform at S in Figure 5.11(b). This would not be desirable for an astable circuit, since S will be the $\bar{Q}$ in the astable circuit, which happens to be one of the outputs. Therefore, R_2, R_2', D_1, and D_1' are added to the circuit shown in Figure 5.14(a). With this modification, the C_x and C_x' are respectively recharged through R_2' and R_2 so that only the waveforms at $\bar{Q}'$ and Q' have slow rise time, not those at $\bar{Q}$ and Q. To be more specific, consider the points at Q' and Q. While C_x' is charging, T_2 is ON and T_1 is OFF; Q' rises slowly, but Q tends to jump to +5 immediately. Of course, if Q and Q' were shorted, Q would rise as slowly as Q'. However, the diode D_1 is isolating Q from Q' since $Q > Q'$ while C_x' is being charged and $Q = Q'$ when C_x' is fully charged, thus D_1 is always cut-off during this period. Now, as T_1 is being turned on, D_1 will always be ON since Q will be 0.2 V and $Q' > Q$ or $Q' \simeq 0.9$, which effectively ties the right-hand side of C_x' to ground so that T_2 will be kept OFF until it is charged to +0.7 V. The operation is the same as that of the one-shot; it will not be repeated here. The waveforms of the points that are interesting to us are shown in Figure 5.14(b). The output waveform at Q has a fast rise time. Since $\bar{Q}$ is the negation of Q, it is not shown in Figure 5.14(b).

In practice this circuit has one drawback. It is possible that both T_1 and T_2 stay saturated at the same time. If so, the circuit will not oscillate until a disturbance is applied to the circuit. The simplest way to get a circuit to start oscillating is to momentarily short the base of either one of the transistors. In order to insure that the circuit will self-start, the D_2 and D_2' are added to the circuit. As long as the parallel value of R_2 and R_C is slightly less than that of R_x, the transistor will be kept in the active region while the capacitors are fully charged. Since the circuit is connected with positive feedback, any natural noise from anywhere can initiate the oscillator, thus the circuit shown is considered a self-started astable multivibrator.

5.3.2. Astable Multivibrator Using 555 Chip

Figure 5.15(a) shows the circuit configuration for an astable multivibrator using a 555 timer. The capacitor C_x is being charged through R_1

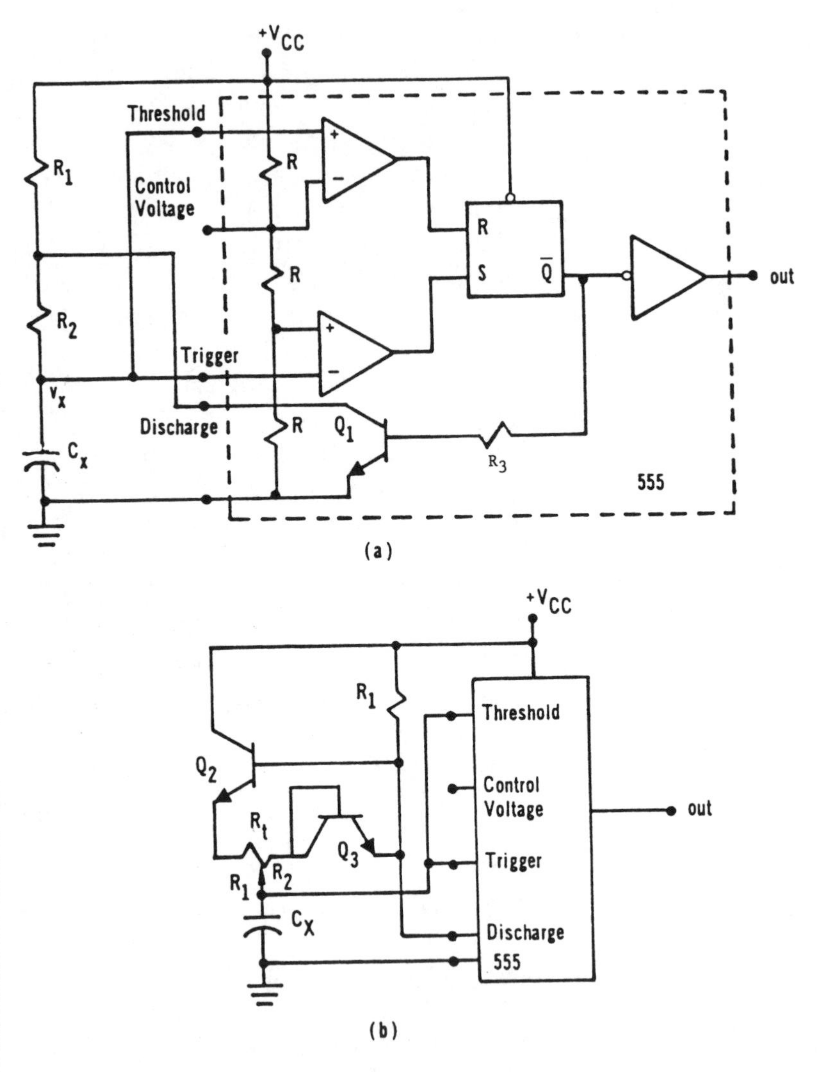

FIGURE 5.15 Astable multivibrator using 555 timer.

$+ R_2$ with a time constant $\tau_c = C_x(R_1 + R_2)$ and discharged through R_2 and Q_1 with a time constant $\tau_d = C_xR_2$. Consider the following three possible cases:

a. For $v_x(t) < \frac{1}{3}V_{CC}$, the flip-flop will have its inputs, R = 0, S = 1, and, according to Table 5.2.1, $\bar{Q} = 0$; Q_1 will be OFF, and C_x will be charged through $R_1 + R_2$.
b. For $\frac{1}{3}V_{CC} < v_x(t) < \frac{2}{3}V_{CC}$, we have R = 0, S = 0, and capacitor C_x will continuously be charged.
c. For $v_x(t) > \frac{2}{3}V_{CC}$, we have R = 1, S = 0, and Q_1 will be ON; C_x will be discharged until $v_x(t) < \frac{1}{3}V_{CC}$. The operation repeats itself and the output yields a square wave clock. It is in HIGH state when C_x

is being charged and LOW state when C_x is being discharged. Following the derivation similar to that described in Section 5.2.3, we have

$$t_H = \text{time while the output is HIGH} = 0.693(R_1 + R_2)C_x$$

$$t_L = \text{time while the output is LOW} = 0.693\,R_2C_x$$

The frequency of the clock is

$$f = \frac{1}{t_H + t_L} = \frac{1.44}{(R_1 + 2R_2)C_x} \tag{5.15}$$

In some applications, one would like to adjust the duty cycle or the ratio of the HIGH and LOW periods of the clock without changing its frequency. To do this, one can use the circuit shown in Figure 5.15(b). Here, the frequency of the clock is

$$f = \frac{1.44}{R_tC_x}$$

where $R_t = R_1 + R_2$, the total resistance of the potentiometer, thus the frequency will remain constant as the charging and discharging time is being controlled by the setting of the potentiometer. Here, transistor Q_2 is turned OFF during discharging period and turned ON during charging. Q_3 is used as a diode which isolates resistance R_2 from the charging loop, therefore we have,

$$t_H = 0.693\,R_1C_x$$

$$t_L = 0.693\,R_2C_x$$

and

$$f = \frac{1.44}{(R_1 + R_2)C_x} = \frac{1.44}{R_tC_x}$$

5.4. SCHMITT TRIGGER CIRCUITS

The Schmitt trigger circuit is another kind of regenerative circuit. Its application is usually as a pulse shaper for improving the rise and fall time of a signal or level detector. The circuit has a predetermined threshold voltage V_H; if the input signal is less than V_H, the output will stay LOW. As soon as the input voltage has reached V_H, the output voltage jumps to HIGH. Thus the circuit has the desirable noise immunity property.

Figure 5.16 shows a basic Schmitt trigger circuit. For BSA purposes, a transistor with high current gain is assumed so that the effect of I_b, the base current, can be neglected. Consider $V_i = 0$; then T_1 will be OFF and T_2 will be ON. We have

$$B = \frac{R_2}{R_{C1} + R_1 + R_2} \cdot 6 = \frac{5}{10} \cdot 6 = 3 \text{ V} \tag{5.16}$$

$$E = B - 0.7 = 2.3 \text{ V}$$

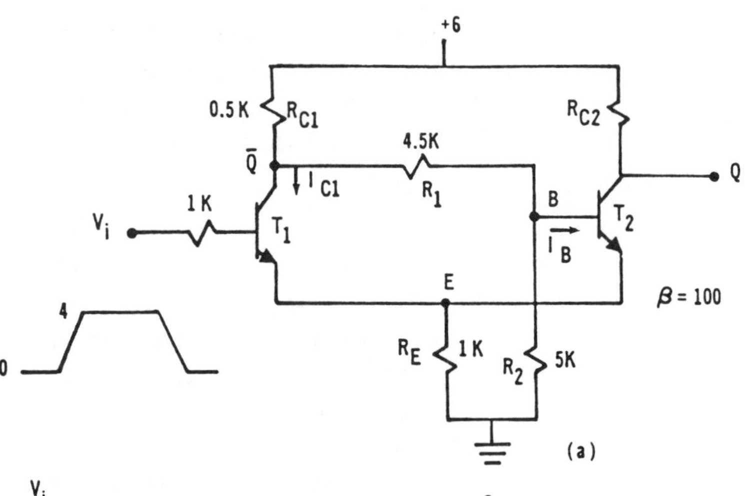

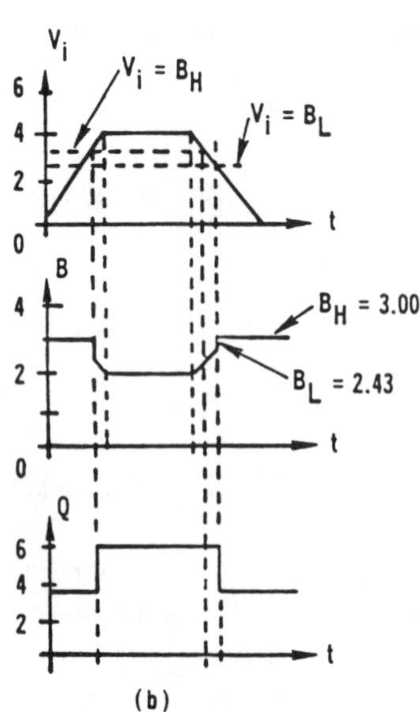

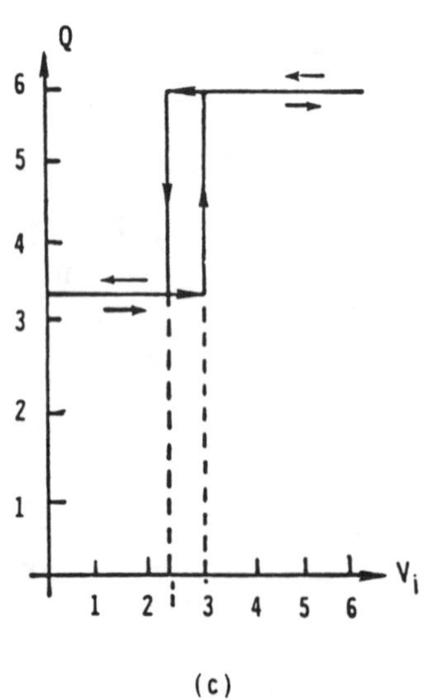

FIGURE 5.16 Basic Schmitt trigger circuit.

Now, as V_i reaches $E + 0.7 = 3$ V, T_1 begins to conduct and I_{C1} causes a voltage-drop across R_{C1}, which results in B decreasing to less than 3 V. As soon as B has dropped below 3 V, T_2 will be OFF. Under this condition, B can be determined by applying the Thevenin Theorem. Let

$$\bar{Q}_{th} \triangleq \text{Thevenin voltage at } \bar{Q}$$
$$R_{th} \triangleq \text{Thevenin resistance at } \bar{Q}$$
$$T_1 \text{ be actively nonsaturated}$$

then

$$\bar{Q}_{th} = 6 - I_{C1}R_{C1}$$
$$R_{th} = R_{C1} = 0.5\text{ K}$$

and

$$B = \frac{R_2}{R_{th} + R_1 + R_2} \cdot \bar{Q}_{th} = \frac{5}{10}[6 - 0.5\, I_{C1}]$$

$$= 3 - 0.25\, I_{C1} \tag{5.17}$$

where $V_i = 3$ V and

$$I_{C1} = \frac{3 - 0.7}{R_E} = \frac{2.3}{1\text{ K}} = 2.3\text{ mA}$$

but when $V_i = 4$, $I_{C1} = 3.3$ mA. Then

$$B(\text{at } V_i = 3\text{ V}) = 3 - 0.25 \times 2.3 = 2.43\text{ V}$$

$$B(\text{at } V_i = 4\text{ V}) = 3 - 0.25 \times 3.3 = 2.17\text{ V}$$

From the above calculation, one can conclude that

$$B = 3\text{ V} \qquad \text{for } V_i = 3^-\text{ V}$$
$$B = 2.43\text{ V} \qquad \text{for } V_i = 3^+\text{ V}$$

However, T_1 will be ON and T_2 will be OFF if $V_i = B + \varepsilon$, and T_1 will be OFF and T_2 will be ON if $V_i = B - \varepsilon$, where $\varepsilon \triangleq$ infinitesimal value. Although B is a function of I_{C1}, according to Equation (5.17), it has only two critical values, i.e.,

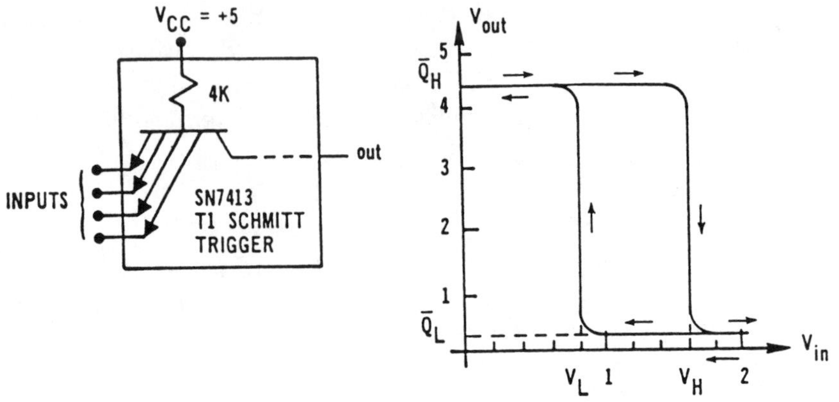

FIGURE 5.17 Schmitt trigger SN7413.

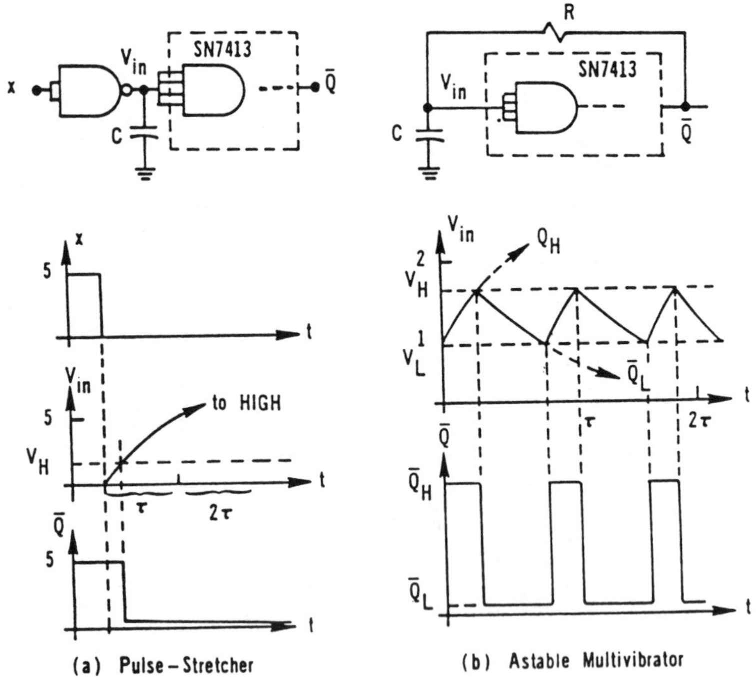

FIGURE 5.18 (a, b) Some applications of Schmitt triggers. (c) Some applications of Schmitt triggers.

$$B_H = 3\ \text{V} \qquad \text{for } V_i = 3^-\ \text{V}$$
$$B_L = 2.43\ \text{V} \qquad \text{for } V_i = 3^+\ \text{V}$$

In other words, $B = B_H = 3$ V if V_i is approaching 3 V from the low value, and $B = B_L = 2.43$ V if V_i is approaching 3 V from the high value. Therefore, (1) if V_i is increasing from low value, T_2 will be turned OFF when $V_i = 3 + \varepsilon$; and (2) if V_i is decreasing from high value, T_2 will be turned ON when $V_i = 2.43 - \varepsilon$. The difference of B_H and B_L is known as the "hysteresis" of the Schmitt trigger. Figure 5.16(b) depicts the cor-

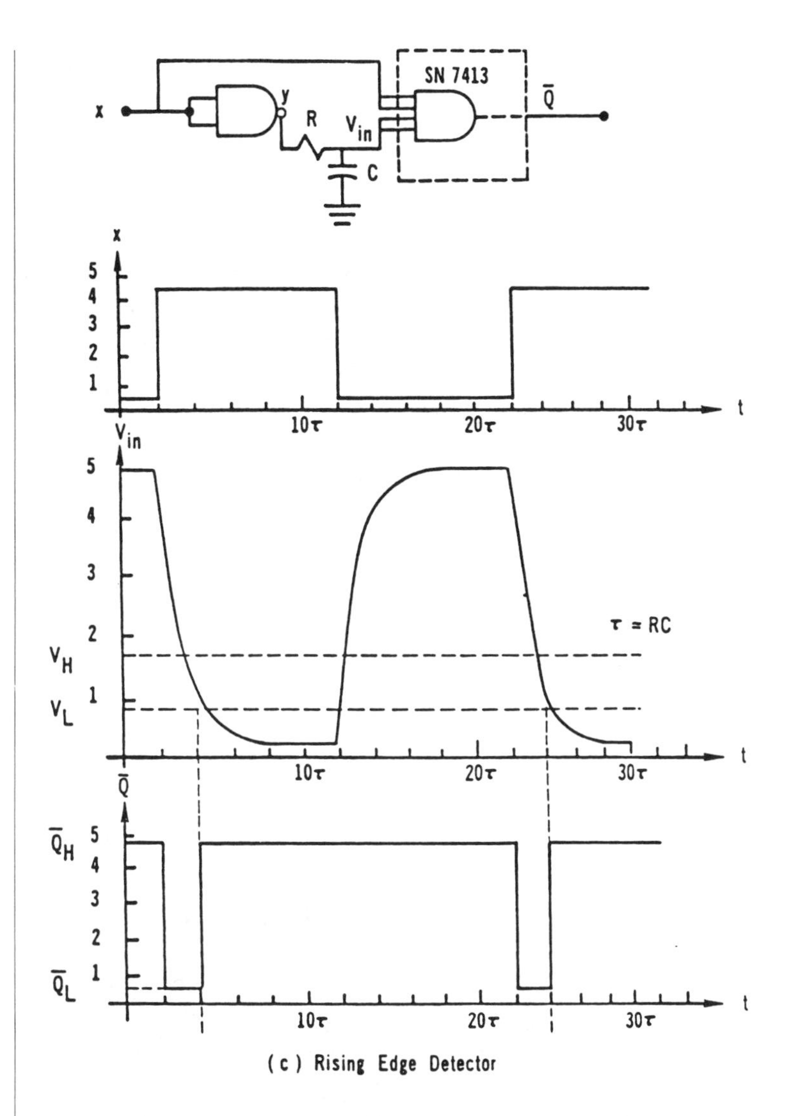

FIGURE 5.18 continued. (a, b) Some applications of Schmitt triggers. (c) Some applications of Schmitt triggers.

responding waveforms at different points of the circuit. Figure 5.16(c) shows its characteristic transfer curve.

The integrated Schmitt trigger circuit is now available on the market. For example, Texas Instruments provides an IC Schmitt trigger circuit registered as SN7413 whose input stage is a four multiemitter transistor, and its input-output transfer curve is shown in Figure 5.17. Figure 5.18(a), (b), and (c) show a few typical applications of the circuit.

In Figure 5.18(a), the circuit functions as a pulse stretcher. The waveforms at different points are shown accordingly. Note that as x is HIGH, V_{in} is LOW, and when x drops to LOW, the capacitor C is being charged to HIGH through the output of the NAND gate. V_{in} rises exponentially and $\bar{Q}$ stays at $\bar{Q}$ = 5 V until V_{in} reaches V_H, the upper threshold of the Schmitt trigger, at which $\bar{Q}$ drops to LOW. The stretched time is determined by the time constant τ which is the product of C and the effective parallel resistance of the output and the input resistance of the NAND gate and Schmitt trigger, respectively. Knowing the input waveform V_{in}, one can plot the output $\bar{Q}$ through the input-output transfer or hysteresis curve shown in Figure 5.17(a).

Figure 5.18(b) is an astable multivibrator using a Schmitt trigger. Consider the instant at which V_{in} is rising from V_L, the lower threshold. According to the hysteresis curve, $\bar{Q}$ stays HIGH but moves toward the right as the arrow shows. During this time, C is being charged through R and the base resistor (4 K) of the multiemitter transistor at the input Schmitt trigger with a time constant

$$\tau \simeq C\left(\frac{R \cdot 4\text{ K}}{R + 4\text{ K}}\right)$$

toward $\bar{Q}_H$. As soon as V_{in} has reached the upper threshold V_H, $\bar{Q}$ drops to LOW ($\bar{Q}_L$) and V_{in} discharges exponentially toward $\bar{Q}_L$ with a time constant

$$\tau' \simeq C\left(\frac{R \cdot R_{in}}{R + R_{in}}\right)$$

where $R_{in} \triangleq$ the input resistance of the Schmitt trigger, which need not be 4 K at this time. During this time $\bar{Q}$ stays LOW until V_{in} reaches V_L from V_L^+ at which time $\bar{Q}$ jumps to $\bar{Q}_H$ according to the hysteresis curve. The cycle repeats and thus the circuit functions as an astable multivibrator. The waveforms at the points of interest are shown in Figure 5.18(b).

Figure 5.18(c) shows another application of the Schmitt trigger circuit, this time as a rising edge detector. The four input terminals are divided into two groups and connected as shown. When x is LOW and V_{in} is HIGH, then $\bar{Q}$ will be HIGH. Now, as x jumps to HIGH, y drops to LOW immediately, but V_{in} is discharged through R exponentially toward LOW. Before V_{in} reaches V_L, all four input terminals of the Schmitt trigger are greater than V_L, and thus $\bar{Q}$ will drop LOW and stay LOW until V_{in} falls to V_L. At this time, as shown in the hysteresis curve, $\bar{Q}$ jumps to HIGH. From here on, as long as x stays HIGH, V_{in} will be LOW and keep $\bar{Q}$ HIGH. As x drops to LOW again, the y responds to HIGH, which will charge the capacitor C through R. Thus V_{in} increases exponentially toward HIGH. Although V_{in} passes through V_H, at which point $\bar{Q}$ would normally drop to LOW, $\bar{Q}$ remains HIGH in this case due to the fact that x is LOW and two of the four inputs of the Schmitt trigger stay LOW, which is lower than V_H. The waveforms at the points of interest are shown in Figure 5.18(c).

5.5. REGISTERS AND COUNTERS

5.5.1. Registers

A register is basically a collection of flip-flops which are logically connected together to perform one or more specific functions, such as latch, buffer, right- or left-shift, parallel-in/parallel-out, series-in/parallel-out, etc. Fortunately, registers with specific functions are available on the market. As a designer, one needs only to be familiar with the functions and make proper decisions. As an example to familiarize the reader with registers, Figure 5.19 shows the logic diagram of a 4-bit bidirectional universal shift register, 74194. Note that there are three distinct subsets of terminals, i.e., the controls, input data, and output data. The control lines contain the function or mode controls, s_0, s_1, clear, and clock. The following table shows the code of the mode controls.

s_0	s_1	Mode
0	0	Clocking of the flip-flop is inhibited
0	1	Shift-left synchronous with rising edge of the clock
1	0	Shift-right synchronous with rising edge of the clock
1	1	Parallel loading of input data

By setting the control mode, one can use this register as a series-in/series-out, parallel-in/parallel-out, series-in/parallel-out, and parallel-in/series-out data manipulator. It is important to point out that the designer should carefully study the timing diagram provided on the data sheet before designing the control signals.

5.2.2. Counters

A counter is also a collection of a set of flip-flops, this time used for counting events or sequencing some controls. Again, there are many kinds of counters available on the market in IC packages. For example, one can purchase some basic counters, such as up/down binary counters, ring counter, decode counter, etc., at a very reasonable cost. More sophisticated ones which offer different kinds of input/output options are also available. Although more sophisticated, they are all

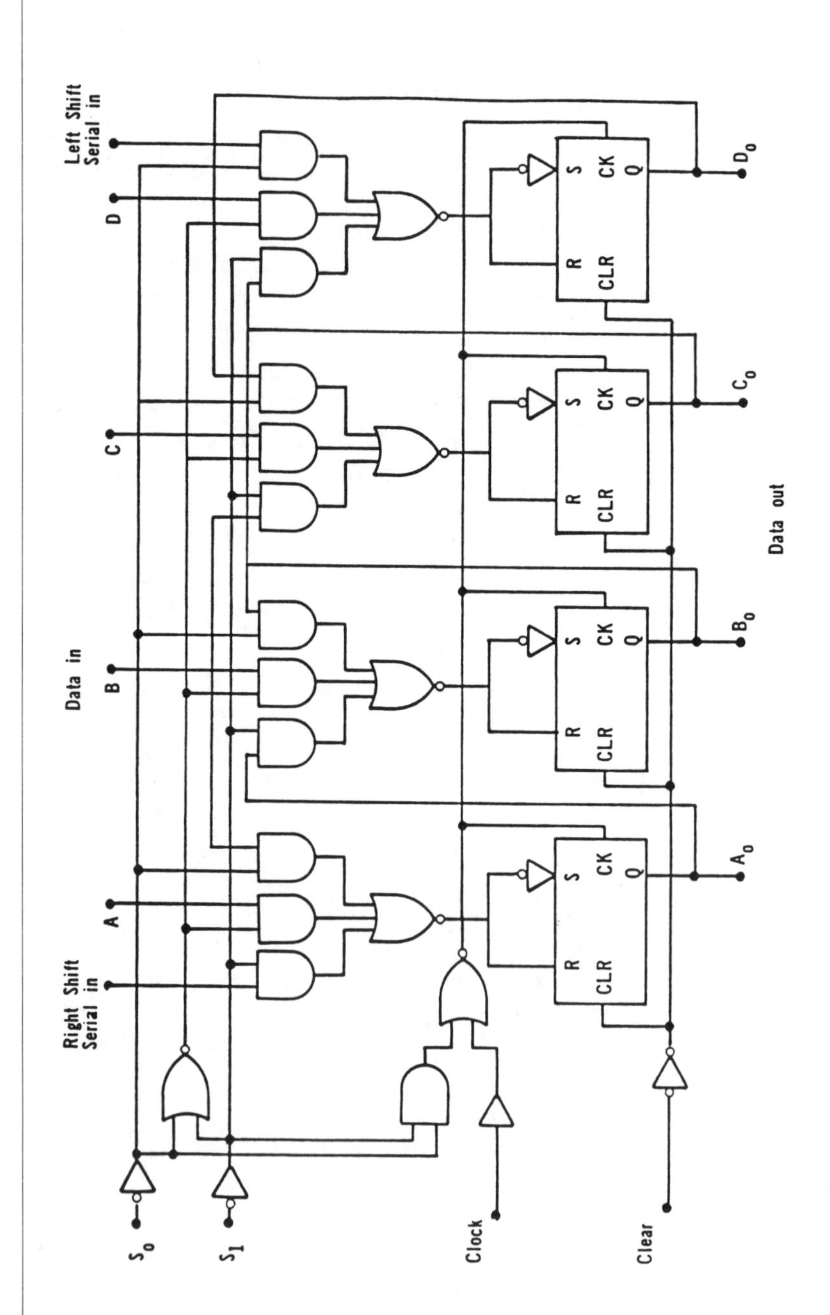

FIGURE 5.19 Logic diagram of a universal shift register.

designed with the basic counters, or flip-flops with gates. To illustrate, a design procedure for a modulo-6 counter using D flip-flops is described in the following example.

Since a count of six is desirable, we need three D flip-flops. The design procedure is as follows.

a. Transition Table

D Flip-Flop			
X	Y	Z	Sequence
0	0	0	0
0	0	1	1
0	1	0	2
0	1	1	3
1	0	0	4
1	0	1	5
1	1	0	6
1	1	1	7

For modulo-6

b. Derivation of Input Equations

Since the sequence for modulo-6 is

$$\text{State } 0 \rightarrow \text{State } 1 \rightarrow \ldots \rightarrow \text{State } 5 \rightarrow \text{State } 0$$

and the input equation for the D flip-flop is

$$Q_{n+1} = (\text{clock}) \cdot D_n$$

the input equation for each flip-flop can be derived based on the transition table and its Karnaugh map. Let the D input for the X, Y, and Z flip-flops be D_x, D_y, and D_z, respectively. Then for the X flip-flop,

X \ YZ	00	01	11	10
0	0	0	1	0
1	1	0	–	–

X_{n+1}

where $X_{n+1} \triangleq$ next sequence output of X flip-flop.

Input Equation $\rightarrow D_x = X Z' + YZ$.

For the Y flip-flop:

YZ \ X	00	01	11	10
0	0	1	0	1
1	0	0	–	–

Y_{n+1}

where $Y_{n+1} \triangleq$ next sequence output of Y flip-flop.
Input Equation → $D_y = X'Y'Z + YZ'$.
For the Z flip-flop:

YZ \ X	00	01	11	10
0	1	0	0	1
1	1	0	–	–

Z_{n+1}

where $Z_{n+1} \triangleq$ next sequence output of Z flip-flop.
Input Equation → $D_z = Z'$.

c. Logic Diagram

Figure 5.20 shows the logic diagram of the modulo-6 counter. Here the edge-triggered D flip-flop is used. This design technique can be used to design counters of modulo-x where x is any positive integer and is $\leq 2^n$, where n is the total number of flip-flops required.

5.6. ENCODERS, DECODERS, MULTIPLEXERS, AND DEMULTIPLEXERS

5.6.1. Encoders

An encoder is a logic network which encodes a specific input line by a set of binary output lines. Keyboard encoding or range selection are typical applications. Figure 5.21 shows the block diagram and truth table of the 10-line to 4-line priority encoder 74147. Note that there are only nine input lines. The zero line of the ten input lines is economically omitted, because logically, when all nine lines are HIGH, this implies that the zero line is LOW.

5.6.2. Decoders

A decoder, as the name implies, is a logic network which functions as the inverse of the encoder. Address decoding network of memory is a typical application. As an example, the 74154 is a device which de-

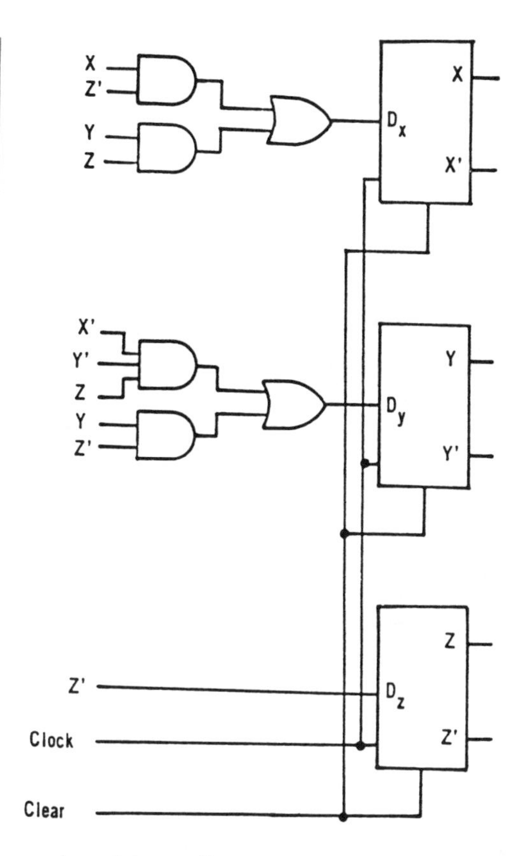

FIGURE 5.20 Logic diagram of modulo-6 counter.

codes 4 input lines into one of 16 mutually exclusive output lines, which performs the inverse function of the 74147 encoder.

5.6.3. Multiplexers

A multiplexer is another type of MSI package. Logically, its time multiplexes an output line to a set of input lines. Electrically, it functions as a rotary switch with the rotating arm as its output. A device such as the 74153 is a dual 4-line to 1-line multiplexer. Two control lines are used to selectively connect one of the four input lines to the output. Figure 5.22 shows the logic diagram of the 74153 and its rotary switch equivalent.

5.6.4. Demultiplexers

A demultiplexer performs the inverse function of a multiplexer. It electrically distributes an input data line to a set of output data lines. It is similar to a rotary switch whose central arm is now used as an input line. Therefore it is a one-to-many data distribution device.

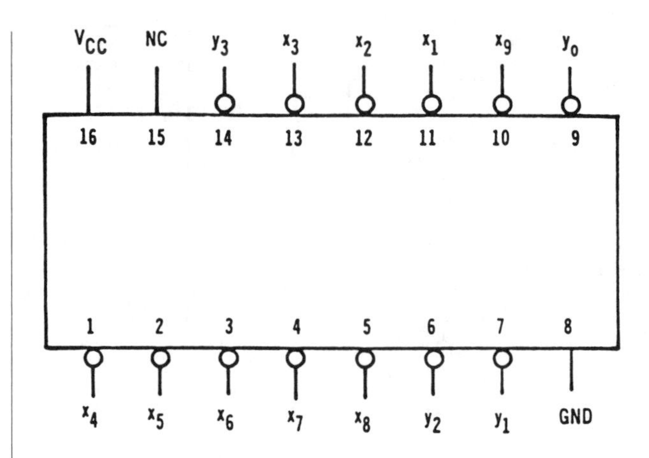

Truth Table

INPUTS									OUTPUTS			
x_1	x_2	x_3	x_4	x_5	x_6	x_7	x_8	x_9	y_3	y_2	y_1	y_0
1	1	1	1	1	1	1	1	1	1	1	1	1
-	-	-	-	-	-	-	-	0	0	1	1	0
-	-	-	-	-	-	-	0	1	0	1	1	1
-	-	-	-	-	-	0	1	1	1	0	0	0
-	-	-	-	-	0	1	1	1	1	0	0	1
-	-	-	-	0	1	1	1	1	1	0	1	0
-	-	-	0	1	1	1	1	1	1	0	1	1
-	-	0	1	1	1	1	1	1	1	1	0	0
-	0	1	1	1	1	1	1	1	1	1	0	1
0	1	1	1	1	1	1	1	1	1	1	1	0

FIGURE 5.21 74147 10-to-4 encoder.

5.6.5. Decoders/Demultiplexers

Since the logic diagrams of decoders and demultiplexers are identical except for how the inputs are defined, devices such as the 74155 can be used as either a decoder or a demultiplexer. By properly defining or labeling the inputs, the 74155 can be used either as a dual 2-line to 4-line decoder or as a dual 1-line to 4-line demultiplexer.

5.7. LINE RECEIVERS AND DRIVERS

So far in this chapter we have presented basic logic elements used within a system. For TTL logic, the binary voltage levels are logic HIGH ≥ 2.4 V and logic LOW ≤ 0.5 V. In many cases, however, we have to deal with data communication, or sending/receiving data through transmission lines. Special devices are therefore required for these applications. There are basically three problems involved, namely, (1)

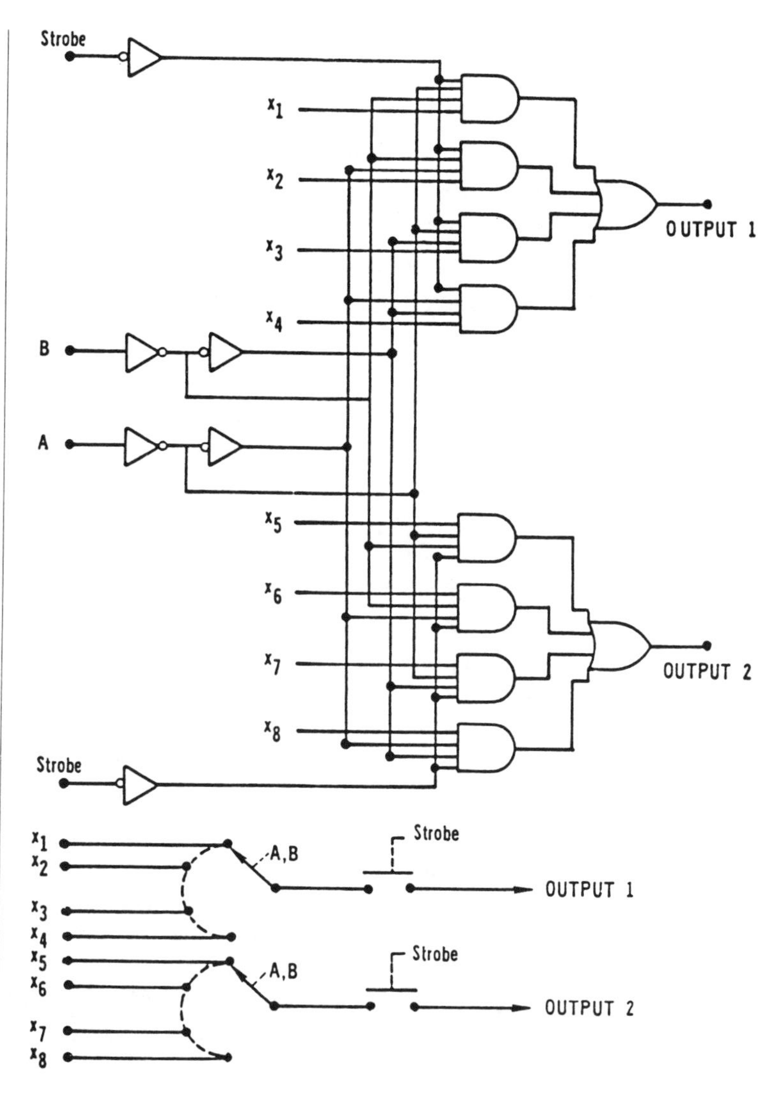

FIGURE 5.22 Logic diagram of multiplexer 74153.

logic level conversions between different systems, (2) impedance matching, and (3) noise immunity and common mode rejection. For example, Signetics Corporation provides the EIA/MIL Line Driver (8T15) and Receiver (8T16) which will satisfy the EIA (Electrical Industry Association) standard RS-232B and C, for which the binary logic levels are ±6 V, therefore for the 8T15 the nominal power supplies for V_{CC} and V_{EE} are ±12 V, respectively. The receiver 8T16 has a desirable hysteresis feature which will receive a ±6 ~ ±12 V signal and convert it into a standard TTL logic level signal. As another example, most of the manufacturers provice devices called transceivers which both receive and transmit conventional logic levels with a built-in tri-state logic gate. Some of the receivers provide the hysteresis property, while others have differential inputs for common mode rejection. Since these devices are normally driving or terminating a transmission line which has low characteristic impedance, the driver should have high input impedance so that the characteristic impedance matching can be achieved by shunting external resistors with appropriate values.

5.8. DEBOUNCERS

In many cases, we have to use a push-button switch to directly excite logic elements. Unfortunately, the push-button switch never produces a clean pulse; instead, it generates a series of noisy pulses which are most unwelcome to the multivibrators and cause errors. It is thus mandatory that all push-button switches be debounced with special circuitry. Figure 5.23(a) shows a very simple debounce circuit which is often used to debounce a manual RESET switch for a microcomputer system. However, the switch must not bounce back while it is being released. Typically, V_{CC} = +5 V, Diode = IN914, R = 6.8 K~10 K, and C = 1~2 μf. Figure 5.23(b) shows a better debouncer which uses two resistors and a cross-coupled flip-flop. Although not absolutely free of bounce, it does serve the purpose in most cases. Figure 5.23(c) shows the circuit of a debouncer guarded by a tri-state switch. Note that a strobe and tri-state control are provided, which will of course assure the "bounce-free" action, if proper timing of the strobe is employed. With the provision of tri-state control, the switch can be disconnected from the output by electric signal or software.

5.9. CONSIDERATION OF INPUT/OUTPUT OF A MODULE

Due to the rapid progress in MSI/LSI technology, the system designer could apparently consider a logic module as a "black box" and design the entire system based on that concept. In practice, however, one would need information on the input/output characteristics of a module and consider compatibility between modules. Fortunately, most data sheets do provide partial input/output equivalent circuits of modules so that the designer can use this information to design an interface circuit between incompatible modules. To clarify this concept, let us consider a typical example. Figure 5.24 shows the input/output equivalent circuits of a typical 745194 4-bit bidirectional universal shift register by Texas Instruments. This shows that at input HIGH state, the

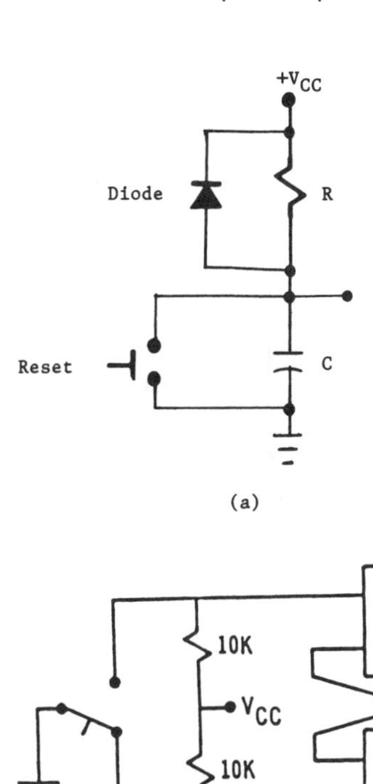

(b) Simple Debouncer Circuit

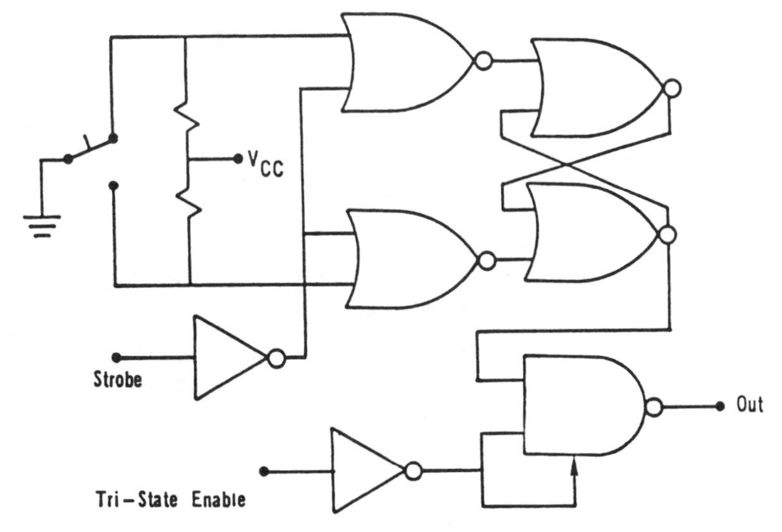

(c) DM75/DM85 44 (Courtesy of National Semiconductor)

FIGURE 5.23 (a) Simple debounced circuit. (b, c) Switch debouncer circuits.

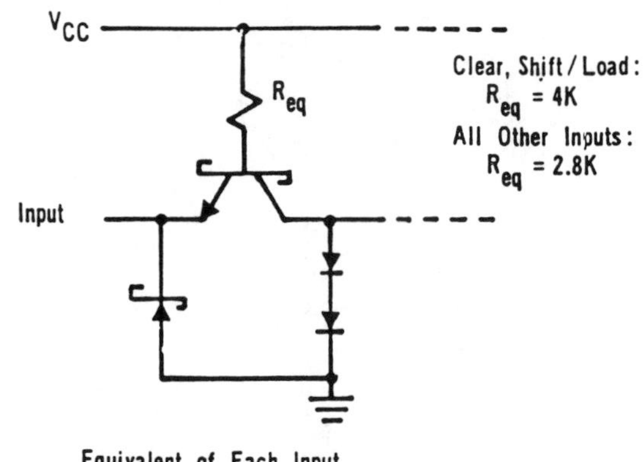

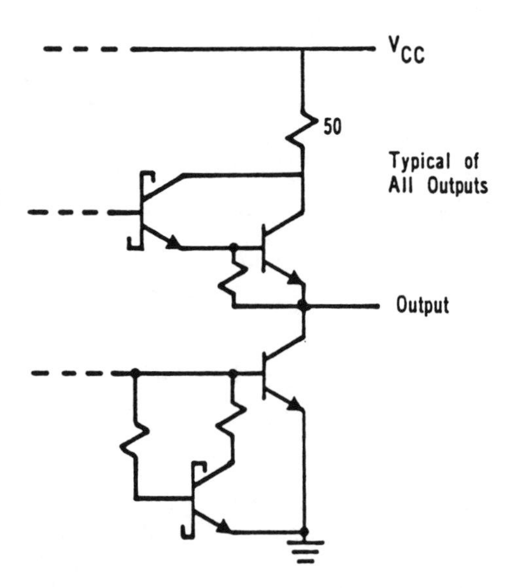

FIGURE 5.24 74S194 4-bit bidirectional universal shift register.

device has very high input impedance and that at input LOW state, its input impedance is approximately equal to the equivalent resistance R_{eq} with current $I = V_{CC}/R_{eq}$ flowing out of the device. The clamping Schottky diode at the input would clamp the negative pulse, and if the driving source resistance is approximately equal to R_{eq}, the device would not operate properly in input LOW state. Consider the output equivalent circuit. This reveals that the device has a totem pole configuration. The upper transistor operates as an emitter-follower which can provide high drain current in output HIGH state, and the 50 Ω resistor in the collector circuit serves as a current limiter if the output is temporarily or accidentally shorted to ground. The lower transistor will sink

TABLE 5.10.1(a) Symbols of Logic Elements and Tables

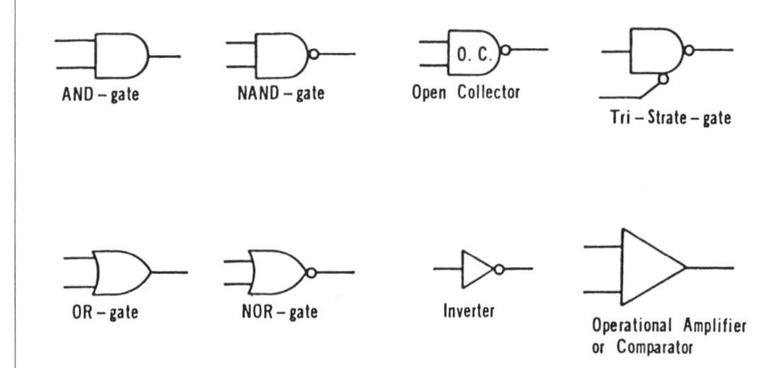

TABLE 5.10.1(b) Symbols of Logic Elements and Modules

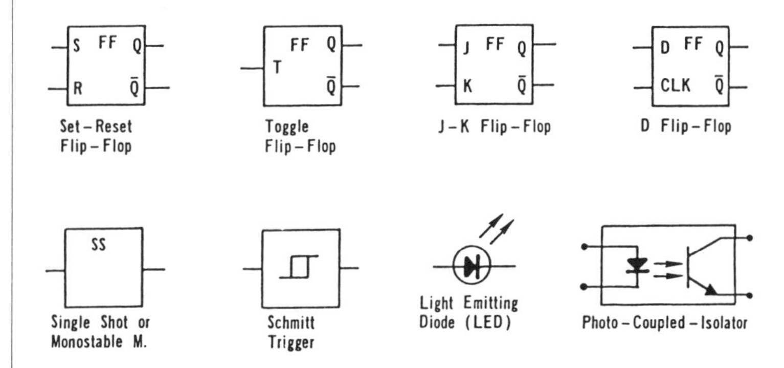

current to its specification when the device is in its output LOW state. This input/output information is extremely important to system designers who are responsible for interfacing two devices, be they SSI (small-scale integrated circuit), MSI (medium-scale integrated circuit) or LSI (large-scale integrated circuit). Incompatibility between the output of a driving device and the input of the device being driven may cause a chaotic result. Designers must apply Thevenin theory to analyze the interface incompatibility.

5.10. A SUMMARY OF SYMBOLS OF LOGIC ELEMENTS AND MODULES

Logic elements and modules have been described in the preceding sections. Although we have been using the most popular symbols which appear in the literature, it is worthwhile to present a summary of the symbols here for convenience. Table 5.10.1(a) and (b) show the most popular logic symbols and their definitions.

Since the early 1970s, a new set of logic symbols has been developed. These new symbols have been appearing in recent data books. Readers who are interested may refer to Explanation of Logic Symbols which begins on page 429.

REFERENCES

1. Millman, J., *Microelectronics: Digital and Analog, Circuits and Systems,* McGraw-Hill, New York, 1979.
2. American Standards Committee, Updating circuit symbols, the graphic language of electronics, *Electronics,* 48(7), 90, April 3, 1975.
3. *TTL Data Book,* Vols. 1 to 4, Texas Instruments, Dallas, 1984—1985.
4. Bell, D. A., *Solid State Pulse Circuits,* 3rd ed., Prentice-Hall, Englewood Cliffs, NJ, 1988.
5. Millman, J. and Grabel, A., *Microelectronics,* 2nd. ed., McGraw-Hill, New York, 1987.
6. Horowitz, P. and Hill, W., *The Art of Electronics,* 2nd ed., Cambridge University Press, New York, 1990.
7. *FACT Data Book,* Fairchild Semiconductor Corporation, 1987.

6 MEMORY SYSTEMS

6.1. INTRODUCTION

In the early 1950s, memory to a logic or digital system designer meant flip-flops or registers, whereas memory systems are only the concern of digital computer designers. As technology has progressed, the cost of the basic memory devices has decreased to the level where a logic designer can now afford to use them to design more sophisticated digital systems. Thus, design of memory systems becomes one of the most important skills that a digital system designer must acquire. In this chapter, we present the structures of memory cells/systems for the magnetic core and semiconductor memories. First, clarifications of confusing terminology are in order. Basically, a "memory system" is a collection of memory cells, where a "cell" is similar to a flip-flop to or from which 1 *bit* of information, either a ONE or a ZERO, can be stored or retrieved. The process of storing digital information is defined as *write,* while that of retrieving digital information is defined as *read.* A memory cell (such as a flip-flop) that holds only one bit of digital information is the basic building element for any memory system. Thus, a memory system is composed of tens, hundreds, thousands, or millions of memory cells. Readers may already be somewhat familiar with memory systems in a digital computer. If so, some terminology used in this chapter might conflict with what you have heard or used before; clarification of those terms is given here.

In a digital computer, the memory system usually consists of a great number of memory cells. Depending on the data width of the CPU (Central Processing Unit) of a given computer, the memory is organized in *strings* of a specific number of contiguous cells assigned with specific names, such as "nibble", "byte", "word", "doubleword", etc., where 4 bits is a nibble, 8 bits or two nibbles is a byte, 16 bits or two bytes is a word, and 32 bits is a doubleword. For example, since the data width of the CPU for an IBM PC/XT is 8 bits wide, its memory system is thus organized around bytes. A good analogy is our cash system, where five pennies make a nickel, ten pennies or two nickels make a dime, etc. Another analogy is the English language where we have letters, words, sentences, paragraphs, etc. However, in English a "word" is composed of a string of letters of variable length, while in computer memory systems a "word" is made up of a fixed number of bits. In general, the capacity of a memory system of a digital computer is specified in kilobytes (1024-bytes). Thus, digital computer designers and users are mostly dealing with bytes when memory systems are considered. As digital system designers, however, we must start at the memory cell or bit level in memory system design. In addition, we must be concerned with the operation of the read/write process in detail. That is, the read/write process mostly deals with which particular memory string we are using and what the digital pattern of that string is. For instance, one byte is composed of 8 bits, thus it can have 256 different digital patterns for representing 256 unique meanings. The unique bit pattern of a byte is called the *content* of that byte. The location of a specific byte is referred to as the *address* of the byte. Currently, memory systems in most digital computers are organized on a byte basis. Thus, for these memory systems, designers and users are dealing with the address and content of a byte they are interested in. However, in digital system design, the designer is not limited to bytes alone. Depending on specific applications the length of a memory string can be 1, 4, 8, or any number of bits. For convenience, in digital system design, the string of a memory is traditionally called *memory-word,* or simply *word* (with variable length of bits), which unfortunately is easily confused with the meaning of "word" in computer technology, where a string of 16 bits is defined as a "word". Therefore, the reader must bear in mind that the memory-word can be any number of bits in width or length. Except for this term, the principle of operation for a memory system, be it for a computer or other digital system, is basically the same. The major operations for a memory system are (1) to specify the address of the memory-word of interest and (2) to issue one of two control signals: READ, for fetching or copying the contents of that word, or WRITE, for storing new contents into that word. We can imagine a memory system is analogous to mailboxes at the post office. Each mailbox has a unique address into which the mailperson places ("WRITES") mail according to the address and from which the addresses takes ("READS") his or her mail. Having described the basic concept of a memory system, we can now classify memory systems according to their store/retrieve technologies into the following categories.

RAM

A RAM is a Random-Access Memory system. In this system, one can store or retrieve data at any location or address at random within the same length of time. In other words, the data access time for this system is independent of data location.

CAM

A CAM is a Content-Addressable Memory system, also known as Associative Memory. In this system, a memory-word, in retrieve process, has no apparent address. Instead, a segment of the content of a memory-word is used as the address of that word. As an example of a typical application, the content of a memory-word for an airline scheduling computer may contain several segments, e.g., time, date, name of the airline, destination, ticket price, etc. One could use the time and destination segments as the addresses for all memory words which match the desired time and destination, and then one could accordingly choose the preferred date and airline.

ROM

ROM is Read-Only Memory. It is a system containing information prestored by the designer so that its content cannot be altered by the user. Therefore, as the name implies, the contents of ROM can only be read. Actually, in our daily life there are many analogies to ROM systems. For example, street signs, road signs, and name plates have information painted or stamped on them and are placed in specific locations for people to "read-only"; they are not supposed to be altered.

RMM

RMM is Read-Mostly Memory. In this system, although the content of a memory-word can be altered, it is mostly meant to be read. In this kind of memory system, the writing process normally takes much longer than the reading process.

6.2. MAGNETIC CORE MEMORY — RAM

6.2.1. Memory Cell

a. Principle of Operation

A memory cell of a magnetic core memory system contains a doughnut-shaped tiny magnetic core with several wires going through the hole as shown in Figure 6.1(a). The magnetic core has a desirable hysteresis characteristic curve for current (i) vs. magnetic flux (ϕ), as shown in Figure 6.1(b). In Figure 6.1(a), there are four wires, where

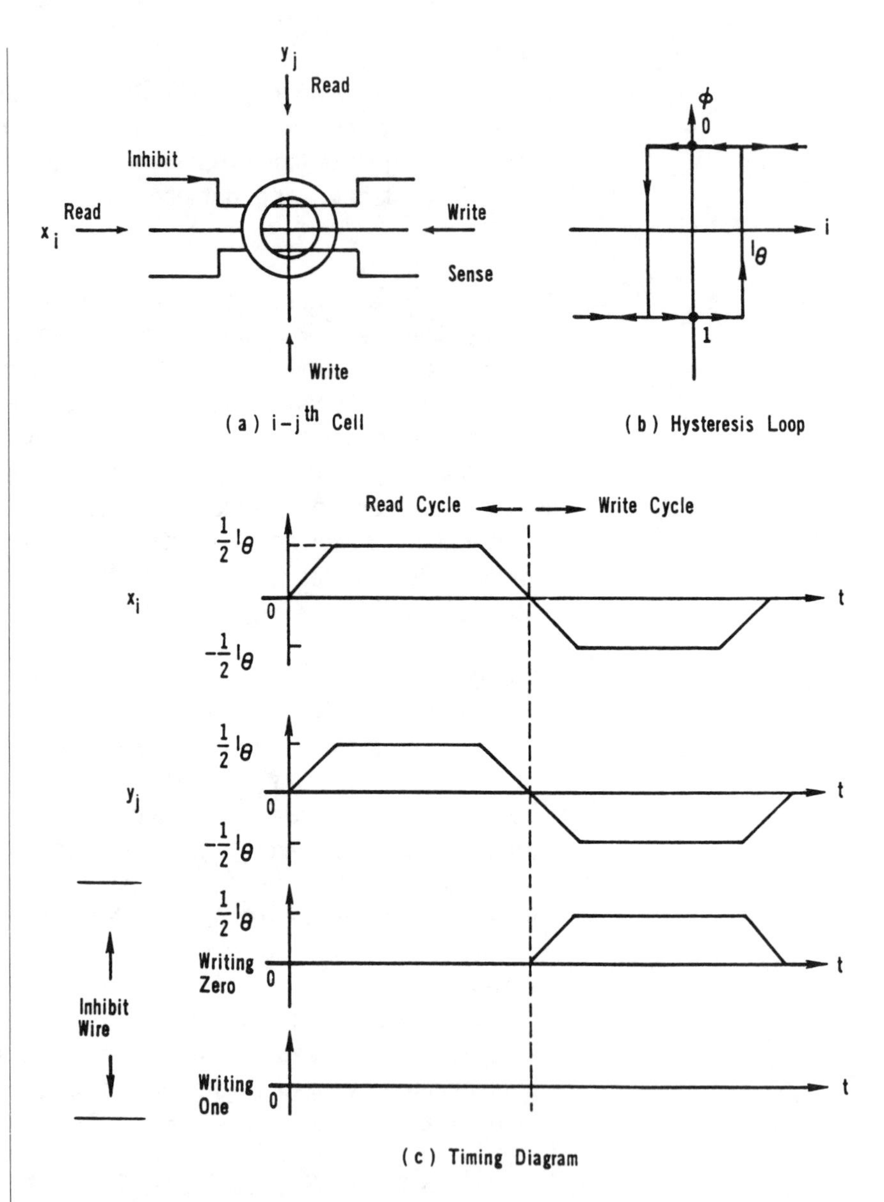

FIGURE 6.1 Magnetic core memory cell.

read, write, and inhibit wires are used for pulse excitation purposes. When excited, a signal will be induced in the sense wire with a magnitude proportional to $\frac{d\phi}{dt}$. Figure 6.1(c) shows the timing diagram of the read/write process. The principle of operation follows.

At quiescent point, no current flows in any of the wires. The core is in one of the two steady states shown in Figure 6.1(b), i.e., [1] or [0]. Consider the core is currently in state [1]. The read cycle starts with a current pulse of $(\frac{1}{2})I_\theta$ in both the x_i wire and the y_j wire (Figure 6.1(c)).

The net current flowing through the core hole will be I_θ which is the threshold of the excitation current (Figure 6.1(b)) so that it will reverse the direction of the magnetic flux in the core. Due to the hysteresis characteristic shown in Figure 6.1(b) by the arrows, the core will not return to [1] but will stay at [0] as the current in the wire becomes zero. This flux reversal will then yield a voltage pulse with a detectable magnitude in the sense wire. Now if the core was in the [0] state, although the total excitation (read) current of x_i and y_j is still I_θ, the core will nevertheless return to the [0] state after the current pulse is gone. There will thus be no significant flux change, and it would induce considerable low voltage in the sense wire. Therefore, based on the sense wire output, the $x_i - y_j$ read currents can be used to determine the original state of the core. Consider the write process. It is important to point out that the core memory is designed in such a way that a write cycle is always preceded by a read cycle. Therefore, it is evident that the core is always in the [0] state after the read cycle and before the write cycle. In reference to Figure 6.1(a) and (c), the write currents in x_i and y_i are always equal in magnitude, but opposite in direction with respect to read currents. Depending on what information is to be stored in the core, one could accordingly set the inhibit current to be logic 1 or 0 as shown in Figure 6.1(c). By setting the inhibit line with a positive current pulse of $(1/2)I_\theta$, the net write current through the hole of the core will be $-(1/2)I_\theta$, which would not be able to reverse the flux direction. As a result, the core will remain in the [0] state. For writing logic 1 in the core, one simply disables the inhibit line and the core will then be driven to the [1] state by the write current pulses.

b. Nonvolatile and Destructive Read-Out

In view of its principle of operation, the magnetic core memory has two important characteristics worth mentioning. First, since the core does not require stand-by power during the quiescent state and the information will remain unchanged even without power, it is a nonvolatile memory device, or a device which is power-failure-proof. Second, as the core always returns to state [0] after being read, it is called a destructive read-out device. That is, the original information is destroyed after it is read. This is, of course, not a desirable feature, therefore a rewrite mechanism after read-out is always required in a core memory system.

6.2.2. Configuration of a 4 × 1 Memory Plane

For clarification, Figure 6.2(a) shows the configuration of a memory plane which has four memory words and the content of each word is one bit wide. The diagram depicts the concept of operation. Figure 6.2(b) shows the concept of a memory map where the address defines the location of each memory-word and the content, either [1] or [0], after the write cycle.

6.2.3. System Configuration

a. 3-D System

Figure 6.3(a) shows the configuration of a 16 × 3 bits 3-D (three-dimensional) core memory system. Here, we have 16 memory-words, and the content of each word is three bits wide. In this diagram, we introduce the terminology of Memory Address Register (MAR) and Memory Data Register (MDR). Here, MAR is a 4-bit register, A_0, A_1, A_2,

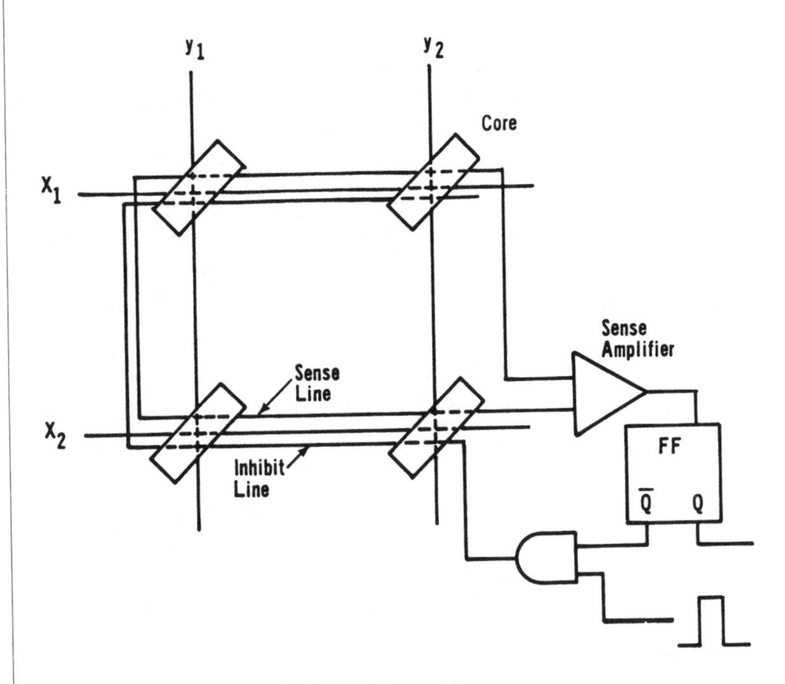

(a) 4 X 1 Memory Plane

ADDRESS				CONTENT
X_1	X_2	y_1	y_2	
1	0	1	0	1/0
1	0	0	1	1/0
0	1	1	0	1/0
0	1	0	1	1/0

(b) Memory Map

FIGURE 6.2
Configuration of a 4 × 1 memory plane.

A_3, where A_0, A_1 define the y-components and A_2, A_3 define the x-components of the address. For example, if the MAR has value $A_0A_1A_2A_3$ = 0011, the right lower corner memory-word will be addressed, and its content will be read and loaded into MDR, then the rewrite cycle follows. As the figure shows, this is obviously a 3-D (three-dimensional) system, i.e., x-address, y-address, and content or data.

Figure 6.3(b) is a general block diagram of a 3-D memory system. Note that there are four blocks, which function as read-write, x-select, y-select, and inhibit drivers, respectively. Those drivers are basically current amplifiers, and in most systems they are made of larger magnetic cores capable of delivering sufficient current to drive through the wires of the memory cells. The diagram, although only depicting the organization of a 3-D memory core system, also shows the major components for other systems such as 2-D and $2\frac{1}{2}$-D described in the next two sections. Furthermore, it illustrates that any memory system, including the semiconductor memory system, can be considered as a black box which interfaces with the outside world with two registers, namely the MAR and MDR, as well as three groups of lines called "buses" which are functionally known as the address-bus, the data-bus, and the control-bus. For an $N \times M$ memory system, there will be N address lines (after decoding) and M data lines. Accordingly, the width of the MAR would be $\log_2 N$ and that of the MDR would be M. The number of lines for the control-bus varies from one system to another. The control-bus carries read/write commands, clock pulses, inhibit pulses, and other information as well.

b. 2-D System

Figure 6.4 is a simplified diagram showing the configuration of a 16 $\times$ 3 bits, 2-D (two-dimensional) system. For simplicity, the inhibit circuitry is not shown. Note that the address has no x-y components; the 16 words, W_0, W_1, ..., W_{15}, can be randomly selected by one of the 16 decoded address lines. Unlike the 3-D system shown in Figure 6.3 where each address line drives 12 cores, in this system each address line drives only three cores.

c. $2\frac{1}{2}$-D System

This system, as the name implies, stays in the middle of the 3-D and 2-D systems based on its cost and performance. Figure 6.5 shows how the 3-D system is evolved into the $2\frac{1}{2}$-D system. For convenience, the 16 $\times$ 3 word 3-D system shown in Figure 6.3 is reproduced in Figure 6.5(a). Conceptually, one could cut the x-address lines at the

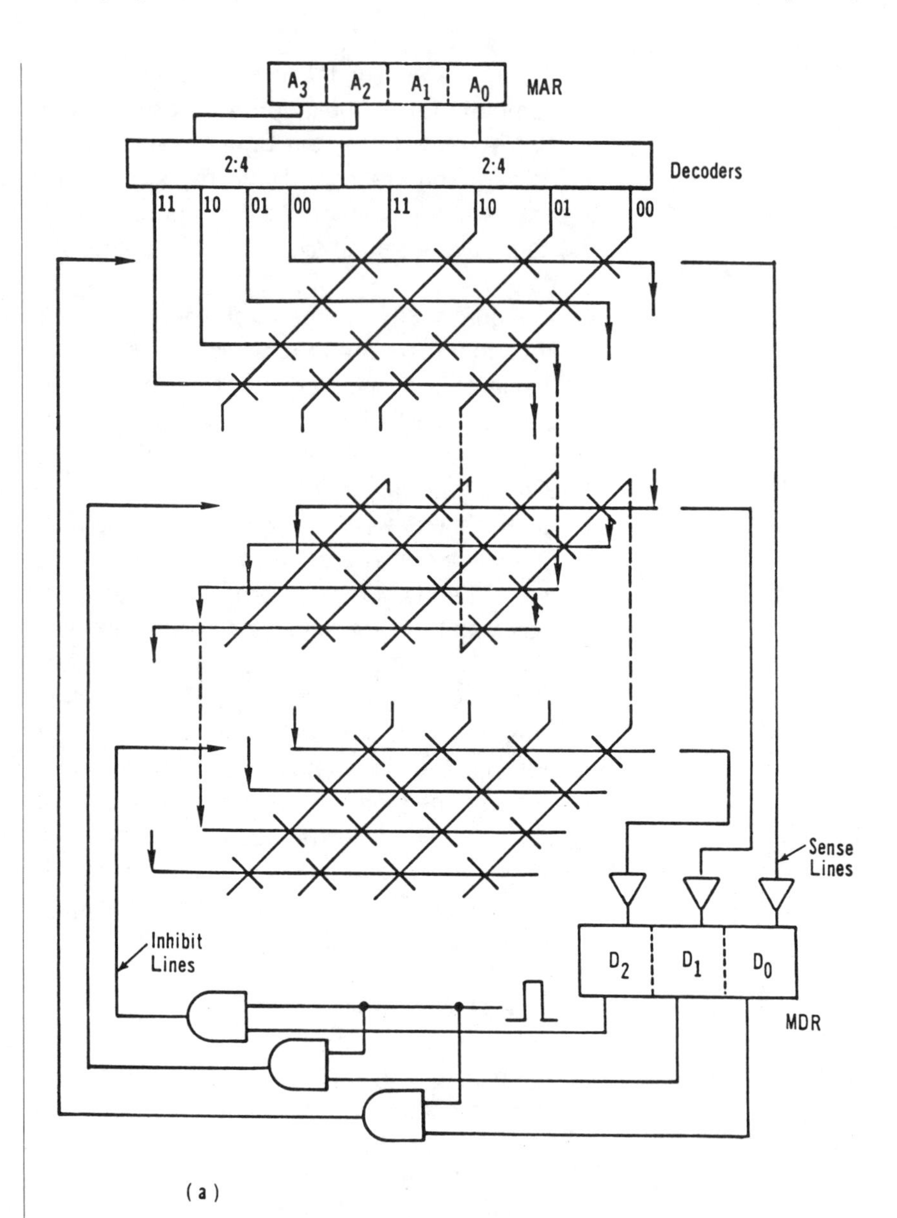

FIGURE 6.3 (a) Configuration of a 16 × 3 words 3-D memory system. (b) A general block diagram of a 3-D memory system.

Xs and then reconfigure it as shown in Figure 6.5(b). Note that the number of y-address lines has been reduced to half the original number, and the number of x-address lines has been doubled, so that the total number of memory-words remains unchanged. Also note that in this example, while each of the y-address lines is still driving 24 cores, the x-address is driving 2 cores by 8 parallel lines.

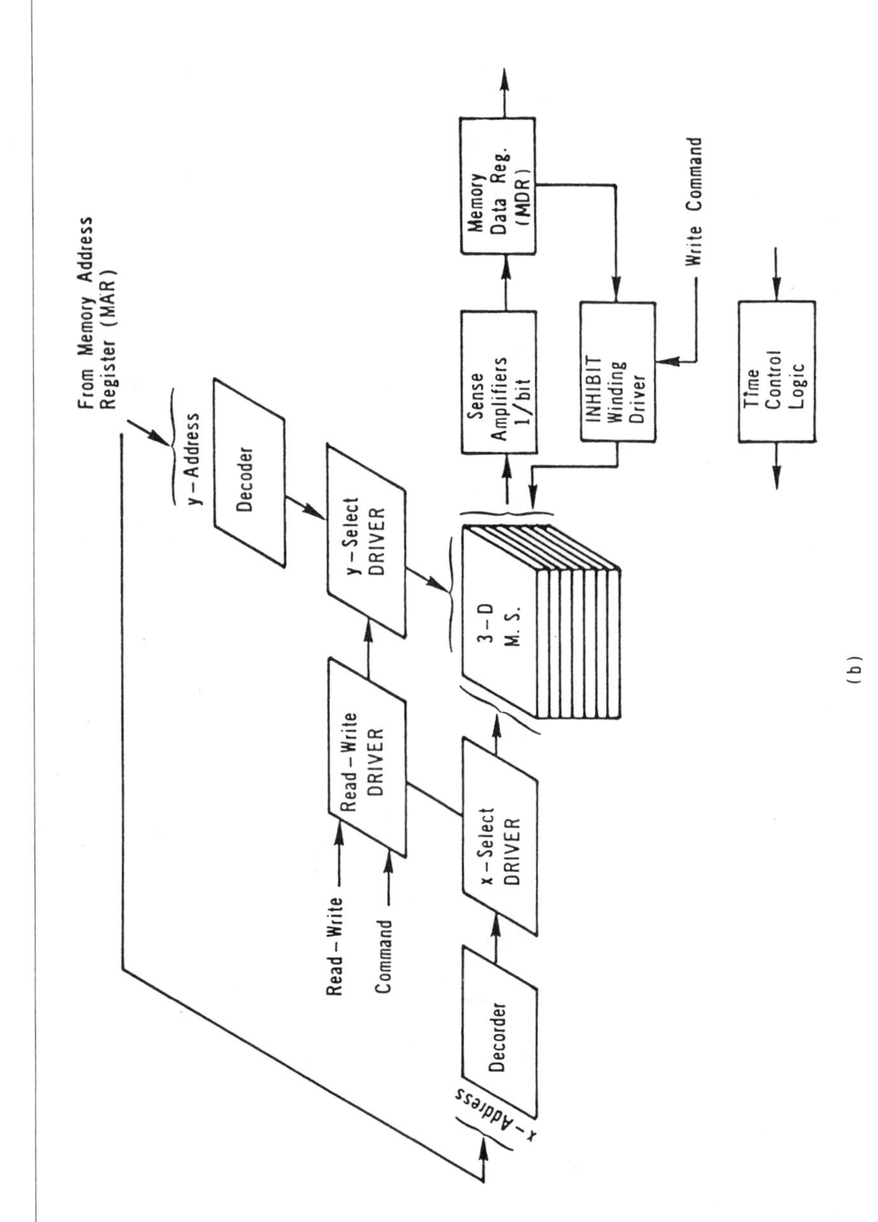

FIGURE 6.3 continued. (a) Configuration of a 16 × 3 words 3-D memory system. (b) A general block diagram of a 3-D memory system.

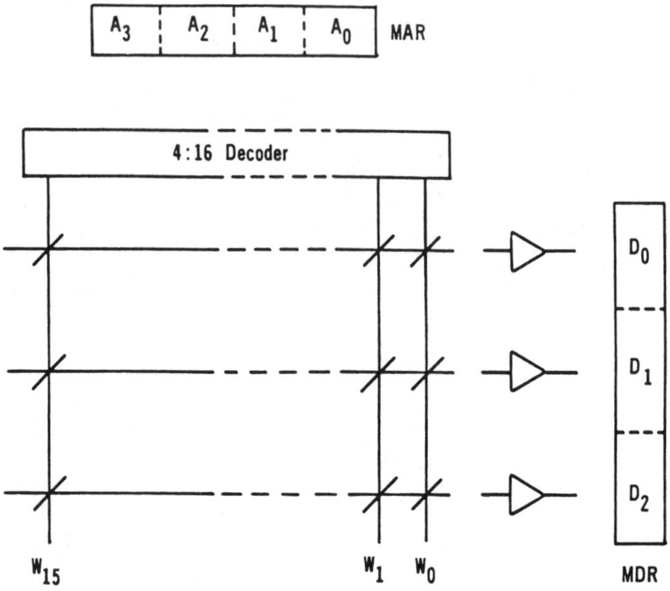

FIGURE 6.4
Configuration of a 16 × 3 words 2-D memory system.

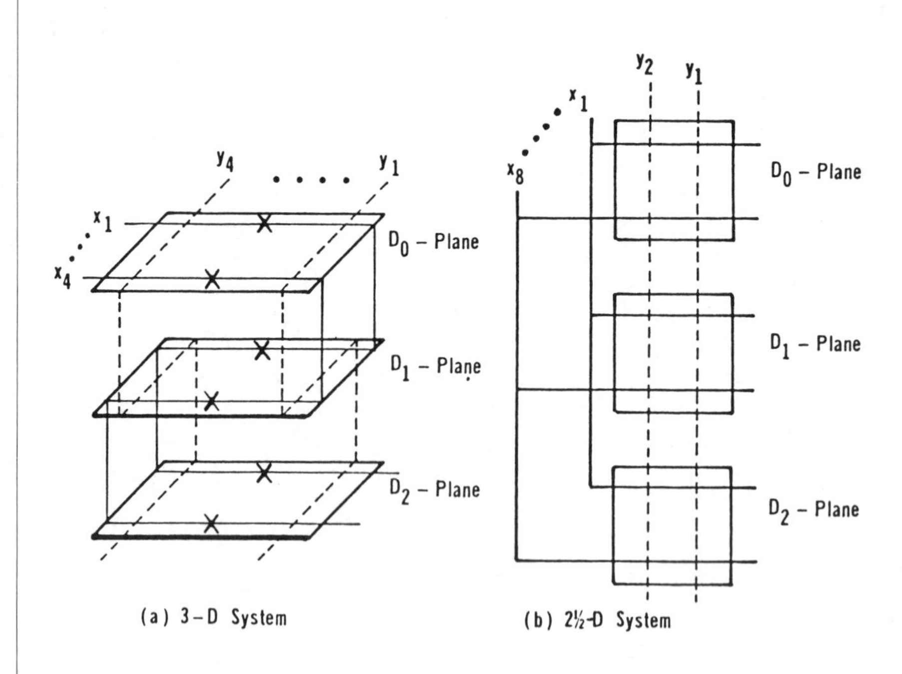

FIGURE 6.5 Evolution of $2\frac{1}{2}$-D system from 3-D system.

6.3. SEMICONDUCTOR MEMORY

6.3.1. Memory Cell

Semiconductor memory, as one may imagine, is a collection of registers which in turn are composed of a set of semiconductor flip-flops. Therefore, a semiconductor memory cell is basically a flip-flop plus the minimum necessary peripheral circuitry for address and data write-in and read-out. To understand how a semiconductor memory system is organized and how it functions, let us examine the circuitry of a few different kinds of memory cells.

a. Static Memory Cell

Figure 6.6(a) to (d) show the circuits of memory cells for different kinds of solid state technologies. Each of the cells basically contains a flip-flop, a word-line (address) , and a pair of bit-lines (data) . Therefore, their operations are similar. For example, consider the flip-flop shown in Figure 6.6(a), for Q_1 ON and Q_2 OFF. The cell is in steady state and the word-line is LOW. In the write process, the Q bit-line is set to HIGH and the $\bar{Q}$ bit-line is LOW. When the word-line goes HIGH, the cell is selected; the B-E junctions connected to the word-line of both Q_1 and Q_2 will be OFF. Since the $\bar{Q}$ bit-line is LOW, Q_2 conducts and its emitter current flows into the $\bar{Q}$ bit-line. In the meantime, Q_1 will be OFF because the Q bit-line is set HIGH. As a result, if we define "ON" as [1], then a [1] is stored in Q_2 as soon as the word-line has returned to LOW and the bit-lines to HIGH. In the read process, the sense amplifier connected to the bit-lines will detect the current if Q_2 has been ON. In summary, the word-line HIGH implies the cell is selected and the bit-lines carry SENSE signal for READ and DRIVE signal for WRITE.

Similarly, one can describe the read/write process for the ECL cell shown in Figure 6.6(b). Here, in steady state, $I_C = I_E$, where I_E is a constant current source. During the read/write process, the word-line goes HIGH, which forces I_C to increase and so $I_C = I_E + I_Q$, where I_Q (the excess current) would flow in the bit-lines. Thus, we again have SENSE for READ and DRIVE for WRITE as described before.

Figure 6.6(c) and (d) show MOS and CMOS cells, respectively. Since MOSFETs are voltage control devices, their configurations are quite straightforward. Q_1 and Q_2 are the resistor equivalents of a conventional flip-flop. Q_5 and Q_6 in both circuits are used as switches. They connect the flip-flop to its corresponding bit-lines. Again, we have word-line HIGH for cell selection and similarly for bit-lines we have SENSE for READ and DRIVE for WRITE. It is evident that the CMOS

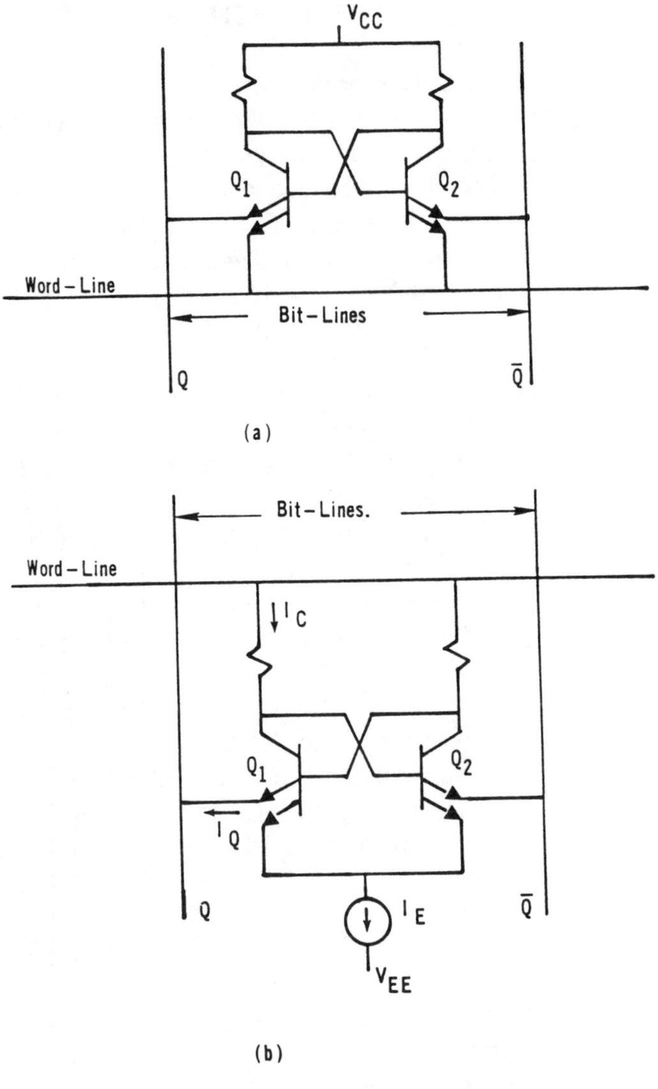

FIGURE 6.6 (a) TTL memory cell. (b) ECL memory cell. (c) MOS memory cell. (d) CMOS memory cell.

memory cell consumes extremely low standby power (see Chapter 4, Section 4.2.8).

b. Dynamic Memory Cell

It is important to point out that the static memory cell just described requires more space in comparison with the dynamic memory cell. In a static memory cell, information is stored in a flip-flop and thus can be retained indefinitely as long as the power is ON. For a dynamic cell, however, the information is retained in a capacitor which may "leak" away as time elapses. Therefore, it requires a refresh-circuit to reinforce the in-

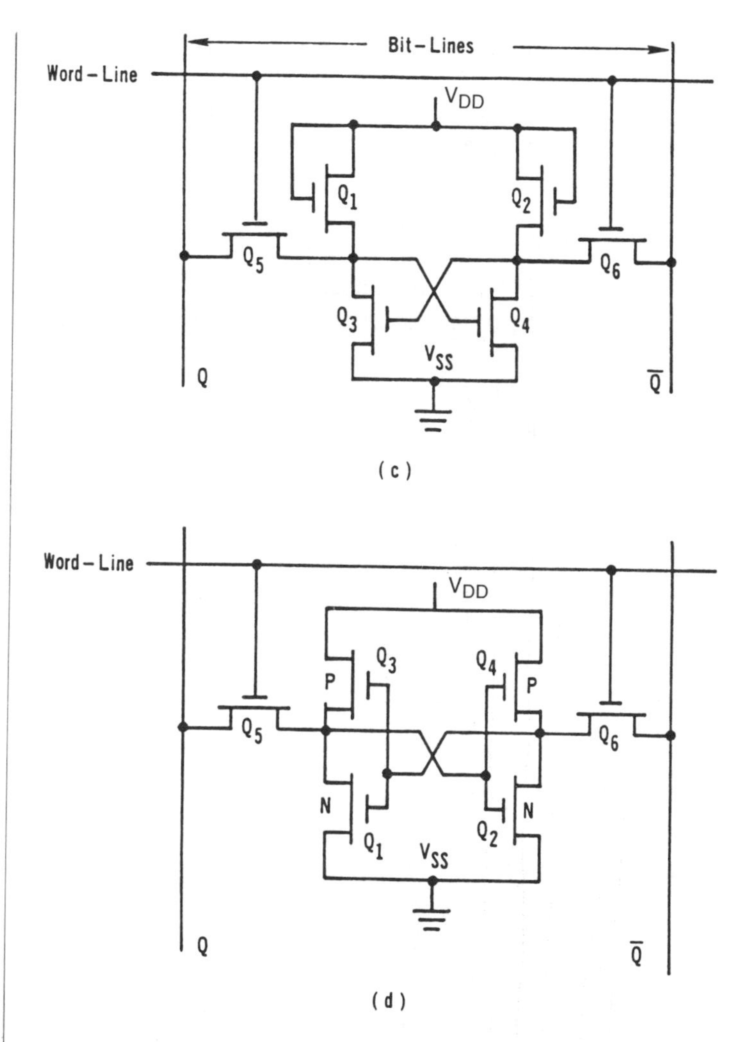

FIGURE 6.6 continued. (a) TTL memory cell. (b) ECL memory cell. (c) MOS memory cell. (d) CMOS memory cell.

formation every 2 ms. Figure 6.7(a) shows an older type of 3-transistor dynamic memory cell. Here, data is stored in the capacitor through the switch Q_1 and read through Q_3. Due to the circuit simplicity of the memory cell, a more sophisticated peripheral circuitry is required in addition to the standard sense/drive circuits for the bit-lines. A brief description of the read/write process follows.

In the read process, the data-out line is first precharged to logic [1], then Q_3 is turned ON by the read-enable (word) line, and the data-out line will be discharged or not discharged through Q_3 and Q_2, depending on the information stored in capacitor C. Thus, the sense amplifier connected to the data-out line can sense the information stored. For writing, the data-in line is activated with the desired information by the drive circuitry connected to it. Q_1 turns ON when the write-enable (word) line is activated and the information in the data-in line is then stored in the capacitor. For the refreshing process, the information in the capacitor is first read to the data-out line and then fed back to the data-in line to write the information just read back to the capacitor. This process is repeated every 2 ms.

Figure 6.7(b) shows a more advanced 1-transistor memory cell. Here, the capacitor is again used as a data storage element. The transistor is selected and activated by the word-line, and it functions as a switch which transfers the information from (or into) the capacitor to (or from) the data-line during the read (or write) process. Due to its simplicity in circuitry, 1 Meg-memory cells, or more, of this kind can be packed on one chip.

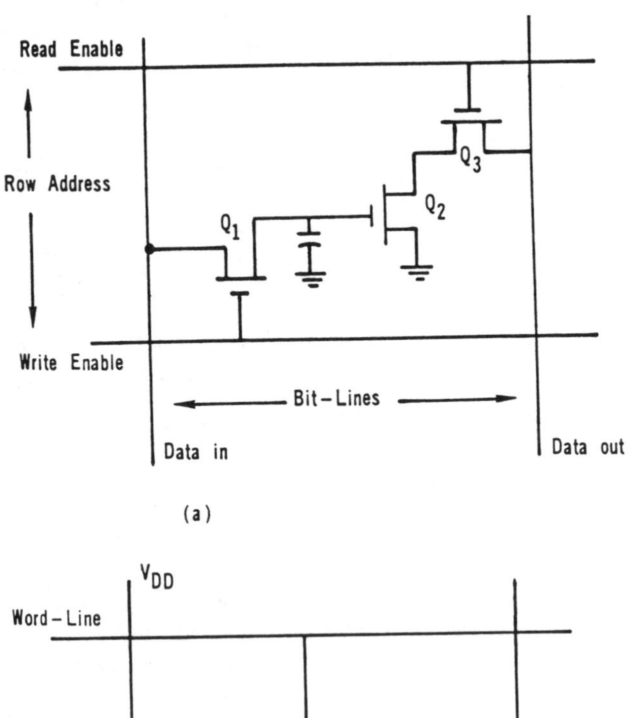

FIGURE 6.7
(a) 3-Transistor memory cell. (b) 1-Transistor memory cell.

c. Volatile and Nondestructive Read-Out

In contrast with the core memory, the semiconductor memory cell is characterized by volatile but nondestructive read-out. While nondestructive read-out is desirable, volatility is not because the information will be lost when the power supply fails. Memory cells of the CMOS type, however, require very low stand-by power, therefore CMOS memory with a battery back-up circuitry can be used for systems which require nonvolatile memory.

6.3.2. System Configuration — Non-RAM

In this section, two special purpose memory systems which can be classified as non-RAM (i.e., not randomly accessible) will be presented. The contents of these memory systems must be accessed following specific rules.

a. LIFO — Last-In First-Out

This type of memory system is also known as a memory-stack, which is well-known in digital computer systems. It is normally used as a temporary storage area for holding information for calling either subroutines or interrupt service routines. As the name implies, information read from this memory is always the latest information stored into it. Recall that a basic memory system (RAM) normally consists of MAR, MDR, read/write control signal, and the main body of the memory. A read signal results in the contents of the location currently specified by MAR being loaded into the MDR. Conversely, a write signal stores the information currently in the MDR into the memory location at the address currently specified by MAR. Thus, one can specify any desired address in the MAR to access the information at that address. That is why it is called random access memory (RAM). However, for the LIFO memory system, the MAR must be either tightly coupled to or replaced by an up/down counter. One can initialize the counter value and then the access process follows. The write signal would decrement the counter or the MAR and then store information from the MDR into memory at the address specified by the counter after the decrement; the read signal would copy the memory content at the address specified by the current value in the counter into the MDR, and then increment the counter. As a result, we have a Last-In-First-Out memory system. In digital computer systems, this kind of write process is conventionally called a *push* and the read process is called a *pop.* In other words, LIFO can only be accessed by incrementing or decrementing the counter to specify the desired address; it cannot be accessed at random.

b. FIFO — First-In First-Out

Basic Principle of Operation

In many applications, we often wish to collect data in a recurrent burst format for data processing. During the burst, we wish to collect the data as fast as they are generated. Between the bursts, we can store the collected data into a mass storage medium such as a floppy or hard disk for future processing. Obviously, in this application, we would need a fast buffer memory to collect the burst data, and therefore, the access time for the mass storage device would not be so critical, providing that the time between bursts is reasonably long. In some cases, however, the requirement for access speed can be just the opposite of what we have just described. That is, the input data can occur infrequently and unpredictably. In that case, we do not want to tie up the mass storage device; we would rather provide a small block of memory as a buffer to collect the trickling-in data and then transfer them into a destination or mass memory at once when the buffer memory is full or nearly full. Figure 6.8(a) depicts the data flow of the process described. We now show as an example how the process can be implemented using a FIFO memory system. Figure 6.8(b) shows the structure of a FIFO system. It is a block of 512 bytes of memory whose address starts at 0 and ends at 511. Two up-counters initialized or reset to 0, depicted by the two arrows, are called *address pointers* that point to the desired addresses at which the contents are to be accessed. For convenience, let us call the hollow arrow the "head" and the solid arrow the "tail". We will use an "access-then-increment" strategy for the read/write process. That is, for the write process, information is written to the location currently pointed at by the "head" which is then incremented by one; in the read process, the contents of the memory cell whose address is currently pointed at by the "tail" is read, then the value of the "tail" is incremented by one. When any one of the pointer values reaches 511, its next increment operation would reset its value to zero, and the process repeats; in this way, the memory block can be used and reused indefinitely, as if we have a memory with a very large capacity. This kind of memory is also called a *circular memory queue.* Figure 6.8(c) demonstrates the circular nature of the concept. However, one may notice that there are two inherent problems in this structure. First, as the "head" catches up with the "tail", the FIFO would be full and further store processes would write over old data which have not yet been read. Second, as the "tail" catches up with the "head", the FIFO would be empty and further read processes would retrieve obsolete data. Therefore, the designer must be aware of these problems and provide some means of detecting whether the

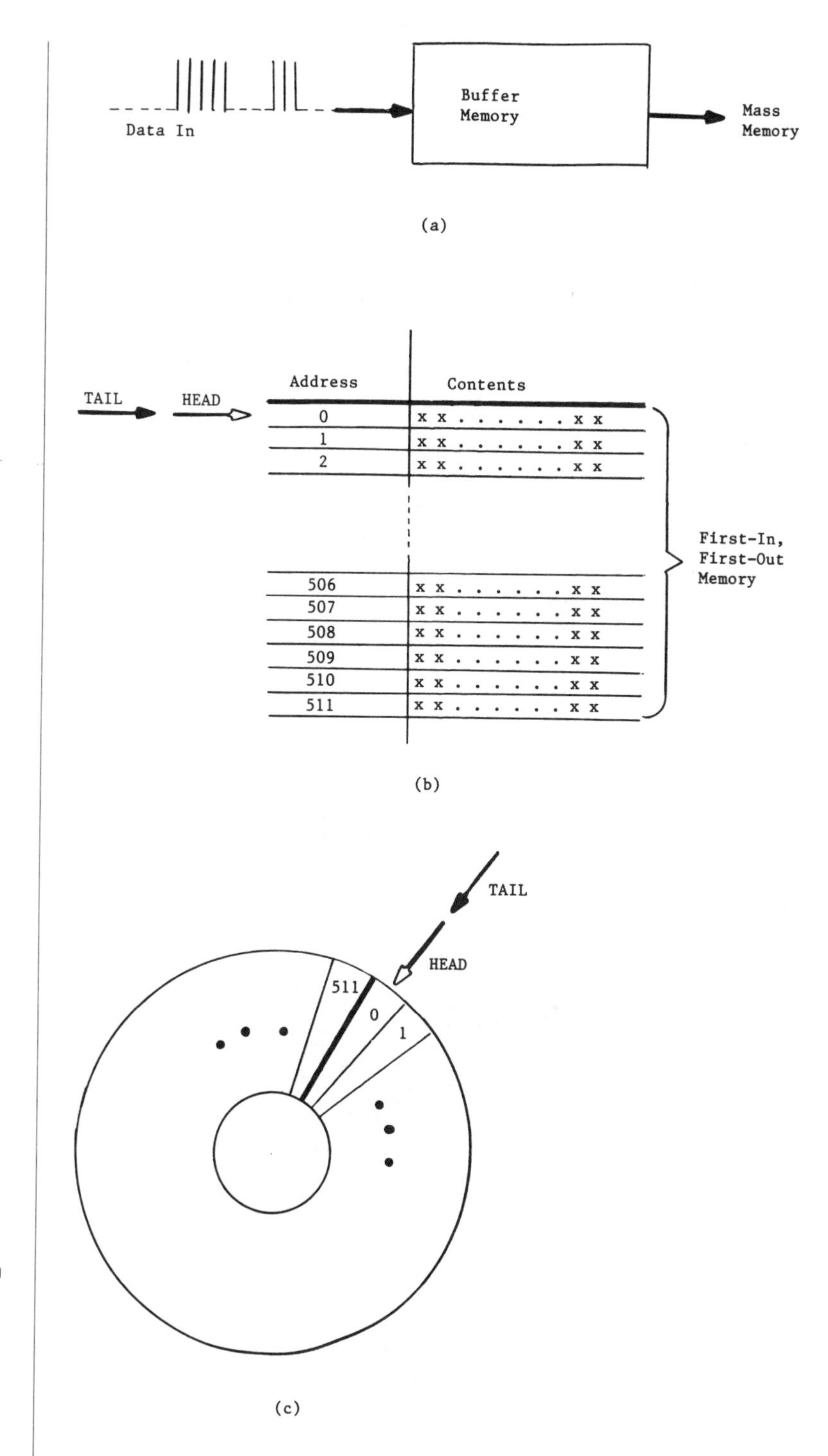

FIGURE 6.8 (a, b) First-in first-out memory. (c) Circular concept of a FIFO memory. (d) 74S225/A 16 × 5 FIFO (courtesy of AMD/MMI).

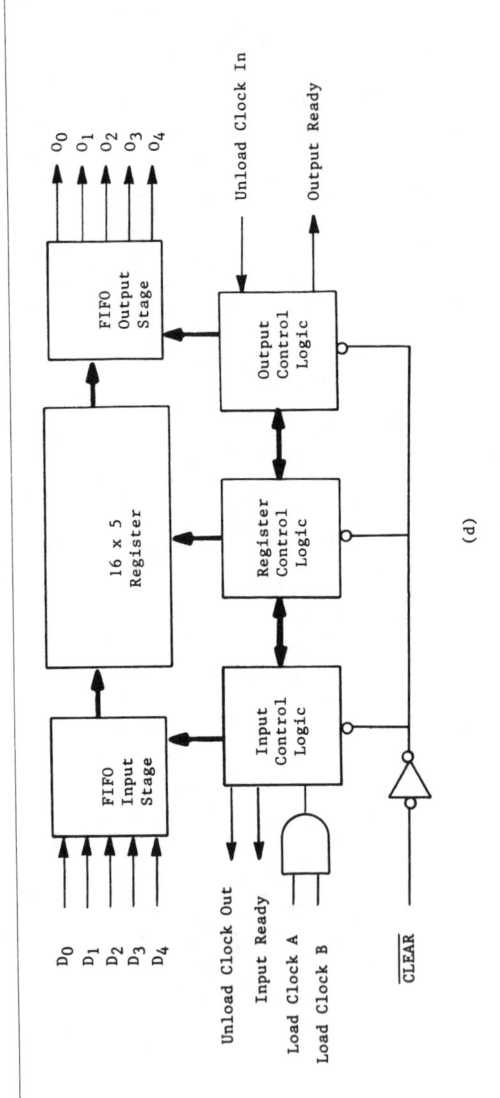

FIGURE 6.8 continued. (a, b) First-in first-out memory. (c) Circular concept of a FIFO memory. (d) 74S225/A 16 × 5 FIFO (courtesy of AMD/MMI).

queue is empty or full. If so designed, the memory would indeed be a first-in first-out memory. As one can see, the FIFO structure is more complicated than that of LIFO. Therefore, FIFO is available on a chip which includes the control logic and indicates whether the buffer is full or empty. A typical FIFO chip is described next.

Integrated Circuit FIFO

Figure 6.8(d) shows the functional block diagram of a typical IC FIFO. It is an Asynchronous First-In First-Out memory chip, labeled as FIFO 16 × 5, 74S225/A. As the figure shows, it has 16 memory words, five bits wide. Here, the low-true CLEAR pin is for reset, or empty the memory; $D_0 - D_4$ and $O_0 - O_4$ are the input and output data lines, respectively. "Input ready" indicates whether the chip is full; "Output ready" denotes if the chip is empty; "Unload clock in" is for the control signal READ; "Load clock A/B" is for the control signal WRITE; and "Unload clock out" indicates that the chip is moving the contents of a word to its next address if there is no unread data at that location. Note that the internal logic of the chip resolves the two inherent problems described in the previous paragraph, i.e., the problems of queue full and queue empty. However, for convenience the chip always moves the information written to it to the location which is available and nearest to the output "door". For example, assume the chip is initialized to "empty". A new word is written into the chip at location 0. Next, since the next address, address 1, is empty, the new word will be shifted to location 1 automatically by internal logic. Eventually, the new word will "fall through" to the last location (in this case, the 15th location) and be ready for unloading or read-out. If the rate of filling-in is equal or slower than that of shipping-out, the chip is like a bottomless bucket and would never fill up. Just as in the waiting room of a doctor's office, there are a number of chairs arranged in a line for the patients to sit while they wait to be seen by the doctor. Then, the first-come, first-served rule governs the order of patient examination. That is why in digital computer terminology the FIFO is known also as a circular data queue.

6.3.3. System Configuration — RAM

In this section we will describe another popular memory system which unfortunately has been named as RAM. Literally, RAM means a memory unit that can be randomly accessed. That is, the contents at one location in the memory unit can be accessed as quickly as the contents at any other location in the same unit. Unfortunately, the memory system which we are about to describe was historically misnamed with this generic name. In the computer field, it really means read/write memory with the random access property. In fact, ROM (Read-Only Memory), which will be described in detail later, can also be accessed randomly, thus logically also belongs to the RAM family. To be precise, we should name the memory described next as Read/Write RAM. Therefore, the reader should be aware that the term RAM as it used in daily conversation refers to memory that is readable/writable and can be accessed randomly. In the following, we describe two different classes of Read/Write RAM that are usually included in all digital computers and in most sophisticated digital system design.

a. Static RAM Memory System (SRAM)

The semiconductor memory systems are normally organized on chip levels. Although in principle the configuration of a semiconductor memory system is similar to the 3-D core memory system described in Chapter 6, Section 6.2.3, it is worthwhile to examine some typical samples. Figure 6.9 shows the logic symbol and block diagram of a 2125 static RAM (SRAM) chip. Each chip has 1024 words and each word is one bit wide. Note that there are ten address-bits, A_0, A_1, ..., A_9, which are subdivided into equivalent x-address and y-address and sometimes called *row-address* and *column-address,* respectively. Since the memory word is one bit wide, there is only one bit data-in (D_{in}) and one bit data-out (D_{out}). $\overline{CS}$ stands for chip-select, negative true. That is, when this line is LOW, the chip is selected. $\overline{WE}$ stands for write-enable, negative true. Therefore, when this line is HIGH, the chip is in read mode. This memory chip has a tri-state output, therefore the output is in high impedance state when the chip is not in read mode. $\overline{CS}$ is actually an address line when multiple chips are used in a system. For example, Figure 6.10(a) and (b) show the 3-D configuration of a 64 K by 8 or 64 K byte memory system using 512 No. 2125 static memory chips. In Figure 6.10(a), there are 64 (8 by 8) subgroups, each of which contains eight chips because the memory word is eight bits wide. Thus, each subgroup contains 1024 $\times$ 8 words. All the address pins of each chip with the same labels are tied together. That is, all A_0s, A_1s, ..., A_9s, and $\overline{CS}$ s, respectively, are tied together within each subgroup. However, for simplicity, in Figure 6.10(a) only the tie-points of $\overline{CS}$ are shown. Figure 6.10(b) shows the 16-bit MAR (Memory Address Register). A_{10}, ..., A_{15} are connected to the 3:8 decoders as shown, whose outputs yield the matrix. Each intersection of the matrix represents an AND gate whose output drives the corresponding $\overline{CS}$. A_0, ..., A_9, respectively, are tied to the corresponding chip-pin with the same label. In this way, each memory-word can be addressed by specifying the value of the MAR. For completeness, an 8-bit MDR is shown in Figure 6.10(b) as a reminder that the system has a data width of 8 bits and thus there are a total of 64 × 8, or 512, #2125 chips.

6.3.4. Dynamic RAM System (DRAM)

a. Principle of Operation

In comparison with the cell of a static RAM system, the dynamic memory cell (Figure 6.7(a) and (b)) is far simpler and its packing density is thus much higher. However, since the dynamic cell uses a capacitor, whose voltage may decay exponentially, as the information storage element, it requires recharging or refreshing. Therefore, the peripheral circuitry of a dynamic RAM chip is more complicated. For easier com-

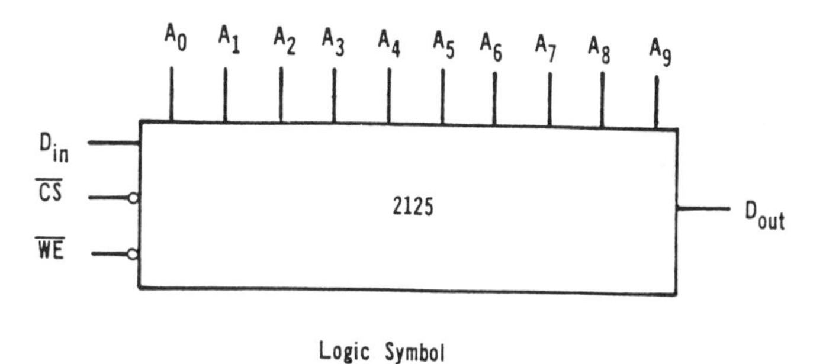

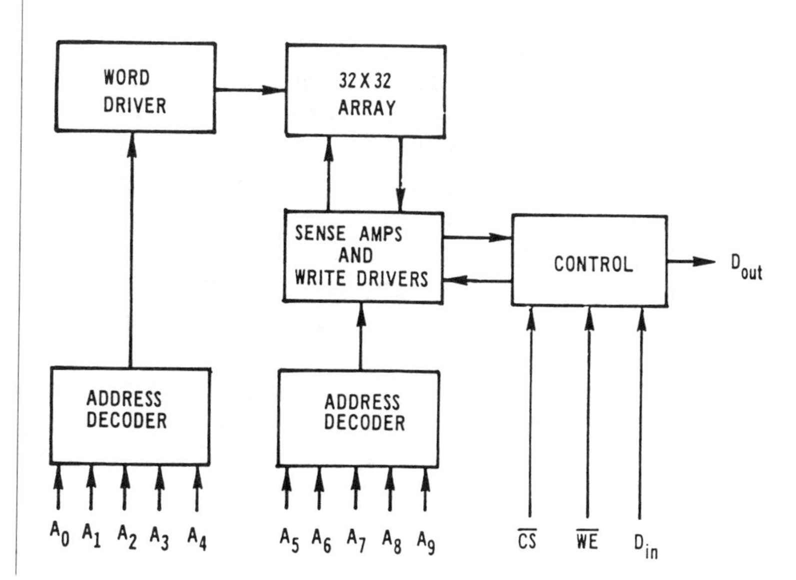

FIGURE 6.9 2125 memory chip.

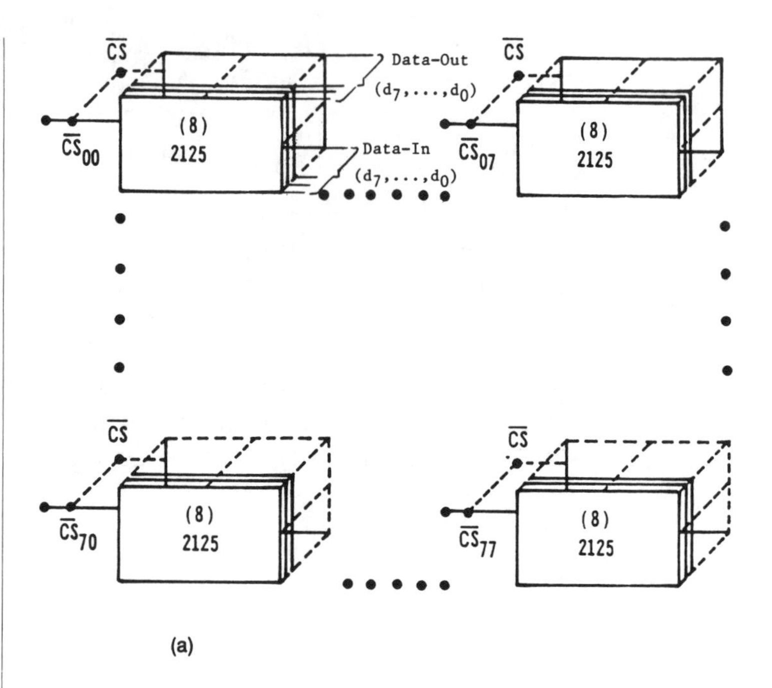

FIGURE 6.10 (a) A 64 K × 8 memory system block diagram. (b) Address map for a 64 K × 8 memory system.

prehension, let us examine a typical 3-transistor dynamic RAM chip shown in Figure 6.11(a). The circuitry of a 3-transistor cell is shown in Figure 6.7(a). This is an m-row by n-column, with word width of one bit, dynamic memory system. In the WRITE process, the x_i and y_j address lines select the ij^{th} cell, and the write command activates the write-enable line. Data provided by the drive circuit in the write column is then stored. In the READ process, the read column is precharged to logic [1]. Selected address lines and the read command activate the desired read-enable line, and the capacitor is charged or not by the precharged read column, depending on whether the data in it is [0] or [1], and thus the sense circuit detects the information. In the REFRESH process, the information in each row is read and fed back to the capacitor through the write column. Therefore, the dynamic memory system is refreshed at a specific interval, i.e., 2 ms, on a row-by-row basis. It is evident that sophisticated timing circuitry is required in a dynamic RAM system. Although straightforward, the peripheral circuitry for a 1-transistor cell, high-density dynamic RAM system would require even more complicated circuitry for the addressing and reading processes. In the 16 K × 1 and 64 K × 4 dynamic RAM chips, for example, row/column address information is time multiplexed and differential amplifiers are used for detection of the considerably weak data signal from storage capacitors.

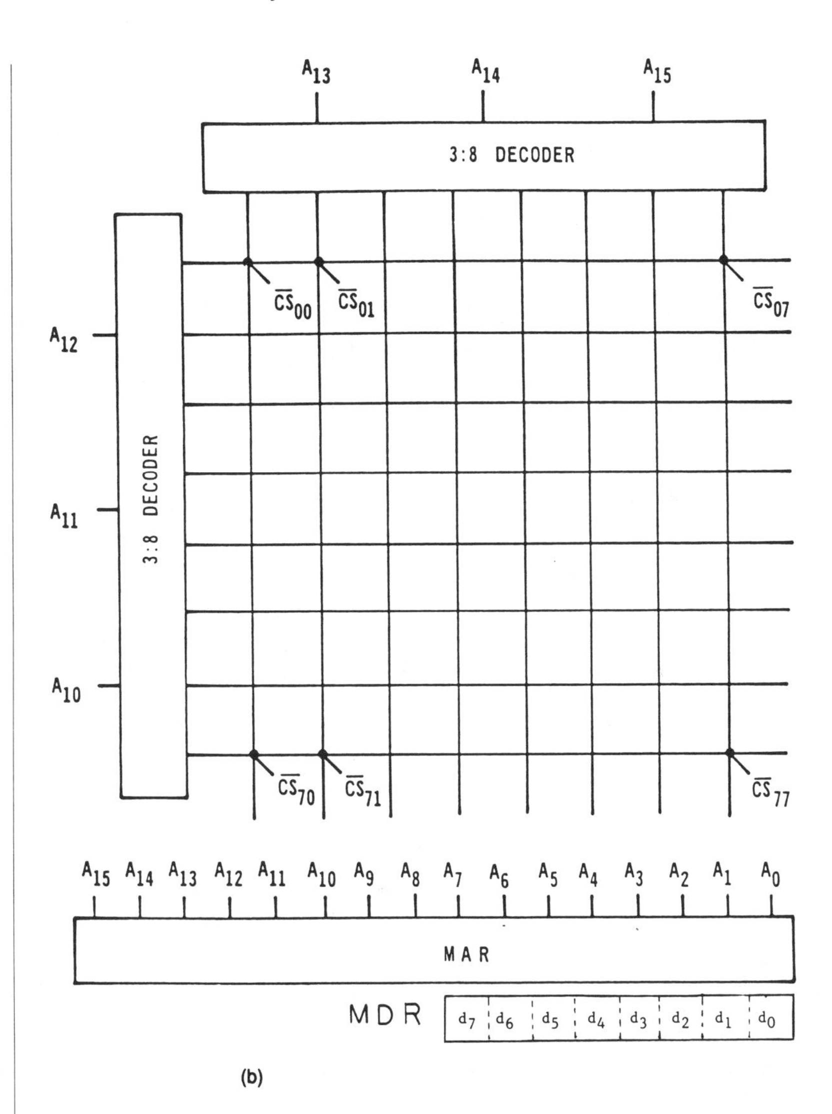

(b)

FIGURE 6.10 continued. (a) A 64 K × 8 memory system block diagram. (b) Address map for a 64 K × 8 memory system.

b. A Typical Dynamic RAM System

As described in Section 6.3.4.a, because of its simplicity in cell structure (especially the single transistor cell), a dynamic memory system (traditionally called DRAM) can be densely packed on a chip. At this point, one megabyte DRAM chips are commercially available; as time progresses, chips of even higher capacity are expected. Therefore, dynamic memory system designers are willing to cope with the inherent unreliability of capacitor-based memory cells by supporting memory chips with sophisticated auxiliary circuitry. Actually, in designing the support circuits for a dynamic memory system, one faces the following design hurdles:

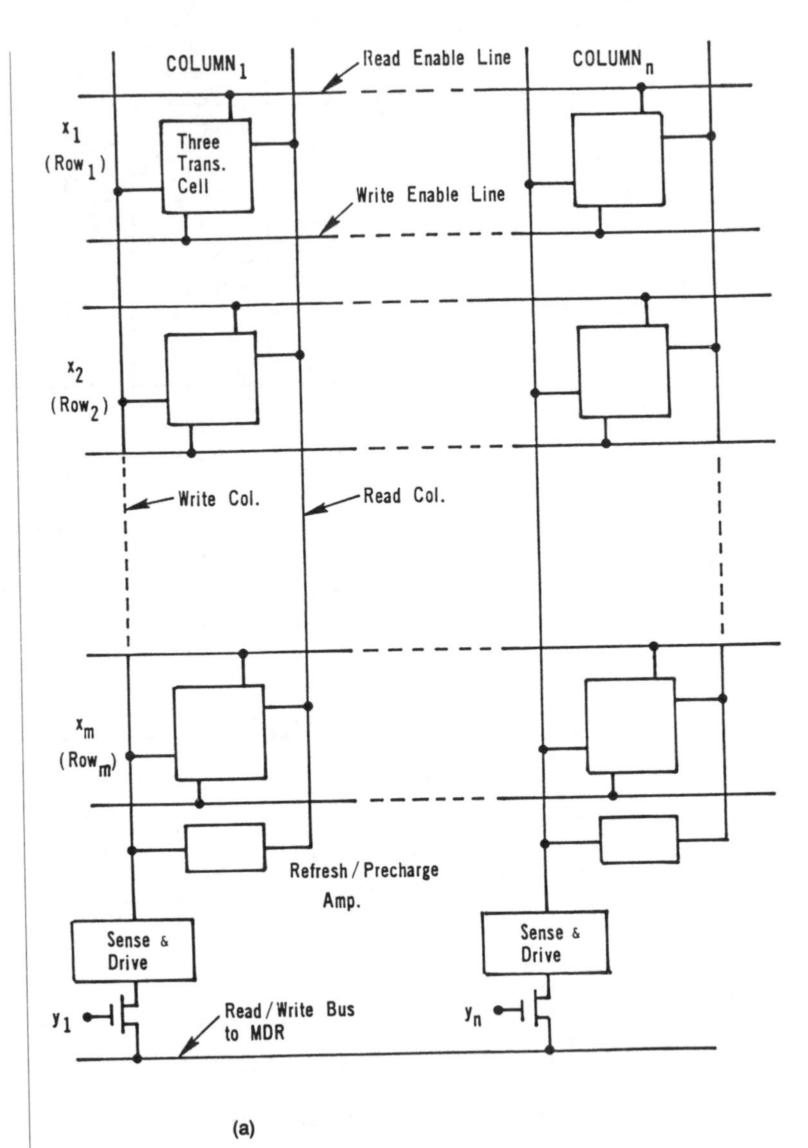

FIGURE 6.11 (a) 3-Transistor dynamic RAM chip. (b) A CPU interfaced to 673102 driving 2 M-bytes of dynamic memory (courtesy of AMD/MMI). (c) A typical dynamic memory system with error detection and correction (courtesy of AMD/MMI).

- Time multiplexing for enabling rows and columns in addressing mode
- Row addressing for the refresh process
- Refresh circuitry
- Timing for the refresh process
- Driving capacitive inputs of DRAM
- Interface between DRAM and its host system
- Error detection and correction for improving data integrity

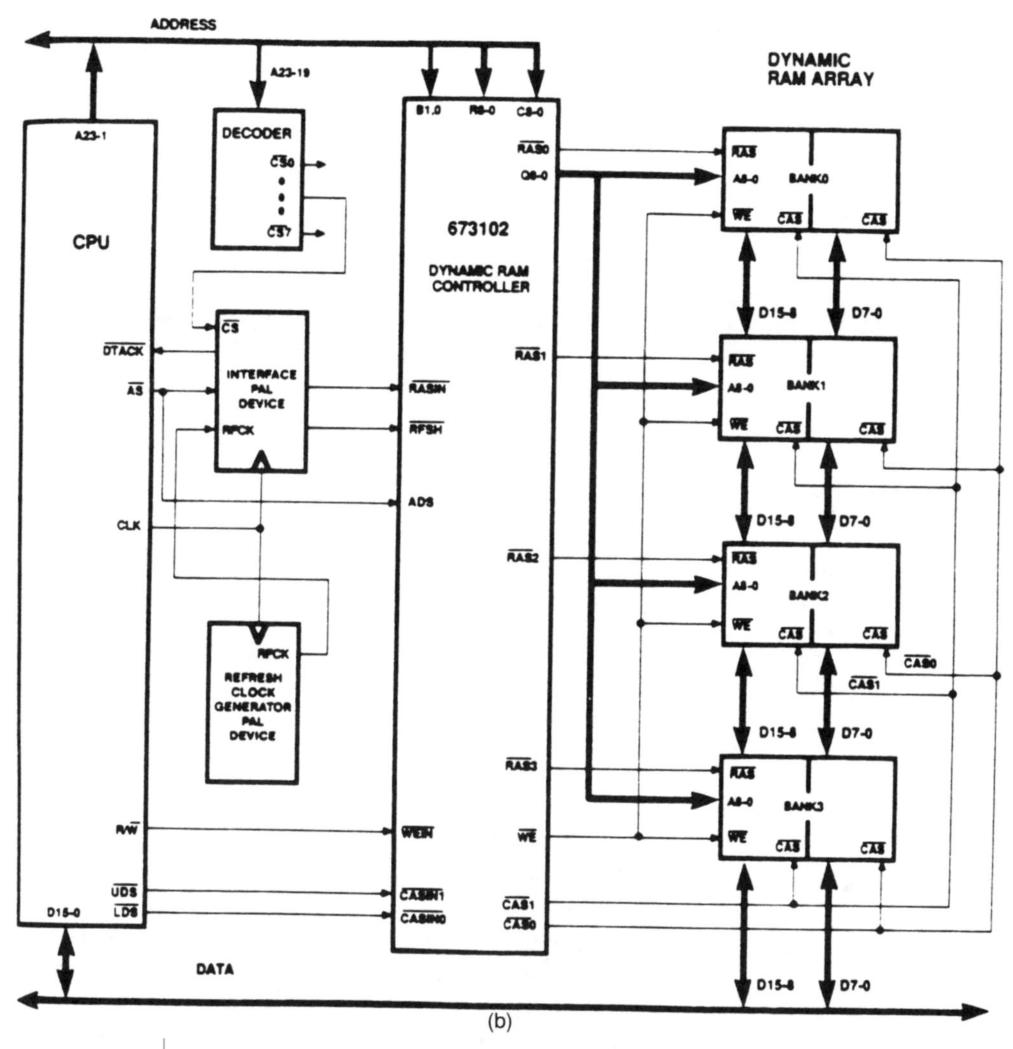

(b)

FIGURE 6.11 continued. (a) 3-Transistor dynamic RAM chip. (b) A CPU interfaced to 673102 driving 2 M-bytes of dynamic memory (courtesy of AMD/MMI). (c) A typical dynamic memory system with error detection and correction (courtesy of AMD/MMI).

Figure 6.11(b) shows a typical dynamic memory system with the dynamic RAM controller, 673102 by AMD/MMI. The system is composed of eight dynamic memory chips, each consisting of 512 × 512 bytes of memory; thus the entire system has 2 Mbytes of total memory. Since the CPU of the computer system to which the DRAM is attached has a data width of 16 bits, this DRAM is structured for 1 Mword; each word is 16 bits wide. Notice that among the CPU's 24 address lines which are capable of directly addressing 16 mega-locations, 20 are connected to the DRAM controller which is responsible for time multiplexing the rows and columns of the dynamic memory chips. In addition, between the controller and CPU one may see a small functional block that is dedicated to the refresh process. It is the controller's responsibility to resolve the potential time conflict between the refreshing process in-

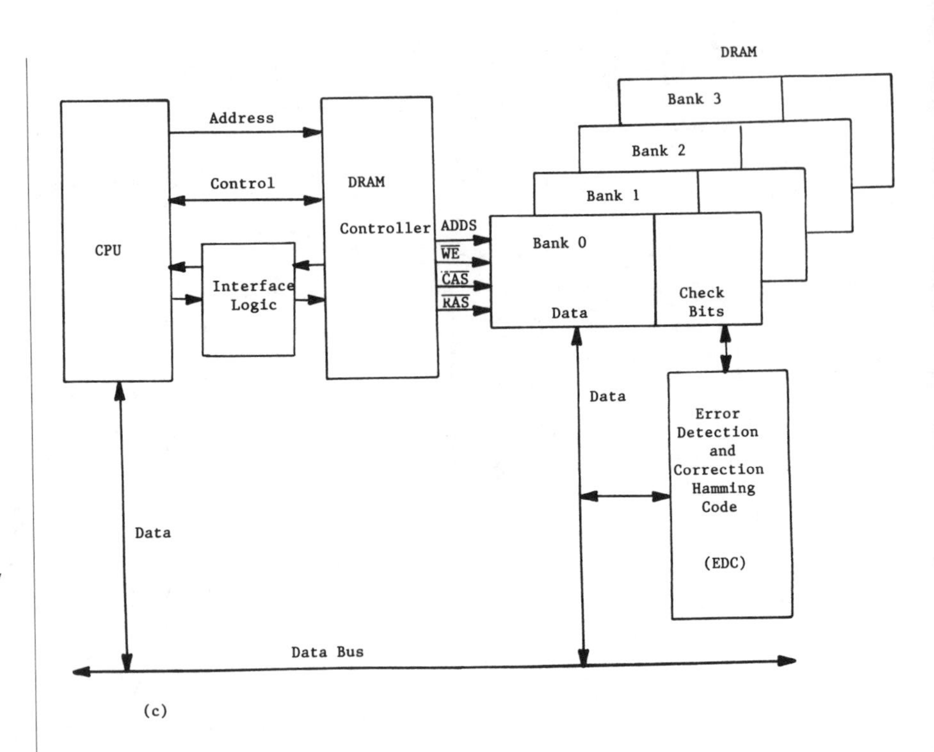

FIGURE 6.11 continued. (a) 3-Transistor dynamic RAM chip. (b) A CPU interfaced to 673102 driving 2 M-bytes of dynamic memory (courtesy of AMD/MMI). (c) A typical dynamic memory system with error detection and correction (courtesy of AMD/MMI).

side the DRAM and the CPU's memory access process outside it. Thus far, it appears that the DRAM controller has solved all the problems inherent in DRAM systems. In practice, one still occasionally suffers errors in accessing DRAM. Figure 6.11(c) shows a more sophisticated DRAM system which consists of a block of hardware dedicated to error detection and correction operations (see References 30 to 32). Specifically, the Hamming Code technique is commonly used for this purpose. Basically it uses the concept of providing a set of redundant bits in conjunction with the data bits to expand the data pattern space. In so doing, it can locate where the error bit is, and then just inverts the binary value of that bit to correct it. As a simple example, consider a block of data four bits wide. In its own data space, it can have only 16 data patterns. Now, let us append two redundant bits, called "check bits" to the block. As a result, we have 64 possible binary patterns to accommodate the 16 unique data patterns. Thus one could assign the 16 real data in this expanded space with maximum-mean distance between them. Accordingly, one could detect the location of the error bit and correct it since each bit has only two values, 1 or 0. That is, if 1 is wrong, the true value must be 0. For an 8-bit data, the Hamming Code requires five check bits to achieve single error detection/correction and double error detection only. Currently, there are a number of DRAM manufacturers who provide Hamming Code error de-

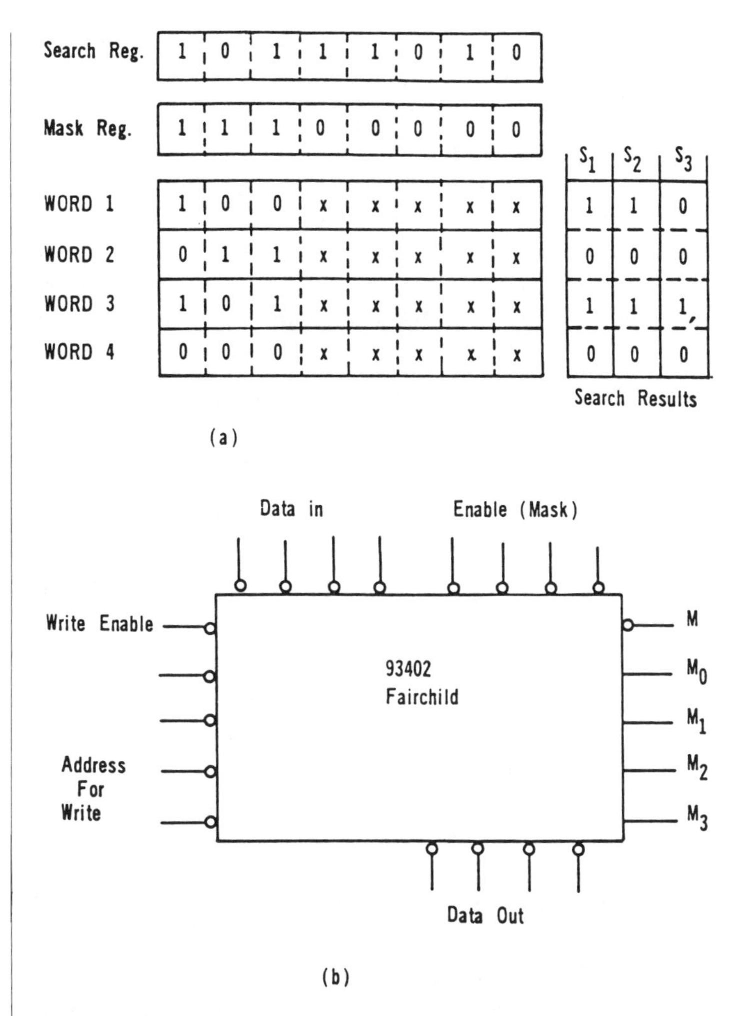

FIGURE 6.12 (a) Matching operation of CAM. (b) A typical 4 × 4 CAM chip (courtesy of Fairchild).

tection/correction (EDC) chips for DRAM designers. Referring to Figure 6.11(c), for example, while the real data on the data-bus is being stored in the DRAM, the EDC generates the proper check bit pattern for each datum. During the read process, the EDC detects and corrects errors, or just sets the error flags of the system. Notice that the width of the memory-word of this system would be wider; that is the price which must be paid for reliable data.

6.4. ASSOCIATIVE OR CONTENT-ADDRESSABLE MEMORY (CAM)

Figure 6.12(a) shows the operation of a CAM. Assume that the system contains 4 × 8 memory-words with the content as shown. Say we want to search for a match word whose three most significant bits are 101. Since we are interested in only the three most significant bits, the *mask register* is set as shown in Figure 6.12(a). The matching process is carried out on a bit-by-bit basis, and there are three search/match cycles. The search results are chronologically shown on the right in Figure 6.12(a). Note that once a bit of a certain word is not matched, it is eliminated by setting its result bit to zero. In Figure 6.12(a), we found that the third memory-word is selected and its entire content is then read to MDR. There are cases in which more than one memory-word may match the content of the search register after the masking operation. In this case, a multiple match resolver is needed. Figure 6.12(b) shows a typical 16 × 4 CAM chip by Fairchild Camera and Instrument Corp., whose outputs labeled as M_0, M_1, M_2, M_3, and $\bar{M}_0$ show the search results and the four enable inputs are for loading desired data into the mask register.

6.5. READ-ONLY MEMORY (ROM)

Read-Only Memory is a special purpose random-access memory. It takes a lot longer to write data into a memory cell than to read data from it. As mentioned earlier, ROM is technically a type of RAM because it is also randomly accessed. However, the term "RAM" is traditionally used now to refer to read/write random-access memory. ROM has two most desirable properties: it is nonvolatile and has nondestructive read-out. It is widely used for logic realization, table look-up, character generation, and other applications. As with the RAM described in the preceding sections, there are many types of ROM. Depending on the type of material used to make up the memory cell, one can classify them as capacitor ROM, core ROM, and semiconductor ROM. Figure 6.13(a), (b), and (c), respectively, show the conceptual diagrams of these three different types. Due to recent rapid progress in solid state technology, semiconductor ROM now dominates all fields of ROM applications. Generally speaking, semiconductor ROM is inexpensive and has faster access speeds than the other types. Depending on the method of its writing process, semiconductor ROM can be further classified as follows:

6.5.1. Nonprogrammable ROM

This type of ROM is permanently programmed during the fabrication process by using a custom-designed mask through a ROM manufacturer. The mask is designed according to the system designer's specifications. It is evident that this kind of ROM is suitable for applications

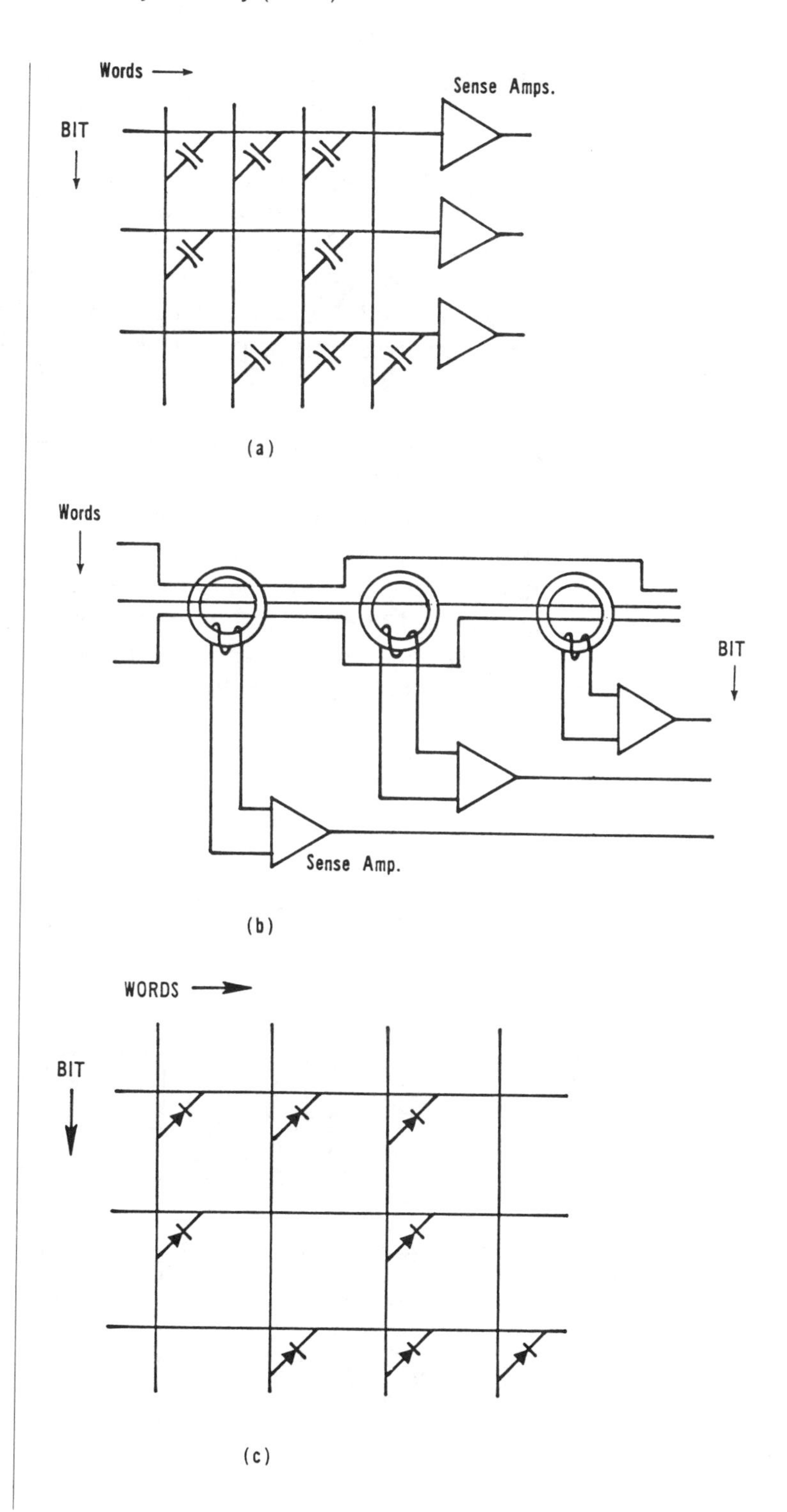

FIGURE 6.13. (a) Capacitor ROM. (b) Core ROM. (c) Semiconductor ROM.

where high volume mass production is expected. Once the mask is made, it is difficult and expensive to change the design.

6.5.2. Field-Programmable ROM (FPROM)

Figure 6.14 shows a field-programmable ROM. In this device, data can be permanently stored by a "burn-in" process. That is, to write logic [0] to a cell, an excess current can be applied to the cell so that the fuse shown in the emitter circuit can be "blown" out and logic [0] is stored. Otherwise a logic [1] remains. Clearly this type of ROM is not reprogrammable.

6.5.3. Ultraviolet Rays Erasable Programmable ROM (EPROM)

For this device, data can be written into it and mass-erased by ultraviolet rays. It is therefore reprogrammable and most desirable for applications of low quantity production. Devices such as the 27010 have a capacity of 128 K × 8 bits per chip, require only +5 V power supply, and have a maximum access time of 200 ns; the programming process requires 15 s. They require 10 to 30 min to mass-erase data by ultravio-

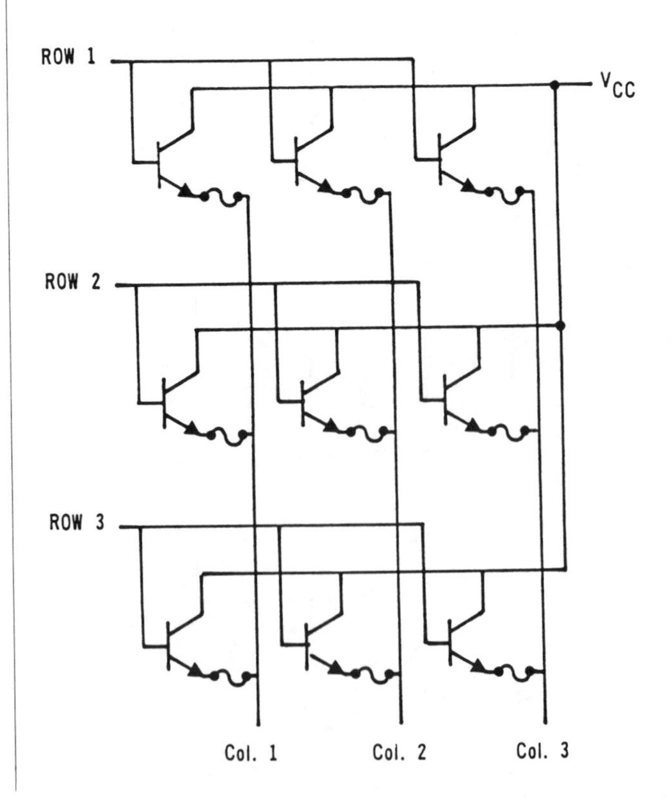

FIGURE 6.14 Field programmable ROM.

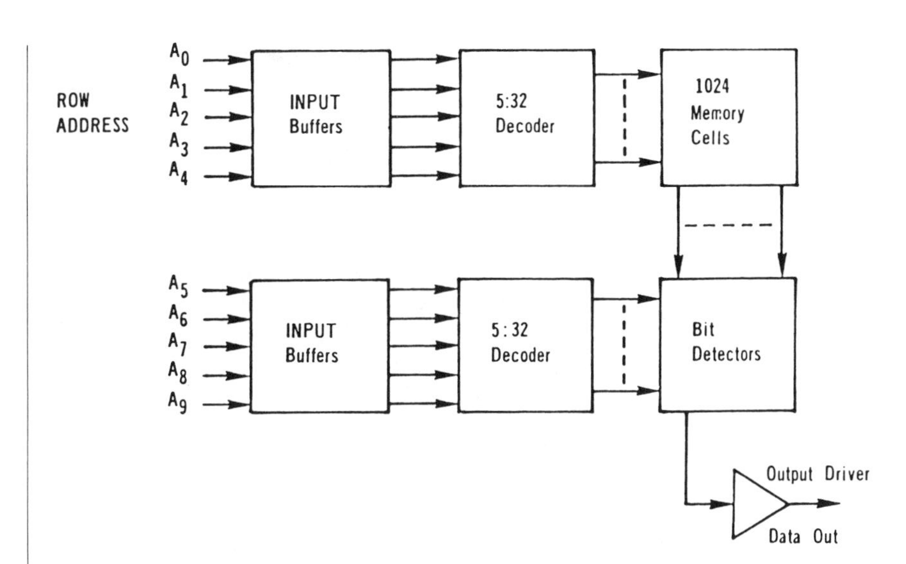

FIGURE 6.15 1024 × 1 ROM system block diagram.

let rays. This type of ROM is ideal for applications in the development and design of systems where short- or medium-term read-only memories are essential. For long-term memory applications, the mask or fusible types of ROM would be more desirable.

6.5.4. Electrically Alterable ROM (E^2ROM)

Note that the types of ROM described so far are "off-line" in the erase/write process and "on-line" in the read process. However, it is most desirable for ROM to be electrically alterable or erasible. Since 1978, this type of ROM has been commercially available from Nippon Electric Corporation (NEC) and General Instrument Corporation, among others. Typically, the 2864A by Intel has a capacity of 8 K × 8. It can be written as if it were a static RAM. During the write process, the chip is isolated from the data-bus at high impedance mode. Its access time is 180 ns.

6.5.5. Read-Only Memory System

Although ROM is normally used for the read process, its system organization is similar to RAM. Again, there are three groups of input/output lines in the system, namely, control, address and data. The address lines are usually subdivided into ROW and COLUMN address lines. In semiconductor ROM, the column address lines, after decoding, are used to control the sense or output circuitry. Figure 6.15 illustrates a typical system organization of a 1024 × 1 ROM.

6.6. MAGNETIC BUBBLE MEMORIES

The concept of magnetic bubble memories has been around since the early 1970s. By the late 1970s, several major semiconductor firms were providing samples of 256 K and one megabyte magnetic bubble memory chips to system designers. Due to its physical properties, a bubble memory system requires special supporting circuitry for which LSI technology has been employed. Therefore, at present bubble memories are sold as a system that includes supporting circuitry instead of just on a single chip. From the system designer's point of view, one may consider a bubble memory system to contain a group of magnetic bubble shift registers complemented with LSI supporting circuitry to carry out such functions as timing, addressing, and read/write data commands. As a typical example, Figure 6.16 shows a block diagram of a magnetic bubble memory system.

6.7. CHARGE-COUPLE DEVICE (CCD)

The charge-couple device is another type of semiconductor memory. The device is basically used as a shift register with an extremely high number of bits. It can be used to fill the gap between mass memory devices such as magnetic disk or tape, and RAM. Figure 6.17 shows its principle of operation. Figure 6.17(a) shows a segment of the device. Assume that the voltages at the electrodes are v_1, v_2, v_3, and v_4, and that $v_3 < v_4 = v_2 = v_1 = \theta < 0$, where θ is the threshold voltage. When the electrode voltage is lower than the threshold of the device, it yields the depletion region. The depth of the region is a function of the applied voltage. When the voltage is sufficiently lower than the threshold voltage, a *potential well* is generated which can re-

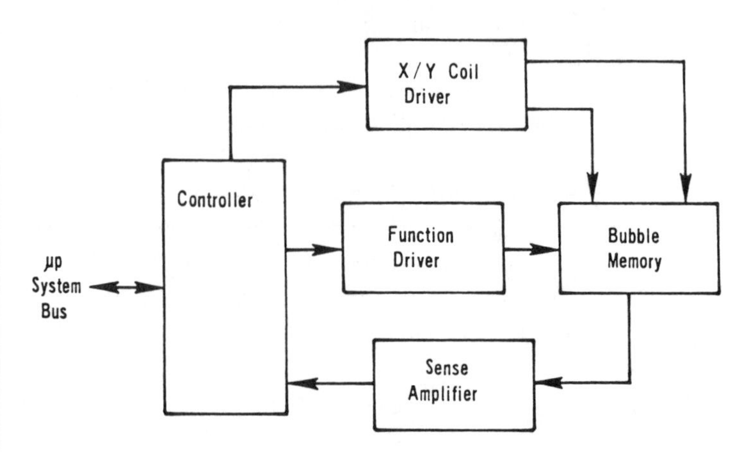

FIGURE 6.16 Bubble memory system block diagram.

ceive or store the positively charged holes. To illustrate the data shifting action, Figure 6.17(b) to (e) shows the data shifting process of a three-phase clocked CCD. Figure 6.17(b) depicts the time waveforms of the voltages at the electrodes for storage and transfer periods. Figure 6.17(c) to (e) shows the shifting of the digital data. In Figure 6.17(c), electrodes tied to ϕ_1 generate the corresponding potential wells. Two of the four wells are holding positive charges, thus the device has the digital information as shown. Entering the shifting mode we have $\phi_2 < \phi_1 < \phi_3$ as shown in Figure 6.17(b). The corresponding depletion diagram is shown in Figure 6.17(d). Here, the potential wells of the electrodes tied to ϕ_2 are deeper than their left-hand neighbors, and as a result the positive charges are transferred to the right as shown. Figure 6.17(e) shows that the status of the device returns to the storage mode after the transfer, and that the corresponding digital data have been shifted one position to the right as compared to Figure 6.17(c). One can see that the fabrication process for CCD is simple and its packing density can be very high. The major applications of these devices are digital memory, shift registers, image processing, and analog and digital signal processing.

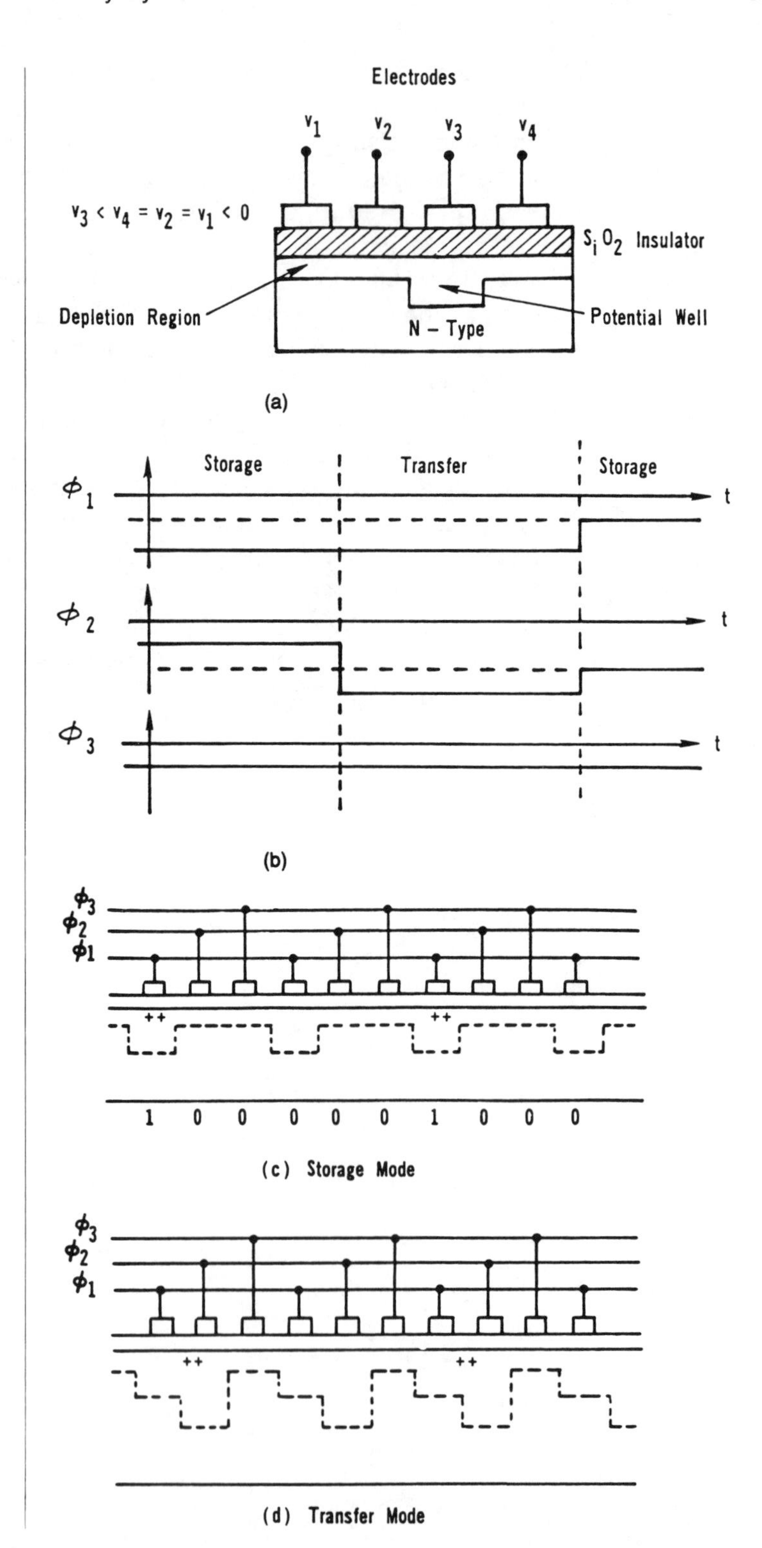

FIGURE 6.17 (a) CCD element. (b) Voltage waveforms of the electrodes. (c, d, e) CCD data shifting operation.

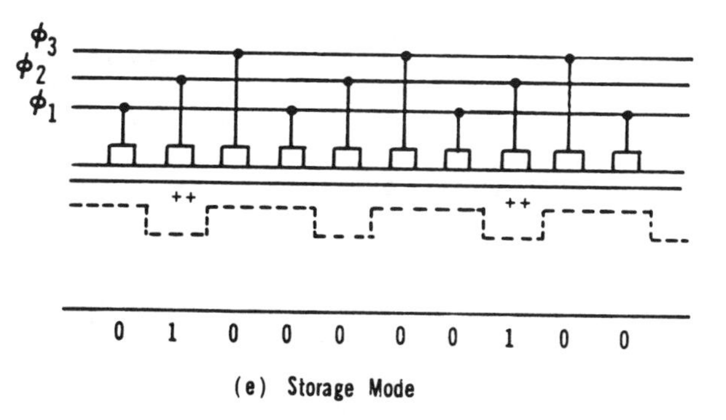

FIGURE 6.17 continued. (a) CCD element. (b) Voltage waveforms of the electrodes. (c, d, e) CCD data shifting operation.

REFERENCES

1. Gilligan, T. J. and Persons, P. B., High-speed ferrite $2\frac{1}{2}$-D Memory, *Fall Joint Computer Conference,* 1965.
2. Gilligan, T. J. and Persons, P. B., Comparison of core memory system organization, *Computer Design,* May 1966.
3. Vadsz, L. L., Chua, H. T., and Grove, A. S., Semiconductor random access memories, *IEEE Spectrum,* 8(5), 40—48, May 1971.
4. Special Issue, Semiconductor memories and optoelectronics, *IEEE J. Solid-State Circuits,* Vol. SC-5, No. 5, October 1970.
5. Moore, G. E., Semiconductor RAMs, *Computer,* 4(2), 6—17, March/April 1971.
6. Rudolph, J. A., et al., With associative memory, speed limits no barrier, *Electronics,* 43(13), 96—101, June 22, 1970.
7. Bartlett, J., et al., Associative memory chips: fast, versatile and here, *Electronics,* 43(17), 96—100, August 17, 1970.
8. Riley, W. B., Ed., *Electronic Computer Memory Technology,* McGraw-Hill, New York, 1971.
9. Hodges, D. A., Ed., *Semiconductor Memories,* IEEE Press, 1972.
10. Altman, L., Charge-coupled devices move in on memories and analog signal processing, *Electronics,* 47(16), 91—101, August 8, 1974.
11. Frankenburg, R. J. and Cross, D., Designer's guide to semiconductor memories, *EDN* 1975 Series: Part 1 (20(14), 22—29, August 5), Part 2 (20(15), 58—65, August 20), Part 3 (20(16), 68—74, September 5), Part 4 (20(17), 62—67, September 20), Part 5 (20(18), 31—35, October 5), Part 6 (20(19), 44—50, October 20), Part 7 (20(20), 59—70, November 5), Part 8 (20(21), 127—137, November 20).
12. Bobeck, A. H., Bonyhard, P. I., and Geusic, J. E., Magnetic bubbles — an emerging new memory technology, *Proc. IEEE,* 63(8), 1176—1195, August 1975.
13. Cohen, M. S. and Chang, H., The frontiers of magnetic bubble technology, *Proc. IEEE,* 63(8), 1196—1206, August 1975.
14. Coe, J. E. and Oldham, W. G., Enter the 16,384-bit RAM, *Electronics,* 49(4), 116—121, February 19, 1976.
15. Sander, W. B., Shepherd, W. H., and Schinelle, R. D., Dynamic I^2L random-access memory competes with MOS designs, *Electronics,* 49(17), 99—102, August 19, 1976.
16. Greene, R., Perlegos, G., Salsbury, P. J., and Morgan, W. L., The biggest erasable PROM yet puts 16,384 bits on a chip, *Electronics,* 50(5), 108—111, March 3, 1977.
17. Proebsting, R., Dynamic MOS RAMs: an economic solution for many system designs, *EDN,* 22(12), 61—66, June 20, 1977.

18. Mohan Rao, G. R. and Hewkin, J., 64-K dynamic RAM needs only one 5-volt supply to outstrip 16-K parts, *Electronics,* 51(20), 109—116, September 28, 1978.
19. Waller, L., Has bubble memory's day arrived?, *Electronics,* 52(3), 80—81, March 29, 1979.
20. Bryson, D., Clover, D., and Lee, D., Megabit bubble memory chip gets support from LSI family, *Electronics,* 52(9), 105—116, April 26, 1979.
21. Capece, R. P., The race heats up in fast static RAMs, *Electronics,* 52(9), 125—135, April 26, 1979.
22. Bisset, S., Bristow, S., and Chen, T. T., Bubble memories demand unique test methods, *Electronics,* 52(10), 117—122, May 10, 1979.
23. Wallace, C., Electrically erasable memory behaves like a fast, nonvolatile RAM, *Electronics*, 52(10), 128—131, May 10, 1979.
24. Welch, T. A., Analysis of memory hierarchies for sequential data access, *Computer,* 12(5), 19—26, May 1979.
25. Halsema, A. I., Bubble memories — a short tutorial, *Byte,* 166—167, June 1979.
26. Juliessen, J. E., Where bubble memory will find a niche, *Mini-Micro Systems,* July 1979.
27. George, P. K. and Reyling, G., Jr., Bubble memories come to the boil, *Electronics,* 52(16), 99—114, August 2, 1979.
28. Lee, D. and Spiegel, P., Ease bubble system design with a few basic guidelines, *EDN,* August 5, 1979.
29. Greene, R. and Louie, F., E-PROM doubles bit density without adding a pin, *Electronics,* 52(17), 126—132, August 16, 1979.
30. McCluskey, E. J., *Logic Design Principles*, Section 1.5, Prentice-Hall, Englewood Cliffs, NJ, 1986, 20—25.
31. AMD/MMI, *System Design Handbook,* 1985, chap. 4.
32. Hamming, R. W., Error detecting and error correcting codes, *Bell System Tech. J.,* 29, 147—160, 1950.
33. Intel, Inc., Memory Components Handbook, 1986.
34. Intel, Inc., Memory Components Handbook, 1988.
35. Intel, Inc., Memory Components Handbook – Supplement, 1988.
36. Texas Instruments, Inc., *MOS Memory Data Book,* 1986.

7 ARITHMETIC LOGIC UNIT (ALU)

Like the memory unit, the arithmetic logic unit (ALU) is another important subsystem in digital system design. In many applications such as numerical control, process control, instrumentation, etc., one would like to carry out binary arithmetic or logic operations within the system. Thanks to large-scale integrated circuit (LSI) technology, different types of ALU chips are now available on the market, from which a proper type can be selected for a specific application. Generally, most of the ALUs available, however, are "ADDER"–based; subtraction and multiplication operations would need additional hardware to implement according to some algorithms. In this chapter, we shall review some basic element of ALU design as well as some of the more popular algorithms. Examples are given and some typical LSI chips described.

7.1. BINARY ADDITION

7.1.1. Basic Element

a. Half-Adder

A half-adder can be defined by the following truth table.

A (Augend)	B (Addend)	S (Sum)	C (Carry)
0	0	0	0
0	1	1	0
1	0	1	0
1	1	0	1

and its switching functions derived from the truth table for Sum and Carry are

$$S = \overline{A}B + B\overline{A} = A \oplus B$$
$$C = AB$$

Figure 7.1(a) shows the logic diagram and symbol of a half-adder.

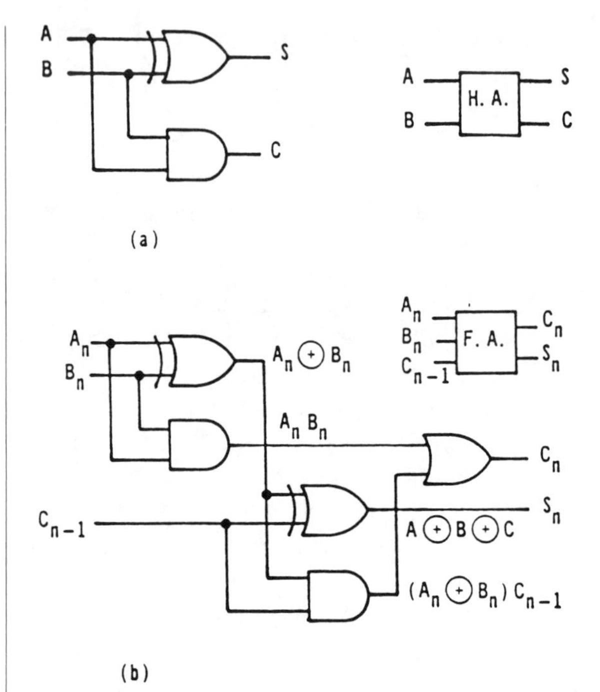

FIGURE 7.1 (a) Logic diagram and symbol of a half-adder. (b) Logic diagram and symbol of a full-adder.

b. Full-Adder

A full-adder can be defined by the following truth table:

A_n	B_n	C_{n-1}	S_n	C_n
0	0	0	0	0
0	0	1	1	0
0	1	0	1	0
0	1	1	0	1
1	0	0	1	0
1	0	1	0	1
1	1	0	0	1
1	1	1	1	1

and its switching functions derived from the truth table are

$$C_n = A'_nB_nC_{n-1} + A_nB'_nC_{n-1} + A_nB_nC'_{n-1} + A_nB_nC_{n-1}$$

$$= A_nB_n + (A_n \oplus B_n)\, C_{n-1}$$

$$S_n = A'_nB'_nC_{n-1} + A'_nB_nC'_{n-1} + A_nB'_nC'_{n-1} + A_nB_nC_{n-1}$$
$$= (A'_nB'_n + A_nB_n)\, C_{n-1} + (A'_nB_n + A_nB'_n)\, C'_{n-1}$$
$$= (A_n \oplus B_n)'\, C_{n-1} + (A_n \oplus B_n)\, C'_{n-1} = A_n \oplus B_n \oplus C_{n-1}$$

where the subscripts n and n–1, respectively, denote the nth and $(n-1)^{th}$ binary bit of a binary number. Figure 7.1(b) shows the logic diagram and symbol of a full-adder.

7.1.2. Multiple-Bit Addition

a. Carry-Ripple-Through Parallel-Adder

The heart of an ALU is a multiple-bit adder which is a collection of full-adders as shown in Figure 7.1(b). Figure 7.2(a) shows the diagram of a "ripple-through" parallel-adder. Note that the output of the nth or most significant bit full-adder unit shown is a function of the $(n-1)^{th}$, $(n-2)^{th}$, ..., 1st full-adder units. Therefore, the final valid output values would not settle after $n\tau$ seconds in the worst case, where τ is the propagation delay of each full-adder unit. This adder is thus called the carry-ripple-through parallel-adder. Due to its simple organization, this type of adder is used in many systems. For systems where speed of operation is essential, however, the $n\tau$ delay may not be desirable. Therefore, another type of adder known as the carry-look-ahead parallel-adder is recommended.

b. Carry-Look-Ahead Parallel-Adder

To avoid the time delay caused by the carry-ripple-through, one can design a logic network where the nth carry bit is independent of the carry bit of any other full-adder unit. The following mathematical derivation will clarify the concept. From Section 7.1.1(b) we have

$$C_n = A_nB_n + (A_n \oplus B_n)\, C_{n-1}$$

Let

$$G_n \triangleq A_nB_n$$
$$P_n \triangleq A_n \oplus B_n$$

Then

$$\begin{aligned} C_n &= G_n + P_n\, C_{n-1} \\ &= G_n + P_n(G_{n-1} + P_{n-1}\, C_{n-2}) \\ &= G_n + P_n\, G_{n-1} + P_n\, P_{n-1}\, C_{n-2} \\ &= G_n + P_n\, G_{n-1} + P_n\, P_{n-1}(G_{n-2} + P_{n-2}\, C_{n-3}) \\ &\ldots\ldots\ldots\ldots\ldots \\ &= G_n + P_n\, G_{n-1} + \ldots + P_n\, P_{n-1} \ldots P_2G_1 + P_n\, P_{n-1} \ldots P_1C_0 \end{aligned}$$

The last line of the derivation reveals that C_n can be so designed that it

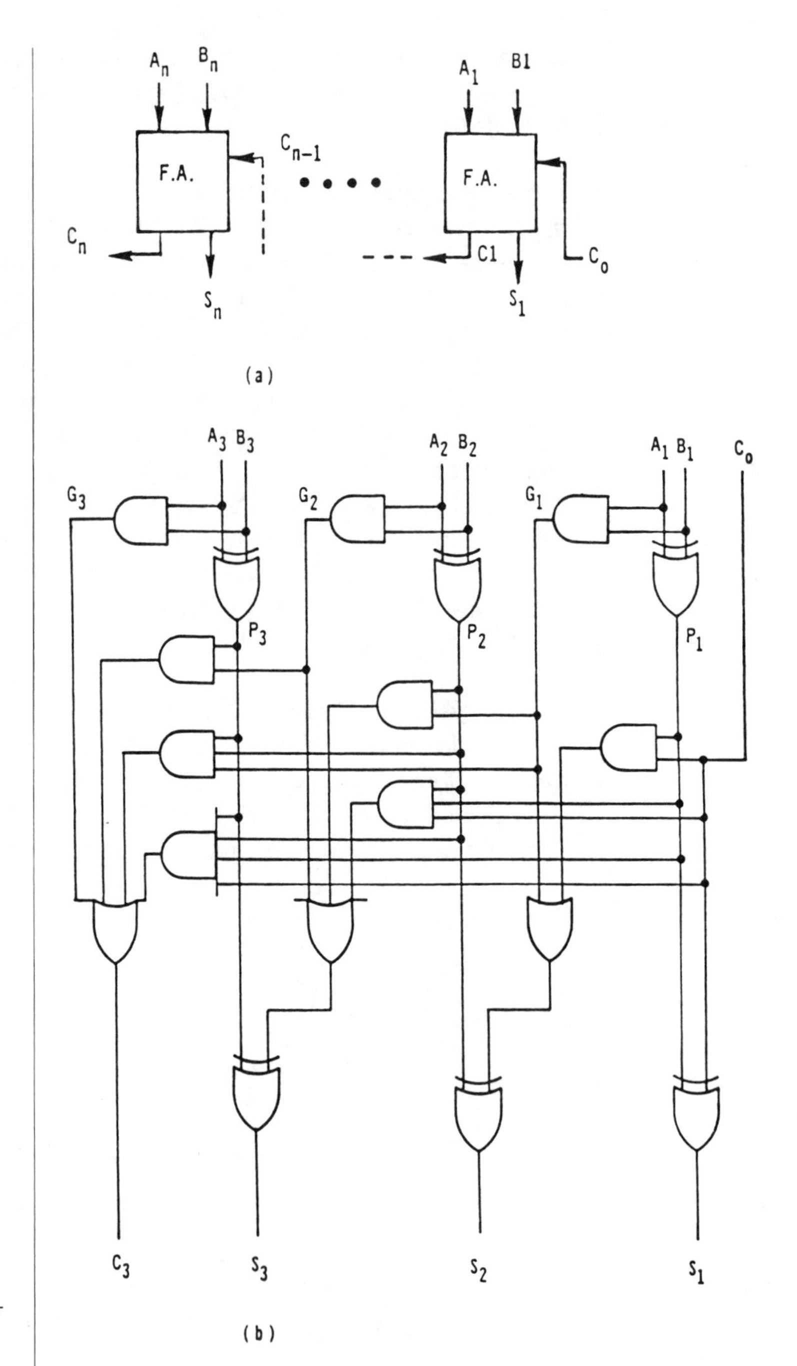

FIGURE 7.2 (a) Ripple through multibit parallel-adder. (b) 3-Bit carry-look-ahead parallel-adder.

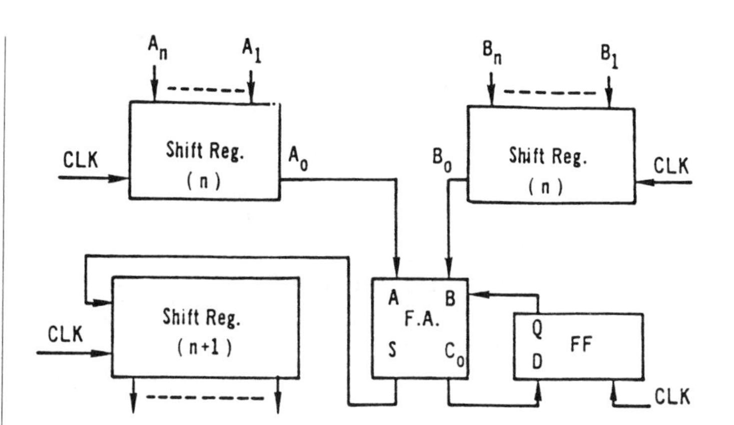

FIGURE 7.3 n-Bit binary serial-adder.

is a function of G_n, P_n, and C_0, which are available at the first level of addition. Figure 7.2(b) shows the logic diagram of a 3-bit carry-look-ahead parallel-adder, which is a direct implementation of the switching function derived. Here we have n = 3.

c. n-Bit Serial-Adder

In many applications, a serial-adder is more appropriate than a parallel-adder if speed of operation is not a major concern. Figure 7.3 shows a logic diagram of a typical serial-adder. The augend and addent are first loaded into a parallel-in/serial-out shift register, respectively. Then the clock pulse will right-shift the registers, and their serially shifted outputs are fed to the full-adder as shown. The sum of the two numbers is then stored in the (n + 1)-bit serial-in/parallel-out register. The D flip-flop is used to delay the carry bit for one clock period and merged into the carry-in input of the full-adder.

d. Accumulator

An accumulator is a device which is able to perform multiple-number additions one at a time and keeps a cumulative total until at the end of the process when it holds the total summation. Obviously it normally contains adders and registers. Figure 7.4 shows a typical block diagram of an accumulator. The D flip-flops are used as a storage register for the sum of the augend or the incoming data A_n, ..., A_1 and the old result; the number in the storage register is then fed back to become the addend of the next addition.

7.2. SUBTRACTION BY COMPLEMENTARY ARITHMETIC

Although one can easily design a subtractor by following the same logical steps for designing an adder, it is normally more convenient to do subtraction using the same hardware used for addition. In other words, an ALU normally contains the basic full-adder elements for add-operation, and it would be convenient if the same hardware elements could also be used to implement subtraction. To do this, a complementary arithmetic is used.

7.2.1. Background

a. *Signed-Binary-Coded Integer*

Conventionally, an n-bit binary coded integer

$$x \triangleq A_{n-1}, A_{n-2}, \ldots, A_0$$

is interpreted as a positive integer with a value of

$$x = A_{n-1}2^{n-1} + A_{n-2}2^{n-2} + \ldots + A_0 2^0$$

For a binary-coded signed-magnitude integer representation, however, the most significant bit, A_{n-1}, is reserved as a sign-bit. That is, when $A_{n-1} = 1$, the number is negative, otherwise it is positive. For example, in signed-magnitude representation,

$$1111 = -7$$

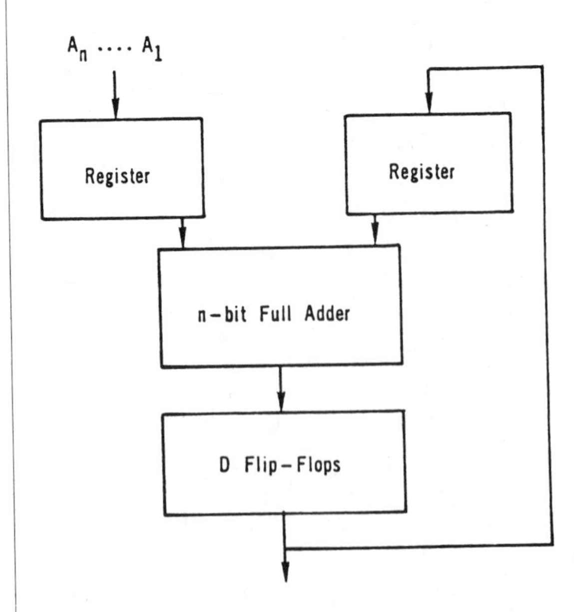

FIGURE 7.4 Block diagram of an accumulator.

and

$$0111 = +7$$

As another example,

$$0000 = 0$$

and

$$1000 = -0$$

Therefore, the range of a 4-bit binary integer is ±7 in which there exists positive and negative representations of zero although ±0 has no mathematical meaning. The binary representation of a number is obviously the natural one in a digital system. It is, however, very inconvenient for human beings. Therefore, the digital designer or digital computer user often uses different ways of representing a binary integer. Octal or hexadecimal are the most common number bases used. For conversion from binary to octal representation, the binary bits are placed in groups of three, while for hexadecimal notation groups of four bits are used. To illustrate these forms of numerical notation, consider an unsigned 8-bit binary integer: $(10100110)_2$. In octal representation this will be $(10,100,110)_2 \rightarrow (246)_8$, and in hexadecimal, $(1010,0110)_2 \rightarrow (A6)_{16}$. The subscripts indicate the radix or base of the number representation. The following table shows the conversion of the three systems.

Binary				Octal		Hexadecimal
0	0	0	0	0	0	0
0	0	0	1	0	1	1
0	0	1	0	0	2	2
0	0	1	1	0	3	3
0	1	0	0	0	4	4
0	1	0	1	0	5	5
0	1	1	0	0	6	6
0	1	1	1	0	7	7
1	0	0	0	1	0	8
1	0	0	1	1	1	9
1	0	1	0	1	2	A
1	0	1	1	1	3	B
1	1	0	0	1	4	C
1	1	0	1	1	5	D
1	1	1	0	1	6	E
1	1	1	1	1	7	F

b. Complement Representations

Two's Complement Representation

The two's complement representation of an n-bit signed-magnitude binary integer is symbolically and mathematically defined as follows. Let

$$Y_2 \triangleq \text{n-bit signed-magnitude binary integer}$$

$$= y_{n-1}, y_{n-2}, \ldots, y_i, \ldots, y_0$$

with $y_i = 1, 0$; $i = 0, 1, \ldots, n-1$. y_{n-1} = sign-bit; $1 \triangleq$ negative, $0 \triangleq$ positive, where the subscript 2 of the uppercase Y indicates base 2 (binary notation). Then, symbolically,

$$Y_2(\overline{2}) \triangleq -y_{n-1}2^{n-1} + y_{n-2}2^{n-2} + \ldots + y_0 2^0 \tag{7.1}$$

where $(\overline{2}) \triangleq$ two's complement.

Other Complement Representations

Similarly, we can define other complement representations symbolically and mathematically as follows.

$$Y_2(\overline{1}) \triangleq \text{one's complement of } Y_2$$

$$Y_2(\overline{1}) \triangleq Y_2(\overline{2}) + 1$$

$$= -y_{n-1}2^{n-1} + y_{n-2}2^{n-2} + \ldots + y_0 2^0 + 1 \tag{7.2}$$

$$Y_8(\overline{8}) \triangleq \text{eight's complement of } Y_8$$

$$Y_8(\overline{8}) \triangleq -y_{n-1}8^{n-1} + y_{n-2}8^{n-2} + \ldots\ y_i 2^i + \ldots + y_0 8^0 \tag{7.3}$$

with $y_i = 0, 1, 2, \ldots, 7$; $i = 0, \ldots, n-1$.

$$Y_8(\overline{7}) \triangleq \text{seven's complement of } Y_8$$

$$Y_8(\overline{7}) \triangleq Y_8(\overline{8}) + 1$$

$$= -y_{n-1}8^{n-1} + y_{n-2}8^{n-2} + \ldots + y_0 8^0 + 1 \tag{7.4}$$

$$Y_{10}(\overline{10}) \triangleq \text{ten's complement of } Y_{10}$$

$$Y_{10}(\overline{10}) \triangleq -y_{n-1}10^{n-1} + y_{n-2}10^{n-2} + \ldots\ y_i 10^i + \ldots + y_0 10^0 \tag{7.5}$$

with $y_i = 0, 1, 2, \ldots, 9$; $i = 0, \ldots, n-1$.

$$Y_{10}(\overline{9}) \triangleq \text{nine's complement of } Y_{10}$$

$$\begin{aligned} Y_{10}(\overline{9}) &\triangleq Y_{10}(\overline{10}) + 1 \\ &= -y_{n-1}10^{n-1} + y_{n-2}10^{n-2} + \ldots + y_0 10^0 + 1 \end{aligned} \quad (7.6)$$

Logical Complement Conversion Procedures

The mathematical definition of a complement number appears complicated. The complement version of a number, however, can be obtained by a simple logical conversion procedure. The following examples will clarify the conversion process.

Let

$$|Y_2| \triangleq \text{the magnitude of a signed binary coded number, or}$$

$$|Y_2| = y_{n-2}, y_{n-3}, \ldots, y_2, y_0$$

then the logical conversion procedure for obtaining one's complement of Y_2 is simply the complement of each individual magnitude-bit, i.e.,

$$\text{L.C. } [Y_2(\overline{1})] = y_{n-1}, \overline{y}_{n-2}, \overline{y}_{n-3}, \ldots, \overline{y}_0 \quad (7.7)$$

where L.C. means "Logical Conversion of". The logical conversion procedure for two's complement is adding one to the magnitude of the one's complement and ignoring the carry if it occurs, i.e.,

$$|Y_2(\overline{2})| = |Y_2(\overline{1})| + 1 \quad (7.8)$$

Note that in logical conversion, the sign-bit is always left untouched.

Example 1: Numerical Example for One's and Two's Complement

A. Logical Conversion Procedure

Consider a binary coded signed-magnitude number,

$$Y_2 = 10\,101\,001$$

whose magnitude, $|Y_2|$, is 0101001. Then, by one's complement logical conversion,

$$Y_2(\overline{1}) = 11\,010\,110 \quad (7.9)$$

From Equation (7.8), we obtain the two's complement of Y_2 through binary addition:

$$\begin{array}{rrl} & 11\ 010\ 110 & \leftarrow \quad Y_2(\bar{1}) \\ +) & 1 & \\ \hline & 11\ 010\ 111 & \leftarrow \quad Y_2(\bar{2}) \end{array}$$

To represent the signed-decimal equivalent of the two's complement of the given Y_2, we have

$$\begin{aligned} Y_2(\bar{2}) &= -|Y_2(\bar{2})| \\ &= -|(1 \cdot 2^6 + 0 + 1 \cdot 2^4 + 0 + 1 \cdot 2^2 + 1 \cdot 2^1 + 1 \cdot 2^0)| \\ &= -|(87)| = -87 \end{aligned}$$

Similarly, the signed-decimal equivalent of the one's complement of the given Y_2 in Equation (7.9) is

$$Y_2(\bar{1}) = -|Y_2(\bar{1})| = -86$$

B. Conversion by Mathematical Definition

Following Equation (7.1), we can convert the given number

$$Y_2 = 10\ 101\ 001$$

into its two's complement

$$\begin{aligned} Y_2(\bar{2}) &= -1 \cdot 2^7 + 0 + 1 \cdot 2^5 + 0 + 1 \cdot 2^3 + 0 + 0 + 1 \cdot 2^0 \\ &= -128 + 32 + 8 + 1 = -87 \end{aligned}$$

From Equation (7.2), $Y_2(\bar{1}) = Y_2(\bar{1}) + 1$, so $-87 + 1 = -86$. This example verifies the agreement of the two conversion methods.

Example 2: Proof of the Equivalence of the Two Conversion Procedures

From Equation (7.7),

$$\text{L.C.}\ [Y_2(\bar{1})] = y_{n-1}, \bar{y}_{n-2}, \bar{y}_{n-3}, \ldots, \bar{y}_0$$

But the bit-complement of a binary variable can be expressed by the following equation: $\bar{y}_i = y_i - 1$, thus

$$
\begin{aligned}
\text{L.C. } [Y_2(\overline{1})] &= y_{n-1}, (y_{n-2} - 1), (y_{n-3} - 1), \ldots, (y_0 - 1) \\
&= -y_{n-1}2^{n-1} + (y_{n-2} - 1)2^{n-2} + \ldots + (y_0 - 1)2^0 \\
&= -y_{n-1}2^{n-1} + y_{n-2}2^{n-2} + \ldots + y_0 2^0 - (2^{n-2} + \ldots + 2^0) \quad (7.10)
\end{aligned}
$$

But,

$$2^{n-1} = (2^{n-2} + 2^{n-3} + \ldots + 2^0) + 1$$

or

$$-(2^{n-2} + 2^{n-3} + \ldots + 2^0) = -2^{n-1} + 1$$

Substituting into Equation (7.10),

$$\text{L.C. } [Y_2(\overline{1})] = -y_{n-1}2^{n-1} + y_{n-2}2^{n-2} + \ldots + y_0 2^0 - 2^{n-1} + 1 \quad (7.11)$$

Note that since the most significant bit of the magnitude is 2^{n-2}, and the sign-bit (2^{n-1}) is to be left alone, the influence of the (-2^{n-1}) term logically disappears. We now can see the equivalence between Equation (7.11) and Equation (7.2).

Example 3: Double Complement Property

To illustrate the double complement property, let us complement the results obtained in Example 1 by logical conversion as follows. We have Y_2 = 10 101 001, and it's one's complement is $Y_2(\overline{1})$ = 11 010 110. Now let us again take the one's complement of $Y_2(\overline{1})$; this gives us 10 101 001, which is the same as the original number. Similarly, since the two's complement of Y_2 is $Y_2(\overline{2})$ = 11 010 111, if we two's complement it again we have

```
Y2(2̄) =   11  010  111 ──┐
                          │  (1̄) Operation
           10  101  000 ←─┘
      +)  ____________1
Y2   ←     10  101  001
```

which is the original number. Therefore, we can conclude that complementing a complemented number will yield the original number.

Example 4: Logical Complement Procedure for Nonbinary Numbers

The logical complement procedure, although without apparent mathematical meaning, has been widely used due to its simplicity in manual operation and hardware implementation. We now consider examples of complementing numbers which are not represented in binary notation.

Consider a signed-magnitude number in base 8. Let

$$Y_8 = y_{n-1}, y_{n-2}, \ldots, y_0$$

where

$$y_i = 0, \ldots, 7 \text{ and } i = 0, 1, n-2, i \neq n-1$$

Then

$$Y_8(\overline{7}) = y_{n-1}, (7 - y_{n-2}), (7 - y_{n-3}), \ldots, (7 - y_0)$$

For example,

given the signed-magnitude octal number $Y_8 = -634$

then its seven's complement is $Y_8(\overline{7}) = -143$

$(\overline{7})$ Operation

and since

$$|Y_8(\overline{8})| = |Y_8(\overline{7})| + 1$$

we have its eight's complement, $Y_8(\overline{8}) = -144$.

Similarly,

given a signed-magnitude decimal number $Y_{10} = -987$

then its nine's complement is $Y_{10}(\overline{9}) = -012$

$(\overline{9})$ Operation

and since

$$|Y_{10}(\overline{10})| = |Y_{10}(\overline{9})| + 1$$

its ten's complement is then $Y_{10}(\overline{10}) = -013$.

Example 5: Complement of a Positive Signed-Magnitude Number

Let $Y_2 = y_{n-1}, y_{n-2}, y_{n-3}, \ldots, y_0$; then $y_{n-1} = 0$ if Y_2 is a positive number. Thus, for $Y_2 > 0$, we have

$$Y_2 = 0, y_{n-2}, y_{n-3}, \ldots, y_0$$

But by the mathematical definition, Equation (7.1),

$$Y_2(\overline{2}) = -0 \cdot 2^{n-1}, y_{n-2}2^{n-2}, + \ldots + y_2 2^0$$

$$= \text{magnitude of } Y_2$$

Therefore, it is evident that the complement of a positive number is the positive number itself.

Example 6

A numerical example for complement number representations of different radixes or bases is shown in the following table.

Binary-coded signed-magnitude	Decimal equivalent	One's complement	Two's complement	Seven's complement	Eight's complement
0 0 0 0	0	0 0 0 0	0 0 0 0	0	0
0 0 1 1	1	0 0 0 1	0 0 0 1	1	1
0 0 1 0	2	0 0 1 0	0 0 1 0	2	2
0 0 1 1	3	0 0 1 1	0 0 1 1	3	3
0 1 0 0	4	0 1 0 0	0 1 0 0	4	4
0 1 0 1	5	0 1 0 1	0 1 0 1	5	5
0 1 1 0	6	0 1 1 0	0 1 1 0	6	6
0 1 1 1	7	0 1 1 1	0 1 1 1	7	7
1 0 0 0	–0	1 1 1 1	0 0 0 0	17	10
1 0 0 1	–1	1 1 1 0	1 1 1 1	16	17
1 0 1 0	–2	1 1 0 1	1 1 1 0	15	16
1 0 1 1	–3	1 1 0 0	1 1 0 1	14	15
1 1 0 0	–4	1 0 1 1	1 1 0 0	13	14
1 1 0 1	–5	1 0 1 0	1 0 1 1	12	13
1 1 1 0	–6	1 0 0 1	1 0 1 0	11	12
1 1 1 1	–7	1 0 0 0	1 0 0 1	10	11

7.2.2. Subtraction by Addition of Two's Complements

a. Procedure for Subtraction by Addition of Two's Complements

Step 1: Convert any negative signed-magnitude numbers into their two's complements.

Step 2: Apply binary addition to the two numbers, including their sign-bits, and ignore the carry-bit if it exists.

Step 3: (a) If the sign-bit of the sum is 1, the result is a negative number in its two's complement representation. To obtain its signed-magnitude representation, it should be two's complemented again.

(b) If the sign-bit of the sum is 0, the result is a positive number and no further operation is necessary.

Example 1: Determine $(18)_{10} - (14)_{10}$

Decimal	Binary-Coded Signed-Magnitude
18	[0] * 1 0 0 1 0
–14	[1] * 0 1 1 1 0

* Enclosed bit denotes sign-bit

Step 1: Two's complement of –14:

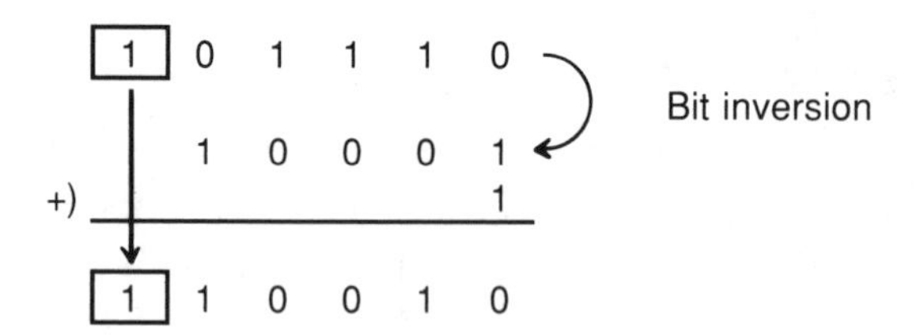

Step 2:

```
   [0] 1 0 0 1 0
+) [1] 1 0 0 1 0
   -------------
   [0] 0 0 1 0 0
```

Step 3: Sum = 4.

Example 2: Determine $(6)_{10} - (14)_{10}$

Step 1:

(6) → [0] 0 0 1 1 0

(–14) → [1] 1 0 0 1 0

Step 2:

```
    [0] 0 0 1 1 0
 +) [1] 1 0 0 1 0
 -----------------
    [1] 1 1 0 0 0   ← negative number in two's complement
```

Step 3:

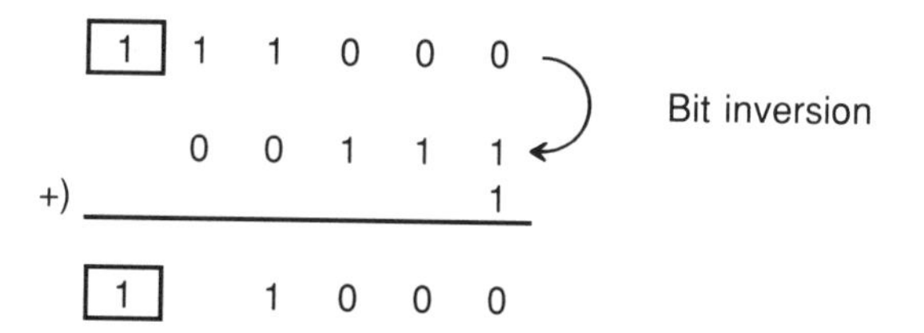

The signed-magnitude result is –8.

Example 3: Determine $(-17)_{10} - (14)_{10}$

Step 1:

```
–17 → [1] 1 0 0 0 1   ⟩ Bit inversion
          0 1 1 1 0
 +)               1
      -----------------
      [1] 0 1 1 1 1

–14 → [1] 1 0 0 1 0
```

Step 2:

```
      [1] 0 1 1 1 1
  +)  [1] 1 0 0 1 0
      -------------
 (1)  [1] 0 0 0 0 1
       ↑
       └──────────── carry-bit should be ignored
```

Step 3:

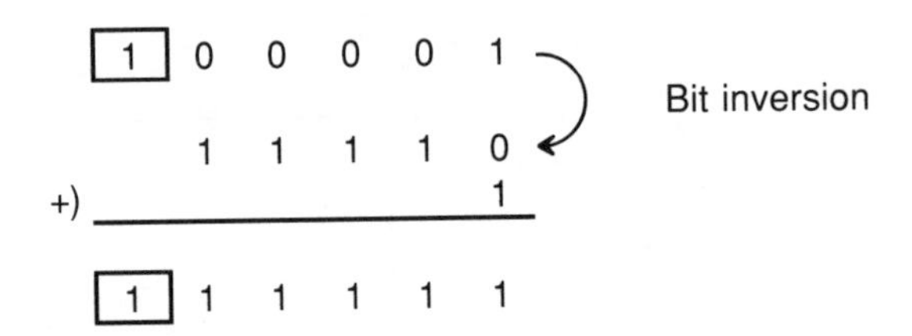

The signed-magnitude result is –31.

b. Overflow Error of Two's Complement Addition

Example 1: Determine $(18)_{10} + (16)_{10}$

Step 1:

```
18 →  [0] 1 0 0 1 0
16 →  [0] 1 0 0 0 0
```

Step 2:

```
      [0] 1 0 0 1 0
  +)  [0] 1 0 0 0 0
      -------------
      [1] 0 0 0 1 0
```

Step 3:

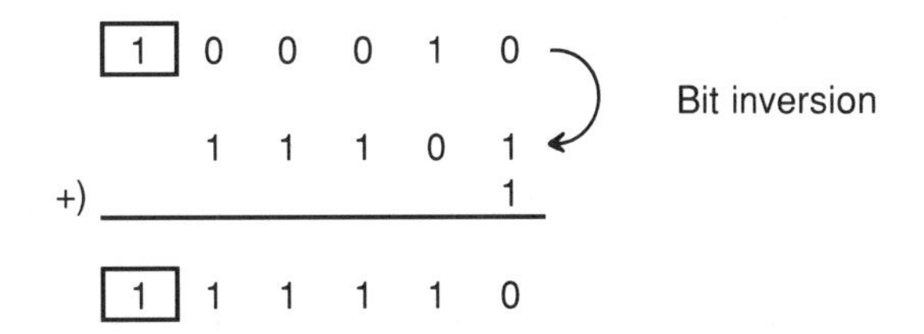

Sum = –30 ← ERROR!.

Example 2: Determine $(-18)_{10} - (16)_{10}$

Step 1:

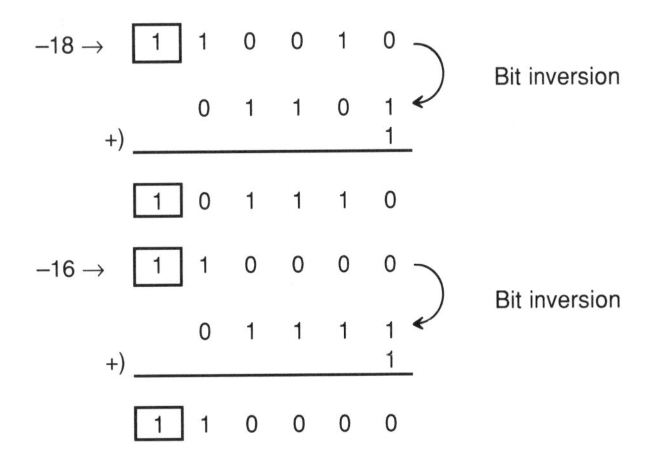

–18 → 1 1 0 0 1 0
0 1 1 0 1
+) 1
1 0 1 1 1 0
Bit inversion
–16 → 1 1 0 0 0 0
0 1 1 1 1
+) 1
1 1 0 0 0 0
Bit inversion

Step 2:

```
      [1] 0 1 1 1 0
  +)  [1] 1 0 0 0 0
      -------------
  (1) [0] 1 1 1 1 0
```

Step 3: Sum = 30 ← ERROR!.

From these two examples, we see that the errors are due to the fact that both of the true answers fall outside of the range ±31 which is the maximum magnitude for a 6-bit ALU; therefore, this kind of error is called "overflow error". It is thus important to design an overflow detection circuit. Let

$$x = \text{Sign-bit of the augend}$$
$$y = \text{Sign-bit of the addend}$$
$$z = \text{Sign-bit of the resulting sum.}$$

Then the following truth table can be constructed.

x	y	z	Overflow
0	0	0	0
0	0	1	1*
0	1	0	0
0	1	1	0
1	0	0	0
1	0	1	0
1	1	0	1**
1	1	1	0

* Addition of two positive numbers yields a negative sum
** Addition of two negative numbers yields a positive sum

According to the truth table, the switching function of the overflow is

$$F = x'y'z + xyz'$$

A detection circuit can then be designed accordingly to caution or "flag" the erroneous results.

c. Shifting of Two's Complement

From Equation (7.1), we have

$$Y_2(\overline{2}) = -y_{n-1}2^{n-1} + y_{n-2}2^{n-2} + \ldots + y_0 2^0$$

To shift the binary bits of $Y_2(\overline{2})$ to the right by one position, we have y_{n-1} moved to the position of 2^{n-2}, and y_{n-2} to that of 2^{n-3}. However, the sign of the number should not be changed. Therefore, in shifting a negative number in two's complement form, the sign-bit, which is always 1, should be "copied", not "moved", to it's right position. For example, to right-shift the number $\boxed{1}\ 0\,1\,1\,0\,1$ three times, we have

$$\boxed{1}\ 0\,1\,1\,0\,1 \overset{RS}{\rightarrow} \boxed{1}\ 1\,0\,1\,1\,0 \overset{RS}{\rightarrow} \boxed{1}\ 1\,1\,0\,1\,1 \overset{RS}{\rightarrow} \boxed{1}\ 1\,1\,1\,0\,1$$

Left Shift (LS)

In the left shift operation, y_{n-2} is being shifted to the 2^{n-1} position; therefore, one can shift the number to the left only if $y_{n-2} = y_{n-1}$, etc. For example, we can shift the following number by two times at the most.

$$\boxed{1}\ 11011 \overset{LS}{\rightarrow} \boxed{1}\ 10110 \overset{LS}{\rightarrow} \boxed{1}\ 01100$$

7.2.3. Procedure for Subtraction by Addition of One's Complements

Step 1: Convert any negative signed-magnitude numbers into their one's complements.

Step 2: Apply binary addition to the two numbers, including their sign-bits. If a carry occurs, add 1 to the resulting sum.

Step 3: (a) If the sign-bit of the sum is 1, apply the one's complement operation again to obtain the signed-magnitude representation of the result.

(b) If the sign-bit of the sum is 0, the result is a positive number and no further operation is necessary.

Example 1: Determine $(34)_{10} - (23)_{10}$

Step 1:

$$34 \rightarrow \boxed{0}\ 1\ 0\ 0\ 0\ 1\ 0$$

$$-23 \rightarrow \boxed{1}\ 0\ 1\ 0\ 1\ 1\ 1$$

$$\boxed{1}\ 1\ 0\ 1\ 0\ 0\ 0$$

Bit inversion

Step 2:

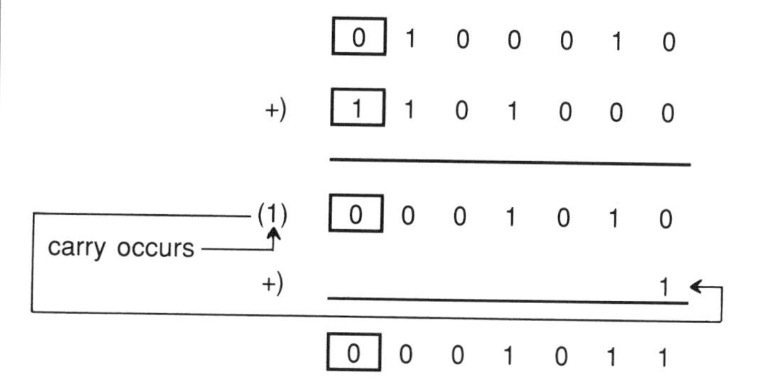

Step 3: Sum = 11.

Example 2: Determine $(17)_{10} - (23)_{10}$

Step 1:

17 →	[0]	0	1	0	0	0	1
–23 →	[1]	1	0	1	0	0	0

Step 2:

	[0]	0	1	0	0	0	1
+)	[1]	1	0	1	0	0	0
	[1]	1	1	1	0	0	1

Step 3:

[1]	1	1	1	0	0	1	Bit inversion
[1]	0	0	0	1	1	0	

Sum = –6, signed-magnitude.

Example 3: Determine $(-22)_{10} - (8)_{10}$

Step 1:

–22 →	[1]	0	1	0	1	1	0	Bit inversion
	[1]	1	0	1	0	0	1	
–8 →	[1]	0	0	1	0	0	0	Bit inversion
	[1]	1	1	0	1	1	1	

Step 2:

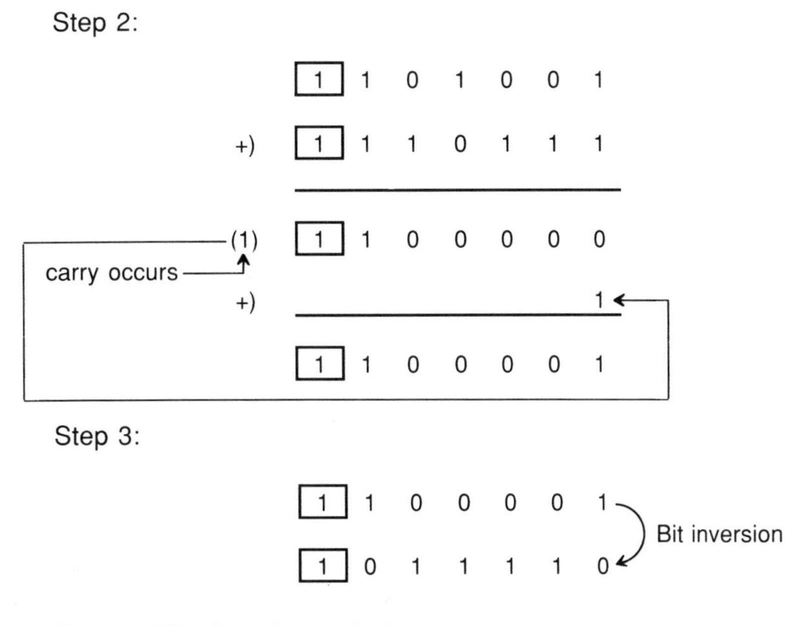

1 1 0 1 0 0 1
+) 1 1 1 0 1 1 1
(1) 1 1 0 0 0 0 0
carry occurs
+) 1
1 1 0 0 0 0 1

Step 3:

1 1 0 0 0 0 1 ⟩ Bit inversion
1 0 1 1 1 1 0

Sum = –30, signed-magnitude.

7.2.4. Subtraction by Addition of Eight's Complements

Following the same logic as above, subtracting two numbers of radix other than one or two can be carried out by addition of its complements. For example, to determine $(715)_8 - (234)_8$, we proceed as follows.

Step 1:

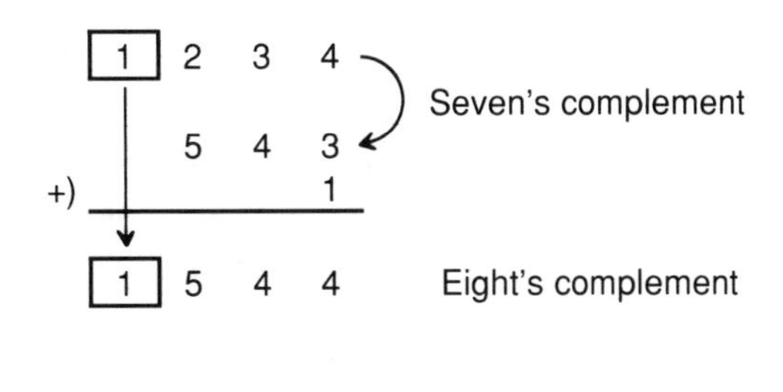

Step 2:

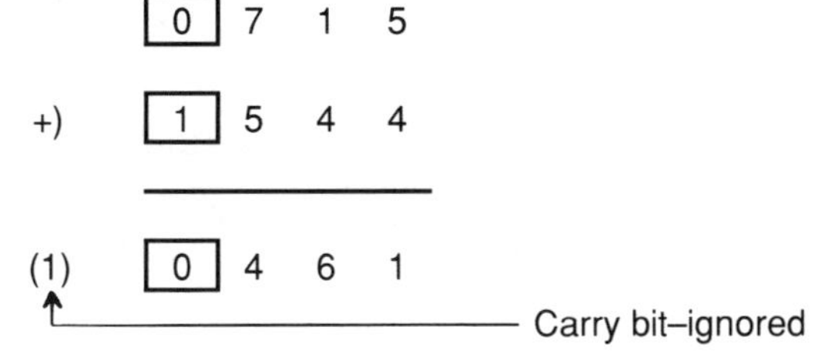

Step 3: Sum = 461_8

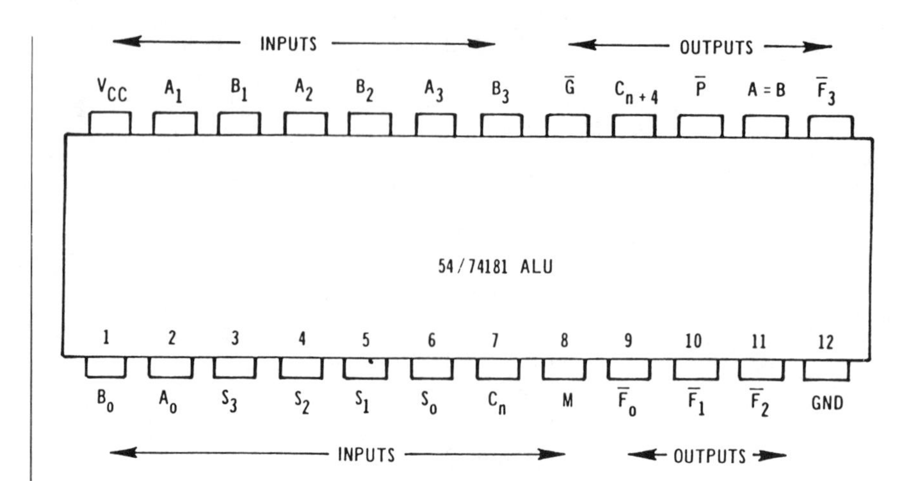

FIGURE 7.5 A typical ALU chip — 54/74181.

7.3. A TYPICAL ALU CHIP

Figure 7.5 shows a typical ALU chip, the 54/74181, which has been commercially available for a number of years. It is a 4-bit ALU. By controlling the M-pin, the chip will operate in logic/arithmetic mode. There are four functional control pins, s_0, s_1, s_2, and s_3, which can provide 16 functions for each type of input data (high or low true). It can also be used as a comparator by referring to the logic levels at C_n and C_{n+4}. Note that both ripple-carry C_n, C_{n+4} and look-ahead-carry $\overline{P}$ and $\overline{G}$ pins are provided. To implement a 16-bit with look-ahead-carry operation, one may use four ALU chips and a look-ahead-carry generator chip (54/74182) which also is available on the market.

7.4. BINARY MULTIPLICATION

7.4.1. Multiplication by Iterative Addition

This method is quite straightforward. It is easily illustrated by a numerical example. For example, one can multiply 23 by 3 by adding 23 three times. This is exactly how it is done with binary numbers. Figure 7.6 shows how this method is implemented with digital hardware. First, the multiplicand and multiplier are loaded respectively into the register and down counter. As the multiply command sets the flip-flop, the system clock enables the ADD operation in the accumulator as well as the count down operation in the down counter, which determines the number of times that the addition of the multiplicand is required. When the down counter reaches zero, the operation is finished.

7.4.2. Multiplication by Shift and Add

The algorithm for this method is the same as we would do manually as shown in the following example.

```
                1 0 1 1 0 1
          ×       1 1 0 0 1
          -----------------
                1 0 1 1 0 1
              0 0 0 0 0 0
            0 0 0 0 0 0
          1 0 1 1 0 1
 +      1 0 1 1 0 1
    -----------------------
    1 0 0 0 1 1 0 0 1 0 1
```

Figure 7.7 shows the exact hardware implementation of the algorithm. Since the diagram is self-explanatory, a detailed description is not given here. Basically, the multiplier controls the action. That is, if [1] occurs, the multiplicand is shifted and added to the accumulator, otherwise, it is just shifted and zeroes, not the multiplicand, are added to the accumulator.

7.4.3. Multiplication of Two's Complement Binary Numbers

a. Booth's Algorithm

Let

$$X \triangleq \text{multiplicand} = (x_{n-1}, x_{n-2}, \ldots, x_0)$$
$$Y \triangleq \text{multiplier} = (y_{n-1}, y_{n-2}, \ldots, y_0)$$

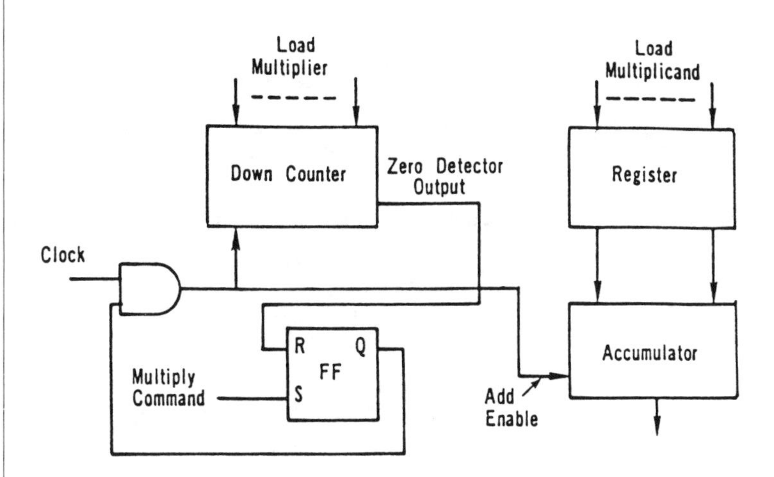

FIGURE 7.6
Multiplication by iterative addition.

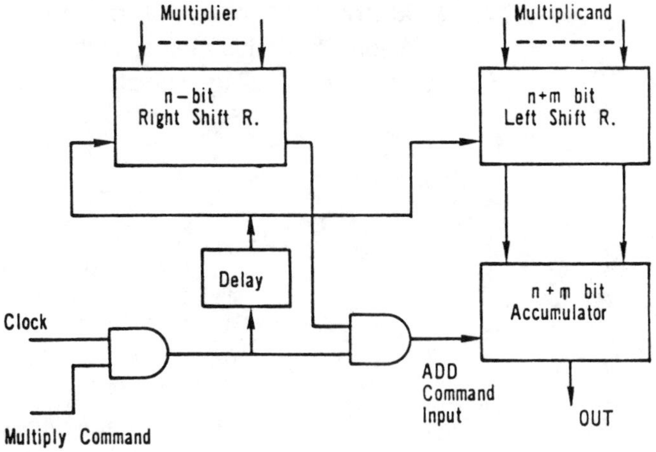

FIGURE 7.7 Multiplication by shifting.

Then

$$Y(\bar{2}) = -y_{n-1}2^{n-1} + y_{n-2}2^{n-2} + \ldots + y_0 2^0 + y_{-1}2^0$$

where $y_{-1} = 0$ is inserted for mathematical derivation. Now, let us add more dummy terms, so we have

$$\begin{aligned}
Y(\bar{2}) &= -y_{n-1}2^{n-1} + y_{n-2}2^{n-2} + (y_{n-2}2^{n-2} - y_{n-2}2^{n-2}) + y_{n-3}2^{n-3} \\
&\quad + (y_{n-3}2^{n-3} - y_{n-3}2^{n-3}) + \ldots + y_1 2 + (y_1 2 - y_1 2) + y_0 2^0 \\
&\quad + (y_0 2^0 - y_0 2^0) + y_{-1}2^0 \\
&= -y_{n-1}2^{n-1} + 2y_{n-2}2^{n-2} - y_{n-2}2^{n-2} + 2y_{n-3}2^{n-3} - y_{n-3}2^{n-3} \\
&\quad + \ldots + 2y_0 2^0 - y_0 2^0 + y_{-1}2^0 \\
&= 2^{n-1}(y_{n-2} - y_{n-1}) + 2^{n-2}(y_{n-3} - y_{n-2}) + \ldots + 2(y_0 - y_1) \\
&\quad + 2^0(y_{-1} - y_0) \\
X \cdot Y &= 2^{n-1}(y_{n-2} - y_{n-1})X + 2^{n-2}(y_{n-3} - y_{n-2})X + \ldots + 2(y_0 - y_1)X \\
&\quad + 2^0(y_{-1} - y_0)X \\
&= p^{n-1} + \ldots + p^i + \ldots + p^0
\end{aligned}$$

where $p^i \triangleq j^{th}$ partial product of $X \cdot Y = 2^j(y_{j-1} - y_j)X$. Note that $2^j(y_{j-1} - y_j)X$ is the product of $(y_{j-1} - y_j)X$ being shifted to the left by j places. However, there are only four possible combinations of a pair of y_js, as shown in the table below.

y_j	y_{j-1}	p_j	Remark
0	0	$2^j(0)X$	Zero being shifted to the left j places
0	1	$2^j(1)X$	X being shifted to the left j places
1	0	$2^j(-1)X$	–X being shifted to the left j places
1	1	$2^j(0)X$	Zero being shifted to the left j places

The following manual operation,

$$
\begin{array}{lr}
 & X \\
\times) & 2^{n-1}(y_{n-2} - y_{n-1}) + 2^{n-2}(y_{n-3} - y_{n-2}) + \ldots + 2^0(y_{-1} - y_0) \\
\hline
 & + 2^0(y_{-1} - y_0)\,X \\
 & + 2(y_0 - y_1)\,X \\
+) & + 2^{n-1}(y_{n-2} - y_{n-1})X \\
\hline
\end{array}
$$

$$
X \cdot Y = \sum_{j=0}^{n-1} 2^j(y_{j-1} - y_j + X)
$$

we see that $2(y_0 - y_1)X$ to the left one place is equivalent to shifting $2^0(y_{-1} - y_0)X$ to the right one place; the last term should not be shifted, however. Hence, Booth's algorithm for binary multiplication can be formulated as follows.

Multiplier bits y_j	Multiplier bits y_{j-1}	Operation
0	0	Shift partial product right 1 bit
0	1	Add X then shift right 1 bit
1	0	Subtract X then shift right 1 bit
1	1	Shift partial product right 1 bit

Example 1: (–9) × 11

Here, we have

	Decimal	Signed-magnitude	+X in two's complement	–X in two's complement
X	–9	–1 0 0 1	[1] 0 1 1 1	[0] 1 0 0 1
Y	+11	1 0 1 1		

Now let the (A) $\triangleq$ the content of the accumulator. Then the multiplication can be shown as follows:

j	y_j	y_{j-1}	Operation	Accumulator (A)
				0 | 0 0 0 0 0 0 0 0
0	1	0	$((A) - X) \rightarrow (A)$	0 | 1 0 0 1 0 0 0 0
			Shift right	0 | 0 1 0 0 1 0 0 0
1	1	1	$((A) + 0) \rightarrow (A)$	0 | 0 1 0 0 1 0 0 0
			Shift right	0 | 0 0 1 0 0 1 0 0
2	0	1	$((A) + X) \rightarrow (A)$	1 | 1 0 0 1 0 1 0 0
			Shift right	1 | 1 1 0 0 1 0 1 0
3	1	0	$((A) - X) \rightarrow (A)$	0 | 0 1 0 1 1 0 1 0
			Shift right	0 | 0 0 1 0 1 1 0 1
4	0	1	$((A) + X) \rightarrow (A)$	1 | 1 0 0 1 1 1 0 1

Inversion

1 0 1 1 0 0 0 1 0

\+ 1

1 0 1 1 0 0 0 1 1

Thus, $XY = [1 + 2 + 32 + 64] = -99$.

Example 2: (–5) × (–7)

$X = -5 = -(0101)_2$ $Y = -7 = -(0111)_2$

$-X = 5 = \boxed{0}\ 0101$ $Y(\bar{2}) = \boxed{1}\ 1001$

$X(\bar{2}) = \boxed{1}\ 0111$

j	y_j	y_{j-1}	Operation	Accumulator (A)
				0 0 0 0 0 0 0 0 0
0	1	0	$((A) - X) \rightarrow (A)$	0 0 1 0 1 0 0 0 0
			Shift right	0 0 0 1 0 1 0 0 0
1	0	1	$((A) + X) \rightarrow (A)$	1 1 1 0 1 1 0 0 0
			Shift right	1 1 1 1 0 1 1 0 0
2	0	0	$(A) \rightarrow (A)$	1 1 1 1 0 1 1 0 0
			Shift right	1 1 1 1 1 0 1 1 0
3	1	0	$((A) - X) \rightarrow (A)$	0 0 1 0 0 0 1 1 0
			Shift right	0 0 0 1 0 0 0 1 1
4	1	1	$(A) \rightarrow (A)$	0 0 0 1 0 0 0 1 1

Thus XY = [1 + 2 + 32] = 35.

b. Booth's Algorithm vs. Conventional Method

Following the description presented in Section 7.4.3.a, the conventional multiplication of two two's complement numbers can be derived as follows. Since

$$X(\bar{2}) = -2^{n-1}x_{n-1} + \sum_{k=0}^{n-2} 2^k x_k = -2^{n-1}x_{n-1} + |X|$$

$$Y(\bar{2}) = -2^{m-1}x_{m-1} + \sum_{j=0}^{m-2} 2^j x_j = -2^{m-1}x_{m-1} + |Y|$$

then

$$X(\overline{2}) \cdot Y(\overline{2}) = (-2^{n-1}x_{n-1} + |X|)(-2^{m-1}x_{m-1} + |Y|)$$

$$= |X||Y| - 2^{m-1}x_{m-1} + |X| - 2^{n-1}x_{n-1} + |Y| + 2^{m+n-2}x_{n-1}y_{m-1}$$

The last equation shown above reveals that if x_{n-1}, y_{m-1} are zero, that is, if X and Y are both positive numbers, the last three terms will disappear and the answer will be correct. If not, the last three terms require more calculation. In Booth's algorithm, however, the correction is not necessary.

c. Features of Booth's Algorithm

(i) Serial-parallel operation
(ii) Faster than shift-and-add multiplication
(iii) No correction is needed for two's complement numbers
(iv) It can easily be implemented with ALU chips and thus applied to signal processing, digital filters, etc.

7.5. BINARY RATE MULTIPLIER

To describe the principle of operation of this device, a 3-bit binary multiplier is shown in Figure 7.8(a), where m_1, m_2, m_3 is the 3-bit multiplier. The output frequency F_{out} is a function of the input frequency F_{in} and the multiplier m_1, m_2, m_3. From the logic diagram in Figure 7.8(a), we have

$$F_o = (m_3Q_1 + m_2\overline{Q}_1Q_2 + m_1\overline{Q}_1\overline{Q}_2Q_3) \cdot \text{pulse}$$

The principle of operation for this network can easily be described by referring to the timing diagram shown in Figure 7.8(b). Note that the number of pulses at the output within the 8-pulse frame of F_{in}, which is determined by the 3-bit counter, is equal to the setting value of m_1, m_2, m_3, i.e.,

$$F_o = \frac{(m_3m_2m_1)_2}{8} F_{in}$$

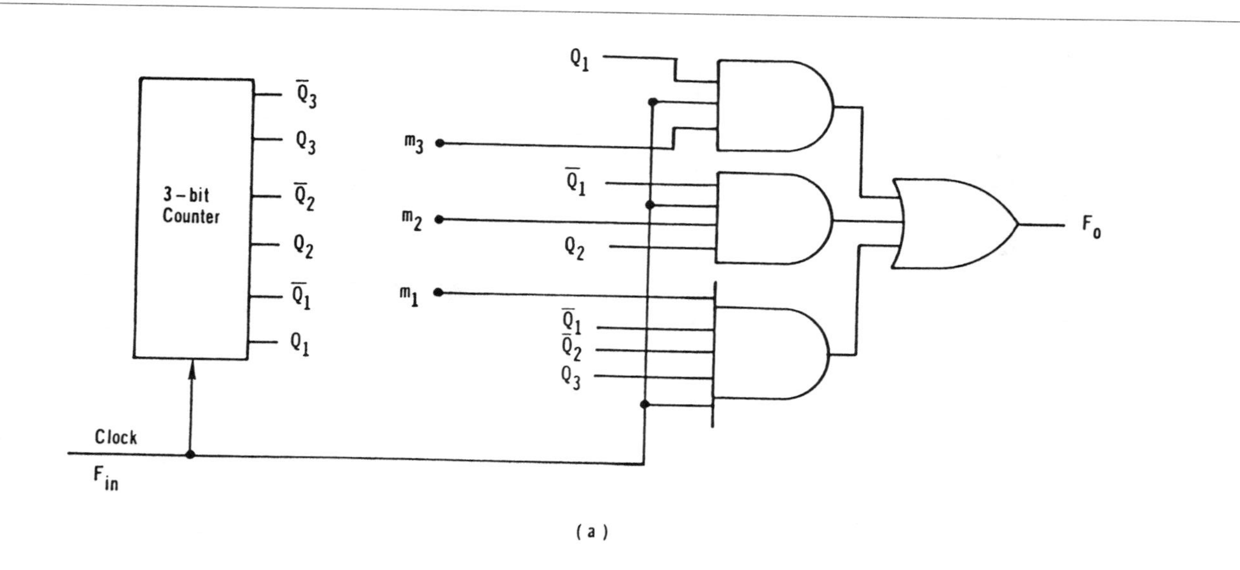

FIGURE 7.8 (a) Binary rate multiplier. (b) Timing diagram of the binary rate multiplier.

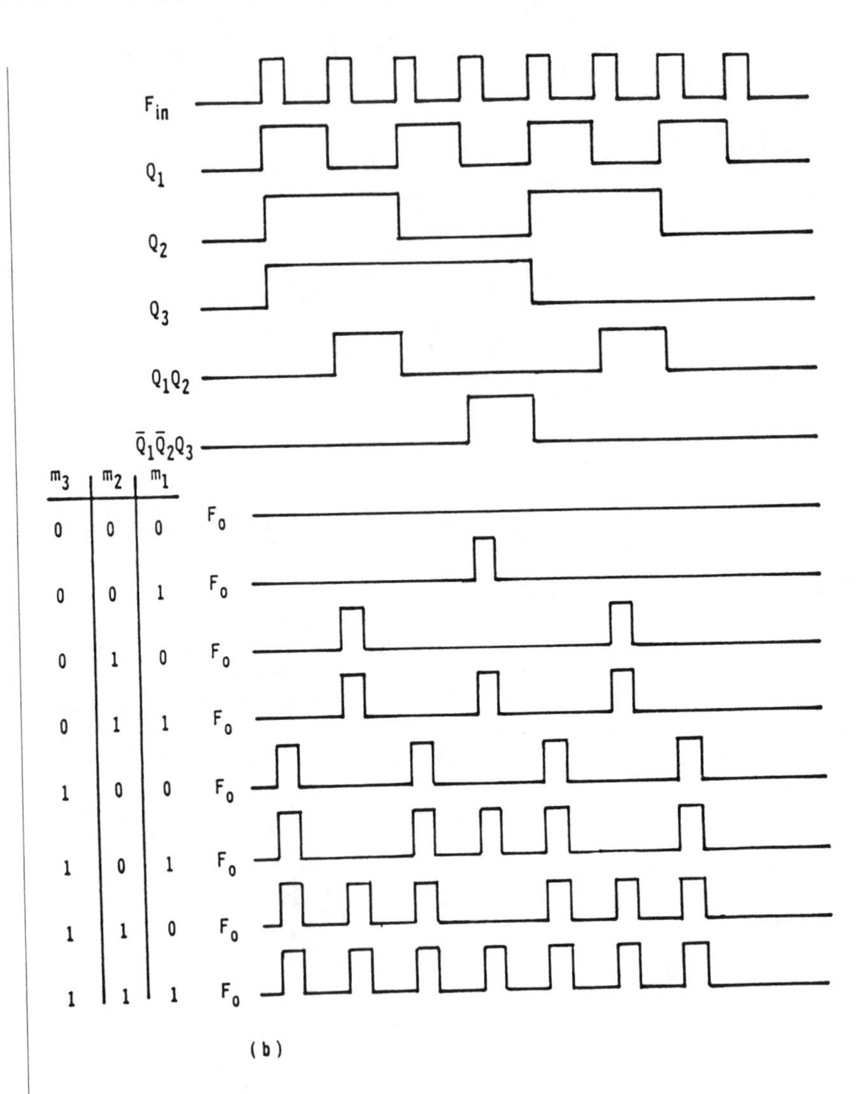

FIGURE 7.8 continued. (a) Binary rate multiplier. (b) Timing diagram of the binary rate multiplier.

In general, if the multiplier is an n-bit binary number and the counter is an n-bit counter, then we have

$$F_o = \frac{(m_n, \ldots, m_1)}{2^n} F_{in}$$

Thus, the output frequency count is always a fraction of the input frequency within a frame of 2^n. It is important to point out the drawback of the nonuniform distribution of the pulse position at the output. For example, as shown in Figure 7.8(b), when $m_3m_2m_1 = 101$, there are three pulses clustered around the center of the eight-pulse frame. This nonuniform pulse position property may not be desirable in some applications.

7.6. BINARY DIVISION

Figure 7.9 shows the hardware implementation of the algorithm known as division by iterative subtraction. It is similar to the manual procedure of long division. A simple numerical example will clarify the algorithm. Let us divide 11 by 2 as follows.

```
2 | 11   |1
     2
   ----
     9   |1
     2
   ----
     7   |1    Total count of 1's is the quotient. Here,
     2                   quotient = 5
   ----
     5   |1              remainder = 1
     2
   ----
     3   |1
     2
   ----
     1    ←   Remainder
```

Note that the up-counter is used to count the total number of subtractions. The dividend has been subtracted by the divisor before the accumulator has a value less than that of the divisor. The digital comparator detects this criterion. When this occurs, the subtract operation is ended by resetting the flip-flop. At this point, we have the quotient in the up-counter and the remainder in the accumulator.

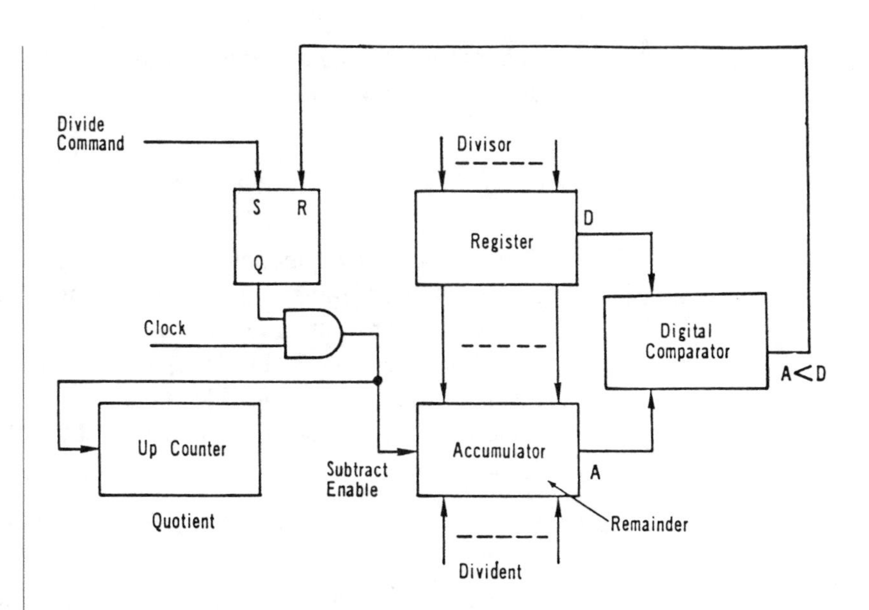

FIGURE 7.9 Logic diagram of a binary divider.

REFERENCES

1. Chu, Yaohan, *Digital Computer Design Fundamentals,* McGraw-Hill, New York, 1962.
2. Rabiner, L. R. and Gold, B., *Theory and Application of Digital Signal Processing,* Prentice-Hall, Englewood Cliffs, NJ, 1975.
3. Mitchell, J. N., Jr., Computer multiplication and division using binary logarithms, *IRE Trans. Electronic Computers,* August 1962.
4. Brubaker, T. A. and Becker, J. C., Multiplication using logarithms implemented with ROM, *IEEE Trans. Computers,* C-24(8), 761—765, August 1975.
5. Peatman, J. B., *The Design of Digital Systems,* McGraw-Hill, New York, 1972.
6. Moshos, G. J., Error analysis of binary rate multiplier, *NASA Tech. Note,* NASA TND-3124, December 1965, Lewis Research Center, Cleveland, OH.
7. Parasuraman, B., Hardware multiplication techniques for microprocessor systems, *Computer Design,* 16(4), 75—82, April 1977.
8. Waser, S., High-speed monolithic multipliers for real-time digital signal processing, *Computer,* 11(10), 19—29, October 1978.
9. *TTL Data Book,* National Semiconductor, 1976.
10. *TTL Data Book,* Texas Instruments, 1977.
11. *Data Manual,* Signetics Corp., 1976.
12. Mick, J. R., *Digital Signal Processing Handbook,* Advanced Micro Devices, Inc., Sunnyvale, 1978.
13. *TTL Data Book,* Vols. I to IV, Texas Instruments, 1984 to 1985.

8 ANALOG-DIGITAL-ANALOG CONVERSION AND DATA ACQUISITION SYSTEMS

8.1. INTRODUCTION

8.1.1. General Consideration of Analog and Digital System Design

As we all know, digital circuits are generally considerably more reliable and less sensitive than analog circuits to variations of circuit components. However, in the real world, output signals generated by the transducers are mostly in analog or continuous forms. To take full advantage of digital technology, one first needs to convert the analog signals into digital and convert them back to analog form afterwards. However, there are a large number of A/D/A (Analog-to-Digital-to-Analog) techniques or devices available to system designers. Their degree of sophistication varies from very simple at low cost to very powerful at high cost. Thus, selection of an appropriate A/D/A conversion technique or device for a system is a nontrivial task for the system designer. Furthermore, consideration of trade-offs between analog and digital techniques becomes an important step to a competent designer. In this chapter, the techniques of A/D/A conversion and major components used to implement them will be described. First, an understanding of the features of analog and digital systems is in order. The following table compares the two in a broad sense and will hopefully clarify any confusion.

Analog system	Digital system
• Directly applicable to the real world	• Requires A/D/A Converters
• Suitable for fast and real-time applications	• Comparatively slower than its analog counterpart
• Simple and low-cost for parallel processing	• Expensive for simple systems, even more so for parallel processing
• Requires frequent and generally complicate system calibration	• More reliable and requires almost no system calibration
• Sensitive to temperature and component aging effects	• Not sensitive to temperature and component aging effects
• Limitation on accuracy; system becomes very expensive if high accuracy is required	• Can be designed for high accuracy applications at reasonable cost
• Difficult to devise an analog random-access memory	• Easy to provide random-access memory at low cost

8.1.2. Typical System Organization of a Digital Data Acquisition System

Figure 8.1(a) shows a typical data acquisition system where A/D/A conversion technologies are used. While the functions of most of the blocks are quite evident, the functions of the analog guard filter, the sample/hold, and sampling clock may require clarification. Since an analog signal is continuously varying, a technique to "freeze" its amplitude for a short period, just long enough to perform A/D conversion, is necessary. The sample/hold block is designed for this purpose. The next question is at what frequency the signal should be "frozen". In other words, what should be the sampling rate of the system? This of course depends on the nature of the analog signal to be processed. If the frequency spectrum of the signal is known, one can use a sampling frequency at least twice as great or greater than the highest frequency component in the frequency spectrum or the bandwidth of that signal. It is not unusual to have a sampling frequency chosen at five times the maximum frequency component of a signal, if the amount of digitized data yielded by the A/D converter is not larger than the digital data processor can manage within a given period of time. It is important for a system designer to be aware of the fact that a so-called "aliasing" problem will occur if this guideline is not observed. Otherwise, the recovered signal after D/A conversion will be distorted and the system will produce errors. The guard filter is used to assure that the incoming signal frequency spectrum is within specification limits; the frequency components higher than the specified maximum frequency are considered as noise, which should be eliminated or attenuated to a negligible level.

What we have just described may appear abstract to those unfamiliar with the field of digital signal processing (DSP). However, as a digital system designer you cannot avoid designing digital systems that must process analog signals. Thus, an understanding of the basic concept behind the system block diagram shown in Figure 8.1(a) is essential.

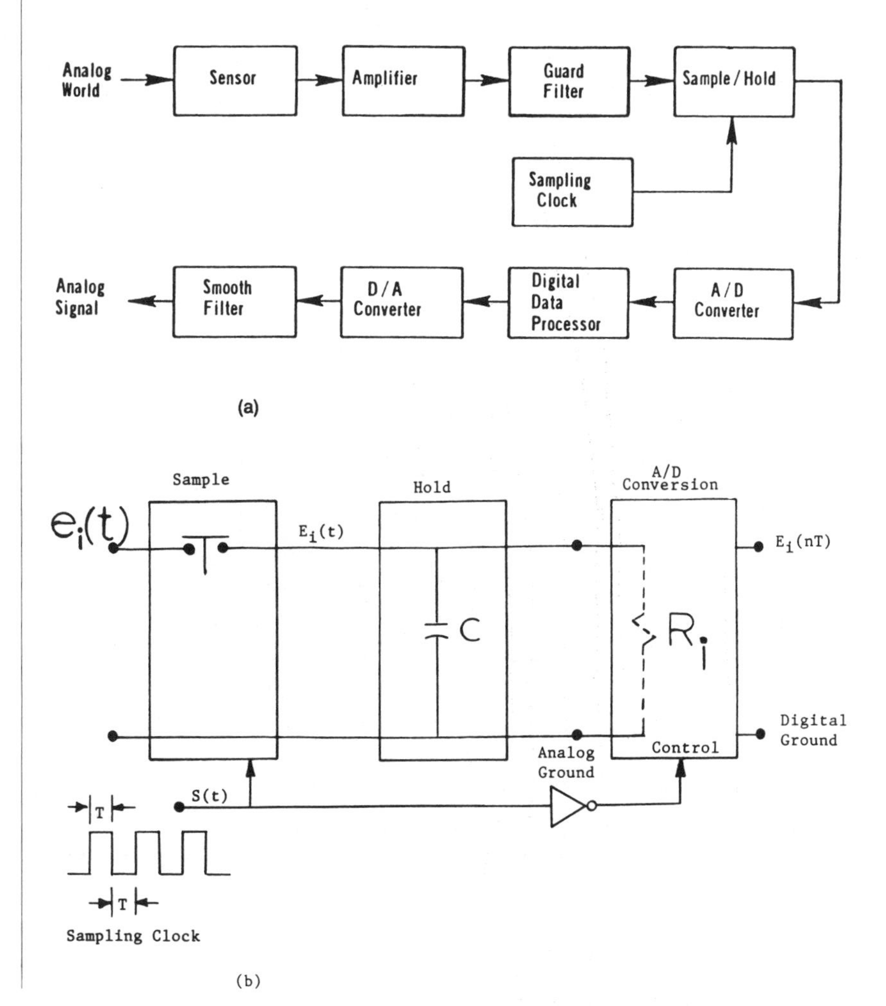

FIGURE 8.1 (a) Block diagram of a typical data acquisition system. (b) Functional diagram of the sample/hold process.

First, let us examine the sample/hold block and the sampling clock. A simple example may clarify the concept. Consider we are interested in temperature variation during the summer in a small town, say, Davis, California. We decide to measure the temperature once a day at 3:00 p.m. during the summer. That is, on the first day at 3:00 p.m. we would measure (sample) the temperature and record (hold) the data until 3:00 p.m. the next day when we would again sample and hold. The process is repeated every day for the whole season. Let us assume that the data collected show that the temperature varied from 90° to 110°F. Therefore, we conclude that the weather in Davis during the summer is around 100°F, very hot! Of course, no one would accept this method of data collection. What about the night-time temperatures — are they 100°F too? It is apparent that measuring or sampling the temperature once a day is not enough. So how do we determine how often the temperature should be measured (sampling rate)? A logical answer is that one must know something about the pattern of temperature variation at the location of interest. For example, if we knew from experience that at Davis the maximum variation could be 30°F within 1 h, then we may decide to sample at least once every one-half hour. This simple example illustrates that the sampling rate must be determined by the nature of the data or the incoming signal. With this in mind, let us consider the details from a system designer's point of view.

Figure 8.1(b) depicts a simplified functional diagram of the sample/hold process in a data acquisition system. This is the circuit which "freezes" the input time-varying voltage for examination at a predetermined rate. Here, $e_i(t)$ is the analog, time-varying, input voltage signal. The sampling switch can be one of the solid-state analog switches which is turned ON and OFF by the sampling clock as shown. The switch is turned ON as the clock is in HIGH state; in the meantime, the capacitor C is being charged to the value of e_i with the time constant R_gC, where R_g is the output resistance of the block (not shown) which precedes the sampling switch plus the ON resistance of the analog switch. Ideally, R_g should be 0 Ω so that the voltage across the capacitor would follow or "track" the input signal with zero time constant. As the clock drops to LOW, say at time $t = 0$, the switch would be OFF. The capacitor would HOLD the voltage value, $E_i(0^-)$, which would be discharged with the time constant, R_iC, during the LOW period of the clock. Here, R_i is the input resistance of the Analog-to-Digital conversion block. In the meantime, the Analog-to-Digital converter begins to convert the voltage value across the capacitor, $E_i(0^-)$, into digital form and yields its digitized output $E_i(nT)$, where $n = 1, 2, \ldots$, and T equals the clock sampling period. Ideally, R_i should be infinitely large so that the discharge time constant would be infinitely large and the voltage $E^i(0^-)$, would stay constant during the A/D conversion time. Thus, the A/D conversion time must be less than T. One may now see why the sample and hold circuit is sometimes called the "track and hold" circuit.

Let us now turn to the next three key points: the sampling rate or the frequency of the sampling clock, the guard filter, and the smooth filter shown in Figure 8.1(a). From the example on temperature measurement given earlier, we learned that the sampling rate is a function of how fast the input signal changes. The system designer must come up with a quantitative value in place of "how fast" in order to determine the frequency of the sampling clock. Fortunately, there is an analytical method for determining this characteristic of the input signal. One examines the spectrum of the input signal, and based on its bandwidth, the highest frequency component, or the high end of the signal in its bandwidth, can be determined. Then, according to the well-known Shannon's criterion, which proposes that the sampling rate must be at least two times greater than the bandwidth or the highest frequency component of the signal in order to minimize the aliasing error. For example, if one wants to digitize the signal of a symphony which has a bandwidth of 20 kHz, then the sampling frequency must be at least 40 kHz. A simple example will further illustrate this criterion. Referring to Figure 8.1(b), let us assume that

$$e_i(t) = A \cos w_i t,$$

and for convenience let A = 1. Also, let the sampling clock be $S(w_o t)$, which is a pulse train with frequency w_o. Then let us examine the signal at the output of the switch with the capacitor disconnected. Notice that as the switch is closed, $E_i(t) = e_i(t)$, and when the switch is open, $E_i(t) = 0$. Equivalently, one may say that $E_i(t) = 1 \cdot e_i(t)$ when the switch is closed and $E_i(t) = 0 \cdot e_i(t)$ as the switch is open. Thus, we can formulate the sampling process as follows:

$$E_i(t) = e_i(t) \cdot S(w_o t)$$

where $S(w_o t)$ is a clock with amplitude of one for HIGH state and zero for LOW state, and oscillates at the frequency of w_o. By Fourier expansion theory, any periodic functions can be expanded into an infinite series of sine/cosine functions. An even periodic function would contain only cosine terms, whereas an odd function has only sine terms. For simplicity, let us assume that the clock is an odd function, and we have

$$S(w_o t) = a_o + \sum_{n=1}^{\infty} b_n \sin(w_o t)$$

Thus,

$$E_i(t) = \cos(w_i t) \cdot S(w_o t)$$

$$= a_o \cos(w_i t) + b_1 \cos(w_i t) \sin(w_o t) + b_2 \cos(w_i t) \sin(2w_o t)$$

$$+ b_3 \cos(w_i t) \sin(3w_o t) + \ldots$$

By applying the trigonometric formula,

$$2 \sin A \cos B = \sin(A + B) + \sin(A - B)$$

we have

$$E_i(t) = a_o \cos(w_i t) + k_1 \sin(w_o + w_i)t + k_1 \sin(w_o - w_i)t + k_2 \sin(2w_o + w_i)t + k_2 \sin(2w_o - w_i)t + \ldots k_i \sin(iw_o + w_i) + k_i \sin(iw_o - w_i) + \ldots$$

where k_i is a function of b_i, with $i = 1, 2, \ldots$.

Notice that there are unwanted components, in all terms except the first, in the above equation after sampling. When the processed signal is to be converted back to analog form by the D/A conversion block shown in Figure 8.1(a), we will need a filter to smooth out the unwanted components resulting from the sampling process. Evidently, this filter must be designed with its cut-off frequency, w_c, such that $w_i < w_c < (w_o - w_i)$. Furthermore, the system must assure that there is no frequency component higher than the highest frequency component of the input signal being applied to the sample/hold circuitry. Therefore, another filter with the bandwidth equal to that of the signal must precede the sample/hold circuitry. This filter is called the "guard filter" or anti-aliasing filter, as shown in Figure 8.1(a). As a trivial example to illustrate the aliasing error, let the bandwidth of an input signal be 100, and the sampling rate be 150, which is less than twice the input signal bandwidth. As a result, we have an unwanted signal component of frequency equal to 150 – 100 = 50, which cannot be filtered out by the smooth filter, whose cutoff frequency is designed to be 100.

Thus, the sampling frequency must be at least equal to or greater than twice the expected bandwidth of the input signal.

Having described a basic digital data acquisition system that consists of A/D and D/A conversion processes as shown in Figure 8.1(a), we now focus on the design details of the essential components of the system.

8.2. DIGITAL-TO-ANALOG CONVERSION

8.2.1. Parallel Conversion

The following equation describes the conversion.

$$v(\text{analog}) = [w_1x_1 + w_2x_2 + \ldots + w_nx_n]\, V_r$$

where $w_1, w_2, \ldots, w_n$ = weight, $x_1, x_2, \ldots, x_n$ = n digital binary (1 or 0) bits, and V_r = reference voltage. The block diagram shown in Figure 8.2(a) realizes this method, where the D/A decoder is usually a resistor network. The following example will clarify the concept.

In reference to Figure 8.2(b), a 4-bit binary number x_4, x_3, x_2, x_1 can be converted into an analog signal v_0. Here, x_4 is the sign-bit, i.e., $x_4 = 1 \Rightarrow$ (implies) the switch is UP, and $x_4 = 0 \Rightarrow$ the switch is DOWN, and x_1 is the least significant bit. Then, since the operational amplifier is in a shunt-to-shunt feedback configuration (see Chapter 2, Section 2.2.3(a)), we have

$$\begin{aligned} v_0 &= (w_1x_1 + w_2x_2 + w_3x_3]\, x_4V_r \\ &= -\left(\frac{R_F}{R_1}x_1 + \frac{R_F}{R_2}x_2 + \frac{R_F}{R_3}x_3\right)x_4V_r \end{aligned}$$

Let $R_1 = 2^3R_F$, $R_2 = 2^2R_F$, $R_3 = 2R_F$, and $V_r = 8V$. Then we have

$$\begin{aligned} v_0 &= -\left(\frac{1}{2^3}x_1 + \frac{1}{2^2}x_2 + \frac{1}{2}x_3\right)x_48 \\ &= -x_4(2^2x_3 + 2^1x_2 + 2^0x_1) \end{aligned} \tag{8.1}$$

8.2.2. Serial Conversion

This operation is described by the mathematical expression

$$v_0(t + T) = \frac{1}{2}\{v_o(t) + x(t + T)V_r\}$$

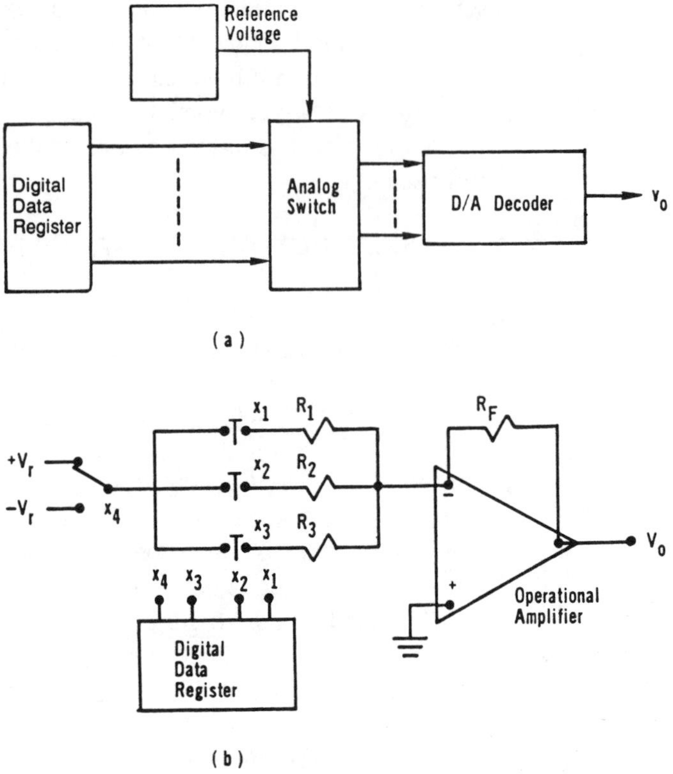

FIGURE 8.2 (a) Block diagram of a parallel D/A converter. (b) A simple 4-bit D/A converter.

where $v_0(t)$ = the output voltage at t, $x(t + T)$ = digital input information, [1] or [0], and V_r = reference voltage. To illustrate, let the input be a 3-bit signal, x_1, x_2, x_3. Let $v_0(t) = v_0(0) = 0$. Then

$$v_0(0 + T) = \frac{1}{2}\{v_0(0) + x_1V_r\}$$

$$v_0(0 + 2T) = \frac{1}{2}\{v_0(T) + x_2V_r\}$$

$$= \frac{1}{2}\left\{\frac{1}{2}x_1V_r + x_2V_r\right\}$$

$$= \left(\frac{1}{2}\right)^2 x_1V_r + \frac{1}{2}x_2V_r$$

$$v_0(0 + 3T) = \frac{1}{2}\{v_0(2T) + x_3V_r\}$$

$$= \left[\left(\frac{1}{2}\right)^3 x_1 + \left(\frac{1}{2}\right)^2 x_2 + \frac{1}{2}x_3\right]V_r$$

In general, for an n-bit digital signal,

$$v_0(0 + nT) = V_r \{2^{-1}x_n + 2^{-2}x_{n-1} + \dots + 2^{-n}x_1\}$$

Figure 8.3(a) shows the system, and Figure 8.3(b) depicts the block diagram for this type of converter.

8.3. ANALOG-TO-DIGITAL CONVERSION

8.3.1. Parallel Conversion

As shown in Figure 8.4, this technique has extremely high conver-

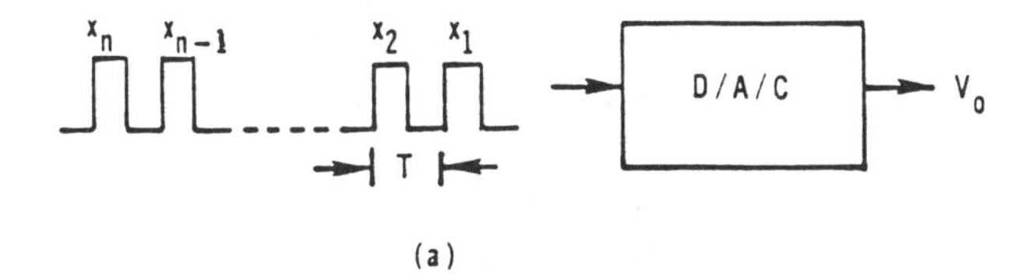

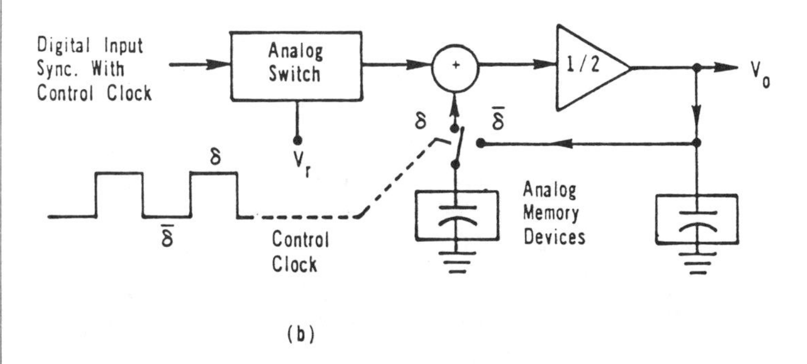

FIGURE 8.3 Serial D/A conversion.

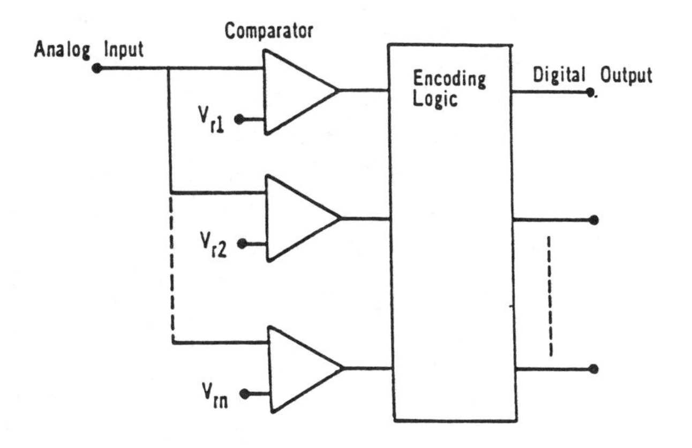

FIGURE 8.4 Parallel A/D converter.

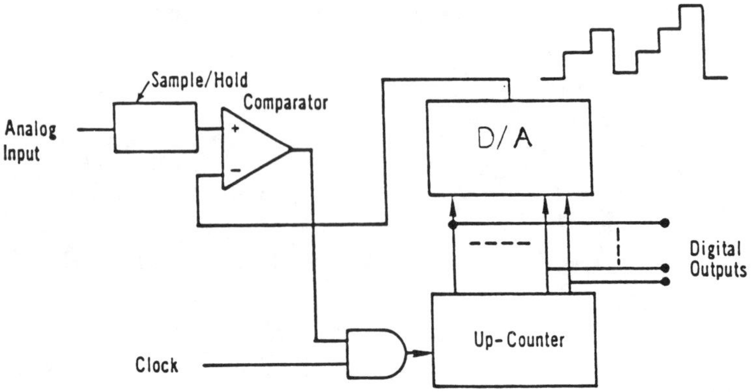

FIGURE 8.5 Counter ramp A/D converter.

sion speed. However, it requires n comparators, although the conversion speed is not a function of the word-width of the digital output. By setting the values of the reference voltages V_{r1} ... to a specific nonlinear pattern, a nonlinear conversion can be achieved. This is most likely to be used in some special applications.

8.3.2. Counter Ramp Converter

As shown in Figure 8.5, the analog signal after sample and hold is input at the positive terminal of the comparator. The counter is originally set to zero, therefore the D/A converter output which inputs at the negative terminal of the comparator is also zero. The output comparator is HIGH, and the clock signal is gated into the up-counter. As a result, there is a ramp signal at the output of the D/A converter. When this output slightly exceeds the incoming analog signal, the comparator output switches to LOW, and the clock signal is blocked from the counter. Thus, the value retained in the counter is the digital equivalent of the analog. This technique is accurate and reliable, since the circuit components used are mostly digital devices. However, the conversion speed is obviously a function of the magnitude of the analog signal and the counter speed. When the input is of full-scale value, it will require a full count of the counter. For example, if an 8-bit A/D or counter is used here, and the clock is 1 MHz, it will require 2^8, or 256 μs to reach full count. It is obvious that during this "ramp" period, no new analog data can be accepted. In other words, this system in this example has a "blind" period of 256 μs which the designer should take into consideration. Normally, there is a digital buffer register at the output. Therefore, as soon as the clock gate is disabled, the data in the up-counter will be transferred to the buffer. The up-counter will then be reset to zero as soon as new data in the sample/hold is ready for conversion. From a system point of view, the converter may be considered as a black box which expects a control pulse, called a "start-to-convert" command from a host system and the converter would generate an "end-of-conversion" pulse when data are ready in the buffer register. Most A/D converters commercially available now have these provisions for system control purposes.

8.3.3. Successive Approximation Converter

Figure 8.6(a) shows the logic diagram of a successive approximation converter. This type of converter is currently the most popular one on the market. In principle, it is similar to counter ramp converters except that it has a shorter "blind period". The ring-counter in Figure 8.6(a) is basically a (N + 1)-bit shift register. Among the N + 1 output bits, one and only one bit at a time is at logical HIGH. The relative timing diagram of the two-phase clock, ϕ_1 and ϕ_2, as well as the partial ring-counter output is shown in Figure 8.6(b). Initially, all flip-flops and ring-counters are set to zero; the output of the D/A converter is thus also zero. The incoming analog signal would cause the output of the comparator z to go LOW. The clock ϕ_1 will set the N-bit of the ring-counter HIGH, and thus the most significant bit (MSB) of the D/A converter is set HIGH. As a result, Y would yield an analog signal with a magnitude of one half of the full scale, or one half of the maximum expected analog signal. At this point, if $X < Y$, then Z goes HIGH and ϕ_2 would reset the MSB. The next ϕ_1 will shift the ring-counter and the (N – 1)-bit goes HIGH; consequently, the next MSB would be set HIGH. Now, if $X > Y$, Z goes LOW which blocks off ϕ_2. As the ring-counter moves downward, this second MSB would remain HIGH. The process continues until the output of the D/A converter matches the value of the unknown analog signal at X. Basically, the algorithm is analogous to a systematic guessing game. The first question is whether the input is greater or less than one half of the full scale. If the answer is "greater than," the next question is whether the input is greater or less than $(\frac{1}{2} + \frac{1}{4})$ of full scale. Of course, just like any other A/D converter, this one also has the provision of accepting a "start-to-convert" command and generating the "end-of-conversion" pulse for system control.

8.3.4. Dual-Slope Integrator Converter

The circuit diagram of this type of converter is shown in Figure 8.7(a). The output waveform of the integrator (see Chapter 2, Section 2.4.1), e_0, is shown in Figure 8.7(b). If the reference voltage V_r is positive, the analog input voltage e_{in} has to be negative. The conversion starts when the s switch is closed and the $\overline{s}$ switch is open. The e_0 begins to rise and the clock gate is enabled. The binary counter

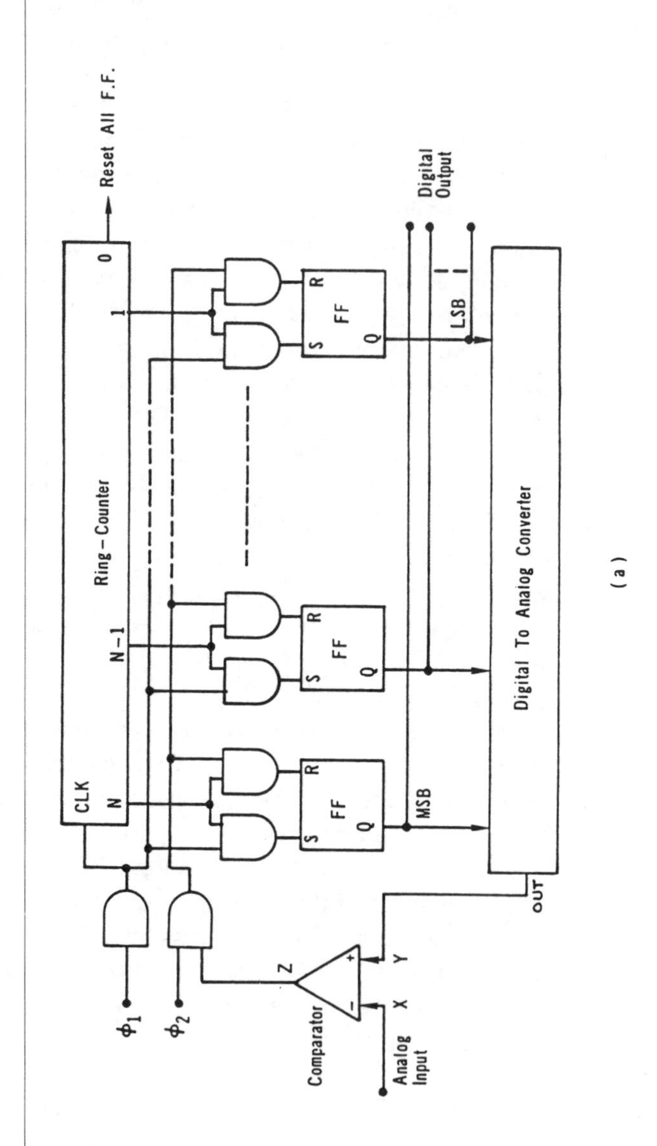

FIGURE 8.6 (a) Successive approximation A/D converter. (b) Timing diagram of the successive approximation A/D converter.

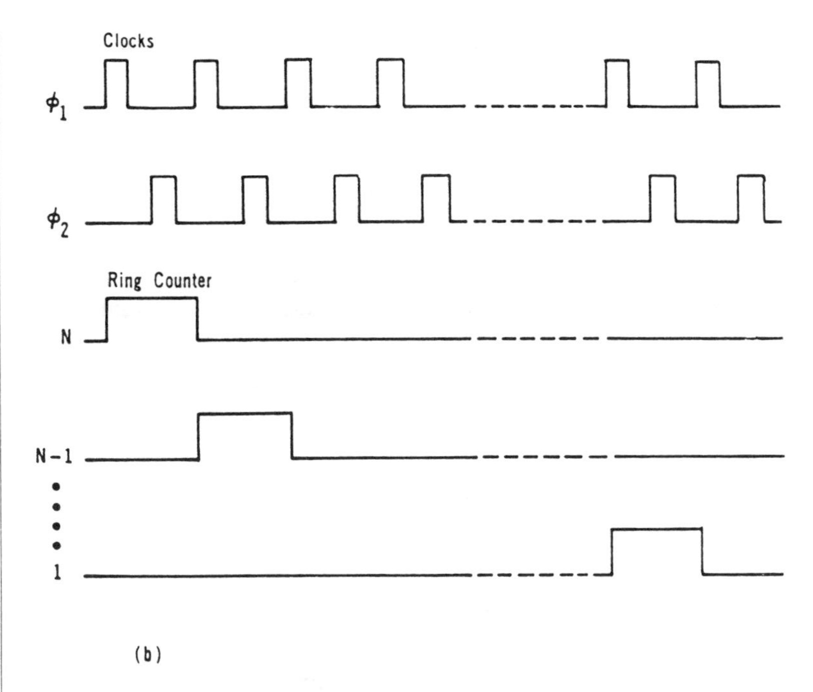

FIGURE 8.6 continued. (a) Successive approximation A/D converter. (b) Timing diagram of the successive approximation A/D converter.

counts upward until all output bits are 1 except the Nth bit, the "carry" or "control" bit. At this point, one more clock pulse would set the control bit to 1 and all of the other bits to 0. The control bit would then close the $\bar{s}$ switch and open the s switch. The V_r with opposite polarity becomes the input and e_0 decreases until it reaches zero or becomes slightly negative. As a result, the clock gate is disabled and the reading of the binary counter is thus the digital value equivalent of the analog input signal. The mathematical derivation follows. In reference to Figure 8.7(b), consider the binary counter set to zero. The s switch is thus closed. Assume that the input $e_{in} < 0$ is applied at $t = T_0$; then we have the integrator output

$$e_0(t) = -\frac{1}{RC}\int_{T_0}^{t} e_{in}\, dt$$

Recall that e_{in} is the output voltage of the sample/hold unit, which can be considered as a constant, thus for $T_0 \le t \le T_1$,

$$e_0(t) = -\frac{1}{RC}\int_{T_0}^{t} e_{in}\, dt = \frac{1}{RC}\,|e_{in}|\,(t - T_0) \tag{8.2}$$

where $|e_{in}|$ is absolute value of e_{in} and T_1 is the time at which the output 1, ..., (N-1)-bits of the counter are all [1] except the carry bit. Assume at this point that

$$e_o(t=T_1) = \frac{1}{RC}|e_{in}|(T_1 - T_0) = \Delta V \quad (8.3)$$

Now, an extra clock pulse would set the carry bit to 1 and the others to 0, which in turn closes the $\bar{s}$ switch and opens the s switch; thus, for $T_1 < t \leq T_2$,

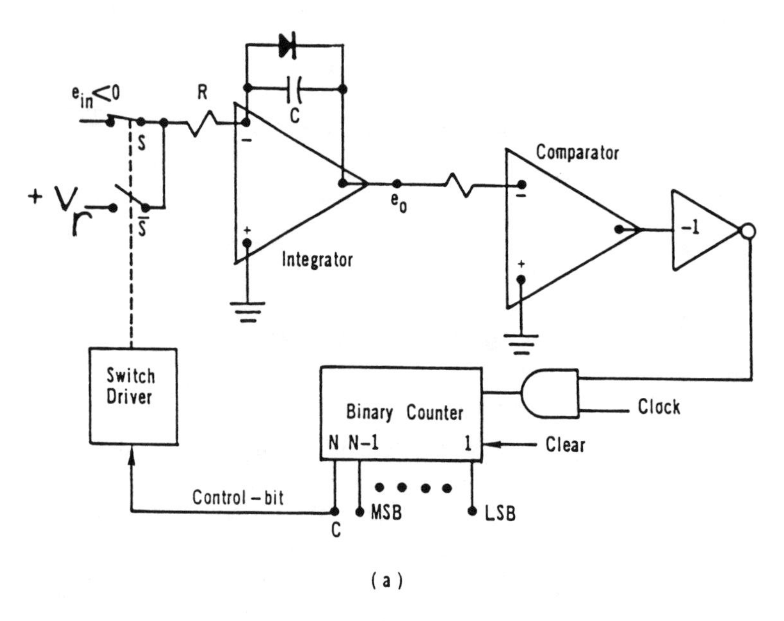

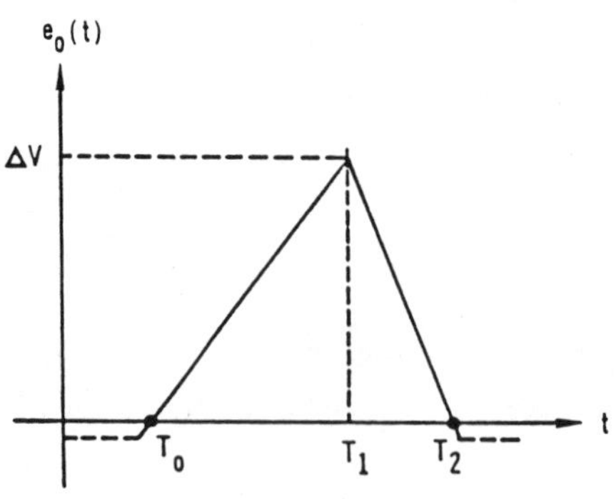

FIGURE 8.7 Dual-slope integrator A/D converter.

$$e_o(t) = -\frac{1}{RC}\int_{T_1}^{t} V_r\,dt + \Delta V$$

$$= -\frac{1}{RC}V_r(t - T_1^+) + \Delta V \tag{8.4}$$

Substituting Equation (8.3) into Equation (8.4),

$$e_o(t) = -\frac{1}{RC}V_r(t - T_1^+) + \frac{1}{RC}|e_{in}|(T_1 - T_0) \tag{8.5}$$

But at $t = T_2$, as shown in Figure 8.7(b), we have $e_o(t{=}T_2) = 0$, which would disable the clock gate and stop the counter. Equation (8.5) becomes

$$-\frac{1}{RC}V_r(T_2 - T_1^+) + \frac{1}{RC}|e_{in}|(T_1 - T_0) = 0$$

or

$$|e_{in}| = V_r\frac{(T_2 - T_1^+)}{(T_1 - T_0)}$$

However, $T_1 - T_0$ is equal to $(2^N - 1)$ clock periods, if the counter is an N-bit counter including the carry bit. Thus,

$$|e_{in}| = K(T_2 - T_1^+) \tag{8.6}$$

where

$$K = \frac{V_r}{T_1 - T_o}$$

But $(T_2 - T_1^+)$ is the counter value at $t = T_2$. Therefore, the digital equivalent of e_{in} is proportional to the value in the counter at $t = T_2$.

Examine Equation (8.6). It is interesting to note that this method of conversion is not a function of the circuit components R and C, nor of the accuracy of the clock rate. It is, however, sensitive to voltage drift of the operational amplifier used as an analog integrator. Furthermore, this A/D converter does not need a D/A converter included in it. It is relatively inexpensive. Most digital meters use this technique.

8.4. KEY ELEMENTS COMMONLY USED IN ANALOG-DIGITAL-ANALOG CONVERSION

8.4.1. Reference Voltage

This element is important in the sense that all analog values yielded are dependent on it. Normally, a zener diode is the basic device used to provide the reference voltage. More sophisticated ones require temperature compensation. For low reference voltage, <1 V, one may use a forward-biased diode as the voltage reference source, or a zener diode followed by an operational amplifier based voltage follower with a voltage gain set to less than 1. For design example, see Chapter 3, Section 3.3.2(e).

8.4.2. Analog Comparator

As described in Section 2.4.4 of Chapter 2, a comparator is a special-purpose operational amplifier. The key parameters to be considered are slewing rate, input/output impedances, settling time, and open-loop gain. The open-loop gain, A, should be greater than the logical swing voltage divided by at least the equivalent voltage of one least significant bit. The slewing rate in volts per second specifies the response speed of the amplifier.

8.4.3. Analog Switch

The desirable characteristics of an analog switch to be used for A/D/A conversion are:

- Switch-on resistance = 0
- Switch-off resistance = ∞
- Switches which do not bounce
- Zero settling time
- High switching speed
- Bilateral

a. Electro-Mechanical Switch

Electro-mechanical switches such as mercury-wet relays satisfy the above requirements. The major drawback of electro-mechanical switches is their switching speed. It would not respond to a high driving frequency. Of course, when the physical size of the analog switch is important, the electro-mechanical switch may not be desirable either.

b. Semiconductor Analog Switch

With the exception of nonlinearity and non-zero switch-on resistance, the CMOS analog switches closely match the desirable characteristics listed above. The non-zero switch-on resistance characteristic of CMOS switches may not be a serious problem in many cases, if the input and output impedances of other associated devices are known; however, the CMOS nonlinearity property needs careful consideration. A thorough understanding of the general functional properties of a CMOS switch is thus important. Figure 8.8(a) shows the schematic of a CMOS analog switch controlled by a CMOS inverter. Nominal voltages are $V_{DD} = +15$ V and $V_{SS} = -15$ V. When C = +15 V, Q_1 is OFF and Q_2 is ON; thus, $\bar{C} = -15$ V. As a result, Q_3 and Q_4 are both conducting. The equivalent channel resistances for Q_3 and Q_4 are functions of the voltage across the channel V_{DS}, which in this application would be the signal voltage, e_{in}, that we wish to switch ON/OFF from R_L. The dotted lines in Figure 8.8(b), respectively, show the channel resistance of each device vs. V_{DS}. Since both Q_3 and Q_4 are conducting, the solid line which is the resultant parallel channel resistance of Q_3 and Q_4 shows the equivalent switch-on resistance, R_{ON}, of the analog switch. Typically, R_{ON} varies between 45 and 55 Ω for $-10 \text{ V} < V_{DS} < +10$ V. If $R_L >> R_{ON}$, then the variation of the parallel channel resistance may become insignificant for processing the analog signal. When C = –15 V, Q_1 conducts and Q_2 is OFF, then $\bar{C} = +15$ V. This will turn off both Q_3 and Q_4, and the analog switch becomes a high impedance device, and e_{in} is then practically disconnected from R_L. The signal voltage e_{in} should not however exceed the range between V_{DD} and V_{SS}. Figure 8.8(c) shows another point of view on how a CMOS analog switch functions by means of Binary State Analysis. As shown, the composite $V_{DS} - I_D$ piece-wise linear curves for the switch ON/OFF mode can be used to plot the response of the device to the input signal e_{in}. Here, the e_{in} shown is a triangular waveform, which is also the voltage across the switch and the load resistor R_L. As the magnitude of e_{in} changes, the load line (assuming $R_L >> R_{ON}$) locus sweeps through the V-I space as shown. Accordingly, the I_D through the switch can be determined graphically as shown. The load line loci, I_D and V_{DS} should satisfy the following loop equation:

$$e_{in} = V_{DS} + I_D R_L$$

For example, when e_{in} = +max, we may determine graphically the corresponding $I_D = I_1$, $V_{DS} = V_1$; similarly, when e_{in} = –max, we have $I_D = I_2$ and $V_{DS} = V_2$. Note that as V_{DS} decreases, the dotted R_L load line moves leftward. Recall that ideally R_{ON} should be negligible with respect to R_L, so that the voltage-drop across the switch, V_{DS}, could be

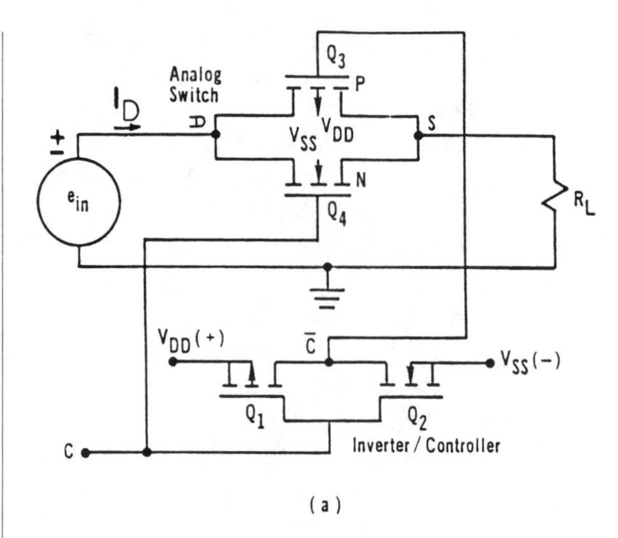

(a)

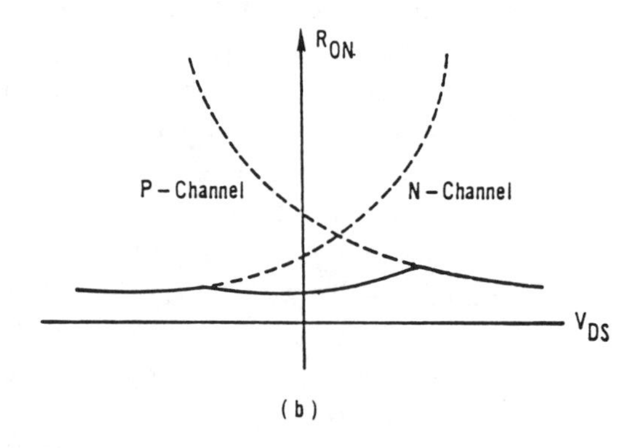

(b)

FIGURE 8.8 (a, b) CMOS analog switch. (c) Graphical analysis of analog switch.

negligible in comparison with the voltage across R_L when the switch is ON.

8.4.4. D/A Decoder Resistor Networks

a. Weight-Resistor D/A Decoder

As shown in Figure 8.9, we have

$$V_0 = -\left(\frac{R_f}{R_0} x_0 + \ldots + \frac{R_f}{R_j} x_j + \ldots + \frac{R_f}{R_{n-2}} x_{n-2}\right) E_{in}$$

$$= x_{n-1} V_R \left(\frac{R_f}{R_0} x_0 + \ldots + \frac{R_f}{R_j} x_j + \ldots + \frac{R_f}{R_{n-2}} x_{n-2}\right)$$

where $x_{n-1}, x_{n-2}, \ldots, x_j, \ldots, x_0$ are digital inputs indicating the switch positions. Note that x_{n-1} is the sign bit; logic [0] = 1, logic [1] = –1. Note also that $x_j = 1; 0,\ j = n–2, \ldots, 0$. For $R_j = (1/2^j)R_f$, we have

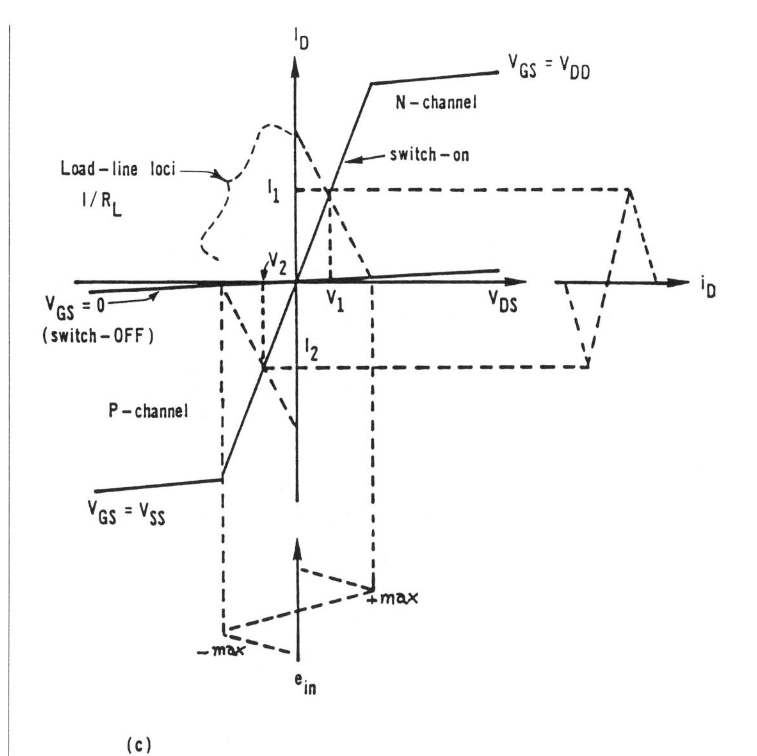

FIGURE 8.8 continued. (a, b) CMOS analog switch. (c) Graphical analysis of analog switch.

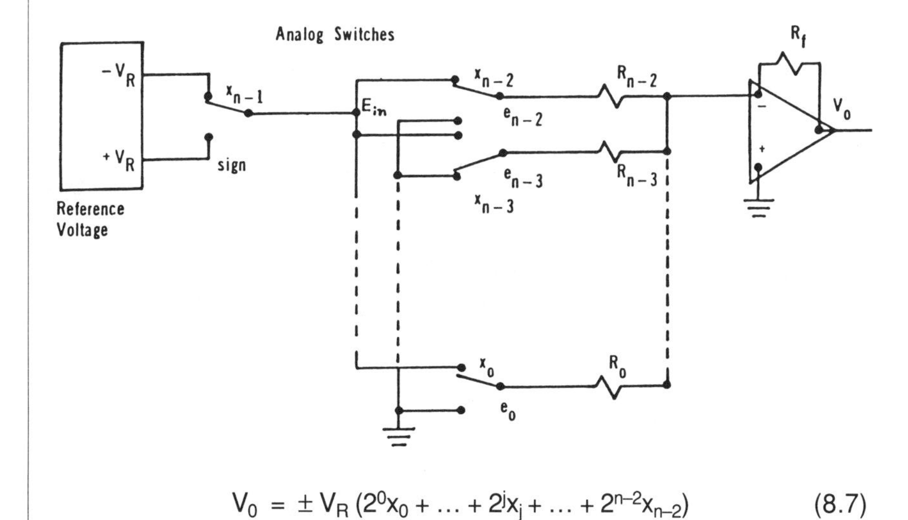

FIGURE 8.9 D/A using weight-resistor D/A decoder.

$$V_0 = \pm V_R (2^0 x_0 + \ldots + 2^j x_j + \ldots + 2^{n-2} x_{n-2}) \tag{8.7}$$

Mathematically, Equation (8.7) appears to have no problem. In practice, however, when n is large, say 12, it may cause problems. To illustrate, let R_{ON} of the analog switch = 50 Ω. In order to make the variation of R_{ON} be insignificant to this system, one may select

$$R_{n-2} = 200 \times R_{ON} = 10\text{ K}$$

But for n = 12,

$$R_f = 2^{n-2} R_{n-2} = 2^{10} \times 10\text{ K}$$

This is obviously not practical. Therefore, this decoder is more appropriate for application in systems containing digital words with a low number of bits.

b. Resistor-Ladder D/A Decoder

The D/A converter using the resistor-ladder as its decoder is shown in Figure 8.10(a). The outstanding feature of this network is that the resistance of each of the three branches at any junction J_i, i = 0, 1, ..., n–2, is always equal to 2R, and the whole network uses only R and 2R as circuit components which in practice is the most desirable feature, regardless of whether the network is to be implemented by discrete components or integrated circuits. The priciple of operation of the resistor-ladder is described next.

Note that the network has the following properties:

i) The total resistance of each branch is 2R.
ii) The current due to each bit is the same, i.e., $I = \frac{V_r}{3R}$, and
iii) The incoming current I at any junction J_i is always split into two equal outgoing currents, $\left(\frac{1}{2}\right)I$, as shown in Figure 8.10(b).

Based on these properties, one can apply the superposition theory to obtain the input current of the operational amplifier.

$$I_t = I\left(\frac{1}{2^{n-2}} x_0 + \frac{1}{2^{n-3}} x_1 + \dots + \frac{1}{2} x_{n-2}\right)$$

$$= \frac{V_r}{3R}\left(\frac{1}{2^{n-2}} x_0 + \frac{1}{2^{n-3}} x_1 + \dots + \frac{1}{2} x_{n-2}\right)$$

But $V_0 = -R_f I_t$, thus

$$V_0 = -\frac{V_r}{3R} \cdot R_f\left(\frac{1}{2^{n-2}} x_0 + \frac{1}{2^{n-3}} x_1 + \dots + \frac{1}{2} x_{n-2}\right)$$

If we let $R_f = 3R$, then

$$V_0 = -V_r\left(\frac{1}{2^{n-2}}x_0 + \frac{1}{2^{n-3}}x_1 + \ldots + \frac{1}{2}x_{n-2}\right)$$

8.4.5. Sample and Hold Elements

a. Circuits

Figure 8.11 shows a typical sample and hold circuit on a commercially available chip. Figure 8.11(a) is a circuit of noninvertible configuration with an overall gain of

$$G = 1 + \frac{R_2}{R_1}$$

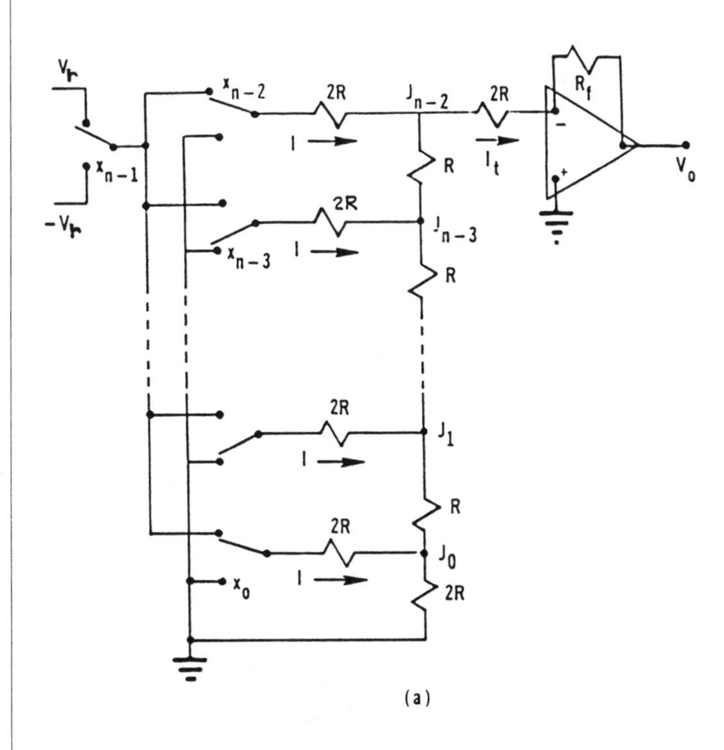

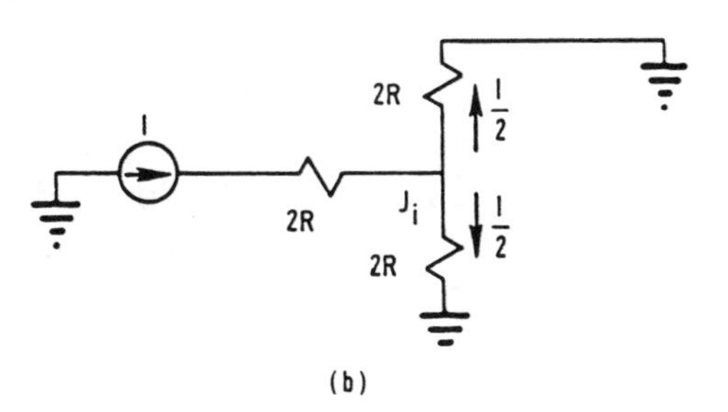

FIGURE 8.10 Resistor ladder D/A decoder.

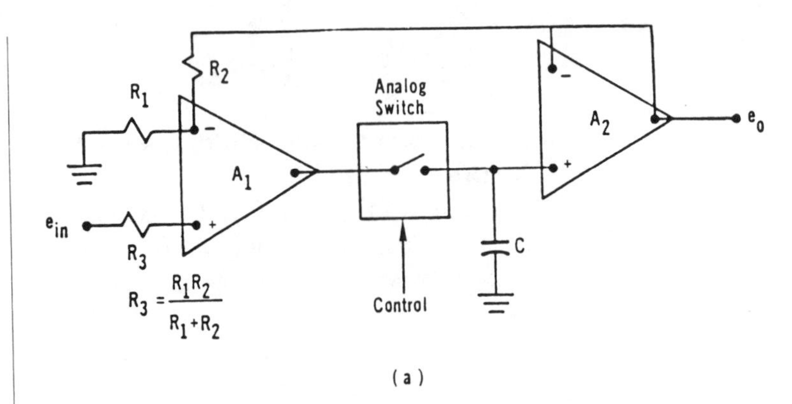

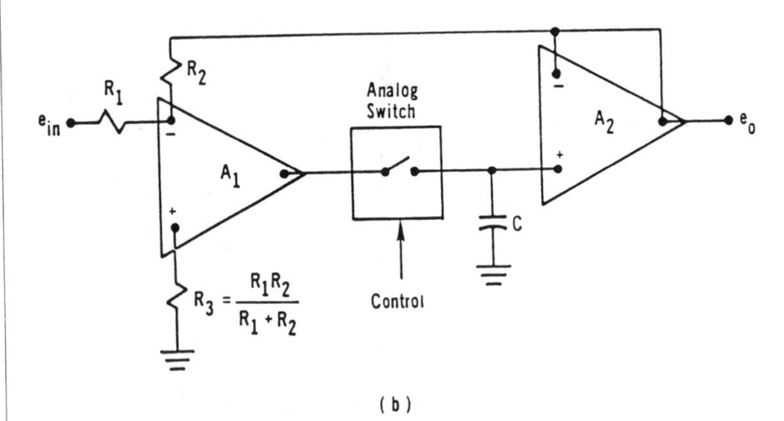

FIGURE 8.11 Sample and hold circuits.

and Figure 8.11(b) is a circuit of invertible configuration with an overall gain of

$$G = -\frac{R_2}{R_1}$$

Operational amplifier A_1, having low output resistance, is used to charge the capacitor, while operational amplifier A_2 is a buffer amplifier which has high input impedance to prevent the information stored in the capacitor from leaking during the A/D conversion process.

b. Definitions

There are two important terms pertaining to the sample and hold circuit which should be defined here.

1. *Aperture time* — the period of delay between the HOLD command and the instant at which the analog switch is really OPEN. A typical aperture time is 50 ns.

2. *Acquisition time* — the time between the SAMPLE command and the instant at which the output voltage e_o settles to ±1% of its final value. A typical acquisition time is 5 μs.

8.5. SPECIAL PURPOSE D/A CONVERSION CIRCUITS

There are two special purpose D/A conversion circuits worthy of mention here.

8.5.1. Inverse D/A Converter

Figure 8.12(a) shows an inverse D/A converter. That is, the output voltage is inversely proportional to the binary-coded input number. Note that the operational amplifier at the output of the resistor network functions as a high input impedance buffer amplifier with unity gain. The equivalent resistance shunting the amplifier input is the total parallel resistance of the D/A decoder network, i.e.,

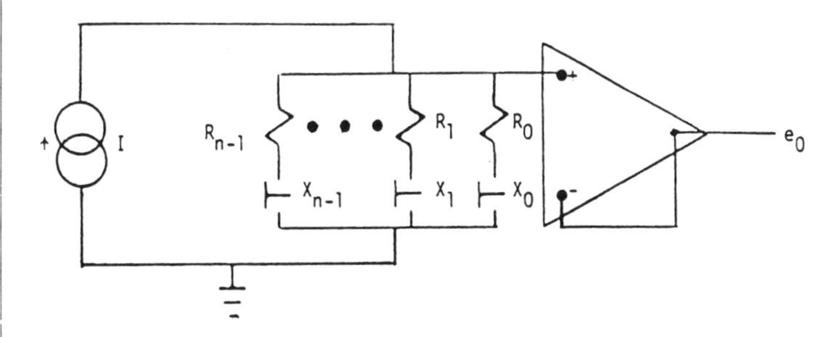

(a) Inverse D/A Converter

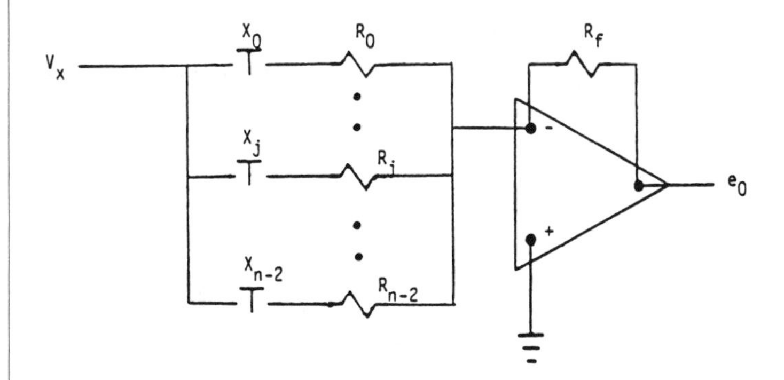

(b) Multiply D/A Converter

FIGURE 8.12 Special purpose D/A conversion circuits.

$$R_t = \frac{1}{\sum_{j=0}^{n-1} \frac{1}{R_j} x_j}$$

Thus,

$$e_o = I R_t = I \frac{1}{\sum_{j=0}^{n-1} \frac{1}{R_j} x_j}$$

where I = a constant current, and $x_j = 1, 0; j = 0, 1, \ldots, n-1$.

8.5.2. Multiply D/A Converter

The circuit shown in Figure 8.12(b) is similar to the parallel D/A converter described in Section 8.2.1, except that V_r in this circuit is no longer a constant. That is, the reference voltage V_r has been changed to V_x, which becomes a variable carrying information. Therefore, in this configuration

$$V_o = V_x (2^0 x_0 + \ldots + 2^j x_j + \ldots + 2^{n-2} x_{n-2})$$

8.6. CODES EMPLOYED BY A/D CONVERTERS

As we know, the function of an A/D converter is to assign a digital code of "one" or "zero" to represent an analog value. There are many kinds of coding systems used for A/D conversion. In this section we describe the three most popular codes using a 4-bit word as an example to represent an analog voltage which ranges from 0 to 1024 mV. In general, one least-significant-bit (LSB) of an n-bit word is equivalent to

$$\frac{\text{Total Voltage Range}}{2^n}$$

For n = 4, and Total Voltage Range = 1024 mV, we have

$$1 \text{ LSB} = 64 \text{ mV}$$

8.6.1. Binary Codes

Tables 8.1 and 8.2, respectively, show a unit polar binary code and a bipolar offset binary code. The latter is called an offset code because the analog equivalent is offset by –512 mV so that the 0 voltage is now in the middle of the binary code.

8.6.2. Complement Codes

One's complement and two's complement codes for bipolar analog voltage are shown respectively in Table 8.3 and Table 8.4. In Table 8.3, the eight negative binary words are the one's complement of the eight positive ones with the most significant bit as sign-bit. They are symmetric and have separate codes for +0 and –0 as shown. In the two's complement code shown in Table 8.4, however, there is only one 0 V equivalent, i.e., 0000 = –32 to +32. The negative codes are the two's complement of the corresponding positive ones, but symmetry with respect to zero is lost in contrast to the one's complement code. Again, the most significant bit is the sign-bit.

8.6.3. Gray Code

Table 8.5 shows the Gray code equivalent of the same analog voltage. The outstanding feature of Gray code is that there is always only one bit difference between adjacent digital words. This is most desirable for coding a rotating shaft. The rule for Gray code is that starting from the LSB column leftward, there are 2, 4, 8, 16, ..., ones alternating with 2, 4, 8, 16, ... zeroes downward, respectively, except where both ends of the column split half the expected number of zeroes, i.e., 2, 4, 8. 16, ... when there is enough space available at the ends to fit in the split number of zeroes.

TABLE 8.1
Unit Polar Binary Code

Analog voltages (mV)	Binary code			
0 to 64	0	0	0	0
64 to 128	0	0	0	1
128 to 192	0	0	1	0
192 to 256	0	0	1	1
256 to 320	0	1	0	0
320 to 384	0	1	0	1
384 to 448	0	1	1	0
448 to 512	0	1	1	1
512 to 576	1	0	0	0
576 to 640	1	0	0	1
640 to 704	1	0	1	0
704 to 768	1	0	1	1
768 to 832	1	1	0	0
832 to 896	1	1	0	1
896 to 960	1	1	1	0
960 to 1024	1	1	1	1

TABLE 8.2
Bipolar Offset Binary Code

Analog voltages (mV)	Binary code
–512 to –448	0 0 0 0
–448 to –384	0 0 0 1
–384 to –320	0 0 1 0
–320 to –256	0 0 1 1
–256 to –192	0 1 0 0
–192 to –128	0 1 0 1
–128 to –64	0 1 1 0
–64 to 0	0 1 1 1
0 to 64	1 0 0 0
64 to 128	1 0 0 1
128 to 192	1 0 1 0
192 to 256	1 0 1 1
256 to 320	1 1 0 0
320 to 384	1 1 0 1
384 to 448	1 1 1 0
448 to 512	1 1 1 1

TABLE 8.3
One's Complement Code

Analog voltages (mV)	One's complement	Signed-magnitude equivalent
–512 to –448	1 0 0 0	–7
–448 to –384	1 0 0 1	–6
–384 to –320	1 0 1 0	–5
–320 to –256	1 0 1 1	–4
–256 to –192	1 1 0 0	–3
–192 to –128	1 1 0 1	–2
–128 to –64	1 1 1 0	–1
–64 to 0	1 1 1 1	–0
0 to 64	0 0 0 0	+0
64 to 128	0 0 0 1	+1
128 to 192	0 0 1 0	+2
192 to 256	0 0 1 1	+3
256 to 320	0 1 0 0	+4
320 to 384	0 1 0 1	+5
384 to 448	0 1 1 0	+6
448 to 512	0 1 1 1	+7

TABLE 8.4
Two's Complement Code

Analog voltages (mV)	Two's complement	Signed-magnitude equivalent
−480 to −544	1 0 0 0	−8
−416 to −480	1 0 0 1	−7
−352 to −416	1 0 1 0	−6
−288 to −352	1 0 1 1	−5
−224 to −288	1 1 0 0	−4
−160 to −224	1 1 0 1	−3
−96 to −160	1 1 1 0	−2
−32 to −96	1 1 1 1	−1
−32 to +32	0 0 0 0	0
32 to 96	0 0 0 1	+1
96 to 160	0 0 1 0	+2
160 to 224	0 0 1 1	+3
224 to 288	0 1 0 0	+4
288 to 352	0 1 0 1	+5
352 to 416	0 1 1 0	+6
417 to 480	0 1 1 1	+7

TABLE 8.5
Gray Code

Analog voltages (mV)	Gray code
−512 to −448	0 0 0 0
−448 to −384	0 0 0 1
−384 to −320	0 0 1 1
−320 to −256	0 0 1 0
−256 to −192	0 1 1 0
−192 to −128	0 1 1 1
−128 to −64	0 1 0 1
−64 to 0	0 1 0 0
0 to 64	1 1 0 0
64 to 128	1 1 0 1
128 to 192	1 1 1 1
192 to 256	1 1 1 0
256 to 320	1 0 1 0
320 to 384	1 0 1 1
384 to 448	1 0 0 1
448 to 512	1 0 0 0

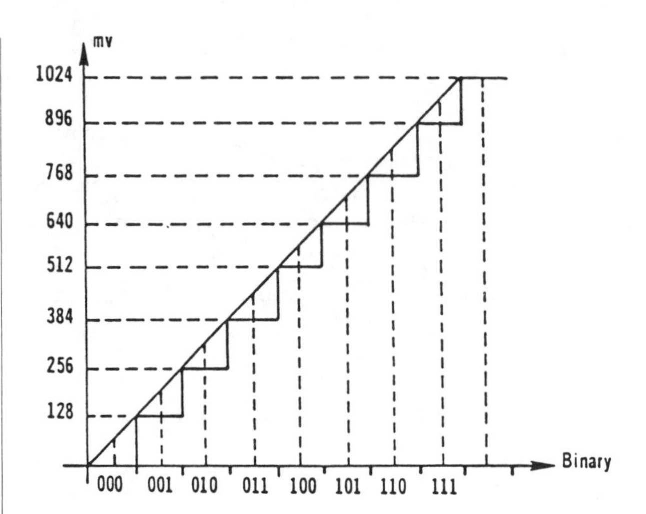

FIGURE 8.13
Quantization error.

8.7. MAJOR ERRORS IN CONVERSION

8.7.1. Quantization Error

Figure 8.13 illustrates the quantization error of a 3-bit binary code for an analog voltage range from 0 to 1024 mV. Note that each binary word represents a range of 128 mV. For example, 001 represents the analog value ranging from 128 to 256 mV; i.e., any analog voltage within the range of 128 to 256 mV will be assigned the value 001. When converting 001 back to analog form, only one value, say 192 mV, will be assigned. Therefore, logically, each binary word inherently has a $\pm 1/2$ LSB error associated with it.

8.7.2. Analog Components Error

This type of error is due to the fact that in reality there are no perfect analog components. Their properties or values change as environmental conditions vary. Therefore, analog errors inherently exist in any system using analog components. Fortunately, most manufacturers of A/D/A converters restrict the analog errors of their products to within a $\pm 1/2$ LSB equivalent, so that the total error will be less than ± 1 LSB.

8.7.3. Aperture Error

Figure 8.14 shows the effect of aperture time on the accuracy of the data obtained. Note that the uncertainty of switching time during the transition of the sample and hold would cause error if the analog signal changed rapidly within this aperture time period. Thus, the aperture error can be estimated by the following equation.

$$\text{Aperture Error} = \frac{\Delta v}{\Delta t} \times T_a$$

where $\frac{\Delta v}{\Delta t}$ is the time slope of the signal as shown, and T_a is the aperture time. Normally, the worst case estimation is employed. That is, the maximum $\frac{\Delta v}{\Delta t}$ of the analog signal is used. For example, if v(t) is the analog signal, then $\left(\frac{\Delta v}{\Delta t}\right)_{max}$ is determined by the equation

$$\frac{d_2 v(t)}{dt^2} = 0$$

8.8. KEY CONSIDERATIONS FOR SYSTEM DESIGN USING A/D/A CONVERTERS

In designing the system shown in Figure 8.1, a designer should first understand the characteristics of the analog signal, i.e., the voltage range, tolerable total error, and the resolution required. Based on this information, the number of bits of the A/D/A converter can be determined. For example, if the range is 1 V and the resolution is 1 mV, then at least a 10-bit A/D/A converter is required, since

$$1 \text{ LSB} = \frac{1 \text{ V}}{2^{10}} = \frac{1}{1024} = 0.977 \text{ mV}$$

Now, if the maximum frequency component of the signal is 100 Hz, or its frequency spectrum or bandwidth ranges from dc to 100 Hz, then a

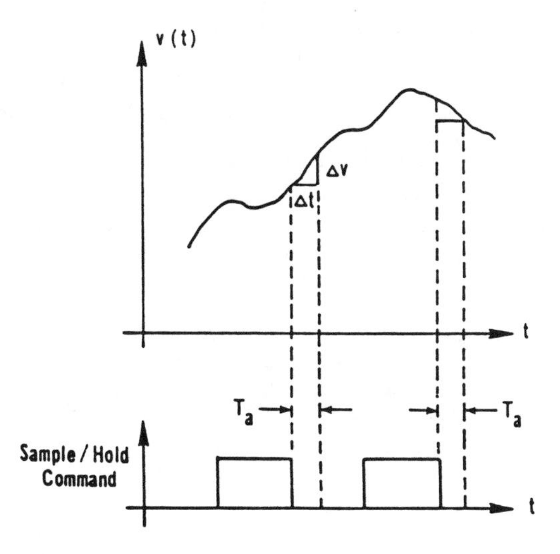

FIGURE 8.14 Aperture error.

low-pass filter cut-off at 100 Hz with sharp attenuation beyond 100 Hz should be used as the guard or anti-aliasing filter. At this point, the sampling frequency of 200 Hz, or preferably more, should be selected. If a sampling frequency of 1000 Hz is chosen, then the maximum hold time would be 500 μs. This figure restricts the maximum conversion time specification of the A/D/A converter. As one may expect, the D/A converter must have the same resolution, e.g., 10-bit, as the A/D converter; the same cut-off frequency value used for the guard filter must also be used for the smooth filter. Finally, one can total the tolerable error of the system. Based on this information, the specifications for aperture time for the sample and hold circuit and that of the total quantization and analog errors for the A/D/A converter can be determined.

REFERENCES

1. Naylor, J. R., Digital and analog signal applications of operational amplifiers I — multiplexers and converters, *IEEE Spectrum,* 8(5), 79—87, May 1971.
2. Naylor, J. R., Digital and analog signal applications of operational amplifiers II — sample/hold modules, peak detectors and comparators, *IEEE Spectrum,* 8(6), 38—46, June 1971.
3. Hoeschele, D. F., Jr., *A/D and D/A Conversion Techniques*, John Wiley & Sons, New York, 1968.
4. Susskind, A. K., *Notes on A/D Conversion Techniques,* MIT Press, Cambridge, MA, 1956.
5. Schmid, H., A practical guide to A/D conversion, *Electronic Design,* 16(22), 49—88, October 24, 1968.
6. Schmid, H., A practical guide to A/D conversion, Parts I to III, *Electronic Design,* 16(25), 49—72, December 5, 1968, 16(26), 57—76, December 19, 1968, and 17(1), 97—112, January 4, 1969.
7. Lin, W. C., A precision A/D converter for nonlinear conversion," *Control Engineering,* 17(4), 92—96, April 1970.
8. Graeme, J. G., Tobey, G. E., and Huelsman, L. P., *Operational Amplifiers: Design and Application,* McGraw-Hill, New York, 1971, chap. 9.
9. Gordon, B. M., Noise Effects on A/D conversion accuracy, Part I, *Computer Design,* 13(3), 65—76, March 1974.
10. Risch, D., Design D/A/D interfaces for your computer, *EDN,* 19(7), 34—40, April 5, 1974.
11. Fogarty, J. D., Know your A/D converter's dynamic range, *Electronic Design,* 23(3), 76—78, February 1, 1975.
12. Guide to Analog CMOS Switches and Multiplexers, Analog Devices, Inc., Norwood, MA, 1974.
13. *Analog/Digital Conversion Notes,* Analog Devices, Inc., Norwood, MA, 1977.
14. Thibodeaux, E., Getting the most out of C-MOS devices for analog switching jobs, *Electronics,* 48(26), 69—74, December 25, 1975.
15. Sheingold, D. H., Ed., *Analog-Digital Conversion Handbook,* Prentice-Hall, Englewood Cliffs, NJ, 1986.
16. *Analog IC Data Book,* Precision Monolithics, Inc., 1988.

17. *Burr-Brown IC Data Book,* Vol. 33, Burr-Brown Corporation, Tucson, AZ, 1989.
18. Application Notes, Burr Brown Corporation:
    AN-130 "Voltage-to-Frequency Converters offer Useful Options in A/D Conversion," 1984.
    AN-90 "Isolation Amplifier Employs Differential Optical Coupler to Achieve Linearity Stability," 1986.

9 NOISE IN DIGITAL SYSTEMS

9.1. INTRODUCTION

Any experienced circuit or system designer has one way or the other experienced the nightmare of noise problems in equipment design. It has been a severe problem to experimenters for centuries and yet has never been appreciated in classrooms. Time after time, manufacturers cannot deliver their products because of their apparently unsolvable noise problems. Although there is no theory or sure way to cope with noise problems, by careful planning plus following a set of guidelines or rules, one may not need to learn things the hard way (or the expensive way). Generally speaking, noise problems in analog systems are quite different from those in digital systems. Electronic systems containing electro-mechanical devices behave differently in comparison with pure electronic systems. Unfortunately, a digital system designer will likely to be asked to design systems involving all three types of circuitry. In this chapter, an attempt will be made to provide the system designer with an analytical background and a set of guidelines to cope with the real problems of the real world.

9.2. EXTERNAL OR RADIATION NOISE

There are many cases where the noise causing erroneous functioning in electronic equipment is external in origin. Analog circuits and regenerative modules within the system (such as operational amplifiers, comparators, one shot, Schmitt triggers, flip-flops) are more sensitive to this type of noise than gates and other passive devices, particularly if high gain or high input impedance amplifiers are used. These noise sources are either high voltage or high current in nature. Thus, noise would be introduced into the system either through electrostatic induction due to the high voltage or through electromagnetic induction due to high current sources. For example, lightning and high voltage switching devices belong to the former category, while welding and starting or stopping large or heavy current motors would belong to the

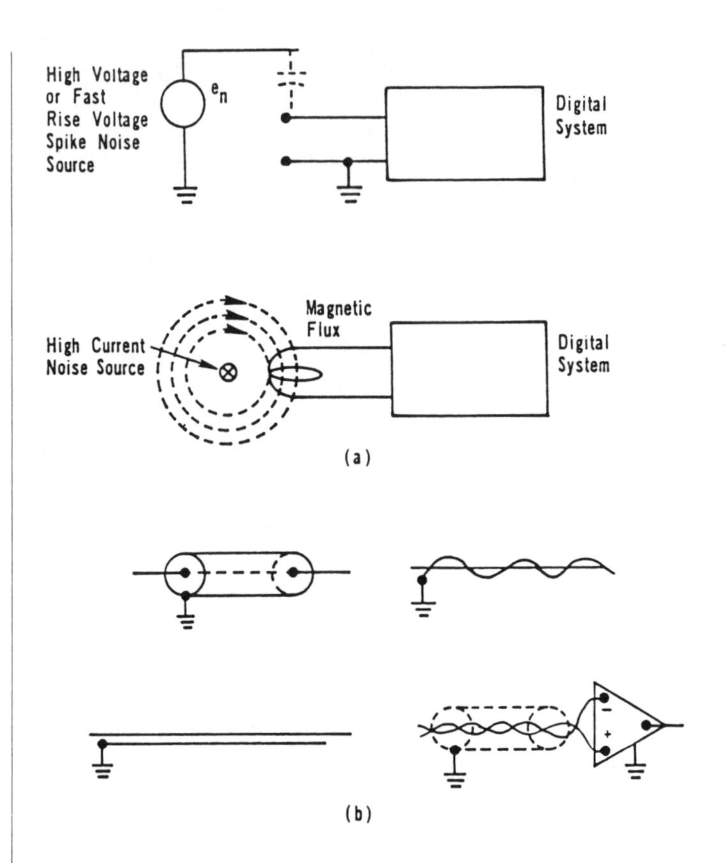

FIGURE 9.1 Shielding techniques.

latter. Figure 9.1(a) shows the concepts of electrostatic and electromagnetic induction. Problems of this sort are mostly solved by proper shielding of the sensitive devices. Figure 9.1(b) suggests some methods of shielding. For electromagnetic noise, some kind of high μ ferromagnetic material should be used for shielding. These materials are expensive; fortunately, high current sources are not as common as high voltage noise sources. For electrostatic noise, some metal such as aluminum makes a satisfactory shield in most cases. The coaxial or twisted-pair wire shielded cable is the most effective among the four shown in Figure 9.1(b). Plain twisted-pair wire is better than parallel grounded wire. As for ribbon cable, if possible one should tie every other conductor to ground. For more critical, low-signal analog circuits, one should consider enclosing the whole section with an aluminum box which should be tied to the system ground with a heavy conductor. If necessary, differential amplifiers should be used in conjunction with two-conductor coaxial cables. Note that the diagram in Figure 9.1(b) has only one end of the shielded wire connected to ground, and the other end is floating. This configuration is recommended only for

for low frequency and low level analog signals. If both ends are grounded, the shielded wire and the ground plane would form a loop which would pick up noise and be amplified as a signal. For digital signals where no amplifier is being used, however, grounding both ends is recommended. A high frequency fast rise-time clock in digital systems usually causes "cross-talk". In this case, it is desirable to shield the clock line using one of the methods shown in Figure 9.1(b). This would prevent the clock from causing cross-talk.

9.3. INTERNAL NOISE

9.3.1. Decoupling Technique

Often in both analog and digital circuits, noise may generate in one circuit and propagate to another when both are sharing the same power supply, as shown in Figure 9.2(a). Here, R_0 is a lump resistance which may include the internal resistance of the power supply and the line resistance (impedance). While Circuit #1 is in operation, it may draw a current pulse from the power supply. This pulse may be transmitted or propagated along the line down to Circuit #2 and become a noise to it. This type of noise is a common problem in both analog and digital circuitry. Therefore, it is recommended that a decoupling capacitor C_d always be included for power supply lines on each circuit

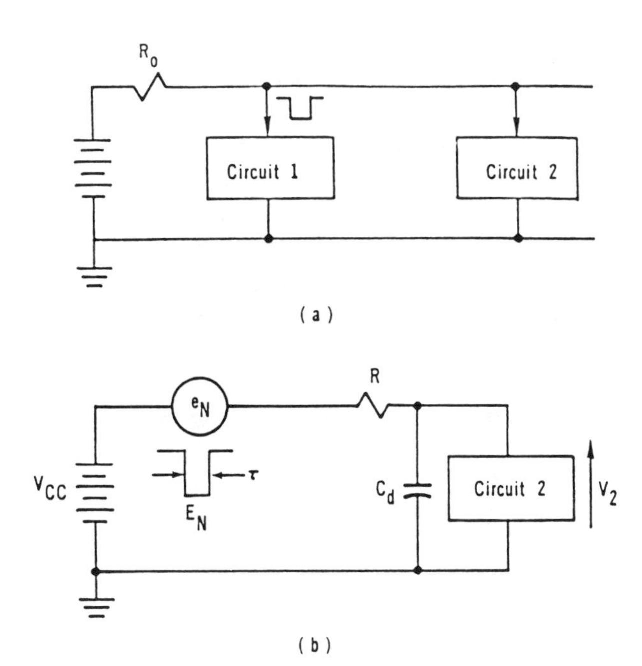

FIGURE 9.2 Noise in power supply line.

board, and that for amplifiers or one shots, Schmitt triggers, or flip-flop chips, one should connect additional C_ds, e.g., of 0.1 μF, directly across its power supplies and ground pins. An example will clarify the problem. As shown in Figure 9.2(b), let the noise pulse produced by Circuit #1 be of amplitude $-E_N$ and have a pulse width of τ. R is the Thevenin resistance of the power supply and Circuit #1. A decoupling capacitor C_d is connected directly across Circuit #2. Then

$$V_2(t) = V_2(\infty) + \left[V_2(0^+) - V_2(\infty)\right] e^{-t/RC_d}$$

$$V_2(\tau) = (V_{CC} - E_N) + \left[V_{CC} - (V_{CC} - E_N)\right] e^{-\tau/RC_d}$$

If

$$RC_d >> \tau$$

then

$$e^{-\tau/RC_d} \approx 1$$

and

$$V_2(\tau) = V_{CC}$$

This means that Circuit #2 will not see the noise if RC_d is much greater than the pulse width τ, e.g., 5 τ. It is, however, not practical to control R; therefore, an electrolytic capacitor of tens or even hundreds of microfarads is generally used for this purpose. It is important to point out that in the very high frequency or switching speed circuitry, a small disc capacitor or 0.1 μF should be connected in parallel with a good sized electrolytic capacitor. The reason for the addition of the small capacitor is that the electrolytic capacitor has inherent inductance which would not fill out radio-frequency, thus the small capacitor used here is sometimes called an RF capacitor.

9.3.2. Grounding Technique

Experience has shown that a well-planned grounding system will eliminate a lot of noise problems in system design, especially for a system which contains a mixture of analog, digital, and electro-mechanical subsystems. As shown in Figure 9.3(a), there are two important points which need to be emphasized, i.e., (1) each subsystem should have its own power supply, and (2) there should be two grounding systems, namely chassis ground and ac power ground. Normally, if each subsystem has its own chassis, then the common-terminal of the power supplies should have one and only one point connected to its individual chassis. The three chassis should then be connected together with heavy high conductance conductor and, finally, electrically tied to the ac ground. Optical coupling devices are strongly recommended for signal coupling between electro-mechanical and other subsystems. If separate power supplies for analog and digital circuitries are not economically feasible, one should at least use separate voltage regulator chips or dc/dc converters which are now commercially available at reasonable cost. Of course, a good decoupling network is still necessary before the voltage regulators. Another problem, as illustrated in Figure 9.3(b), is that there sometimes may be too many devices sharing the same "return-line." In this case,

$$i_t = i_s + i_1 + \ldots + i_j + \ldots + i_n$$

If z_0 is the impedance of the return wire segment as shown, the input signal of D_1 would be $e_s - i_t z_0$ instead of pure e_s, the signal voltage. Figure 9.3(c) shows the "center point" configuration for dealing with this kind of problem. This same recommendation holds for the power distribution layout for the back-plane wiring.

9.3.3. Example

In Sections 9.2 and 9.3 we presented techniques such as power supply decoupling, shielding, and grounding for minimizing the operational errors in a digital system caused by internal and external noise sources. While the decoupling technique has been discussed and illustrated with a numerical example in Section 9.3.1, others were stated as general rules. It may be helpful to consider a practical example for illustration purposes. Say we want to design a mobile robot system. This system would definitely consist of three essential subsystems: analog, digital, and electro-mechanical. Figure 9.3(d) depicts a simplified functional system diagram of the mobile robot, where the shielding and grounding techniques are explicitly shown. We must remember that, although they are not shown, there are decoupling capacitors with values of tens or hundreds of microfarads across the power supply lines (at the entering points), and that 0.1 μF disk capacitors are uniformly distributed, say one every 3 to 6 in., along power lines on each IC board. In addition, local decoupling capacitors of 0.1 μF have also been installed across the power/ground pins of all sensitive chips such as flip-flops, one shots, etc. For simplicity's sake, the

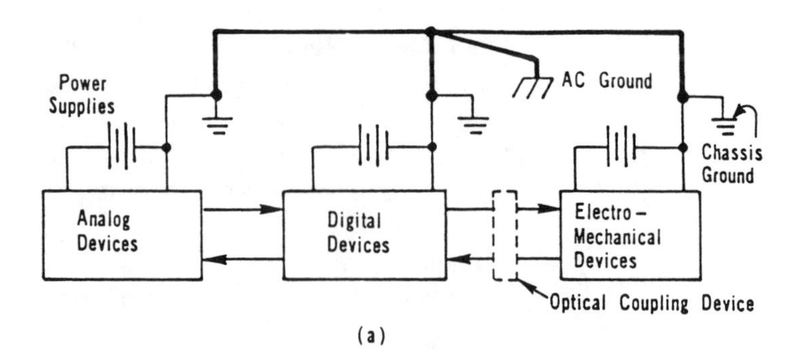

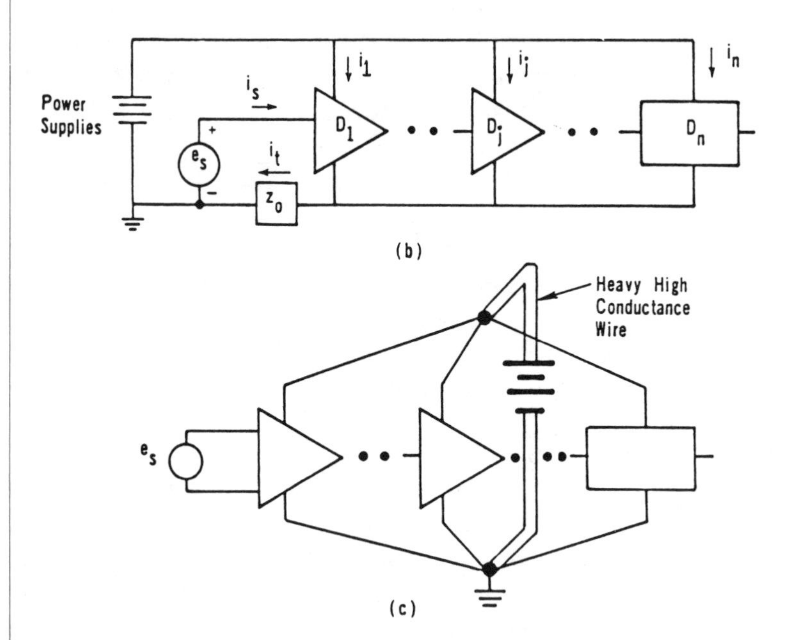

FIGURE 9.3 (a, b, c) Grounding technique. (d) Simplified functional diagram of a mobile robot.

mobile robot is assumed to serve only one function, namely, cleaning the floor of a specific area. If there are obstacles in the way, it avoids them by moving around them. The functional description of the three subsystems as shown in Figure 9.3(d) follows.

In this system, an ultrasound ranging device (see Reference 11 of this chapter for details) is used as an ultrasound transmitter and receiver. As the analog subsystem shown in the diagram, the controller enables the driver to apply a burst of 45 kHz, 300 V electric signals on the device, which in turn transmits a burst of 45 kHz ultrasound in the air. As this signal encounters an obstacle, an echo is generated which is then sensed by the ultrasound receiver after some delay time, which indicates the distance between the robot and the obstacle.

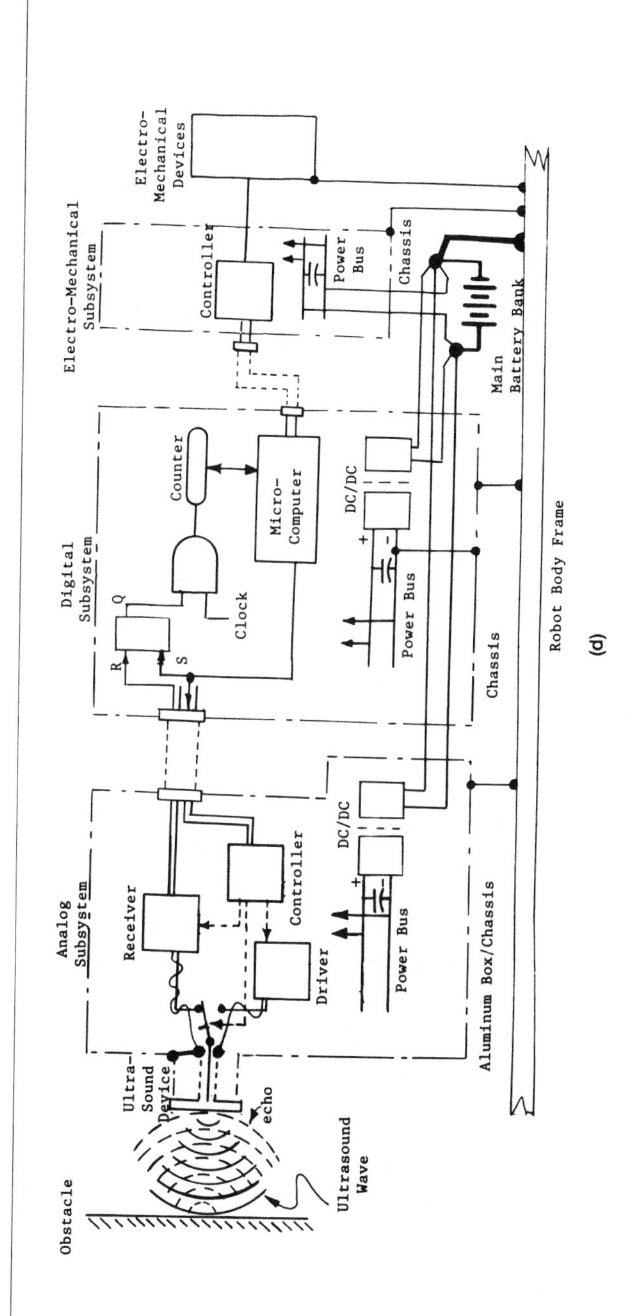

FIGURE 9.3 continued. (a, b, c) Grounding technique. (d) Simplified functional diagram of a mobile robot.

Functionally, the microcomputer in the digital subsystem periodically issues a command signal which sets the R-S flip-flop and enables the counter in the digital subsystem as shown; meanwhile, the command also enables the ultrasound transmitter to emit its signal. The echo received back is amplified by the high-gain operational amplifier in the receiver. As a result, a pulse is issued and sent to reset the flip-flop shown in the digital system, which in turn disables the counter as well as interrupts the microcomputer. The interrupted microcomputer obtains the distance between the robot and the detected obstacle by reading the value in the counter and accordingly makes a decision on what action should be taken. It may order the electro-mechanical subsystem to move wisely to avoid the obstacle in its way. Let us turn now to the noise minimization techniques employed in this mobile robot system.

a. Shielding

Examine the analog subsystem and notice that the ultrasound device is connected to the symbolized switch controlled by the control block by a shielded coaxial cable. Since normally the echo signal is weak, a high-gain (operational) amplifier must be used. This is the most noise-sensitive spot in the whole system. The noise radiated by the motors, power relays, and switches in the electro-mechanical subsystem would definitely be picked up by this cable and amplified if the physical layout of the system components were not carefully planned. Thus, this cable must be as short as possible and physically as far away as possible from the electro-mechanical subsystem. The dashed lines that enclose the electro-mechanical controller and the digital subsystems denote their chassis; however, for the analog subsystem, the dashed lines denote the aluminum enclosure or box for shielding the analog subsystem from the noise generated by the electro-mechanical subsystem. The signal and signal return pins of the ultrasound device connected to the coaxial cable must be covered by the aluminum box, otherwise a double-shielded coaxial cable must be used, with its outer shield connected to the body frame. In Figure 9.3(d), we assume that the coaxial cable can be covered by the aluminum box.

The cable connecting the analog and digital subsystems would not be as critical as that for the ultrasound device. It can be a ribbon cable which carries the transmission command signal from the microcomputer, and the pulse sent from the receiver for disabling the counter. These signals are pretty strong (3 to 5 V), thus the cable can be a long one if necessary, but the unused conductors of the cable must be tied together with the return conductors of the signals (command and the disable pulse). These return conductors are also called "digital ground lines". The connection between the microcomputer and the controller for the electro-mechanical subsystem can be a long cable, but the two subsystems must be interfaced with optical-coupled devices so that the systems are electrically isolated from each other.

b. Power Supplies

Note that all three subsystems are powered by one rechargeable storage battery bank. However, two *isolated* and *regulated* dc/dc converters (note: a dc/dc converter is a device which converts a direct current voltage from high to low, or low to high) are used to independently power the analog and digital subsystems, respectively. Although they all draw power from the main battery bank, most dc/dc converters on the market whose power transferring between input and output is transformer- or magnetic-coupled, and there is no direct electrical connection between them. As a result, each subsystem essentially has its own isolated power supply.

c. Grounding

In the diagram, all of the chassis (shown by dashed lines) must be bolted to the metal (steel) body-frame of the robot. (Note: in the case of equipment powered by ac house current, the chassis should be tied to the third (ground) conductor of the ac wall socket, the ac ground.) Since the switch that time-multiplexes between the ultrasound driver and receiver is functionally shown, it may not be physically close as implied in the diagram. Actually, the switch, the output of the driver, and the input of the receiver amplifier are very close together. Although normally the common terminal of the output side of the dc/dc converter is tied to the chassis as depicted in the digital subsystem, in the analog system we found experimentally that the ground (common) side of the amplifier input is a better place to be tied to the chassis as shown. It is important to point out that in this analog subsystem, only one of the two places (the common side of the amplifier or the common of the output of the dc/dc converter), but not both, may be tied to the chassis. In many critical applications, such as medical instrumentation where transducers yield very low output signals (i.e., a fraction of a microvolt), but have high output impedance (i.e., 1 MΩ or greater), a separate battery power supply would be required. When a separate battery is used, the analog subsystem can be completely "sealed" (i.e., shielded) and "float" from the system ground (i.e., have no direct connection to the system ground).

9.4. TRANSMISSION LINE REFLECTION

9.4.1. Introduction

A transmission line need not be a wire which is miles long. Generally speaking, if the propagation delay of a segment of a wire is greater than the transition time of a pulse to be sent, it can be considered to be a transmission line, where line reflection may result in ringing, overshooting, and producing an erroneous signal. A brief review of transmission line theory follows.

Figure 9.4(a) shows the equivalent circuit of a segment of transmission line, where L is the inductance in henry per unit length of wire, and C is the capacitance in farads per unit length of wire. Both L and C are functions of the width, thickness, and spacing of conductors, and the dielectric constant of the insulation materials. The time delay per unit length $= \sqrt{LC} = \tau$. The characteristic impedance of a transmission line is defined as $z_0 = \sqrt{\frac{L}{C}}$.

Figure 9.4(b) shows the equivalent circuit of a transmission line with sending voltage V_s and a load impedance z_L. If the delay time of the transmission line is 1 T, then the reflection voltage for t = T at the receiving end is

$$V_r(t{=}T) = \frac{z_L - z_0}{z_L + z_0} V_s = \rho V_s \tag{9.1}$$

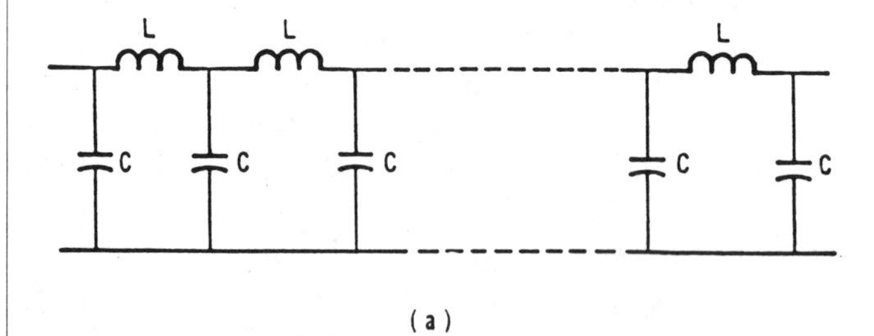

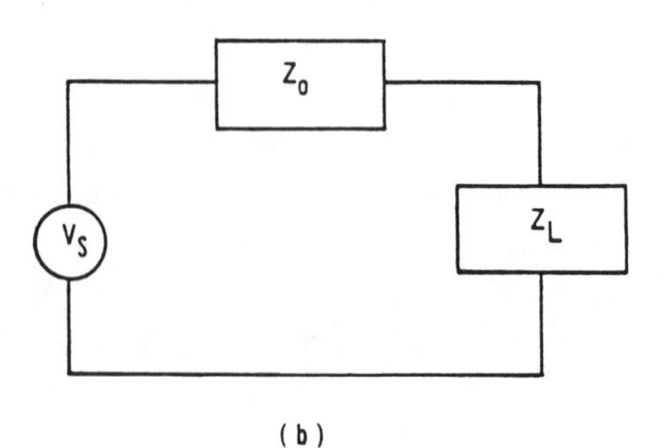

FIGURE 9.4 Equivalent circuits of a transmission line.

where

$$\rho \triangleq \text{reflection coefficient} \triangleq \frac{z_L - z_0}{z_L + z_0} \tag{9.2}$$

9.4.2. Transmission Line Terminated by Linear Devices

As a numerical example, consider an open-ended transmission line driven by a 1 V source which has an output impedance $R_s = 25\ \Omega$. As shown in Figure 9.5(a), we have $z_L = \infty$. The switch S is closed at $t = 0$. We are interested in the time waveforms of the driving voltage, $V_D(t)$ and the receiving voltage $V_R(t)$ at the driving and receiving ends of the transmission line which has a delay time of T. Basically, we are studying the transient response of the transmission line to a step function. Figure 9.5(b) shows a graph describing the reflection process of the transmission line. The lower line represents the time axis of $V_D(t)$ and the upper line, that of $V_R(t)$. The triangle-shaped line with arrows depicts the traveling process of the step function. That is, at $t = 0^+$, $V_D(0^+)$ starts to travel down the line and reaches the end of the line when $t = T$. Therefore, $V_R(0 < t < T)$ is the receiving end voltage before the traveling wave arrives at the end of the line, so we have $V_R(0 < t < T) = 0$. At $t = 2T$, the reflection wave would arrive at the driving end and reflect again. The wave is then reflected back and forth as the arrow line shows. It is clear that the magnitudes of $V_D(t)$ and $V_R(t)$ will change whenever $t = 2T, 4T, \ldots$ and $t = T, 3T, 5T, \ldots$, respectively. Eventually, $V_D(\infty) = V_R(\infty) = 1$ V. To plot $V_D(t)$ and $V_R(t)$, we have to determine the values at these critical times. It is important to point out that all the voltages at the critical times are composed of three components, namely, (1) the original voltage V_1, (2) the incoming voltage V_2, and (3) the reflected voltage V_3. Calculation of these voltages follows.

Reflection coefficients:

$$\rho_D(\text{driving end}) = \frac{R_s - z_0}{R_s + z_0} = \frac{25 - 100}{25 + 100} = -0.6$$

$$\rho_R(\text{receiving end}) = \frac{z_L - z_0}{z_L + z_0} = \frac{\infty - 100}{\infty + 100} = 1$$

For $t = 0^+$:

Since at point D, as far as the driving source is concerned, its load impedance is z_0, regardless of how the transmission line is terminated at the receiving end, since for transient analysis the source does not "see" what is at the other end of the transmission line. It only sees the

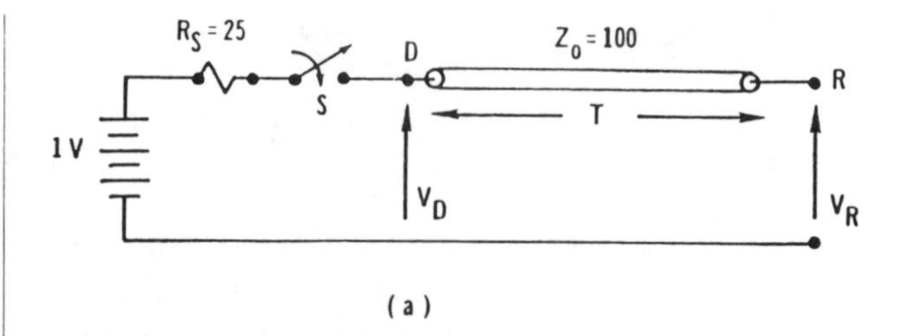

FIGURE 9.5 Transmission line terminated by linear device.

equivalent circuit shown in Figure 9.4(a). Thus, we have $V_D(0^+) = V_1 + V_2 + V_3$, which can be determined as follows:

$$V_1 = \frac{z_0}{R_s + z_0} 1_v = \frac{100}{25 + 100} \times 1 = 0.8$$

$$V_2 = 0$$

$$V_3 = 0$$

$$V_D(0+) = 0.8\ \text{V}$$

Similarly, we have

For t = T

$$V_1 = 0$$

$$V_2 = 0.8$$

$$V_3 = \rho_R \times 0.8 = 1 \times 0.8 = 0.8$$

$$V_R(T) = 0 + 0.8 + 0.8 = 1.6$$

For t = 2T

$$V_1 = V_D(0^+) = 0.8$$

$$V_2 = 0.8$$

$$V_3 = \rho_D \times 0.8 = -0.6 \times 0.8 = -0.48$$

$$V_D(2T) = 0.8 + 0.8 - 0.48 = 1.12$$

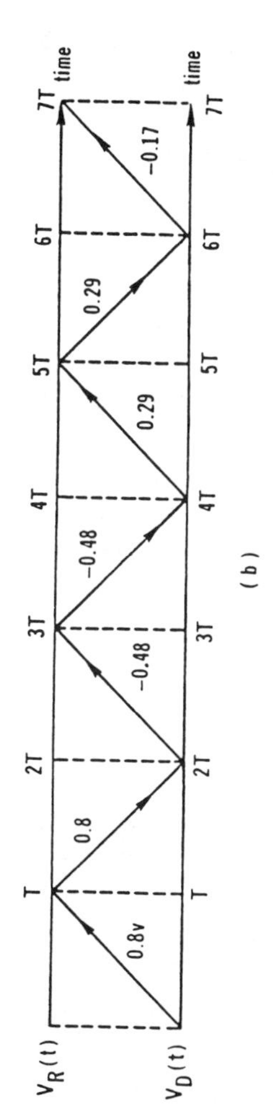

FIGURE 9.5 continued. Transmission line terminated by linear device.

For t = 3T

$$V_1 = V_R(T) = 1.6$$

$$V_2 = -0.48$$

$$V_3 = -0.48$$

$$V_R(3T) = 1.6 - 0.48 - 0.48 = 0.64$$

For t = 4T

$$V_1 = V_D(2T) = 1.12$$

$$V_2 = -0.48$$

$$V_3 = (-0.6) \times (-0.48) = 0.288$$

$$V_D(4T) = 1.12 - 0.48 - 0.288 = 0.928$$

For t = 5T

$$V_1 = V_R(3T) = 0.64$$

$$V_2 = 0.29$$

$$V_3 = 0.29$$

$$V_R(5T) = 0.64 + 0.29 + 0.29 = 1.22$$

For t = 6T

$$V_1 = V_D(4T) = 0.93$$

$$V_2 = 0.29$$

$$V_3 = (-0.6) \times 0.29 = -0.17$$

$$V_D(4T) = 0.93 + 0.29 - 0.17 = 1.05$$

For t = 7T

$$V_1 = V_R(5T) = 1.22$$

$$V_2 = -0.17$$

$$V_3 = -0.17$$

$$V_R(7T) = 1.22 - 0.17 - 0.17 = 0.88$$

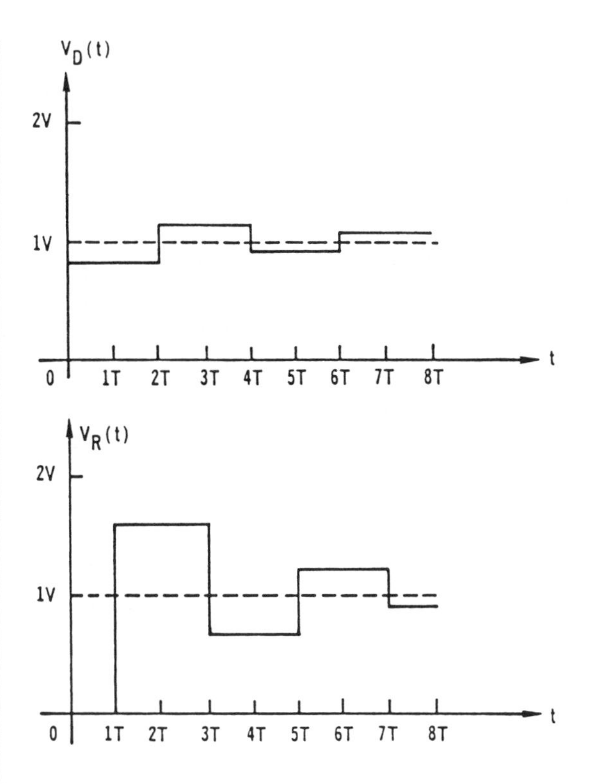

FIGURE 9.6 Waveforms of driving and receiving voltages.

Figure 9.6 shows the plots of $V_D(t)$ and $V_R(t)$ according to the calculated results obtained.

9.4.3. Transmission Line Terminated by Nonlinear Devices

As described in the preceding sections, the analysis of transmission lines terminated by linear devices appears to be straightforward. In digital systems, however, almost all transmission lines are driven and terminated by nonlinear devices such as TTL gates, etc. In this section we illustrate how the piece-wise linear graphical technique can be used for analysis of TTL driven and terminated transmission lines. Consider the transmission line which is driven and terminated by TTL gates shown in Figure 9.7(a). It is evident that when the driver is in logic [1] or HIGH state, the input of the receiving gate is also high; therefore, practically no current is flowing in the line. When the driver is in logic [0] or LOW state, it will sink current from the receiver. For purposes of analysis, let us conveniently stipulate that the current flow out of the receiver is positive, and the flow into the receiver is negative. First let us determine the input V-I curve of the receiver; as shown in Figure 9.7(a), one can assume that a variable dc testing voltage source is used, and it starts at +5 V and decreases until reaching the safety margin at some negative voltage. The corresponding I_i and V_i readings can be used to construct the input V-I curve. For example, we see from Binary State Analysis (BSA) that the base-emitter diode junction will not conduct until $V_i \leq 2.1$ V. The current increases and reaches $\frac{5-0.7}{2.8K} = 1.5$ mA when $V_i = 0$. As V_i continuously decreases, the current increases until $V_i \leq -0.7$ V. At this point, B_1, the clamping diode conducts, and therefore $I_i = I + I'$ and increases rapidly. The V_i–I_i curve is shown in Figure 9.7(b). Following the same logic, the output V-I curves for logic [1] and [0], respectively, are shown in Figures 9.8(a) and (b). In the case of logic [1], Q_5 is OFF and Q_3 and Q_4 are ON. Since it is a Darlington circuit, Q_3 and Q_4 are in the linear region, and the base current, I_B, of Q_3 is negligible. The Thevenin output voltage then is approximately equal to $5 - 1.4 = 3.6$ V, where the drop of 1.4 V is contributed by the B-E junctions of Q_3 and Q_4. Therefore, the current I changes to negative direction or flows outward when the testing voltage $V \leq 3.6$ V. When $V = 0$, $I = \frac{5-0.9}{58} = 70$ mA, where the 0.9 V drop is due to the C-E junction of Q_3 (0.2 V) (since we assume

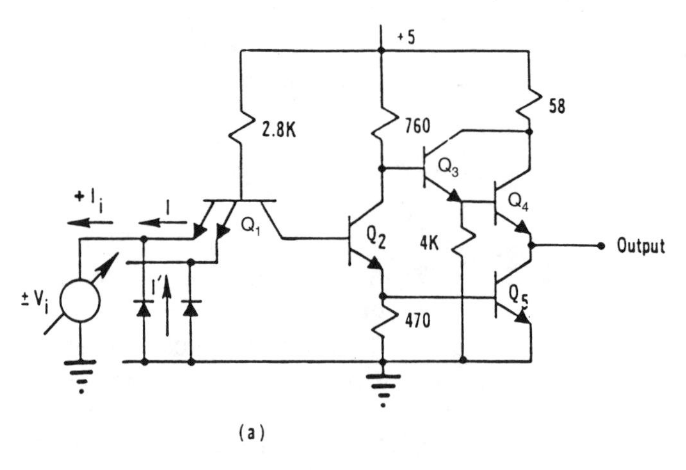

(a)

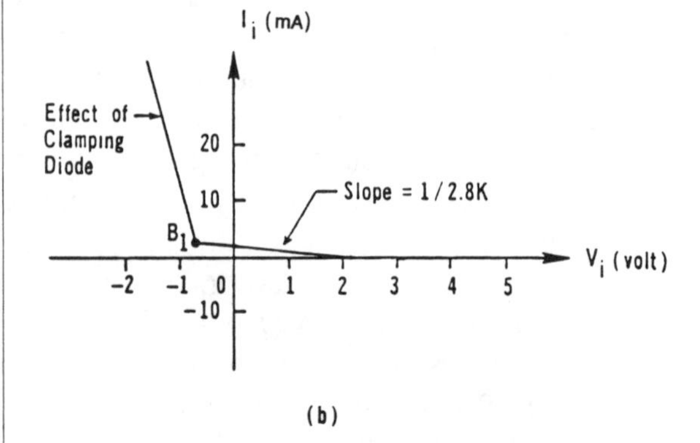

(b)

FIGURE 9.7 TTL input V-I curve.

Q_3 is saturated by now) and the B-E junction of Q_4 (0.7 V). As the source voltage, V, decreases to –0.7 and less, the B-C junction of Q_5 begins to be forward biased and current will flow from ground through the 470 Ω resistor. The output V-I curve for logic [1] is shown in Figure 9.8(a), where B_1 is the break point as the B-C junction of Q_5 conducts. Now let us consider the output of the driver in the logic [0] case shown in Figure 9.8(b). Here, Q_4 is OFF. A constant current I_b flows into the base of Q_5, thus Q_5 conducts. For $V \geq 0$ V, the V-I curve will simply be the I_C-V_{CE} curve at $I_B = I_b$, which is one of the output, or C-E V-I, curve family of a transistor that we are familiar with. When $V \leq 0.7$ V, both the B-E junction of Q_4 and the B-C junction of Q_5 begin to conduct and the current increases rapidly. The break point is B_1 as shown in Figure 9.8(b).

The composite V-I curves of two output states and one input of a transmission line terminated by these TTL gate are shown in Figure 9.9(a). Figure 9.9(b) shows the circuit configuration where the transmission line has a characteristic impedance of 100 Ω. Figure 9.9(a) shows the two steady state points P_1 and P_0, where P_0 is the intersection of input curve and output LOW curve, while P_1 is the intersection of the input curve and the output HIGH curve; that is, for steady state, $V_D(\infty) = V_R(\infty)$. Referring to Figure 9.9(b), consider the driver is in HIGH state and is to be switched to LOW. We know that the initial point will be P_1 and the final point will be P_0 as shown in Figure 9.9(a). The question is, what is the operational path in the composite V-I curves during the transition from logic HIGH to logic LOW. From Figure 9.9(b), we have

$$V_R(t) - I(t)Z_0 = V_D(t) \tag{9.3}$$

For switching from HIGH to LOW, at $t = t_0$, $V_R(t)$ cannot change instantly; thus, we could consider that $V_R(t)$ at this moment is a constant. By differentiating Equation (9.3), we have

$$0 - \frac{dI(t)}{dt} Z_0 = \frac{dV_D(t)}{dt}$$

or

$$\frac{dI(t)}{dV_D(t)} = -\frac{1}{Z_0}$$

That is, the direction of the operational path is $-\frac{1}{Z_0}$. When $t = t_0 + T$, the travelling pulse arrives at the receiving end and $V_R(t)$ changes

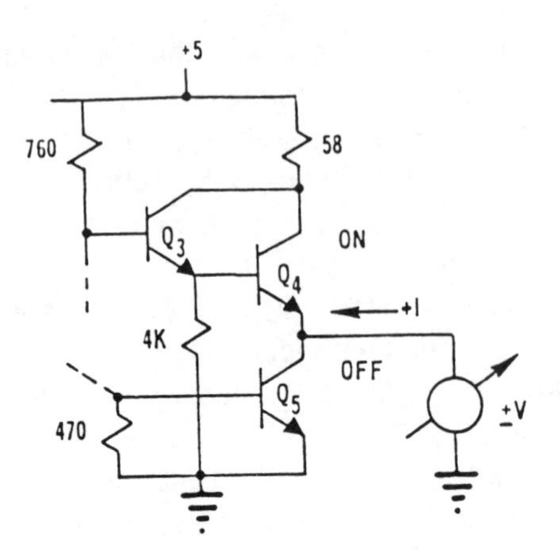

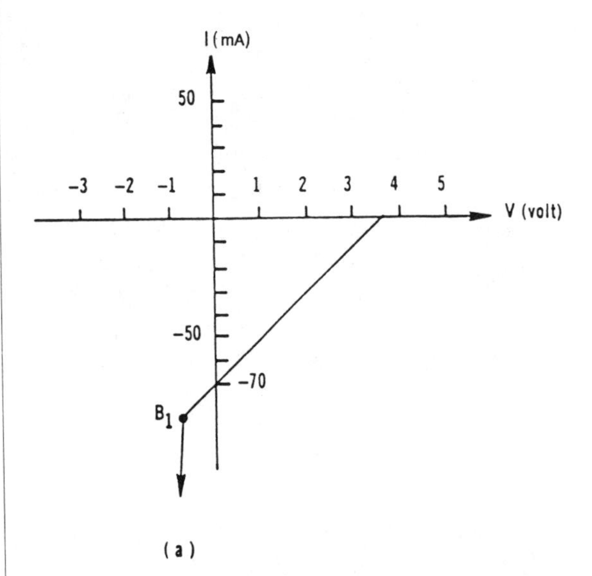

FIGURE 9.8 (a) TTL output V-I curve for logic 1 (HIGH). (b) TTL output V-I curve for logic 0 (LOW).

suddenly while $V_D(t)$ can be considered as a constant at this time; thus, by differentiating Equation (9.3) again, we have

$$\frac{dV_R(t)}{dt} - \frac{I(t)}{dt} Z_0 = 0$$

or

$$\frac{dI(t)}{dV_R(t)} = \frac{1}{Z_0}$$

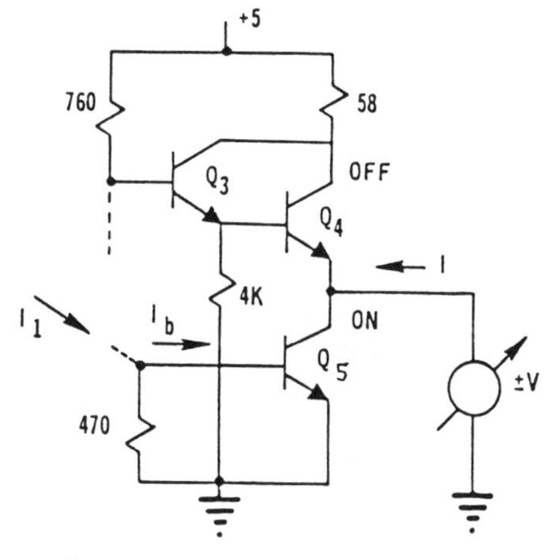

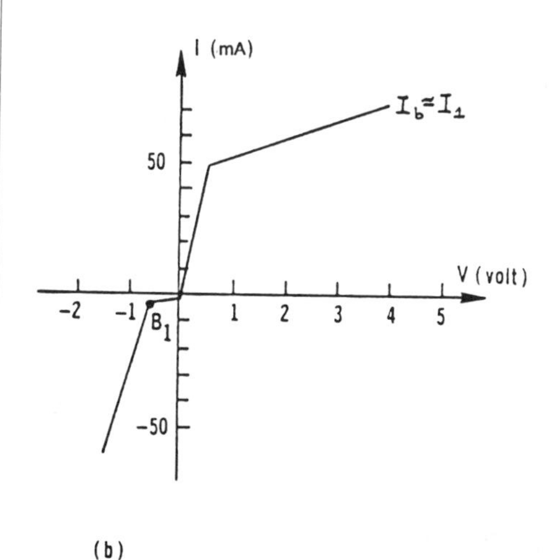

FIGURE 9.8 continued. (a) TTL output V-I curve for logic 1 (HIGH). (b) TTL output V-I curve for logic 0 (LOW).

which has a positive slope. Since we now know the direction of the operational paths and the starting/ending points, as well as the V-I curves, we can conclude that the reflecting pulse will travel back and forth between input curve and output LOW curve as shown in the dotted lines in Figure 9.9(a). The voltage magnitudes of $V_D(t)$ and $V_R(t)$ at t_0, T, 2T, 3T, ... can be read from the graphs; their waveforms are shown in Figure 9.9(c). Notice that in Figure 9.9(c), we see –1 V at the receiving end. By the same principle, one can analyze the response of the transmission line for switching from LOW to HIGH.

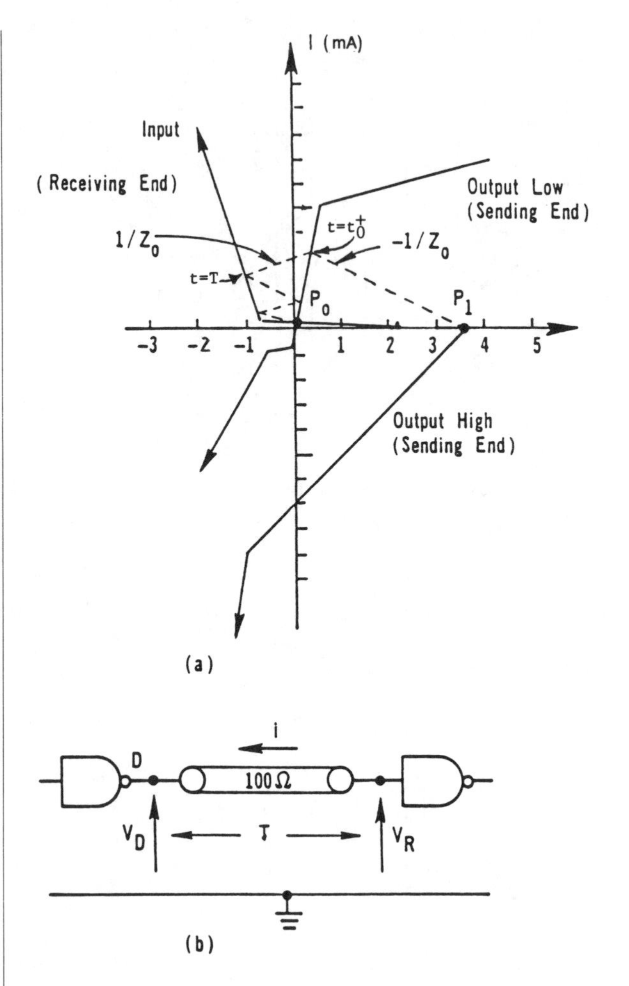

FIGURE 9.9 (a, b) Transmission line terminated by TTLs. (c) V_D and V_R waveforms of TTL terminated transmission line.

9.4.4. Impedance Matching for Transmission Line

The reflection coefficient, Equation (9.2),

$$\rho = \frac{z_L - z_0}{z_L + z_0}$$

tells us that if the load impedance z_L can be made nearly or exactly equal to z_0, the reflection voltage can be reduced or eliminated. This concept has indeed been implemented in practice. Figure 9.10(a) shows a practical way to achieve this. Note that the driver is an open collector gate which has very high output impedance when in HIGH state. At the receiving end, the input impedance of a TTL gate is very high in HIGH state, and several thousand ohms in LOW state, for the TTL shown in Figure 9.7(a) it will be 2.8 K. Let us assume that the re-

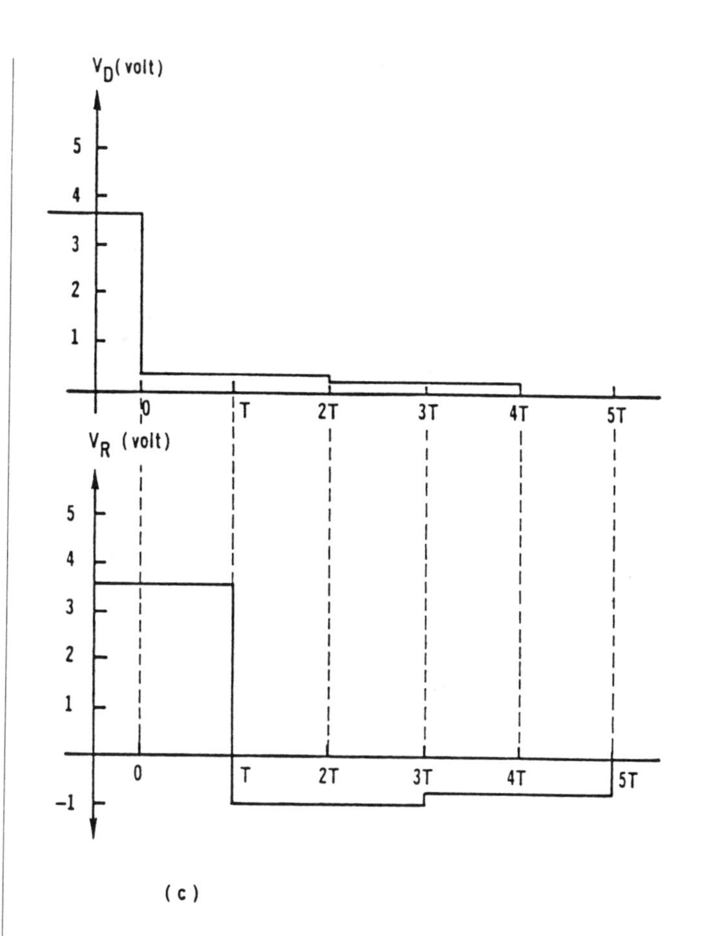

FIGURE 9.9 continued. (a, b) Transmission line terminated by TTLs. (c) V_D and V_R waveforms of TTL terminated transmission line.

ceiving end gate is also a TTL whose circuit is shown in Figure 9.7(a).

Compare the transmission line configurations shown in Figure 9.9(b) and Figure 9.10(a). Note that for the latter, two identical resistor networks have been added at both ends. The purpose of these resistor networks is to modify the z_L at both ends so that they will nearly match the z_0 of the transmission line. Let us follow the graphical analysis procedure described in Section 9.4.3 and plot the time waveform of the driving end and receiving end voltages during the HIGH to LOW transition period.

a. Equivalent Circuit

By applying Thevenin's Theory to the resistor network alone, we can calculate the Thevenin voltage and resistance, V_{th} and R_{th}, respectively, as follows:

$$V_{th} = 5 \times \frac{390}{180 + 390} = 3.4 \text{ V}$$

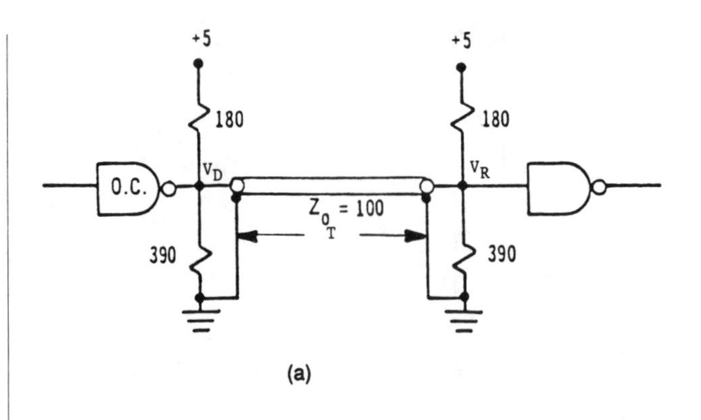

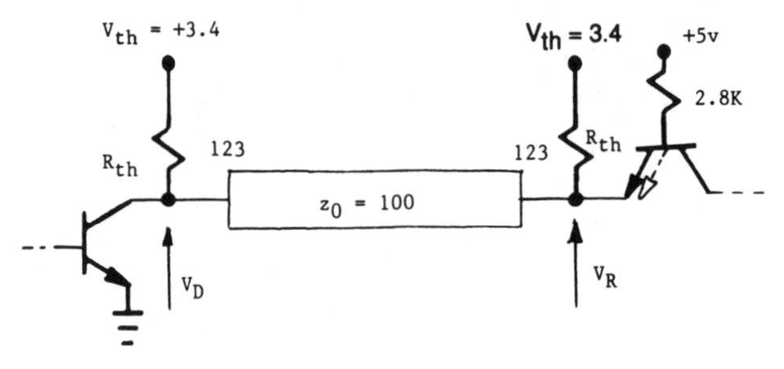

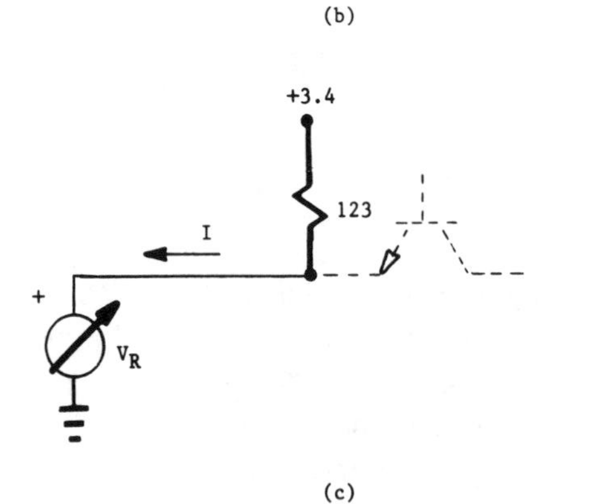

FIGURE 9.10 (a) Matching transmission line impedance. b) Equivalent circuit. (c) Receiving end. (d) Receiving end V-I curve. (e) Driving end at logic HIGH state. (f) Driving end at logic LOW state. (g) Driving end V-I curves. (h) Graphical analysis for switching from logic HIGH to logic LOW state. (i) Voltage time waveforms for driving/receiving end.

$$R_{th} = \frac{180 \times 390}{180 + 390} = 123\ \Omega$$

Thus, the equivalent circuit of Figure 9.10(a) is shown in Figure 9.10(b).

b. V-I Curve for the Receiving End

At the receiving end, there are two circuits in parallel: the Thevenin equivalent circuit of the resistors and the input of the TTL gate. It ap-

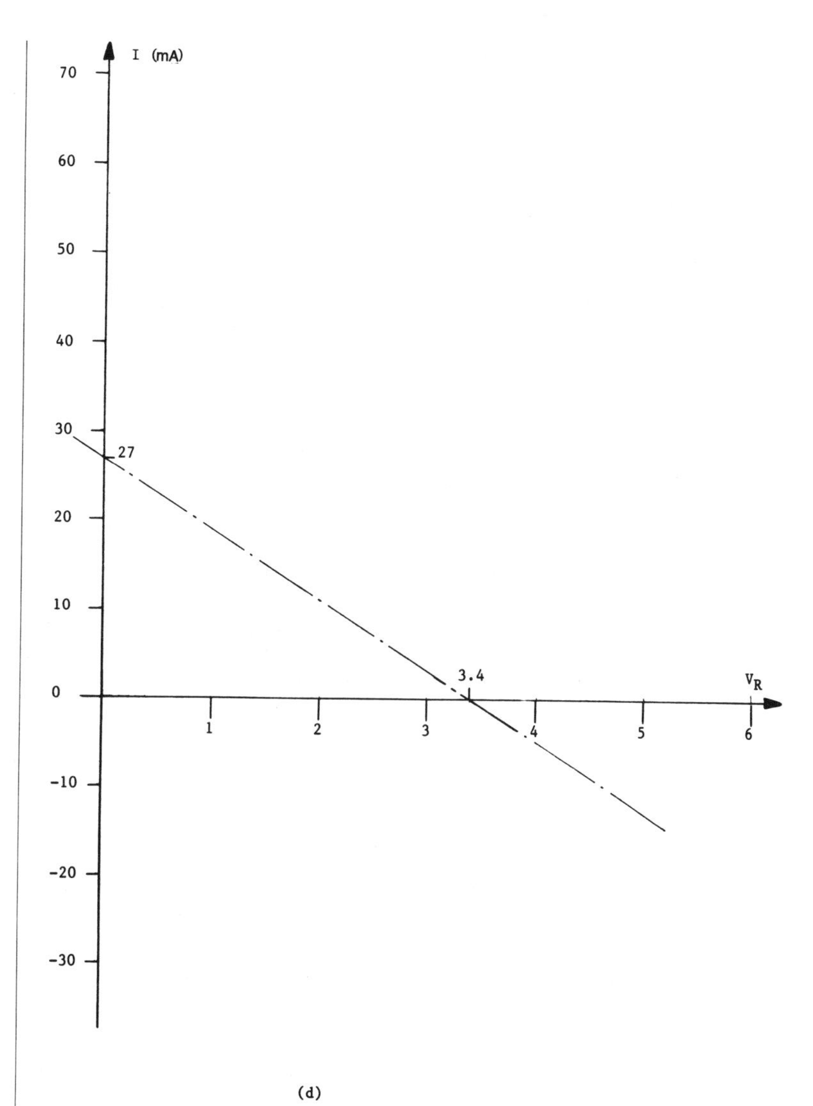

(d)

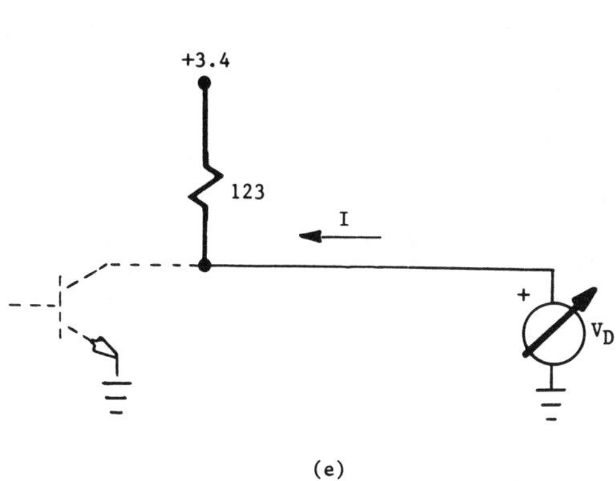

(e)

FIGURE 9.10 continued. (a) Matching transmission line impedance. b) Equivalent circuit. (c) Receiving end. (d) Receiving end V-I curve. (e) Driving end at logic HIGH state. (f) Driving end at logic LOW state. (g) Driving end V-I curves. (h) Graphical analysis for switching from logic HIGH to logic LOW state. (i) Voltage time waveforms for driving/receiving end.

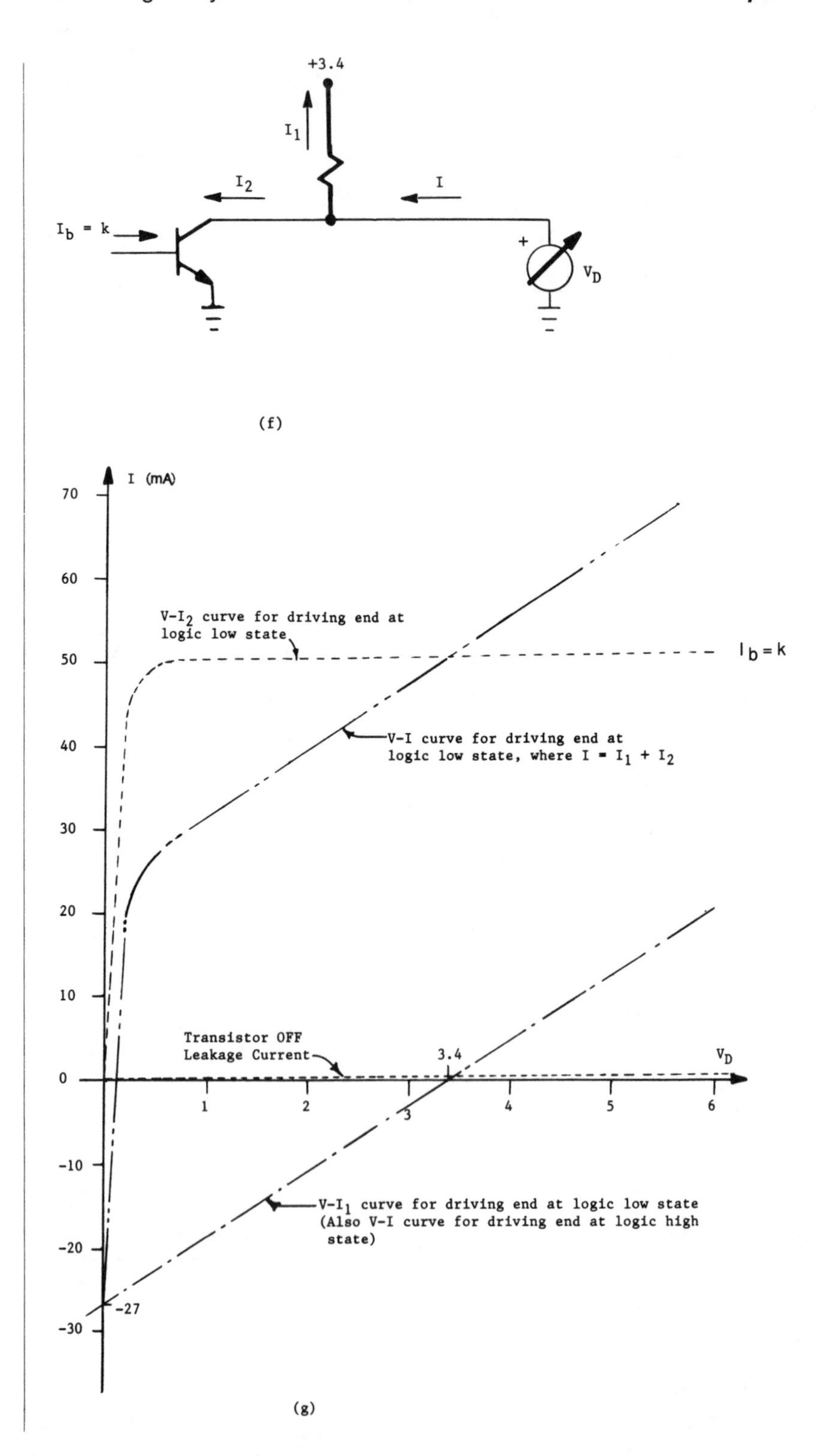

FIGURE 9.10 continued. (a) Matching transmission line impedance. b) Equivalent circuit. (c) Receiving end. (d) Receiving end V-I curve. (e) Driving end at logic HIGH state. (f) Driving end at logic LOW state. (g) Driving end V-I curves. (h) Graphical analysis for switching from logic HIGH to logic LOW state. (i) Voltage time waveforms for driving/receiving end.

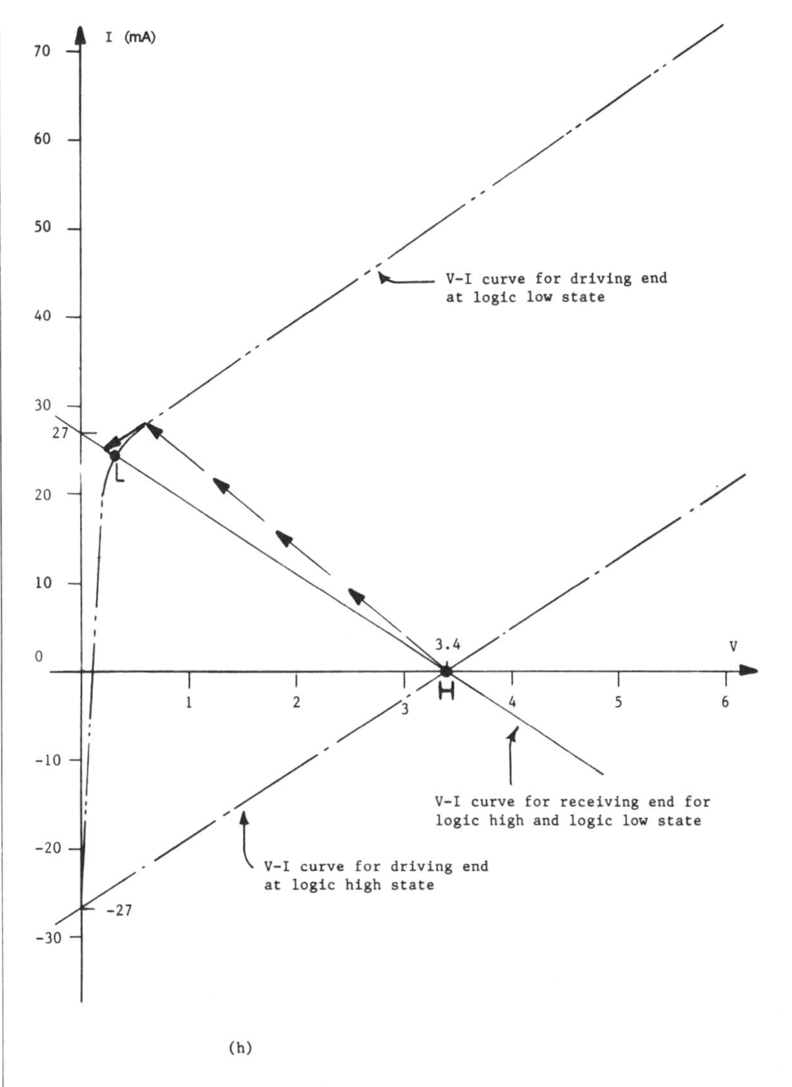

FIGURE 9.10 continued. (a) Matching transmission line impedance. b) Equivalent circuit. (c) Receiving end. (d) Receiving end V-I curve. (e) Driving end at logic HIGH state. (f) Driving end at logic LOW state. (g) Driving end V-I curves. (h) Graphical analysis for switching from logic HIGH to logic LOW state. (i) Voltage time waveforms for driving/receiving end.

pears that for deriving the V-I curve of this end one must consider the composite circuit. Fortunately, we can neglect the effect of the TTL gate here. This is because of the fact that when the driving end is HIGH the B-E junction of the transistor is OFF, thus its input impedance would be very large and its parallel effect is negligible. Consider the driver in LOW state. The input impedance would be 2.8 KΩ which again can be neglected in comparison with R_{th}. Therefore the circuit at the receiving end for deriving the V-I curve can be simplified as shown in Figure 9.10(c). Here, we have

$$V_R = 3.4 - 123 \times I$$

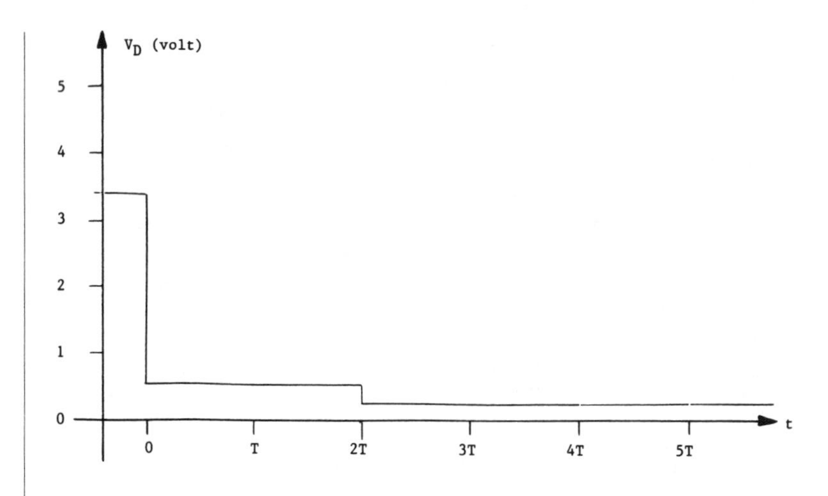

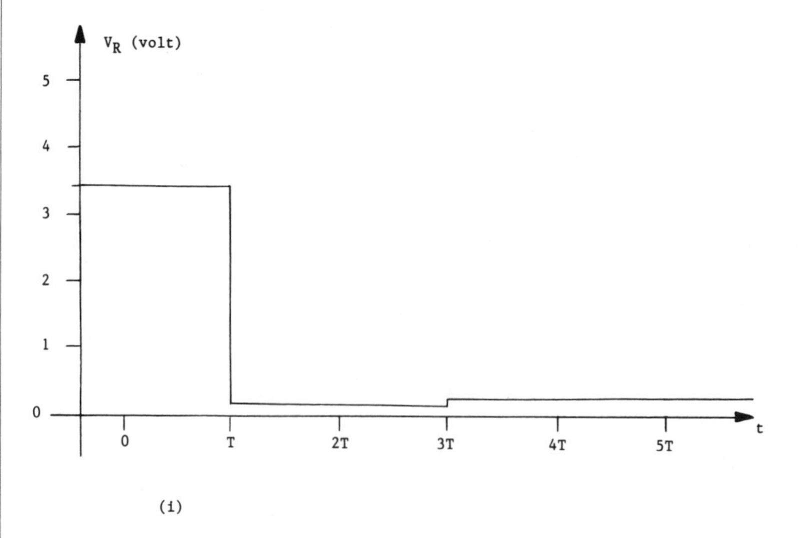

FIGURE 9.10 continued. (a) Matching transmission line impedance. b) Equivalent circuit. (c) Receiving end. (d) Receiving end V-I curve. (e) Driving end at logic HIGH state. (f) Driving end at logic LOW state. (g) Driving end V-I curves. (h) Graphical analysis for switching from logic HIGH to logic LOW state. (i) Voltage time waveforms for driving/receiving end.

for I = 0, V_R = 3.4 V, and for V_R = 0, I = 27 mA. Figure 9.10(d) shows the V-I curve for the receiving end.

c. V-I Curve for the Driving End

Logic High State

Figure 9.10(e) depicts the equivalent circuit of the driving end at its logic HIGH state. Recall that at high state, the output transistor (shown in dotted lines) of the open collector (O.C.) gate is OFF, where only the very low leakage current exists (see the nearly horizontal dotted line close to the X-axis in Figure 9.10(g)), and it can be ignored. The circuit can then be reduced to a single resistor circuit similar to that described for the receiving end analysis. However, since initially at the receiving end we have designated the outward current as positive, the inward current at the driving end must also be considered positive for consistency. Thus the loop equation here is different in sign in comparison with that of the receiving end. That is,

$$V_D = 123 \times I + 3.4\ V$$

and the V-I curve can be determined as follows: for $I = 0$, $V_D = 3.4$ v, and for $V_D = 0$, $I = -27$ mA. The V-I curve is shown by the dashed line (dash-dot-dash) in Figure 9.10(g); this is a mirror image of the curve for the receiving end shown in Figure 9.10(d).

Logic Low State

As shown in Figure 9.10(f), in logic low state the driving end current, I, has two significant components, I_1 and I_2. Thus, $I = I_1 + I_2$, where I_1 is identical to what we have described for the logic HIGH state and thus we can use the same equation by replacing I with I_1, or

$$V_D = 123 \times I_1 + 3.4\ V$$

whose V-I curve is plotted in Figure 9.10(g). As expected, this V-I_1 curve coincides with that V-I of the logic HIGH state. Thus, keep in mind that this V-I curve now represents two identical curves: the current I as the circuit is in logic HIGH state (transistor off) and I_1 as the logic LOW state (transistor on). As for I_2, the problem is slightly more complicated. Since it is a transistor in the ON state, its V-I curve is nonlinear as shown by the dotted line with $I_b = k$, where k is the value of base current determined by the IC designer for that O.C. gate. With the analysis just described, we can now determine the composite V-I curve for the driving end by adding I_1 with I_2 point by point graphically. As a result, we obtain the composite V-I curve for logic low state as shown in Figure 9.10(g) with the dash-dot-dot-dash line. Finally, we obtain two useful V-I curves for logic HIGH state and logic LOW state for the driving end shown in dash-line, respectively, in Figure 9.10(g).

d. Graphical Analysis for Switching from Logic High to Logic Low

By overlaying FIgure 9.10(g) (dropping the dotted lines) over Figure 9.10(d), we obtain Figure 9.10(h). Note that here we have only three V-I curves. The reason for this is that the curves for logic high and logic low for the receiving end are identical, whereas for the driving end the V-I curves for logic high and logic low are different. Referring to Figure 9.10(h), we can find that the logic high curve of the driving end intersects with the logic high curve of the receiving end at point H, while the logic low curve of the driving end intersects with the logic low curve, which happens to be the same as the logic high curve of the receiving end at point L. Those are two steady state operating points for the circuit. Thus, for switching from HIGH to LOW, we follow the arrow path with the slopes $(-\frac{1}{z_0})$ and then $(\frac{1}{z_0})$ from H and eventually arrive at L. By reading the voltage values at t = 0, T, 2T, ..., we can plot the time waveforms for the driving and receiving ends, respectively, as shown in Figure 9.10(i). By comparing Figure 9.10(i) with Figure 9.9(c) we can easily see the effect of matching z_L to z_0. Examine Figure 9.10(h) again. If one is interested in the current time waveform of the circuit, one can read the current values at the critical times from Figure 9.10(h) during the transition period. It is left to the reader to practice the graphical analysis technique by plotting the current time waveform and by following a similar procedure to obtain the time waveforms for switching from LOW to HIGH for the same configuration.

9.5. CROSS-TALK

When two unshielded signal lines are closely placed in parallel, the undesirable cross-talk problem may occur. To illustrate, let one line carry a fast clock signal with fast rise/fall edges, and adjacent as well as parallel to it is a line connected to the input of a one shot chip. The latter would pick up negative and positive going spikes in synchronous with the falling and rising edge of the clock on the other line, respectively. The spike train induced from the clock line appears to be a derivative of the clock signal. As one may expect, the one shot connecting to it would be triggered by the spikes and cause errors, which are known as cross-talk errors. To minimize this kind of problem, one may shield the clock line with coaxial cable or twist pair lines. As another alternative, one may reroute the clock line in such a way that it is perpendicular or far away from all noise sensitive lines. Still another inexpensive way, but not as effective as the ones just described, is to route the clock (or other trouble-making lines that carry sharp pulses) close to and in parallel with the power supply lines on the IC board. This works because power lines, whether positive or common, can be considered as ground lines as far as noise problems are concerned.

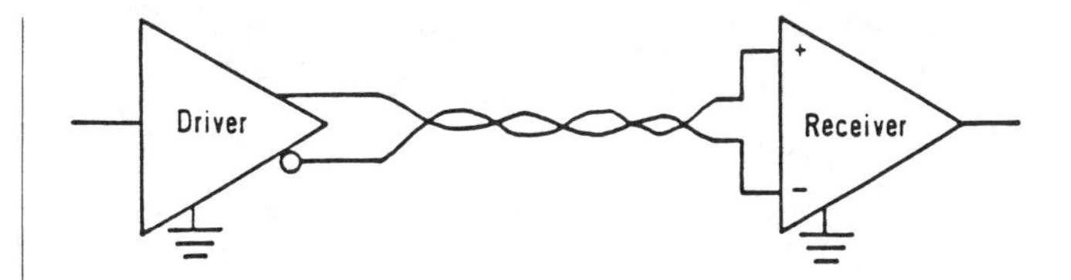

FIGURE 9.11 Differential transmission line driver and receiver.

9.6. DIFFERENTIAL LINE DRIVER AND RECEIVER

For long-distance transmission lines and noisy environments, the system designer should consider using the differential line driver and receiver as shown in Figure 9.11 These devices are now commercially available at a very reasonable cost. Note that the driver simultaneously generates a pair of signals with opposite phase, and the receiver output is the difference of its input pair. Therefore, it has the desirable common mode rejection feature which minimizes the radiation type noise induced in the transmission line. The input impedance of the receiver is normally fairly high; thus, a transmission line impedance matching network can be inserted at the receiving and sending ends to reduce the reflection voltage.

REFERENCES

1. Ott, H. W., *Noise Reduction Techniques in Electronic Systems,* 2nd ed., John Wiley & Sons, New York, 1988.
2. Jones, J. P., *Causes and Curves of Noise in Digital Systems,* Computer Design Publishing Co., W. Concord, MA, 1964.
3. Heniford, B., Noise in 54/74 TTL Systems, TI Application Report, CA-108.
4. Blood, W. R., MECL System Design Handbook, 4th ed., Motorola Semiconductor, Inc., Phoenix, AZ, 1983.
5. Morris, R. L. and Miller, J. R., Ed., *Designing with TTL Integrated Circuits,* McGraw-Hill, New York, 1971.
6. Morrison, R., *Grounding and Shielding Techniques in Instrumentation,* John Wiley & Sons, New York.
7. DeFalco, J. A., Reflection and crosstalk in logic circuit interconnections, *IEEE Spectrum,* 7(7), 44—50, July 1970.
8. Cushman, R. H., Make the most of noise — correlate it, Parts I and II, *EDN,* 16(5), 29—35, March 1, 1971 and 16(8), 39—45, April 15, 1971.
9. Balph, T., Drive 50 ohm transmission lines with TTL and ECL gate outputs, *EDN,* 24(2), 92—94, January 20, 1979.
10. Matick, R. E., *Transmission Lines for Digital and Communication Networks,* McGraw-Hill, New York, 1969.
11. Tam, Shirley Y., Polaroid Ultrasonic Ranging System Handbook: Application Notes and Technical Papers, Polaroid Corp., Cambridge, MA, 1980.

10 REVIEW OF SEQUENTIAL MACHINE DESIGN FUNDAMENTALS

10.1. BASIC CONCEPTS

What we have presented thus far are the basic building elements for our final goal: the design of a complete digital system. Those basic elements can either be hardware or analysis/design techniques. Among the design techniques that we have discussed, the logic design techniques are mostly for the so-called *combinational logic* design. That is, when a logic module or a black box that we have designed, whose output(s) is exclusively a function of its present input(s), we would name it as a combinational logic module. By "module" we mean any subsystem that implements one or more logical functions. Therefore, a digital system can consist of one or more logic modules. As a trivial example, the tri-state gate described in Chapter 4, Section 4.2.5(b) is a combinational logic module. It can be described by the following switching function:

$$\mathrm{OUT} = \bar{E}X$$

Note that the output, OUT, responds only to what the input logic values are at any moment. In general, a combinational logic module does not consist of any memory elements such as flip-flops. In contrast, a sequential logic module is one whose output(s), is a function not only of the present input(s), but also the preceding or historical states of the system. A washing machine is an example of a very simple sequential machine, whereas a digital computer is a very sophisticated sequential machine built on a number of sequential logic modules. To summarize, a sequential logic module consists of combinational logic blocks and memory devices. Its output is a function of the present input and present and past memory states.

In our daily lives we can easily find examples of sequential logic systems or modules. As a very general example of a sequential logic process, our next action (the "output") depends on our recollection (the "memory" state) of what we have recently done, plus what situation we are presently in (the "input"). Specifically consider our education system. As one enters college, one normally follows the *sequence* of freshman-sophomore-junior-senior. Here we have four obvious states in a restricted order. To enter the freshman state, one must apply for and be granted admission to a college or university. To enter the sophomore state, one must have been in freshman state for at least a year and have completed certain course requirements, and so on. Expressing this sequence in terms of sequential machine modules, we (the system) must check one's record (the memory) if he/she has been a freshman, then read in (input) whether he/she has passed the required courses; if so, we may declare (output) that he/she can become a sophomore, otherwise he/she must remain a freshman. Specifically, there are three essential elements, i.e., input, memory, and output, plus a rule for the orderly sequence of a sequential machine or sequential logic module.

According to their manner of operation, sequential machines are generally classified into two categories: synchronous and asynchronous. The former usually consists of a system clock. The system changes state only at the rising or falling edge, or during the high state of the system clock. By contrast, the latter changes state immediately after the inputs have changed. Although they appear faster and more efficient at first glance, asynchronous sequential machines require careful time analysis. Similar to the TV show "Mission Impossible", timing is a critical design factor. On the other hand, synchronous sequential machines do not care how the input variables behave between the clock pulses as long as they are in their stable states at the moment before the clock pulse arrives. Therefore, synchronous sequential machines are simpler to design. In practice, synchronous sequential systems satisfy most system specifications. Our society as whole is synchronous, since our activities are synchronized by "standard" or "daylight" time (Pacific in the case of the West Coast of the U.S.).

10.2. SEQUENTIAL MACHINE MODELS AND FUNCTIONAL DESCRIPTIONS

As mentioned in the preceding section, there are three essential elements in any sequential machine, namely, inputs, memory and output. These elements and how they are configured or interconnected are shown in Figure 10.1(a) and (b). According to its system configuration, a sequential machine can be placed into one of two major classes or models: (1) Mealy machine and (2) Moore machine. Figure 10.1(a) depicts a simplified functional system block diagram of the former, while Figure 10.1(b) shows the latter. Here, $X(i) = (x_1, x_2, \ldots, x_n)$ denotes the binary input vector at time $t = i$; $Z(i) = (z_1, z_2, \ldots, z_m)$ is the binary output vector at time $t = i$; $Q(i) = (q_1, q_2, \ldots, q_k)$ is the binary vector code for the present (the "i^{th}") state of the machine; and $D(i+1) = (d_1, d_2, \ldots, d_k)$ is the binary vector code of the next, or "$(i+1)^{th}$" state. To illustrate, we may define the mathematical symbols for the education system described as follows.

$$X(i) = (\text{application, admission, pass all required courses})$$

(Here, for example, $X(i) = (1,1,0)$ implies that one has filed application and received admission, but not yet passed all required courses).

$$Z(i) = (\text{freshman, sophomore, junior, senior})$$

$$D(i+1) = (d_1,d_2)$$

where d_1,d_2 = 00, 01, 10, 11, respectively, denote freshman, sophomore, junior, and senior

$$Q(i) = (q_1,q_2)$$

where q_1,q_2 = 00, 01, 10, 11 again denote freshman, sophomore, junior, and senior. Notice that the only difference between the two models is the connection of the input lines or X(i). For the Mealy machine, the output vector is a direct function of the ith input vector and the ith state of the machine, while for the Moore machine it is only a direct function of the ith state of the machine, although the ith state itself is still a function of the ith input vector. However, in both machine models, there are an input logic layer, a memory layer, and an output logic layer. Both input and output logic layers are realized or implemented by combinational logic. Note that the output of the memory layer yields Q(i), the ith or present state of the machine; the output logic layer yields Z(i), the ith or present output vector; and the input logic layer having X(i) and Q(i) as its inputs yields D(i+1), the $(i+1)^{th}$ or next state of the machine. The D(i+1) is waiting at the "door" of the memory layer. As soon as the clock goes high, D(i+1) is loaded into the memory layer, and what was the "next state" then becomes the "present state" of the machine at the output. This new present state then immediately shows

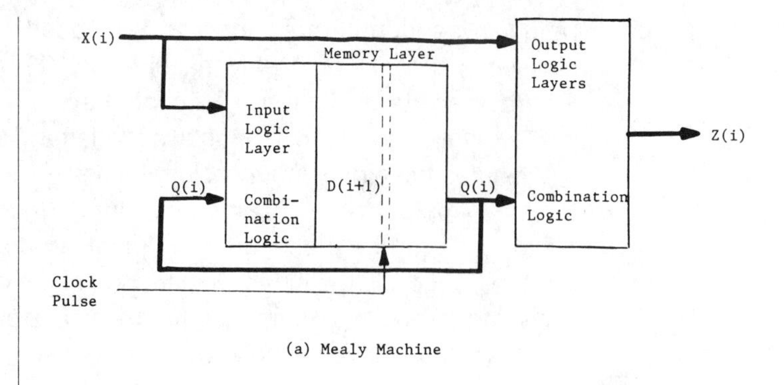

(a) Mealy Machine

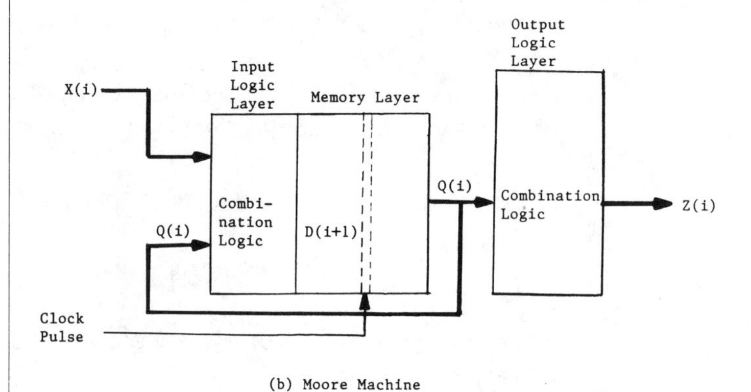

(b) Moore Machine

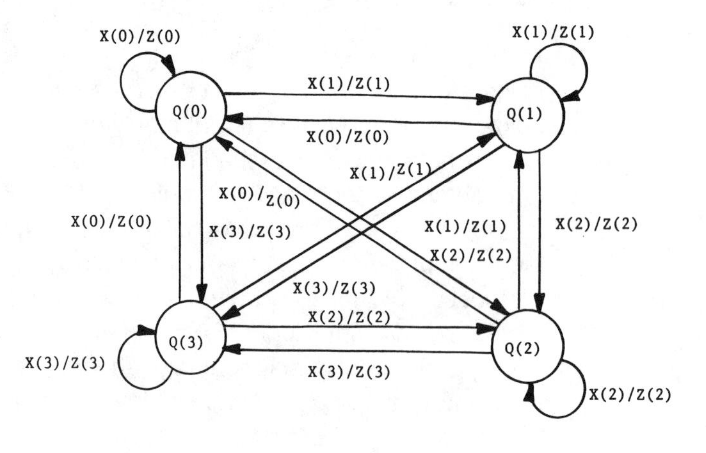

(c) Mealy Machine

FIGURE 10.1 (a, b) Sequential machine models. (c, d) Sequential machine state graphs. (e, f) Sequential machine state graph (continued). (g, h) State transition tables.

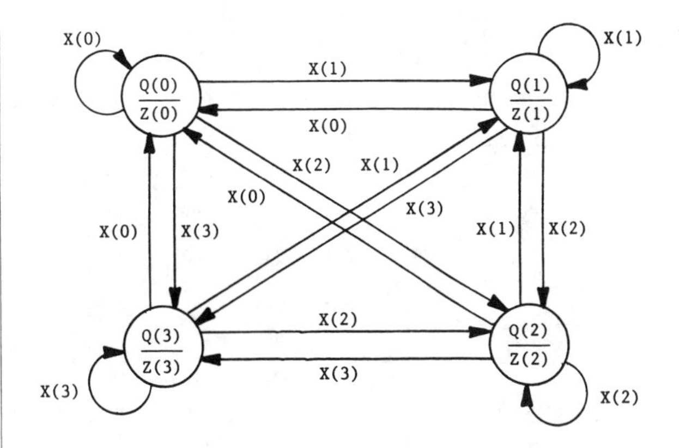

(d) Moore Machine

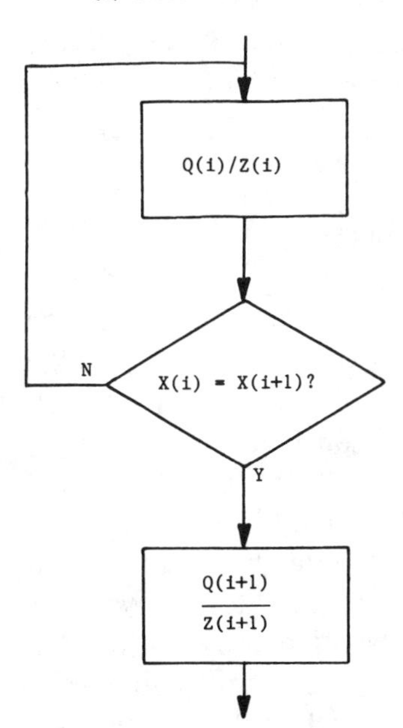

(e) Flow Chart Description

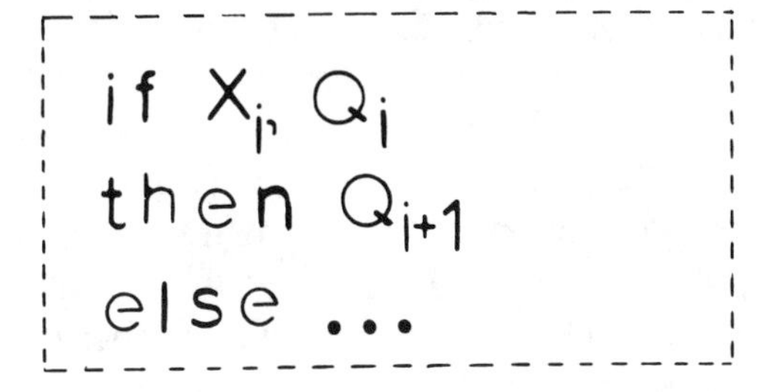

(f) If/Then/Else Description

FIGURE 10.1 continued. (a, b) Sequential machine models. (c, d) Sequential machine state graphs. (e, f) Sequential machine state graph (continued). (g, h) State transition tables.

Present	Next State				Present Output Z(i)			
State Q(i)	X(0)	X(1)	X(2)	X(3)	X(0)	X(1)	X(2)	X(3)
Q(0)	Q(0)	Q(1)	Q(2)	Q(3)	Z(0)	Z(1)	Z(2)	Z(3)
Q(1)	Q(0)	Q(1)	Q(2)	Q(3)	Z(0)	Z(1)	Z(2)	Z(3)
Q(2)	Q(0)	Q(1)	Q(2)	Q(3)	Z(0)	Z(1)	Z(2)	Z(3)
Q(3)	Q(0)	Q(1)	Q(2)	Q(3)	Z(0)	Z(1)	Z(2)	Z(3)

(g) Mealy machine

Present	Next State				Present
State Q(i)	X(0)	X(1)	X(2)	X(3)	Output Z(i)
Q(0)	Q(0)	Q(1)	Q(2)	Q(3)	Z(0)
Q(1)	Q(0)	Q(1)	Q(2)	Q(3)	Z(1)
Q(2)	Q(0)	Q(1)	Q(2)	Q(3)	Z(2)
Q(3)	Q(0)	Q(1)	Q(2)	Q(3)	Z(3)

(h) Moore machine

FIGURE 10.1 continued. (a, b) Sequential machine models. (c, d) Sequential machine state graphs. (e, f) Sequential machine state graph (continued). (g, h) State transition tables.

at the door of the input logic layer, and joins the input vector to cause the input logic layer to repeat the process all over again. As mentioned, the input/output logic layers are combinational logic networks. However, the memory layer is made of a group of any types of flip-flops, which defines the binary code of the present or ith state of the machine. For convenience, we use D-flip-flops here.

Figures 10.1(c) and (d), respectively, show the state graph (diagram) for Mealy and Moore machines. This is an explicit way of describing the operational sequence of a sequential machine system. Note that in contrast to the system block diagram, the state graph shows the details of the state transition path or sequence. It clearly shows where the system "goes" next if a specific condition occurs.

Figure 10.1(e) and (f) show two more ways to describe the operation sequence of a sequential module. Figure 10.1(e) is simply a flow chart such as programmers have been using for decades. Here, the first rectangular block plus the diamond block are equivalent to one state-node in a state graph. Similarly, as in Figure 10.1(f), a state-node in a state graph can be described with the familiar IF-THEN-ELSE statement. Thus, one may have the choice of using one of the four methods of Figure 10.1(c), (d), (e), or (f) to describe the operation sequence of any sequential machine.

Figure 10.1(g) and (h) depict, respectively, the state transition table corresponding to the state graphs shown in Figure 10.1(c) and (d). By comparing Figure 10.1(g) with Figure 10.1(c), and Figure 10.1(h) with

Figure 10.1(d), we quickly find that they convey identical information. Therefore, one has the choice of using one or both of the two descriptions.

10.3. GUIDELINES FOR SEQUENTIAL MACHINE DESIGN

A complete digital system generally consists of transducer(s), multiplexer, filter, sample/hold device, A/D converter, data processor, D/A converter, plus a system controller (an underground hero). If the system is microprocessor based, then the functions of the data processor and the system controller are carried out by an appropriate microprocessor. Otherwise, the controller would be a custom-designed unit which controls the data path and the operation sequence of the whole system. Thus, the design of a digital system can generally be divided into two major tasks: (1) design of the data paths and (2) design of the controller. The controller is nothing but a sequential machine. Following are the proposed guidelines for the design of a sequential machine.

- Analyze the system specifications.
- Select the machine model (Mealy, Moore) for system implementation.
- Define the input variables and the input binary vector.
- Define the output variables and the output binary vector.
- Develop a state graph (state diagram) and state transition table, followed by the state minimization process.
- Assign the internal states.
- Encode the internal states.
- Develop the excitation and output matrix for the input logic layer and the memory (flip-flop) layer.
- Design the output logic layer.
- Hardware realization or implementation.

10.4. DESIGN EXAMPLES

In this section we will present two design examples. The first example is a simple one so that its design process need not clearly or explicitly show each of the design steps proposed in the preceding section. However, in the second example, one may clearly see most of the steps.

10.4.1. Example 1: Design of a BCD Counter

For illustration purposes, we will show the design in two parts (1) based on the Moore machine model and (2) based on the Mealy machine model. Let us follow the proposed guideline as follows.

Since BCD stands for Binary Coded Decimal, it means that we need four binary bits to represent (code) one decimal digit which may be 0, 1, ..., 9. Note that the equivalent four bit codes for 10, ..., 15 are not used in BCD code, so we may consider them as "don't care" conditions which we can take advantage of in the logic minimization process for combinational logic design later.

As stated above, we will use the Moore machine model for solution (1) and the Mealy machine model for solution (2).

We need only one input variable, the clock pulse. Thus, the input vector for t = i, X(i) = (x_1), where x_1 = 1 implies a clock pulse, otherwise no pulse.

Define the output vector with the binary coded decimals. Thus, the output vector Z(i) = (z_1, z_2, z_3, z_4) will use four LEDs, one for each binary bit to indicate the binary code. That is, binary output 0001 is the code for decimal 1, and binary output 0111 is the code for decimal 7, etc.

Finally, develop the state diagram.

Solution (1): The Moore Machine Model Realization

Figure 10.2(a) shows the Moore state graph and the state transition table. Since ten is obviously the minimum number of states for this system, we can skip the state minimization step.

Figure 10.2(b) shows the tables for the encoded state and output vectors, respectively. Obviously, for the state table we need at least four flip-flops to encode the 10 stable states; we choose four D flip-flops for this purpose. Thus, we have d_3, d_2, d_1, d_0 waiting at the doors or entrances to the memory layer, which encodes the next states and q_3, q_2, q_1, q_0 as the output of the memory layer, for encoding the present states.

Figure 10.2(c) shows the Karnaugh maps derived from Figure 10.2(b) for the design of characteristic equations or the excitation tables for the D flip-flops. As a result, we obtain the four switching functions (combinational logic) for the d_i's of the four flip-flops, which encode the next states of the system. Thus, we have completed the design of the input logic layer. Note that in each Karnaugh map, there are six "don't care" entries marked with " – " in the lower right corner of each entry, which we have used to minimize the combinational logic design of the input layer.

Since we only need the binary codes for display, the output logic layer design is a trivial task. We just use the negation outputs of the D flip-flops to drive the LEDs. It is important to point out here that as far as a logic designer is concerned, their design job appears complete at this point. However, for a digital system designer, one further step must be considered. That is, can the flip-flops directly drive the LEDs? For this task, one must select the right LEDs and then the proper flip-flops that can SINK the current of the LEDs without burning the flip-flops, otherwise a driver should be inserted between the LEDs and the output of each flip-flop.

Figure 10.2(d) depicts the hardware implementation of the design. It is advisable for the reader to check the schematic shown against the diagram shown in Figure 10.1(b) for identifying the corresponding essential system components. In addition, it may be beneficial for the reader, by referring to the schematic diagram, to assume that the system is at any one of the ten states, and that that state is the present state. Then he/she might imagine that a clock pulse has just arrived at the input and thus logically analyze the system and verify if the next state is indeed the expected one.

Solution (2): The Mealy Machine Model Realization

To make the problem a little more interesting, let us assume that the customer requests that the display of the counter should be changed to a true decimal numerical figure instead of the 4 bits of binary code displayed by the four LEDs. Obviously, all we have to do is to improve the output logic layer and the display devices of our present design. However, let us further assume that we would like to take this opportunity to illustrate how the Mealy machine model can also be used to design this system. Recall that the only difference between the Mealy machine and Moore machine models is their system configurations, state graphs, and state transition tables. Thus, we will use the same vector notation defined in the preceding solution for the input/output, present and next states, etc. Since the system configuration of a Mealy machine has been shown in Figure 10.1(a), we wil not repeat it here; however, the Mealy machine state graph and state transition table are shown in Figure 10.2(e). As for the display device, we decide to use one 7-segment display device in addition to the four LEDs. However, the 7-segment device has seven inputs, while in our old design, the output vector yielded from the memory layer is a pure 4-bit binary-coded decimal, which must be decoded and followed by a driver to drive the 7-segment display. Thus, it appears that we need to

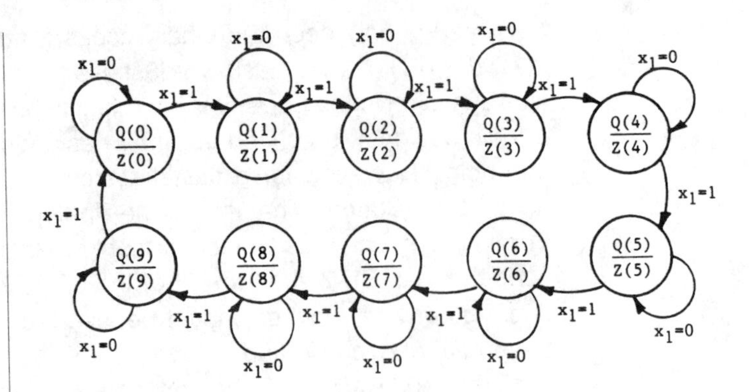

BCD Counter State Graph (Moore Model)

Present State Q(i)	Present Output Z(i)	Next State	
		$x_1=0$	$x_1=1$
Q(0)	Z(0)	Q(0)	Q(1)
Q(1)	Z(1)	Q(1)	Q(2)
Q(2)	Z(2)	Q(2)	Q(3)
Q(3)	Z(3)	Q(3)	Q(4)
Q(4)	Z(4)	Q(4)	Q(5)
Q(5)	Z(5)	Q(5)	Q(6)
Q(6)	Z(6)	Q(6)	Q(7)
Q(7)	Z(7)	Q(7)	Q(8)
Q(8)	Z(8)	Q(8)	Q(9)
Q(9)	Z(9)	Q(9)	Q(0)

BCD Counter State Transition Table (Moore Model)

(a)

FIGURE 10.2 (a) State graph and state transition table. (b) State coding and output coding. (c) Design of input logic layer. (d) Schematic diagram for BCD counter (Moore machine). (e) State graph and state transition table. (f) Schematic diagram for BCD counter with decimal display (Mealy model). (g) SN74LS49 numerical designations and resultant displays. (h) SN74LS49 function table (courtesy of Texas Instruments). (i) SN74LS49 Logic diagram (courtesy of Texas Instruments).

design a more complicated output logic layer. Fortunately, there are many kinds of BCD to 7-segment decoder/drivers (Figure 10.2(h, i)) available on the market, so we do not have to "reinvent the wheel". Here, we choose to use the SN74LS49 decoder/driver in our design. Figure 10.2(f) shows the schematic for our new system. Figure 10.2(g), (h), and (i), respectively, show the numerical designation and resultant displays, the function table, and the logic diagram for the SN74LS49 BCD to 7-segment decoder/driver. Note that in Figure 10.2(f), the input clock pulse, followed by a one-shot, is connected to the blanking input pin of the 7-segment display. The output of the one-

State Coding and State Transition:

PRESENT STATE ($x_1 = 0$)					NEXT STATE ($x_1 = 1$)				
State	Binary Code				State	Binary Code			
Q(i)	q_3	q_2	q_1	q_0	D(i+1)	d_3	d_2	d_1	d_0
0	0	0	0	0	1	0	0	0	1
1	0	0	0	1	2	0	0	1	0
2	0	0	1	0	3	0	0	1	1
3	0	0	1	1	4	0	1	0	0
4	0	1	0	0	5	0	1	0	1
5	0	1	0	1	6	0	1	1	0
6	0	1	1	0	7	0	1	1	1
7	0	1	1	1	8	1	0	0	0
8	1	0	0	0	9	1	0	0	1
9	1	0	0	1	0	0	0	0	0

Output Coding:

OUTPUT VECTOR	CODING			
Z(i)	z_3	z_2	z_1	z_0
Z(0)	0	0	0	0
Z(1)	0	0	0	1
Z(2)	0	0	1	0
Z(3)	0	0	1	1
Z(4)	0	1	0	0
Z(5)	0	1	0	1
Z(6)	0	1	1	0
Z(7)	0	1	1	1
Z(8)	1	0	0	0
Z(9)	1	0	0	1

(b)

FIGURE 10.2 continued. (a) State graph and state transition table. (b) State coding and output coding. (c) Design of input logic layer. (d) Schematic diagram for BCD counter (Moore machine). (e) State graph and state transition table. (f) Schematic diagram for BCD counter with decimal display (Mealy model). (g) SN74LS49 numerical designations and resultant displays. (h) SN74LS49 function table (courtesy of Texas Instruments). (i) SN74LS49 Logic diagram (courtesy of Texas Instruments).

shot is a short negative going pulse. During its low state, the display will be blanked out. As a result, it will steadily show the present state, or the current count of the counter as a decimal number. Note that for this configuration, the output logic layer yields the output vector at t = i, Z(i) = (z_a, z_b, …, z_g), which now has seven components, and it is a direct function of the present state and the input clock pulse.

10.4.2. Example 2: Traffic Light Controller

Design a traffic light controller for the intersection of a main street and a side street. As shown in Figure 10.3(a), there are two push-buttons (P_1 and P_2) and two car sensors (S_1 and S_2) in the side street.

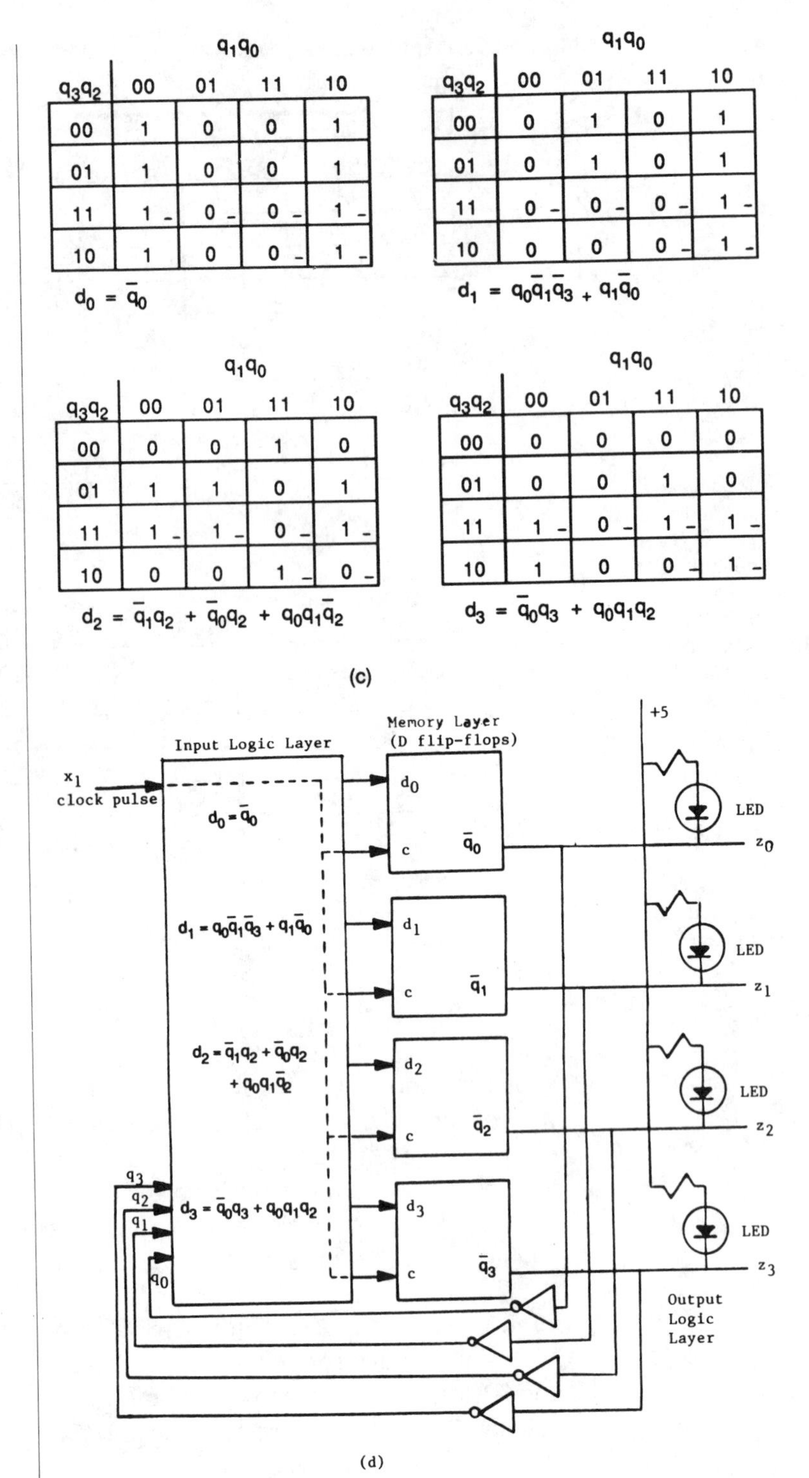

FIGURE 10.2 continued. (a) State graph and state transition table. (b) State coding and output coding. (c) Design of input logic layer. (d) Schematic diagram for BCD counter (Moore machine). (e) State graph and state transition table. (f) Schematic diagram for BCD counter with decimal display (Mealy model). (g) SN74LS49 numerical designations and resultant displays. (h) SN74LS49 function table (courtesy of Texas Instruments). (i) SN74LS49 Logic diagram (courtesy of Texas Instruments).

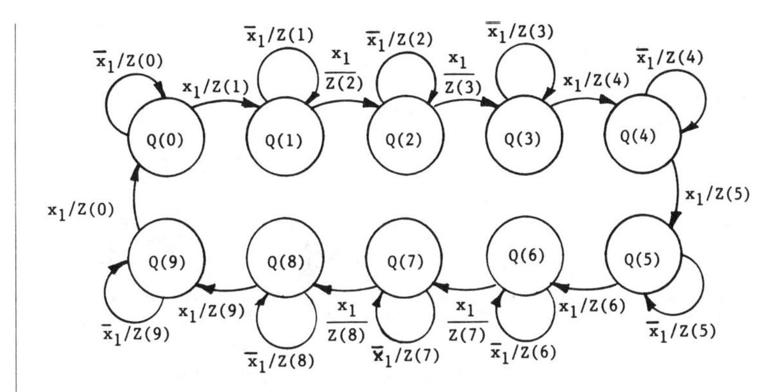

BCD Counter State Graph (Mealy Model)

Present State Q(i)	Next State		Present Output Z(i)	
	$x_1 = 0$	$x_1 = 1$	$x_1 = 0$	$x_1 = 1$
Q(0)	Q(0)	Q(1)	Z(0)	Z(1)
Q(1)	Q(1)	Q(2)	Z(1)	Z(2)
Q(2)	Q(2)	Q(3)	Z(2)	Z(3)
Q(3)	Q(3)	Q(4)	Z(3)	Z(4)
Q(4)	Q(4)	Q(5)	Z(4)	Z(5)
Q(5)	Q(5)	Q(6)	Z(5)	Z(6)
Q(6)	Q(6)	Q(7)	Z(6)	Z(7)
Q(7)	Q(7)	Q(8)	Z(7)	Z(8)
Q(8)	Q(8)	Q(9)	Z(8)	Z(9)
Q(9)	Q(9)	Q(0)	Z(9)	Z(0)

BCD Counter State Transition Table (Mealy Model)

(e)

FIGURE 10.2 continued. (a) State graph and state transition table. (b) State coding and output coding. (c) Design of input logic layer. (d) Schematic diagram for BCD counter (Moore machine). (e) State graph and state transition table. (f) Schematic diagram for BCD counter with decimal display (Mealy model). (g) SN74LS49 numerical designations and resultant displays. (h) SN74LS49 function table (courtesy of Texas Instruments). (i) SN74LS49 Logic diagram (courtesy of Texas Instruments).

According to statistics, very few cars or pedestrians go across the main street. The system should be designed so that the green light for the main street (Gm) is mostly ON, and the green light for the side street (Gs) is to be turned on only when the push-buttons (P_1 and P_2) or the car sensors are enabled. The yellow lights are designed to be ON for only 10 s. Figure 10.3(b) shows the design procedure in nine steps for this example.

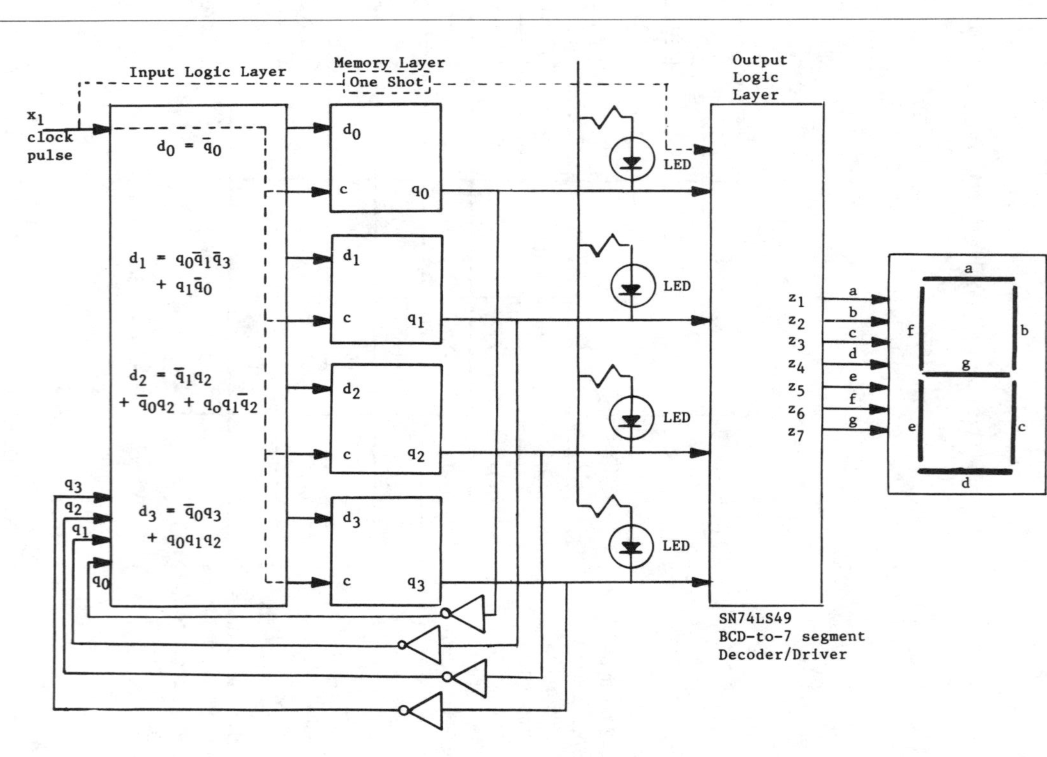

FIGURE 10.2 continued. (a) State graph and state transition table. (b) State coding and output coding. (c) Design of input logic layer. (d) Schematic diagram for BCD counter (Moore machine). (e) State graph and state transition table. (f) Schematic diagram for BCD counter with decimal display (Mealy model). (g) SN74LS49 numerical designations and resultant displays. (h) SN74LS49 function table (courtesy of Texas Instruments). (i) SN74LS49 Logic diagram (courtesy of Texas Instruments).

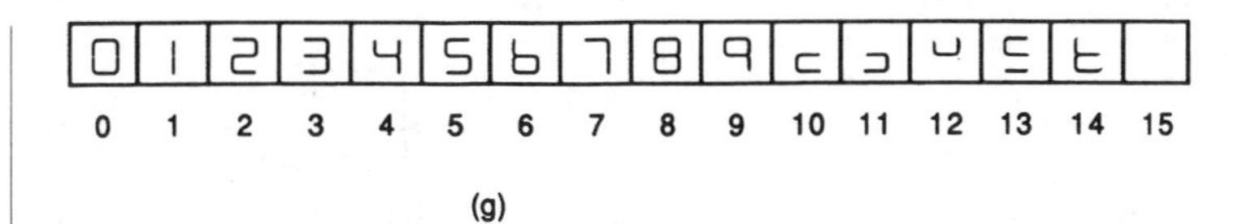

(g)

DECIMAL OR FUNCTION	INPUTS					OUTPUTS							NOTE
	D	C	B	A	$\overline{BI}$	a	b	c	d	e	f	g	
0	L	L	L	L	H	H	H	H	H	H	H	L	1
1	L	L	L	H	H	L	H	H	L	L	L	L	
2	L	L	H	L	H	H	H	L	H	H	L	H	
3	L	L	H	H	H	H	H	H	H	L	L	H	
4	L	H	L	L	H	L	H	H	L	L	H	H	
5	L	H	L	H	H	H	L	H	H	L	H	H	
6	L	H	H	L	H	L	L	H	H	H	H	H	
7	L	H	H	H	H	H	H	H	L	L	L	L	
8	H	L	L	L	H	H	H	H	H	H	H	H	
9	H	L	L	H	H	H	H	H	L	L	H	H	
10	H	L	H	L	H	L	L	L	H	H	L	H	
11	H	L	H	H	H	L	L	H	H	L	L	H	
12	H	H	L	L	H	L	H	L	L	L	H	H	
13	H	H	L	H	H	H	L	L	H	L	H	H	
14	H	H	H	L	H	L	L	L	H	H	H	H	
15	H	H	H	H	H	L	L	L	L	L	L	L	
BI	X	X	X	X	L	L	L	L	L	L	L	L	2

H = High level, L = Low level, X = irrelevant

Notes:
1. The blanking input ($\overline{BI}$) must be open or held at a high logic level when output functions 0 through 15 are desired.
2. When a low logic level is applied directly to the blanking input ($\overline{BI}$), all segment outputs are low regardless of the level of any other input.

(h)

FIGURE 10.2 continued. (a) State graph and state transition table. (b) State coding and output coding. (c) Design of input logic layer. (d) Schematic diagram for BCD counter (Moore machine). (e) State graph and state transition table. (f) Schematic diagram for BCD counter with decimal display (Mealy model). (g) SN74LS49 numerical designations and resultant displays. (h) SN74LS49 function table (courtesy of Texas Instruments). (i) SN74LS49 Logic diagram (courtesy of Texas Instruments).

Step 1: Engineering Interpretation/System Specifications

Lights	Main street	Side street
Green	G_m	G_s
Yellow	O_m	O_s
Red	R_m	R_s

Since there are only four states for the output, we may define the output states vector as follows:

States	G_m	O_m	R_m	G_s	O_s	R_s
Z_1	1	0	0	0	0	1
Z_2	0	1	0	0	0	1
Z_3	0	0	1	1	0	0
Z_4	0	0	1	0	1	0

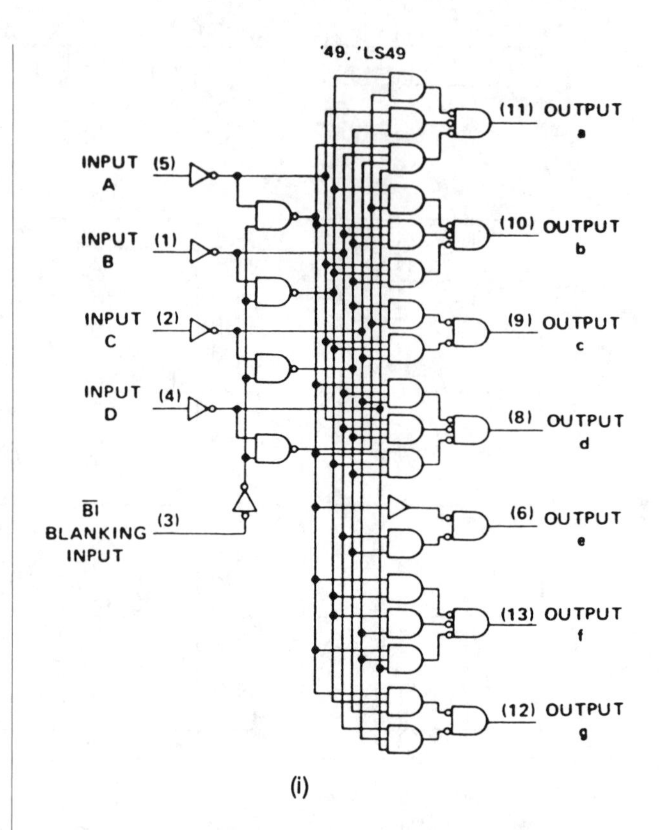

(i)

FIGURE 10.2 continued. (a) State graph and state transition table. (b) State coding and output coding. (c) Design of input logic layer. (d) Schematic diagram for BCD counter (Moore machine). (e) State graph and state transition table. (f) Schematic diagram for BCD counter with decimal display (Mealy model). (g) SN74LS49 numerical designations and resultant displays. (h) SN74LS49 function table (courtesy of Texas Instruments). (i) SN74LS49 Logic diagram (courtesy of Texas Instruments).

Figure 10.3(c) shows the system block diagram. Note that two one-shots are used for timing purposes. The 60-s one-shot is used to hold the push-button signals for the pedestrian (two debouncers may be used here), and the 10-s one-shot is used to hold the ON-time of the yellow light. Both yellow lights, O_m and O_s, share the same one-shot. The light controller is shown in the form of a Mealy sequential logic model, where we have

$$\text{INPUT vector } X(i) = (x_1, x_2, t)$$
$$\text{OUTPUT vector } Z(i) = (G_m, O_m, R_m, G_s, O_s, R_s)$$
$$\text{FEEDBACK vector } Q(i) = (q_1, q_2, \ldots, q_k)$$

Step 2: State Diagram (Graph)

Figure 10.3(d) shows the primitive Mealy state diagram. For example, when we have the input vector

$$X(0) = (x_1, x_2, t) = (0, 0, 0)$$

and the output vector

$$Z = Z(1) = (1, 0, 0, 0, 0, 1)$$
$$= \text{Green on main street, Red on side street}$$

If one or both of the inputs, i.e., the push-button or car sensor, are activated, the machine jumps to second state which will result in activating the yellow light so that t = 1 and the controller enters the third state. Similarly, the rest of the states are defined as shown. However, while the controller is in its sixth state, at which we have Z(4) as output, we do allow the pedestrian or car on the side street to have priority. Therefore, the controller will jump back to the fourth state and let the car or pedestrian go through the side street. One could easily make the controller provide more options and generate more states, but we would rather keep the problem reasonably simple so that the main purpose of design and logic implementation will remain at tutorial level.

Step 3: State Transition Table (PRIMITIVE)

	(x_1, x_2, t)							
States	**000**	**010**	**110**	**100**	**101**	**111**	**011**	**001**
1	①, Z(1)	2, Z(2)	2, Z(2)	2, Z(2)	–	–	–	–
2	–	②, Z(2)	②, Z(2)	②, Z(2)	3, Z(2)	3, Z(2)	3, Z(2)	–
3	–	4, Z(3)	4, Z(3)	4, Z(3)	③, Z(2)	③, Z(2)	③, Z(2)	–
4	5, Z(4)	④, Z(3)	④, Z(3)	④, Z(3)	–	–	–	–
5	⑤, Z(4)	–	–	–	6, Z(4)	6, Z(4)	6, Z(4)	6, Z(4)
6	1, Z(1)	4, Z(3)	4, Z(3)	4, Z(3)	⑥, Z(4)	⑥, Z(4)	⑥, Z(4)	⑥, Z(4)

Note that a circle denotes that the controller is in its stable state.

Step 4: State Minimization

	(x_1, x_2, t)							
States	**000**	**010**	**110**	**100**	**101**	**111**	**011**	**001**
1, 2	①, Z(1)	②, Z(2)	②, Z(2)	②, Z(2)	3, Z(2)	3, Z(2)	3, Z(2)	3, Z(2)
3	–	4, Z(3)	4, Z(3)	4, Z(3)	③, Z(2)	③, Z(2)	③, Z(2)	–
4, 5	⑤, Z(4)	④, Z(3)	④, Z(3)	④, Z(3)	6, Z(4)	6, Z(4)	6, Z(4)	6, Z(4)
6	1, Z(1)	4, Z(3)	4, Z(3)	4, Z(3)	⑥, Z(4)	⑥, Z(4)	⑥, Z(4)	⑥, Z(4)

Step 5: Internal State Assignment (Omit Output Entries)

Define:

States $1, 2 \rightarrow q_1$
State $3 \rightarrow q_2$
States $4, 5 \rightarrow q_3$
State $6 \rightarrow q_4$

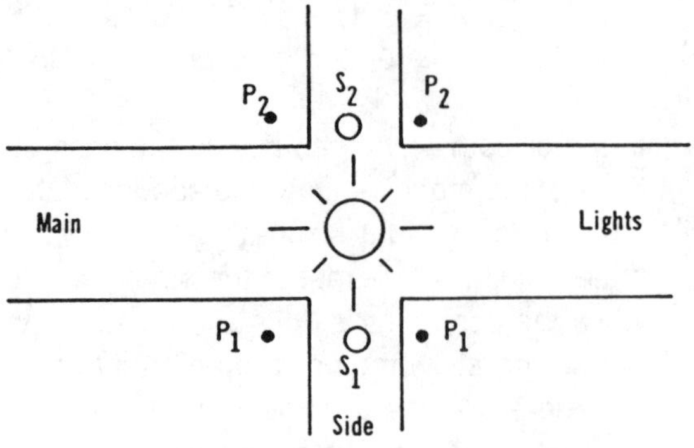

FIGURE 10.3 (a) Street map. (b) Design procedure. (c) System block diagram of traffic light controller. (d) Primitive Mealy state diagram. (e) Hard-wired logic realization.

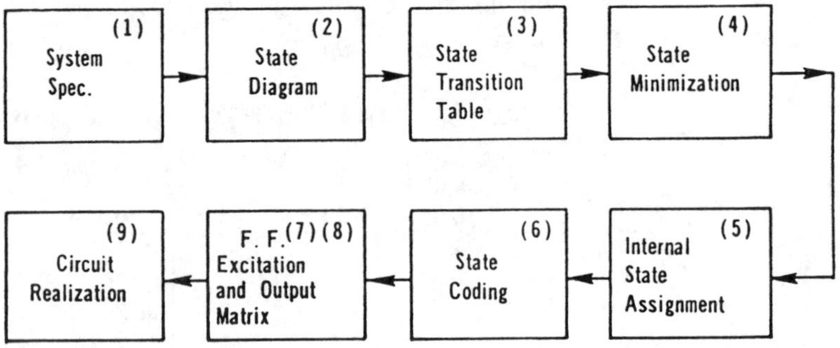

States	000	010	110	100	101	111	011	001
q_1	(q_1)	(q_1)	(q_1)	(q_1)	q_2	q_2	q_2	q_2
q_2	–	q_3	q_3	q_3	(q_2)	(q_2)	(q_2)	–
q_3	(q_3)	(q_3)	(q_3)	(q_3)	q_4	q_4	q_4	q_4
q_4	q_1	q_3	q_3	q_3	(q_4)	(q_4)	(q_4)	(q_4)

Step 6: State Coding

$$q_1 \triangleq 00, \quad q_2 \triangleq 01, \quad q_3 \triangleq 11, \quad q_4 \triangleq 10$$

Present states	(x_1, x_2, t) 000	010	110	100	101	111	011	001
00	00	00	00	00	01	01	01	01
01	11	11	11	11	01	01	01	01
11	11	11	11	11	10	10	10	10
10	00	11	11	11	10	10	10	10

Note that in this table, entries = *next states*.

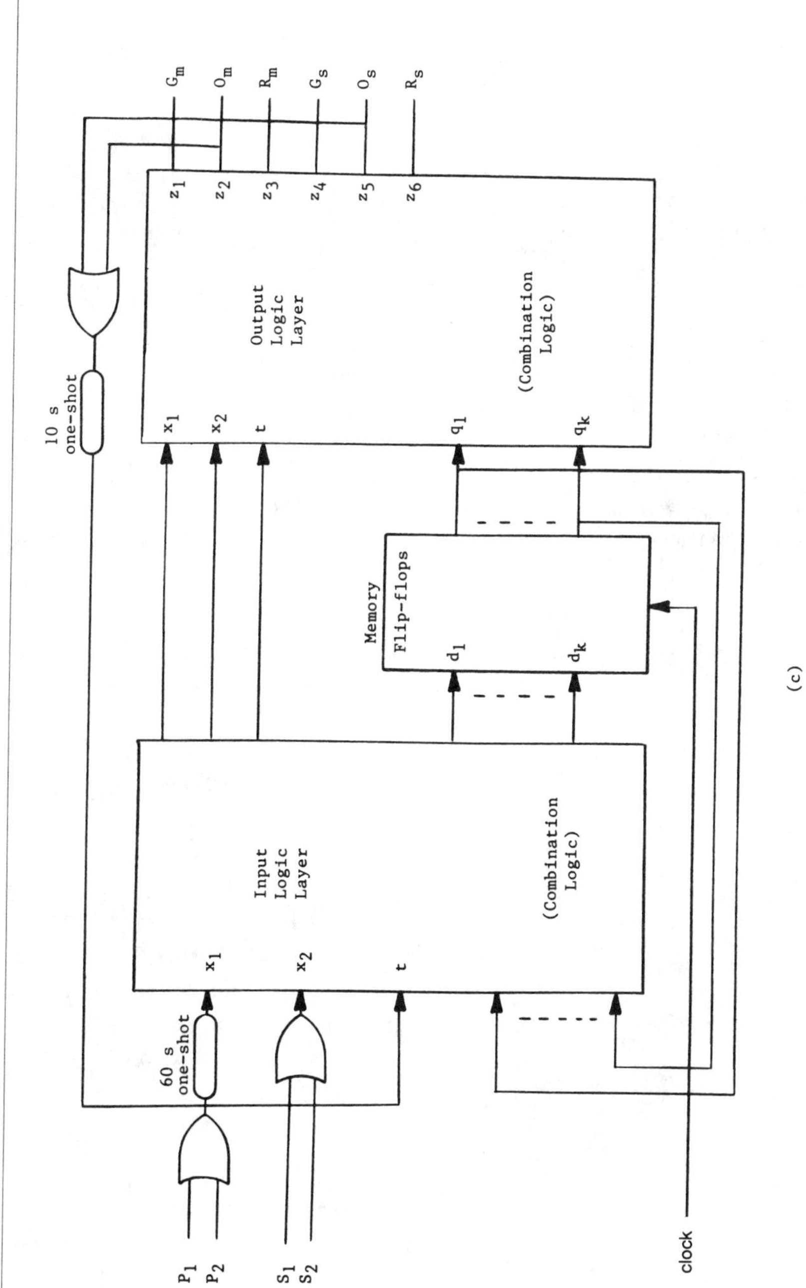

FIGURE 10.3 continued. (a) Street map. (b) Design procedure. (c) System block diagram of traffic light controller. (d) Primitive Mealy state diagram. (e) Hard-wired logic realization.

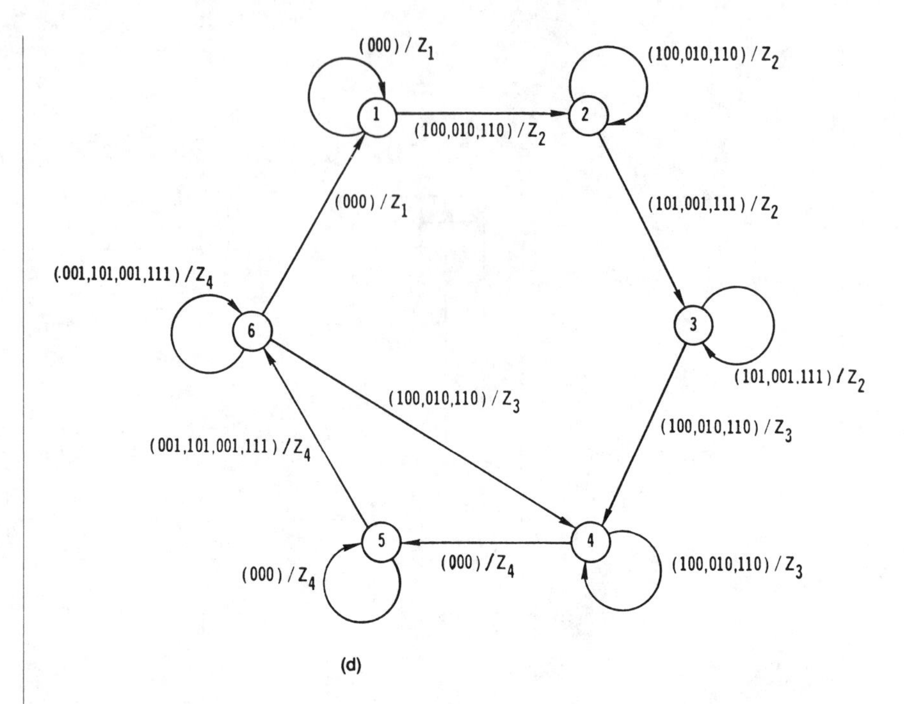

(d)

FIGURE 10.3 continued. (a) Street map. (b) Design procedure. (c) System block diagram of traffic light controller. (d) Primitive Mealy state diagram. (e) Hard-wired logic realization.

Step 7: Flip-Flop Excitation Matrix

Since there are four states, two flip-flops are needed. Let us assign

$$\text{the left column} \rightarrow q_1(i),\ d_1 \triangleq q_1(i+1)$$
$$\text{the right column} \rightarrow q_2(i),\ d_2 \triangleq q_2(i+1)$$

where

$$q_1(i) \triangleq \text{the present state of the first flip-flop}$$
$$d_1 \text{ or } q_1(i+1) \triangleq \text{the next state of the first flip-flop}$$
$$q_2(i) \triangleq \text{the present state of the second flip-flop}$$
$$d_2 \text{ or } q_2(i+1) \triangleq \text{the next state of the second flip-flop}$$

Then, we have

$q_1q_2(i)$	(x_1, x_2, t) 000	010	110	100	101	111	011	001
00	0	0	0	0	0	0	0	0
01	1	1	1	1	0	0	0	0
11	1	1	1	1	1	1	1	1
10	0	1	1	1	1	1	1	1

for $d_1 = q_1(i+1)$, and

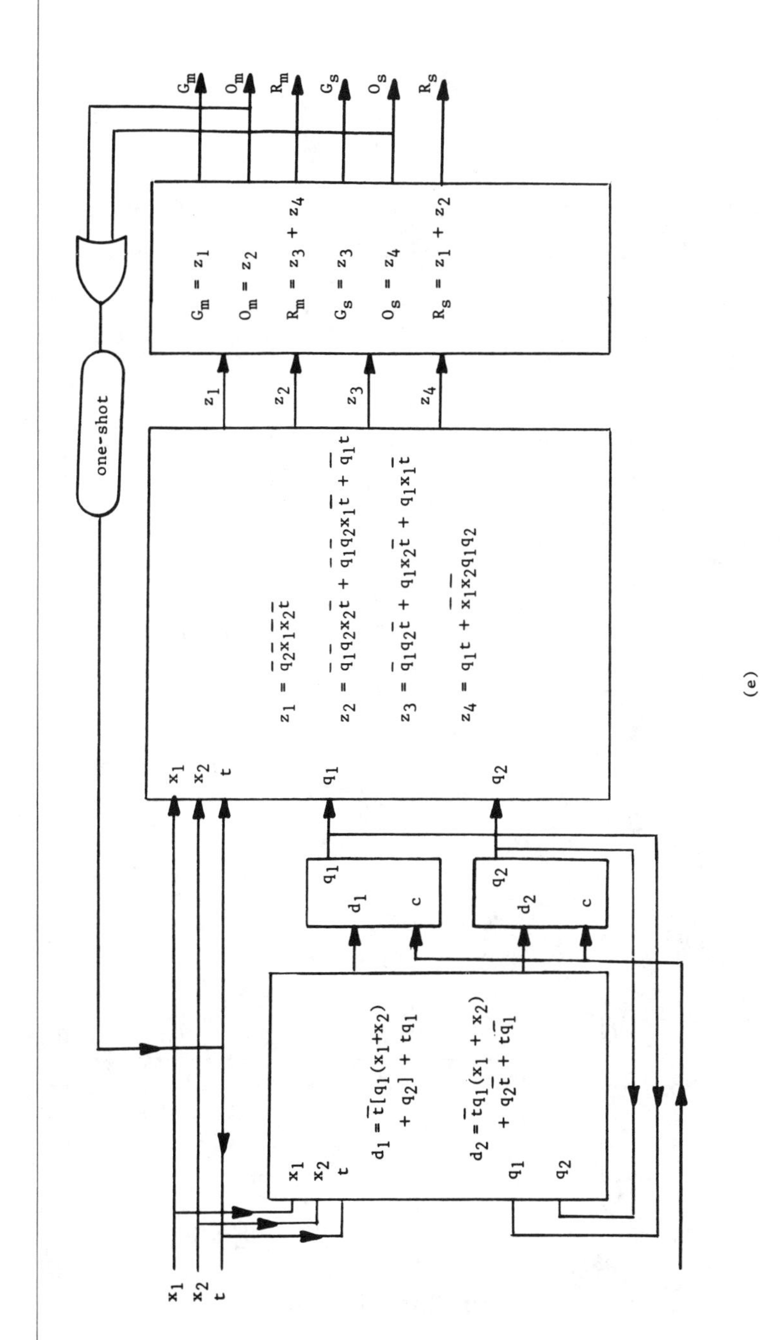

FIGURE 10.3 continued. (a) Street map. (b) Design procedure. (c) System block diagram of traffic light controller. (d) Primitive Mealy state diagram. (e) Hard-wired logic realization.

$q_1q_2(i)$ \ (x_1, x_2, t)	000	010	110	100	101	111	011	001
00	0	0	0	0	1	1	1	1
01	1	1	1	1	1	1	1	1
11	1	1	1	1	0	0	0	0
10	0	1	1	1	0	0	0	0

for $d_2 = q_2(i+1)$. Let us use a D flip-flop which has the following characteristics:

$Q(n)$	$D(n^+)$	$Q(n+1)$
0	0	0
0	1	1
1	0	0
1	1	1

D Q
Clock

Hence, we can derive the matrix for the D flip-flop:

$q_1q_2(i)$ \ (x_1, x_2, t)	000	010	110	100	101	111	011	001
00	0	0	0	0	0	0	0	0
01	1	1	1	1	0	0	0	0
11	1	1	1	1	1	1	1	1
10	0	1	1	1	1	1	1	1

$$d_1(n^+) = q_2\bar{t} + q_1x_2\bar{t} + q_1x_1\bar{t} + q_1t$$

$$= \bar{t}\left[q_1(x_1 + x_2) + q_2\right] + tq_1$$

Similarly,

$q_1q_2(i)$ \ (x_1, x_2, t)	000	010	110	100	101	111	011	001
00	0	0	0	0	1	1	1	1
01	1	1	1	1	1	1	1	1
11	1	1	1	1	0	0	0	0
10	0	1	1	1	0	0	0	0

$$d_2(n^+) = q_2\bar{t} + q_1x_2\bar{t} + q_1x_1\bar{t} + t\bar{q}_1$$

Step 8: Output Matrix

From Step 4 we have

$q_1q_2(i)$ \ (x_1, x_2, t)	000	010	110	100	101	111	011	001
00	z_1	z_2	z_2	z_2	z_2	z_2	z_2	z_2
01	z_3	z_3	z_3	z_3	z_2	z_2	z_2	z_2
11	z_4	z_3	z_3	z_3	z_4	z_4	z_4	z_4
10	z_1	z_3	z_3	z_3	z_4	z_4	z_4	z_4

Thus,

$$z_1 = \bar{q}_2\bar{x}_1\bar{x}_2\bar{t}$$
$$z_2 = \bar{q}_1\bar{q}_2x_2\bar{t} + \bar{q}_1\bar{q}_2x_1\bar{t} + \bar{q}_1t$$
$$z_3 = \bar{q}_1q_2\bar{t} + q_1x_2\bar{t} + q_1x_1\bar{t}$$
$$z_4 = q_1t + \bar{x}_1\bar{x}_2q_1q_2$$

From the definition of the output states, we have

$$G_m = Z_1\bar{Z}_2\bar{Z}_3\bar{Z}_4 \rightarrow Z_1, \text{ since } Z_1, Z_2, Z_3, Z_4 \text{ are mutually exclusive}$$
$$O_m = \bar{Z}_1Z_2\bar{Z}_3\bar{Z}_4 \rightarrow Z_2$$
$$R_m = \bar{Z}_1\bar{Z}_2Z_3Z_4 \rightarrow Z_3 + Z_4$$
$$G_s = \bar{Z}_1\bar{Z}_2Z_3\bar{Z}_4 \rightarrow Z_3$$
$$O_s = \bar{Z}_1\bar{Z}_2\bar{Z}_3Z_4 \rightarrow Z_4$$
$$R_s = Z_1Z_2\bar{Z}_3\bar{Z}_4 \rightarrow Z_1 + Z_2$$

Step 9: Circuit Realization

Figure 10.3(e) shows the hard-wired logic design.

REFERENCES

1. Peatman, J. B., *The Design of Digital Systems,* McGraw-Hill, New York, 1972.
2. Krieger, M., *Basic Switching Circuit Theory,* Macmillan, New York, 1967.
3. Friedman, A. D., *Logic Design of Digital Systems,* Computer Science Press, New York, 1975.
4. Blakeslee, T. T., *Digital Design with Standard MSI and LSI,* 2nd ed., John Wiley & Sons, New York, 1979.
5. Roth, C. H., Jr., *Fundamentals of Logic Design,* 3rd ed., West Publishing, St. Paul, MN, 1985.
6. Johnson, E. L. and Karim, M. A., *Digital Design – A Pragmatic Approach,* PWS Engineering Publishers, Boston, 1987.
7. Comer, D. J., *Digital Logic and State Machine Design,* Holt, Rinehart and Winston, New York, 1984.
8. McCluskey, E. J., *Logic Design Principles,* Prentice-Hall, Englewood Cliffs, NJ, 1986.
9. Cheung, J. Y. and Bredeson, J. G., *Modern Digital System Design,* West Publishing, St. Paul, MN, 1990.

11 RANDOM LOGIC DESIGN USING PROGRAMMABLE LOGIC ELEMENTS

11.1. GENERAL BACKGROUND

Random logic, sometimes called control logic, is a logic network that has no definite pattern, in contrast to array or tree logic networks. In any digital system design, one is always involved in designing some subsystem which controls the signal paths and schedules the time or sequence order of logical events. The design is usually customized and irregular in fabrication, therefore it is an inherently time-consuming task and prone to fabrication errors. In this chapter, we shall describe the techniques of realizing random logic using regular or programmable logic elements which have been commercially available for a number of years. A brief review of the logical design concept follows.

The conventional procedure for the logic design of a switching network or a digital system is shown in Figure 11.1. To clarify the concept, a simple design example is in order. Following the procedure shown in Figure 11.1, one can show how a 1-bit full-adder is designed. Figure 11.2 illustrates the design in three steps: (a), (b), and (c). However, the final step of hardwired logic implementation connecting "discrete" logic elements, such as AND and EXCLUSIVE-OR gates, in a customized fashion can be replaced by programmed logic elements such as MULTIPLEXER or programmable logic elements such as ROM, Programmable-Logic-Array (PLA), etc. Thanks to LSI technology, a logic designer now has a choice of realizing his design by one of the five methods, namely hard-wired or discrete logic, multiplexers, ROM, PLA, or a microprocessor.

11.2. DESIGN WITH MULTIPLEXERS

A typical 4:1 multiplexer is shown in Figure 11.3. It is originally designed for multiplexing digital data. However, it is important to point out that a designer can use or "borrow" it to perform an unintended task,

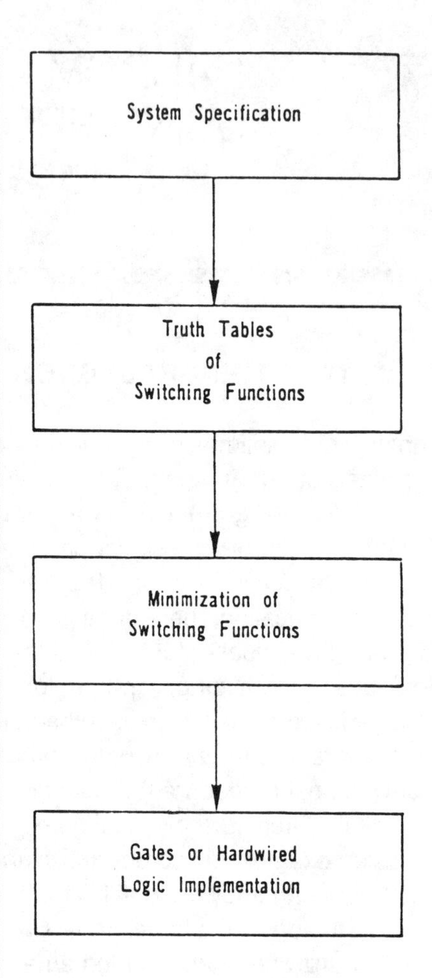

FIGURE 11.1 Logic design procedure.

such as realization of some general switching functions of several variables.

Referring to Figure 11.3 again, note that the device is really a realization of the following switching function:

$$F = \overline{S}_1\overline{S}_0 I_0 + \overline{S}_1 S_0 I_1 + S_1\overline{S}_0 I_2 + S_1 S_0 I_3 \quad (11.1)$$

This equation reveals that the device has all four combinations of two switching variables, i.e., S_0 and S_1. Thus, it can potentially realize 2^{2n} switching functions, where n=2 in this example. Recall (Chapter 5, Figure 5.22) that a multiplexer can be considered as an electronically controlled rotary switch. For a 4:1 multiplexer, it is a 4-to-1 rotary switch with its arm controlled by S_1S_0 signal. Some design examples follow.

x_1 Augend	x_2 Addend	x_3 Carry – in	z_1 Sum	z_2 Carry – out
0	0	0	0	0
0	0	1	1	0
0	1	0	1	0
0	1	1	0	1
1	0	0	1	0
1	0	1	0	1
1	1	0	0	1
1	1	1	1	1

(a) Truth Table of a Full Adder

Before Minimization:

$$z_1 = \bar{x}_1\bar{x}_2x_3 + \bar{x}_1x_2\bar{x}_3 + x_1\bar{x}_2\bar{x}_3 + x_1x_2x_3$$
$$z_2 = \bar{x}_1x_2x_3 + x_1\bar{x}_2x_3 + x_1x_2\bar{x}_3 + x_1x_2x_3$$

After Minimization:

$$z_1 = x_1 \oplus x_2 \oplus x_3$$
$$z_2 = x_1x_2 + (x_1 \oplus x_2)x_3$$

(b) Minimization Process

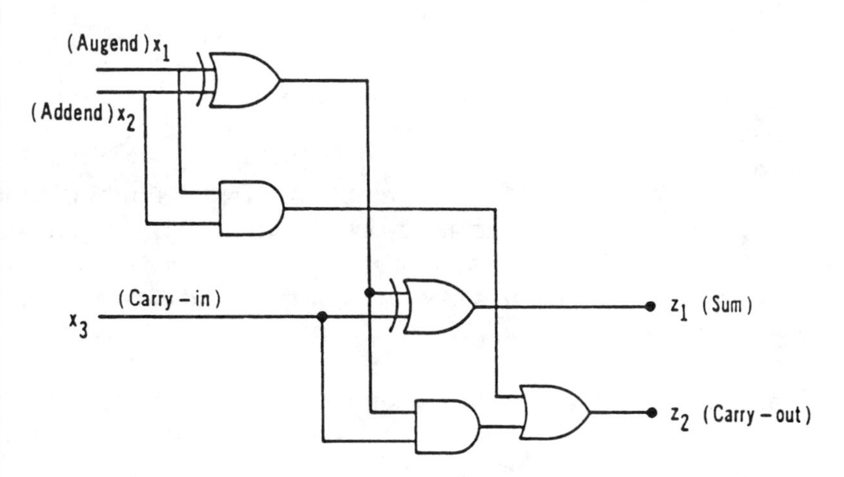

(c) Hardwired Logic Implementation

FIGURE 11.2 Conventional combinational logic design procedure (full-adder).

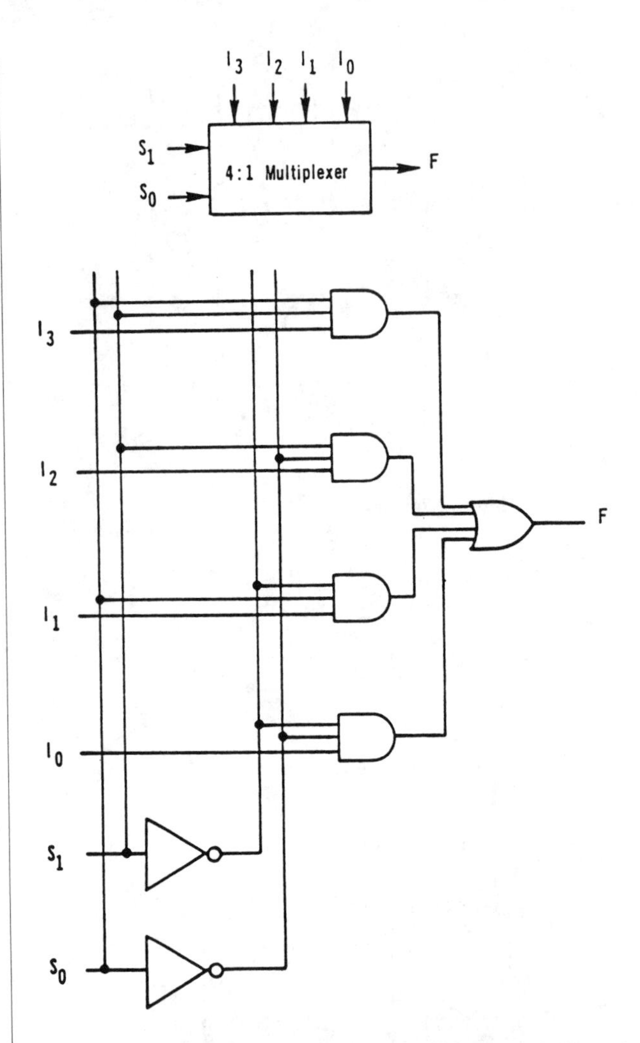

FIGURE 11.3 4:1 digital multiplexer.

11.2.1. Example 1: Two Switching Variables

For realization of $F = x_1\overline{x}_2 + \overline{x}_1x_2$, with reference to Equation (11.1). one simply ties I_0 and I_3 to ground, or logic [0]; I_1 and I_2 to V_{CC}, or logic [1]; and x_1 to S_0, x_2 to S_1, and the device will realize the given switching function.

11.2.2. Example 2: Three Switching Variables

By following a similar procedure, one can use the same element to realize a switching function of three variables.

Problem: Realization of $F = x_1\overline{x}_2 + \overline{x}_1x_2x_3$

Solution

Step 1: Let us define a "template" in Karnaugh Map format for this 4:1 multiplexer as shown in Figure 11.4(a).

Step 2: Expand the switching function back to the expression before minimization, i.e., $F = x_1\overline{x}_2x_3 + x_1\overline{x}_2\overline{x}_3 + \overline{x}_1x_2x_3$. Then, construct its corresponding Karnaugh Map as shown in Figure 11.4(b).

Step 3: Obtain Figure 11.4(c) by overlaying (b) on (a).

Step 4: Specify connection of multiplexer according to the shaded pattern, whose result is shown in Figure 11.4(d). To be more specific, for 11.4(c) we proceed as follows.

(a) Since the whole columns of I_0 and I_3 are unshaded, we tie I_0 and I_3 to ground.

(b) Since the whole column of I_1 is shaded, we tie I_1 to V_{CC} or logic [1].

(c) Since the lower section of the I_2 column is shaded, we tie I_2 to x_3. *Note*: if the upper section of the I_2 column were shaded, then I_2 would be tied to $\overline{x}_3$.

(d) Finally, we tie x_1 to S_0 and x_2 to S_1.

Step 5: Verify the results of Step 4, through Equation (11.1). We have

$$F = 0 + x_1\overline{x}_2 + \overline{x}_1x_2x_3$$

11.2.3. Example 3: Adders

By following the procedure proposed in Example 2, one can implement the full-adder shown in Figure 11.2 by the same multiplexer used in Examples 1 and 2 as follows. First, the Karnaugh Map for the Sum,

$$S = \overline{x}_1\overline{x}_2x_3 + \overline{x}_1x_2\overline{x}_3 + x_1\overline{x}_2\overline{x}_3 + x_1x_2x_3$$

is

x_3 \ x_2x_1	00	01	11	10
0	0	1	0	1
1	1	0	1	0

Therefore, we have

$x_1 = S_0$, $x_2 = S_1$, $x_3 = I_0$

$\overline{x}_3 = I_1$ $\overline{x}_3 = I_2$ $x_3 = I_3$

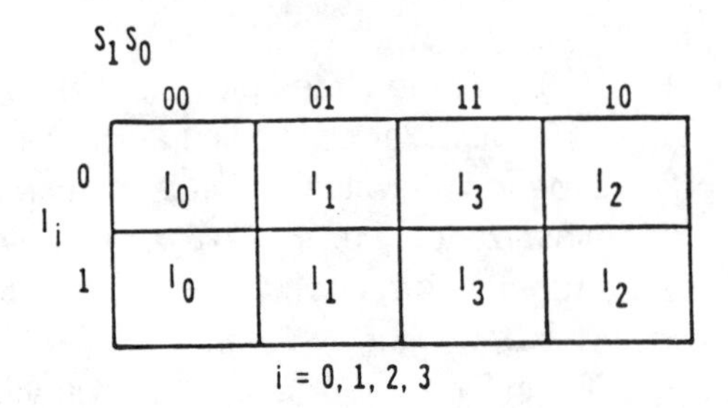

(a) Multiplexer Template

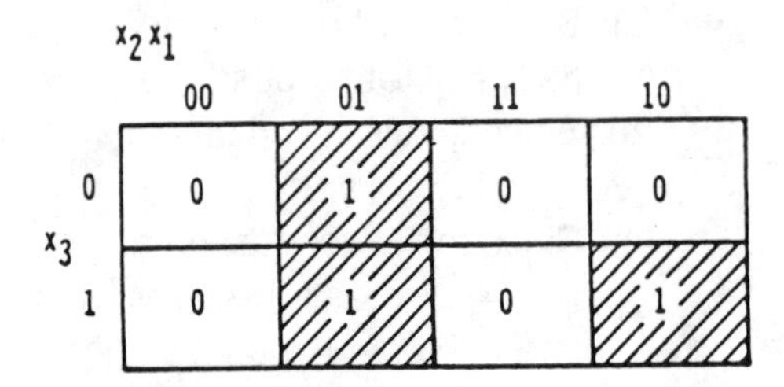

(b) Karnaugh Map of the Switching Function

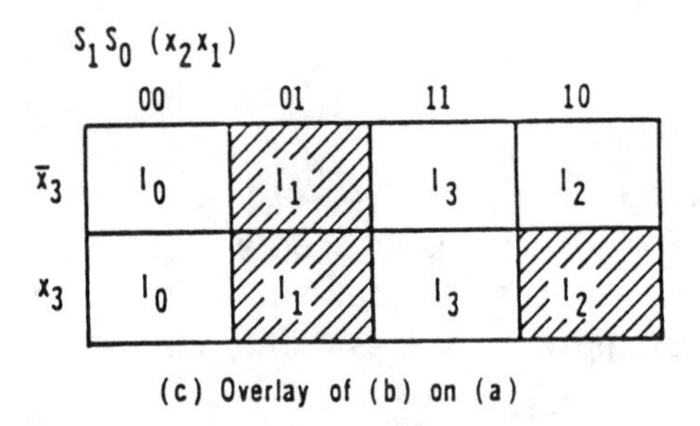

(c) Overlay of (b) on (a)

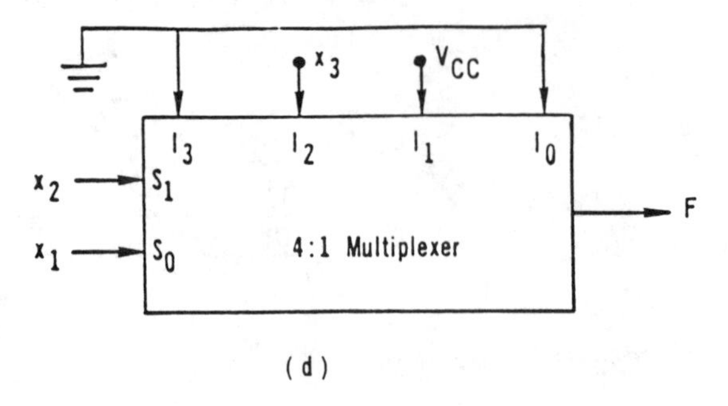

(d)

FIGURE 11.4 (a, b, c, d) Procedure for using multiplexer.

The Karnaugh map for Carry

$$C = \overline{x}_1x_2x_3 + x_1\overline{x}_2x_3 + x_1x_2\overline{x}_3 + x_1x_2x_3$$

is

x_3 \ x_2x_1	00	01	11	10
0	0	0	1	0
1	0	1	1	1

Therefore, we have

$x_1 = S_0$, $x_2 = S_1$, $0 = I_0$
$x_3 = I_1$ $x_3 = I_2$ $1 = I_3$

The circuit diagram is shown in Figure 11.5.

At this point, one may ask what is the advantage of using multiplexers as opposed to using gates to implement a full-adder? For a simple logic network, the hard-wired logic implementation using gates may still be the best one. In this example we merely use the design of the full-adder as a vehicle to illustrate the concept. However, if more complicated switching functions are to be implemented, one would gain reliability, simplify assembly operation in production lines, and reduce

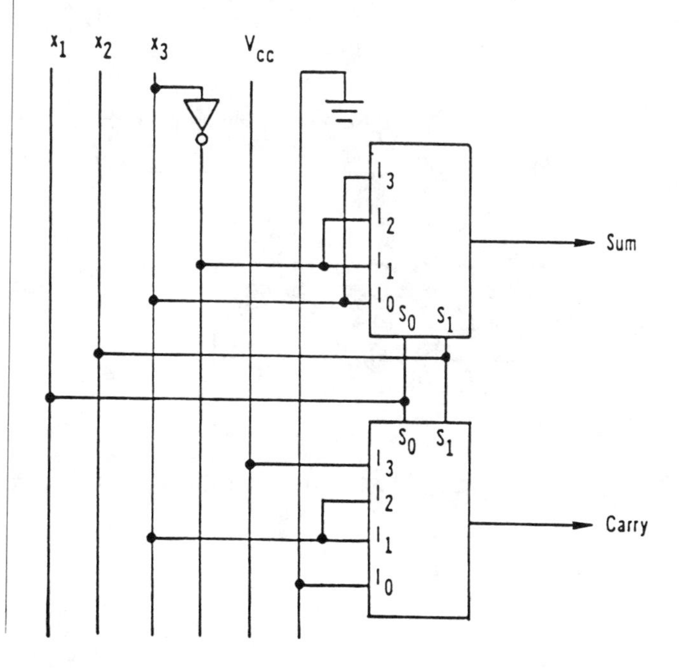

FIGURE 11.5
Implementation of a full-adder with multiplexers.

maintenance expense by using multiplexers. The next example shows how a five-variable switching function can be realized by a 4-control, 16:1 multiplexer.

11.2.4. Example 4: Five Switching Variables

Problem: Realization of a five-variable switching function which has the following Karnaugh Maps:

x_4x_3 \ x_2x_1	00	01	11	10
00	0	1	0	0
01	1	0	1	1
11	1	1	0	0
10	0	1	1	0

$x_5 = 0$

x_4x_3 \ x_2x_1	00	01	11	10
00	1	1	0	1
01	0	0	0	1
11	0	0	0	1
10	1	1	0	1

$x_5 = 1$

Solution:

As the proposed procedure, the Karnaugh Map template of the 4-control, 16:1 multiplexer can be constructed as follows.

S_4S_3 \ S_2S_1	00	01	11	10
00	I_0	I_1	I_3	I_2
01	I_4	I_5	I_7	I_6
11	I_{12}	I_{13}	I_{15}	I_{14}
10	I_8	I_9	I_{11}	I_{10}

$I_j = 0$

S_4S_3 \ S_2S_1	00	01	11	10
00	I_0	I_1	I_3	I_2
01	I_4	I_5	I_7	I_6
11	I_{12}	I_{13}	I_{15}	I_{14}
10	I_8	I_9	I_{11}	I_{10}

$I_j = 1$

Next, overlay the Karnaugh Map of the switching function on the template to get

x_4x_3 \ x_2x_1	00	01	11	10
00	0		0	0
01		0		
11			0	0
10	0			0

$I_i = \overline{x}_5$

x_4x_3 \ x_2x_1	00	01	11	10
00			0	
01	0	0	0	
11	0	0	0	
10			0	

$I_i = x_5 = 1$

Accordingly,we can define the connection as follows:

TABLE 11.1
Verification Table for Example 4

Switching function						MUX controls				MUX		
F	x_5	x_4	x_3	x_2	x_1	S_3	S_2	S_1	S_0	Inputs		Output (W)
0	0	0	0	0	0	0	0	0	0	I_0	x_5	0
1	0	0	0	0	1	0	0	0	1	I_1	1	1
0	0	0	0	1	0	0	0	1	0	I_2	x_5	0
0	0	0	0	1	1	0	0	1	1	I_3	0	0
1	0	0	1	0	0	0	1	0	0	I_4	$\bar{x}_5$	1
0	0	0	1	0	1	0	1	0	1	I_5	0	0
1	0	0	1	1	0	0	1	1	0	I_6	1	1
1	0	0	1	1	1	0	1	1	1	I_7	$\bar{x}_5$	1
0	0	1	0	0	0	1	0	0	0	I_8	x_5	0
1	0	1	0	0	1	1	0	0	1	I_9	1	1
0	0	1	0	1	0	1	0	1	0	I_{10}	x_5	0
1	0	1	0	1	1	1	0	1	1	I_{11}	$\bar{x}_5$	1
1	0	1	1	0	0	1	1	0	0	I_{12}	$\bar{x}_5$	1
1	0	1	1	0	1	1	1	0	1	I_{13}	$\bar{x}_5$	1
0	0	1	1	1	0	1	1	1	0	I_{14}	$\bar{x}_5$	0
0	0	1	1	1	1	1	1	1	1	I_{15}	0	0
1	1	0	0	0	0	0	0	0	0	I_0	x_5	1
1	1	0	0	0	1	0	0	0	1	I_1	1	1
1	1	0	0	1	0	0	0	1	0	I_2	x_5	1
0	1	0	0	1	1	0	0	1	1	I_3	0	0
0	1	0	1	0	0	0	1	0	0	I_4	$\bar{x}_5$	0
0	1	0	1	0	1	0	1	0	1	I_5	0	0
1	1	0	1	1	0	0	1	1	0	I_6	1	1
0	1	0	1	1	1	0	1	1	1	I_7	$\bar{x}_5$	0
1	1	1	0	0	0	1	0	0	0	I_8	x_5	1
1	1	1	0	0	1	1	0	0	1	I_9	1	1
1	1	1	0	1	0	1	0	1	0	I_{10}	x_5	1
0	1	1	0	1	1	1	0	1	1	I_{11}	$\bar{x}_5$	0
0	1	1	1	0	0	1	1	0	0	I_{12}	$\bar{x}_5$	0
0	1	1	1	0	1	1	1	0	1	I_{13}	$\bar{x}_5$	0
1	1	1	1	1	0	1	1	1	0	I_{14}	$\bar{x}_5$	1
0	1	1	1	1	1	1	1	1	1	I_{15}	0	0

$$
\begin{array}{llll}
x_1 = S_0, & x_2 = S_1, & x_3 = S_2, & x_4 = S_3 \\
I_0 = x_5, & I_1 = 1, & I_2 = x_5, & I_3 = 0 \\
I_4 = \bar{x}_5, & I_5 = 0 & I_6 = 1, & I_7 = \bar{x}_5 \\
I_8 = x_5, & I_9 = 1, & I_{10} = x_5, & I_{11} = \bar{x}_5 \\
I_{12} = \bar{x}_5, & I_{13} = \bar{x}_5, & I_{14} = x_5, & I_{15} = 0
\end{array}
$$

This result is verified by the verification table (Table 11.1). Note that the W column represents the corresponding outputs of the multiplexer which are totally in agreement with the values of F, the desired outputs of the switching function.

11.3. DESIGN WITH ROM AND PROM

Using ROM to realize switching functions is another way of implementing random logic with programmable logic elements. As described in Chapter 6, ROM is a special type of memory. It has an address and a content for each memory word which consists of a fixed number of memory cells, but the content or the information is prestored in the memory cells in the design phase. During operation, the user has no way to change its contents. All the user can do is find what information has been stored in a particular memory word at its unique address by specifying that address and issuing a READ command to retrieve it. Following are three design examples whose problem specifications are already no stranger to the readers. We will illustrate how one can use this new ROM technique to solve old design problems.

11.3.1. Example 1: Adder Design With ROM

As we know, there are many ways to realize a ROM in circuitry. For demonstration purposes we will begin with a simple diode-matrix ROM to implement parallel/serial full-adders. Figures 11.6(a) and (b), respectively, show the schematic diagram of the ROM for realizing parallel and serial full-adders. Recall that for the diode-matrix ROM, the data must be prewired according to the designer's specification. Here, the information to be stored is the adder's switching functions to be realized. However, during operation, the user cannot change its contents. As shown, both Figure 11.6(a) and (b) have eight 2-bit memory words addressed by 000, 001, ..., 111, respectively. Since the principles of operation for both types of adder are similar, we will for convenience focus our discussion on the parallel-adder only. For clarification, let us describe the schematic diagram shown in Figure 11.6(a) in a table form (also known as a "memory map"), as shown in Figure 11.6(c). Notably, the truth table in Figure 11.2(a) and the ROM memory map shown in Figure 11.6(c) are identical, and surprisingly, the ROM realizes the switching functions that are not yet minimized. Here, each address is equivalent to a minterm of the switching function. If one READs the ROM through the whole address space, one would find that the readings of the first memory cell for the whole read process constitutes the switching function z_1; similarly, that of the second cell constitutes the switching function z_2. Thus, we have z_1 and z_2 in minterm format (remember, each address equivalently represents one minterm) as follows.

$$\begin{aligned} \text{SUM:} \quad z_1 &= (m_0 + m_1 + \ldots + m_7)\ \text{(read pulse)} \\ \text{CARRY:} \quad z_2 &= (m_0 + m_1 + \ldots + m_7)\ \text{(read pulse)} \end{aligned}$$

where (read pulse) means "as the READ command pulse is applied", and m_i is the expression for the ith minterm.

Referring to Figure 11.6(a), for z_1 there are only four out of the eight cells holding logic [1] information, and similarly for z_2; the rest of the cells are holding logic [0]. Therefore, the results of the two switching functions can be reduced by eliminating the cells with [0] and expressed in the familiar "sum of products" format as follows:

$$\text{SUM:} \quad z_1 = \bar{x}_1\bar{x}_2x_3 + \bar{x}_1x_2\bar{x}_3 + x_1\bar{x}_2\bar{x}_3 + x_1x_2x_3 \qquad (11.2)$$

$$\text{CARRY:} \quad z_2 = \bar{x}_1x_2x_3 + x_1\bar{x}_2x_3 + x_1x_2\bar{x}_3 + x_1x_2x_3 \qquad (11.3)$$

Note that these results are identical to the switching functions shown in Figure 11.2(a) before minimization. Referring to Figure 11.6(a) or (c), we basically see the structure of an 8×2 ROM which would have a 3-bit MAR and a 2-bit MDR connected to it to function as a memory system. However, in logic design terms, we may consider the three address bits in the MAR, labeled as x_1, x_2, and x_3, as the three switching variables, and the two data bits in the MDR, labeled as z_1 and z_2, as the switching functions of the three switching variables. Note further that in Figure 11.6(c) the address column shows eight memory locations which is nothing but the complete set of the eight minterms for the three switching variables, while the content column shows the truth table for the two switching functions, z_1 and z_2. In this case they realize the SUM and CARRY for the full-adder.

To illustrate the operation of the ROM, let us assume that at $i = 0$, the input vector $x_1x_2x_3 = 001$; in this case, according to the equations (before minimization) shown in FIgure 11.2(a), $z_1 = 1$ and $z_2 = 0$. Now referring to Figure 11.6(a) or (c), that input vector would point to the address location 001 for $i = 0$ and a READ control signal would load the contents of that location, in this case 10, into the MDR. As a result, we would get $z_1 = 1$ and $z_2 = 0$, which would produce the same result as the equations in Figure 11.2(a).

Let us look at this same discussion from a memory designer's viewpoint instead of a logic designer's viewpoint. Figure 11.6(d) shows the 8×2 memory structure whose schematic is shown in Figure 11.6(a)

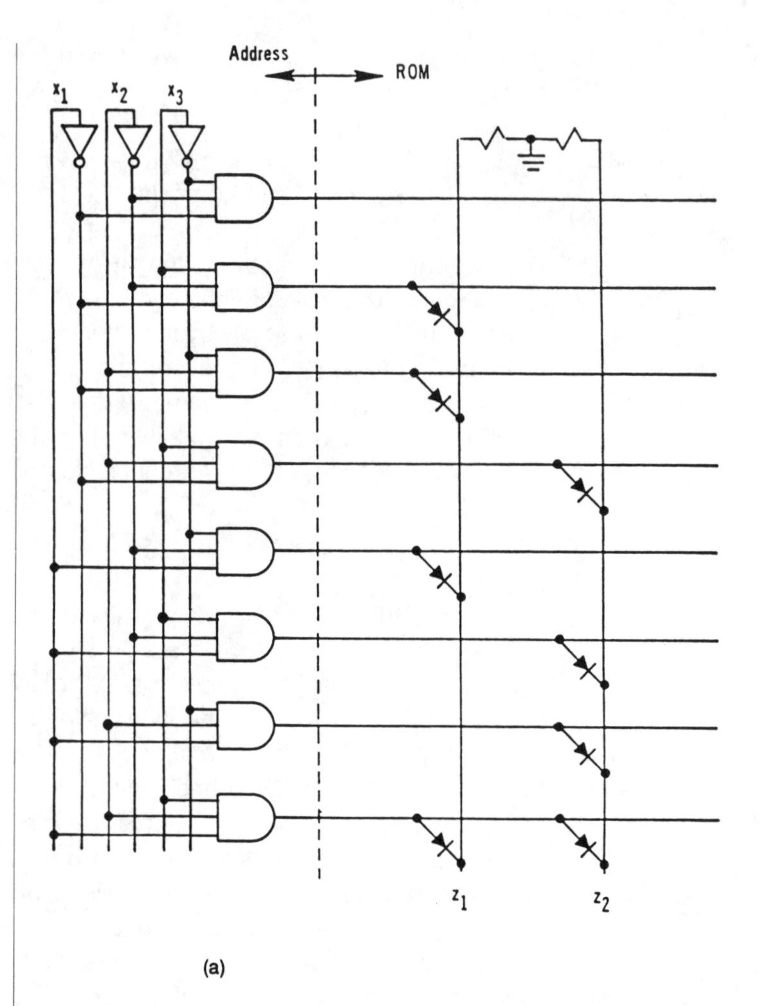

FIGURE 11.6 (a) ROM realization of a full-adder. (b) ROM realization of a serial-adder. (c) ROM memory map for a parallel full-adder. (d) Parallel full-adder using 8 × 2 ROM. (e) BCD counter using 4 × 4 ROM. (f, g) ROM realization of traffic light controller.

and whose memory map is shown in Figure 11.6(c). Here we have all the components for a memory subsystem, namely, the MAR, address decoder, memory cells, MDR, and the READ and WRITE control lines. Since the system shown is a ROM, we do not need the WRITE control line; it is thus shown by a hollow dot arrow to remind us that it is needed for a RAM. Notice that the eight outputs of the decoder specify the eight unique addresses of the eight 2-bit data words. For a detailed schematic of the decoder, see Figure 11.6(a), where the AND network serves as the address decoder. Meanwhile, they are also the eight product terms or the complete set of the minterms of the three switching variables. As memtioned before, if one observes the data in the MDR while reading the ROM, from addresses 000, 001, ..., 111, one could verify that the data bits of the MDR indeed implement the whole functions of the SUM and CARRY as we intended. It is important

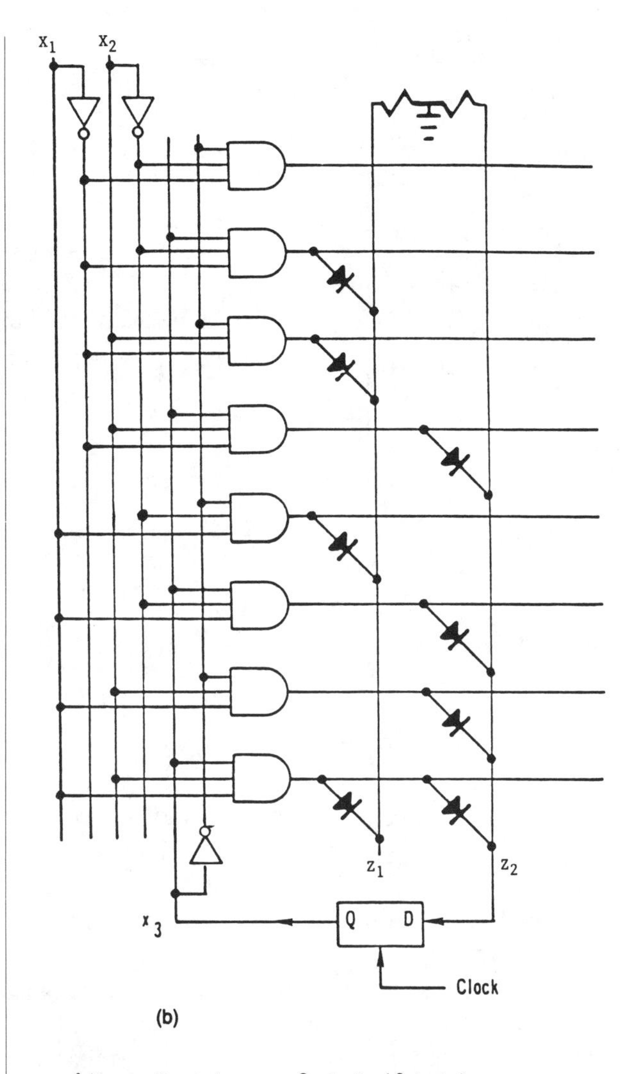

(b)

Address (Inputs)			Contents (Outputs)	
x_1	x_2	x_3	z_1	z_2
0	0	0	0	0
0	0	1	1	0
0	1	0	1	0
0	1	1	0	1
1	0	0	1	0
1	0	1	0	1
1	1	0	0	1
1	1	1	1	1

(c)

FIGURE 11.6 continued. (a) ROM realization of a full-adder. (b) ROM realization of a serial-adder. (c) ROM memory map for a parallel full-adder. (d) Parallel full-adder using 8 × 2 ROM. (e) BCD counter using 4 × 4 ROM. (f, g) ROM realization of traffic light controller.

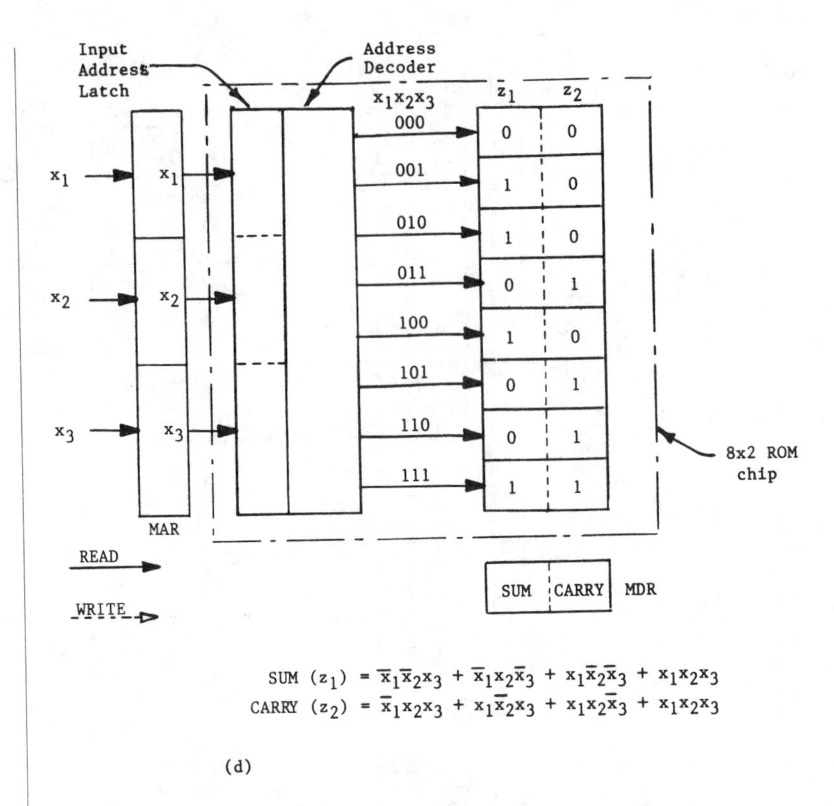

FIGURE 11.6 continued. (a) ROM realization of a full-adder. (b) ROM realization of a serial-adder. (c) ROM memory map for a parallel full-adder. (d) Parallel full-adder using 8 × 2 ROM. (e) BCD counter using 4 × 4 ROM. (f, g) ROM realization of traffic light controller.

to point out here that the address decoder produces all possible minterms (product terms or AND terms), and the contents of the memory cells define the SUM (OR terms) and CARRY (OR terms), while the MDR and READ command together implement the Boolean (binary) OR operations.

Recall from the fundamentals of logic design that any combinational logic circuit can be realized with a switching function in the sum of products format. This means that we can conclude that any combinational logic circuit can be realized by a ROM that has enough memory capacity to implement all the switching functions required for that logic circuit. Specifically, if one wants to implement, for example, four switching functions having three switching variables, one could use a 8 × 4 ROM to implement them by properly loading the desired contents into that ROM.

11.3.2. Example 2: BCD Counter

As another example, let us implement the BCD counter which we used as a design example in Chapter 10, Section 10.4.1, by means of ROM. Note that here we have four input switching variables and four

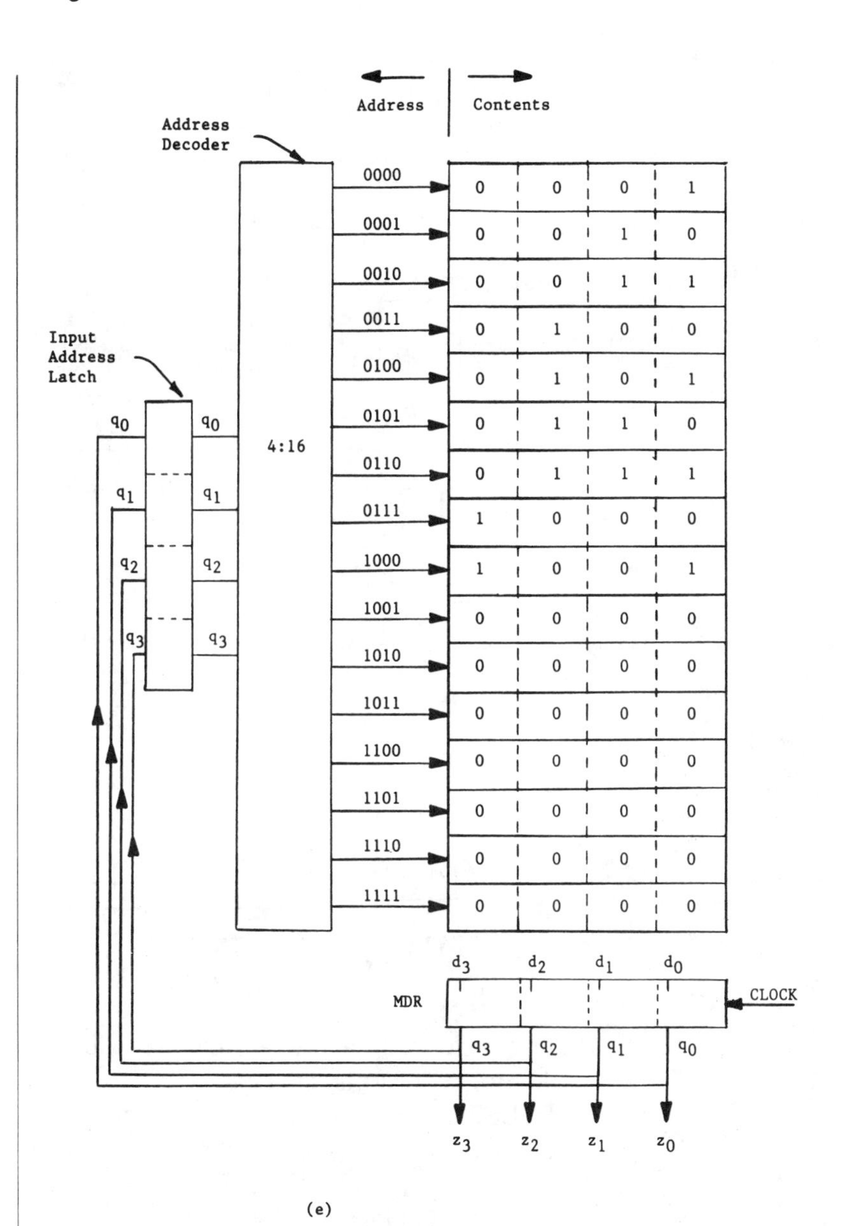

(e)

FIGURE 11.6 continued. (a) ROM realization of a full-adder. (b) ROM realization of a serial-adder. (c) ROM memory map for a parallel full-adder. (d) Parallel full-adder using 8 × 2 ROM. (e) BCD counter using 4 × 4 ROM. (f, g) ROM realization of traffic light controller.

switching functions to be implemented, which means that we need a 16 × 4 ROM. Figure 11.6(e) shows the schematic for the realization of this BCD counter. Recall that a BCD counter is a sequential machine. Here, we use the MDR (not included in the ROM chip) as the memory layer, and the input address latch plus the address decoder inside the ROM chip as the input logic layer for the sequential machine. Examine Figure 11.6(e) vs. Chapter 10, Figure 10.2(b). You will quickly notice

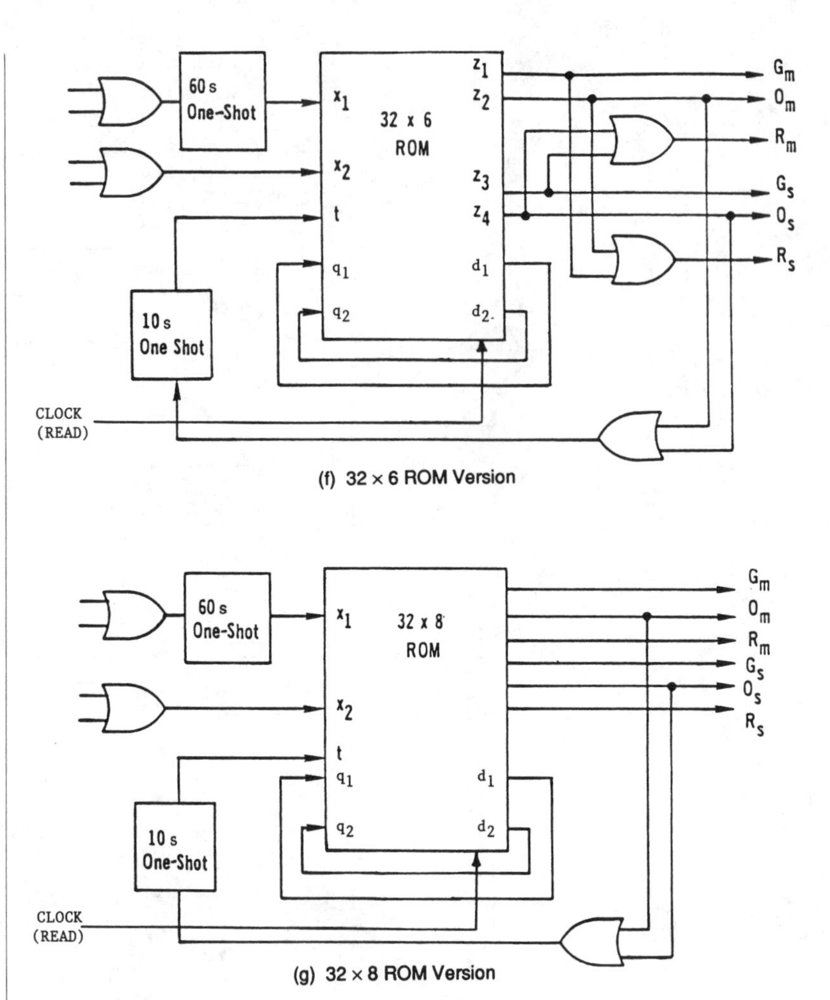

(f) 32 × 6 ROM Version

(g) 32 × 8 ROM Version

FIGURE 11.6 continued. (a) ROM realization of a full-adder. (b) ROM realization of a serial-adder. (c) ROM memory map for a parallel full-adder. (d) Parallel full-adder using 8 × 2 ROM. (e) BCD counter using 4 × 4 ROM. (f, g) ROM realization of traffic light controller.

that the former is almost a "Xerox copy" of the latter. Note that in Figure 11.6(e) the MDR and the input latch are holding the binary code of the PRESENT STATE and the decoder outputs show the same pattern as the entries of the present state table, while the contents of the ROM similarly is a copy of the entries for the next state table shown in Chapter 10, Figure 10.2(b). It is interesting to note that in this implementation, the address decoder which yields the minterms or the AND (product) terms, together with the memory cells or the contents of the memory, implement the input logic (combinational logic) layer of this sequential machine. However, in this implementation the logic minimization step has been skipped. Finally, let us verify the design by initializing the MDR with 0000. As a result, the input address latch/decoder of the chip will point to the 0000 address, the contents of which is 0001 as shown. The next CLOCK or READ signal reads these contents into the MDR and the decoder output in turn points to address 0001. The process repeats until the last count, 9. One more READ, and the MDR is given the contents of address 9 which are 0000, thus the counter returns to 0000.

11.3.3. Example 3: Traffic Light Controller

From Step 7 of the example described in Chapter 10, Section 10.4.2, we obtain the next state information, d_1d_2, in conjunction with the inputs x_1x_2, t and q_1q_2. With the same input condition, x_1x_2, t and q_1q_2, we obtain the corresponding outputs Z_i, i = 1, 2, 3, 4, from the output matrix in Step 8 of the same example. Accordingly, we can construct Table 11.2, which is a composite of the $\{x_i\}$ $\{q_k\}/\{d_k\}$ tables of Step 7 and the output table of Step 8. For example, for $q_1q_2x_1x_2t$ = 00000, from the tables in Step 7 we read $d_1 = 0$ and $d_2 = 0$; by the same token, from the output table or matrix for $q_1q_2x_1x_2t$ = 00000 in Step 8, we read $z_1 = 1$. Because z_1, z_2, z_3 and z_4 are mutually exclusive, we have $z_1 = 1$ and $z_2 = z_3 = z_4 = 0$. As we know that all programmable logic is based on the address-content configuration, we could simply let the left half of Table 11.2 be the addresses and the right half be the contents. Since d_1d_2 contribute to the decision making of the next state, or the next address in this case, they must accordingly be connected to q_1q_2. Therefore, the system diagram for implementing the controller by ROM is as shown in Figure 11.6(f) with two data bits of the MDR (not shown) as the feedback, and four data bits as the output driver. The ROM used is a 32-word ROM, and each word is six bits wide. Recall that the Zs are used to drive the light through the simple combinational logic shown in Step 8. We can integrate this logic into the ROM by increasing the width of the ROM from 6 to 8 bits. Figure 11.6(g) shows the 32 × 8 ROM version with two data bits of the MDR (not shown) as the feedback, and six data bits of it as output driver. The detailed contents of both versions are shown in Table 11.3.

11.3.4. Design with PROM

In the preceding section, we described, using three examples, how random logic circuitry can be realized with ROM. As we know, normally ROM is not user-programmable and is expensive for the designer to reprogram. When a designer has completed his or her design, or decided what the contents of each address in ROM should be, the details of that design are given to the chip manufacturer to make a mask for the ROM chips. After this point, the designer cannot change the design without paying a heavy price. To avoid this serious prob-

TABLE 11.2

q_1	q_2	x_1	x_2	t	d_1	d_2	Z_1	Z_2	Z_3	Z_4
0	0	0	0	0	0	0	1	0	0	0
0	0	0	0	1	0	1	0	1	0	0
0	0	0	1	0	0	0	0	1	0	0
0	0	0	1	1	0	1	0	1	0	0
0	0	1	0	0	0	0	0	1	0	0
0	0	1	0	1	0	1	0	1	0	0
0	0	1	1	0	0	0	0	1	0	0
0	0	1	1	1	0	1	0	1	0	0
0	1	0	0	0	–	–	–	–	–	–
0	1	0	0	1	–	–	–	–	–	–
0	1	0	1	0	1	1	0	0	1	0
0	1	0	1	1	0	1	0	1	0	0
0	1	1	0	0	1	1	0	0	1	0
0	1	1	0	1	0	1	0	1	0	0
0	1	1	1	0	1	1	0	0	1	0
0	1	1	1	1	0	1	0	1	0	0
1	0	0	0	0	0	0	1	0	0	0
1	0	0	0	1	1	0	0	0	0	1
1	0	0	1	0	1	1	0	0	1	0
1	0	0	1	1	1	0	0	0	0	1
1	0	1	0	0	1	1	0	0	1	0
1	0	1	0	1	1	0	0	0	0	1
1	0	1	1	0	1	1	0	0	1	0
1	0	1	1	1	1	0	0	0	0	1
1	1	0	0	0	1	1	0	0	0	1
1	1	0	0	1	1	0	0	0	0	1
1	1	0	1	0	1	1	0	0	1	0
1	1	0	1	1	1	0	0	0	0	1
1	1	1	0	0	1	0	0	0	1	0
1	1	1	0	1	1	0	0	0	0	1
1	1	1	1	0	1	1	0	0	1	0
1	1	1	1	1	1	0	0	0	0	1

lem, one could use PROM (described in Chapter 6) instead of using plain ROM. By using PROM one can easily change, improve, or upgrade the design. Consider the three design examples just presented using ROM. We implemented the full-adder with 8×2 ROM, the BCD counter with 16×4 ROM, and the traffic light controller with 32×8 ROM. It appears that one would need a custom-made ROM to realize each of these designs. But we may instead use one type of PROM, say 32×8 PROM, to implement any of them. The only thing we need to do is to program or load the specific contents for the design in question into a 32×8 PROM and plug that PROM into the socket for that circuit. We can even use the same hardware or circuit layout to implement all of the three designs as long as we plug the PROM with the right contents. This is because in contrast to the circuit layout for random logic, the wiring or circuit layout of a ROM or PROM is systematic and regular (consisting only of the lines for addresses, data, and read-control), and is independent of its contents. Think of all the labor costs for assembling the hardware of random layout circuitry which we save by this method. Cost conscious readers may ask whether we have wasted the unused memory space in the 32 × 8

TABLE 11.3
ROM Maps

Address					**Content (8-bits)**								**Content (6-bits)**					
q_1	q_2	x_1	x_2	t	d_1	d_2	G_m	O_m	R_m	G_s	O_s	R_s	d_1	d_2	Z_1	Z_2	Z_3	Z_4
0	0	0	0	0	0	0	1	0	0	0	0	1	0	0	1	0	0	0
0	0	0	0	1	0	1	0	1	0	0	0	1	0	1	0	1	0	0
0	0	0	1	0	0	0	0	1	0	0	0	1	0	0	0	1	0	0
0	0	0	1	1	0	1	0	1	0	0	0	1	0	1	0	1	0	0
0	0	1	0	0	0	0	0	1	0	0	0	1	0	0	0	1	0	0
0	0	1	0	1	0	1	0	1	0	0	0	1	0	1	0	1	0	0
0	0	1	1	0	0	0	0	1	0	0	0	1	0	0	0	1	0	0
0	0	1	1	1	0	1	0	1	0	0	0	1	0	1	0	1	0	0
0	1	0	0	0	–	–	–	–	–	–	–	–	–	–	–	–	–	–
0	1	0	0	1	–	–	–	–	–	–	–	–	–	–	–	–	–	–
0	1	0	1	0	1	1	0	0	1	1	0	0	1	1	0	0	1	0
0	1	0	1	1	0	1	0	1	0	0	0	1	0	1	0	1	0	0
0	1	1	0	0	1	1	0	0	1	1	0	0	1	1	0	0	1	0
0	1	1	0	1	0	1	0	1	0	0	0	1	0	1	0	1	0	0
0	1	1	1	0	1	1	0	0	1	1	0	0	1	1	0	0	1	0
0	1	1	1	1	0	1	0	1	0	0	0	1	0	1	0	1	0	0
1	0	0	0	0	0	0	1	0	0	0	0	1	0	0	1	0	0	0
1	0	0	0	1	1	0	0	0	1	0	1	0	1	0	0	0	0	1
1	0	0	1	0	1	1	0	0	1	1	0	0	1	1	0	0	1	0
1	0	0	1	1	1	0	0	0	1	0	1	0	1	0	0	0	0	1
1	0	1	0	0	1	1	0	0	1	1	0	0	1	1	0	0	1	0
1	0	1	0	1	1	0	0	0	1	0	1	0	1	0	0	0	0	1
1	0	1	1	0	1	1	0	0	1	1	0	0	1	1	0	0	1	0
1	0	1	1	1	1	0	0	0	1	0	1	0	1	0	0	0	0	1
1	1	0	0	0	1	1	0	0	1	0	1	0	1	1	0	0	0	1
1	1	0	0	1	1	0	0	0	1	0	1	0	1	0	0	0	0	1
1	1	0	1	0	1	1	0	0	1	1	0	0	1	1	0	0	1	0
1	1	0	1	1	1	0	0	0	1	0	1	0	1	0	0	0	0	1
1	1	1	0	0	1	0	0	0	1	1	0	0	1	0	0	0	1	0
1	1	1	0	1	1	0	0	0	1	0	1	0	1	0	0	0	0	1
1	1	1	1	0	1	1	0	0	1	1	0	0	1	1	0	0	1	0
1	1	1	1	1	1	0	0	0	1	0	1	0	1	0	0	0	0	1

the adder. Indeed, this is true in the case of full-adder implementation, since it requires only an 8×2 ROM, and we use a 32×8 ROM. Fortunately, in comparison with the labor costs, the cost of the wasted memory space in such cases is minimal, except in those cases where the memory space wasted is on the order of hundreds or more. In such conditions, however, we may consider using the PLA or PAL devices introduced in Sections 11.4 and 11.5.

11.4. DESIGN WITH PLA

11.4.1. Introduction

Consider the design shown in Figure 11.6(e). Here, there are 16 minterms generated by the address decoder. Notice that the four switching functions, d_0, d_1, d_2, and d_3, use only 10 out of the 16 total available minterms. As a result, 6 out of 16 memory words are wasted. In this case, fortunately, we have only four input switching variables. In practice, one may expect more input variables. For example, suppose we are designing a system which needs ten input variables and five switching functions with EPROM. We would need a 1024×5 PROM to implement this design. However, in looking through the standard PROM chips available on the market, we may find that the closest chip to our specifications is 1024×8. Evidently, unless we want to go to the great expense of ordering custom-made 1024×5 chips, we would have to settle for the 1024×8 chips, with which we would be wasting 3 bits in each memory word. In addition, this chip would have 1024 available minterms, which means that each switching function could have up to 1024 product terms and 1023 OR operations. Obviously, this is unlikely in real practice. Similarly, as in the case of the BCD counter, we may need only a small subset of the 1024 available minterms. Unfortunately, in the case of ROM and PROM, all the minterms are there regardless of whether the designer needs them or not. It is desirable to have PROM-like devices that have a fairly high number of input variables but provide only a sufficient number of minterms for the designer to use. It would be even better if the values of these minterms are programmable in such a way that the designer can define the needed minterms within the limited number available and leave the unwanted minterms out of the picture. One would also wish that the number of OR-terms in any switching function is programmable by the designer. Thanks to technological advances, this kind of device is now available. They are called Programmable Logic Arrays (PLA). A brief description of these devices and some application examples follow.

11.4.2. Fundamentals of PLA

Figure 11.7(a) shows a conceptual PMOS based n-input NAND gate. The lower n transistors in parallel function as switches, and the upper transistor functions as the load resistor shared by the lower n transistors. Because it is a PMOS based device, its power sources and logic levels are specified as follows:

$$V_{DD} < V_{GG} < 0v$$
$$\text{Logic } [1] = 0v, \text{ Logic } [0] = V_{DD}$$

thus, positive logic is used here. If the input $x_i = 1$, the ith transistor will be OFF, otherwise it is ON. The reader can use Binary State Analysis to verify that the circuit is indeed a NAND gate as shown.

Figure 11.7(b) shows two NAND gates configured in a small array pattern. Figure 11.7(c) is a direct translation of the array shown in (b) into our familiar logic symbols. Figure 11.7(d) is the proposed symbolic notation for the same logic array. Figure 11.8(a) shows an array with two levels of NAND gate. Its equivalent switching functions are shown on the right-hand side of the figure. Figure 11.8(b) is the equivalent logic symbol representation. As we know, two levels of NAND gate in cascade will yield the familiar sum of products format, as shown in Figure 11.8(c). As a result, we can generalize the representation of the logic array in block form as shown in Figure 11.8(d), and in matrix form as in Figure 11.8(e). The presence or absence of a black dot at the intersection of the lines in (e) denotes whether the transistors linking them are intact or blown. That is, the array is structured like a two-level ROM. The conventional symbols for AND and OR shown here serve as a reminder that the upper and lower sections, respectively, are AND and OR logic matrices. Since these logic matrices are ROM-like in structure, the two layers of block form shown in Figure 11.8(d) are respectively called AND-ROM (for the products) and OR-ROM (for the sum) layers. However, the notation shown in Figure 11.8(e) conveys more information about the arrays, since it is a direct translation of the switching functions that the logic array implements. It is important to mention that most PLA arrays nowadays are NMOS based, since NMOS transistors can be packed more densely and have higher operation speed than PMOS. For NMOS based logic arrays, we expect the following.

$$V_{DD} = V_{GG} = +V$$
$$\text{Logic } [1] = +V, \text{ Logic } [0] = 0$$

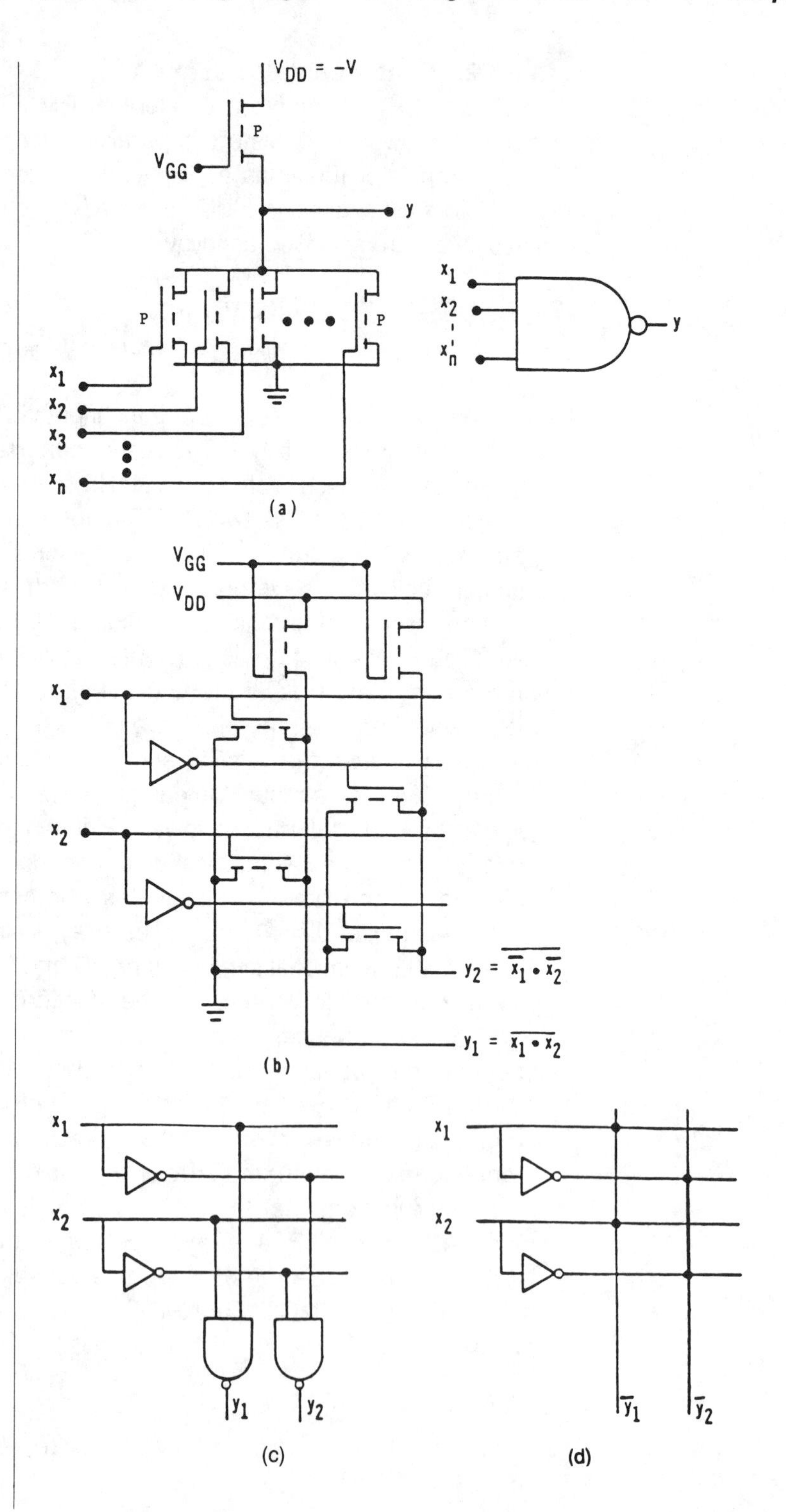

FIGURE 11.7 (a, b) PMOS NAND gate. (c) Gate notation. (d) Logic array notation.

It is also worthy to note that currently there are PLAs available that are implemented with CMOS or ECL technology.

11.4.3. Design Examples

a. Full-Adder

Figure 11.9 shows the PLA implementation of a parallel full-adder in AND/OR matrix structure. Note that there are only seven vertical lines depicting the required minterms. Since the switching functions do not need the 000 term, it has been eliminated from the AND matrix. Figure 11.10 shows the PLA implementation of a serial full-adder. Since in detail its AND/OR matrix is similar to that of the parallel full-adder, the figure shows only the AND/OR block diagram.

b. Four-Digit Binary Counting on Any Specific Sequence

Figure 11.11(a) shows a specific counting sequence for a four-digit counter that we want to implement with a PLA. In fact, the design presented here is good for any specified sequence. However, for illustration purposes we choose the sequence shown. Note that here we need four flip-flops for a four-digit binary counter; thus, 16 minterms are possible. For comparison, let us design the counter by following two different routes, i.e., after logic minimization and before logic minimization.

Solution 1 — After Minimization

Figure 11.11(b) shows the logic minimization process by means of Karnaugh maps, according to the counting sequence specified in Figure 11.11(a). Figure 11.11(c) shows the design implementation with PLA which includes four D flip-flops.

Solution 2 — Before Minimization

As described before, if we implement the design with ROM or PROM, we would normally realize the logic before minimization and the memory spaces with the addresses (minterms) not used in any of the switching functions would be wasted. Since with PLAs we can program or select the minterms or addresses, it would be interesting to see the result if the minimization step is not employed in the design process. From Figure 11.11(a), we find that we need only 12 minterms for the problem specification; accordingly, implementation before minimization is shown in Figure 11.11(d). Compare the final designs of Figure 11.11(c) and (d). The former requires two more memory word spaces. That is, for this problem, the minimization process is not a desirable route to take. In summary, we may conclude that this

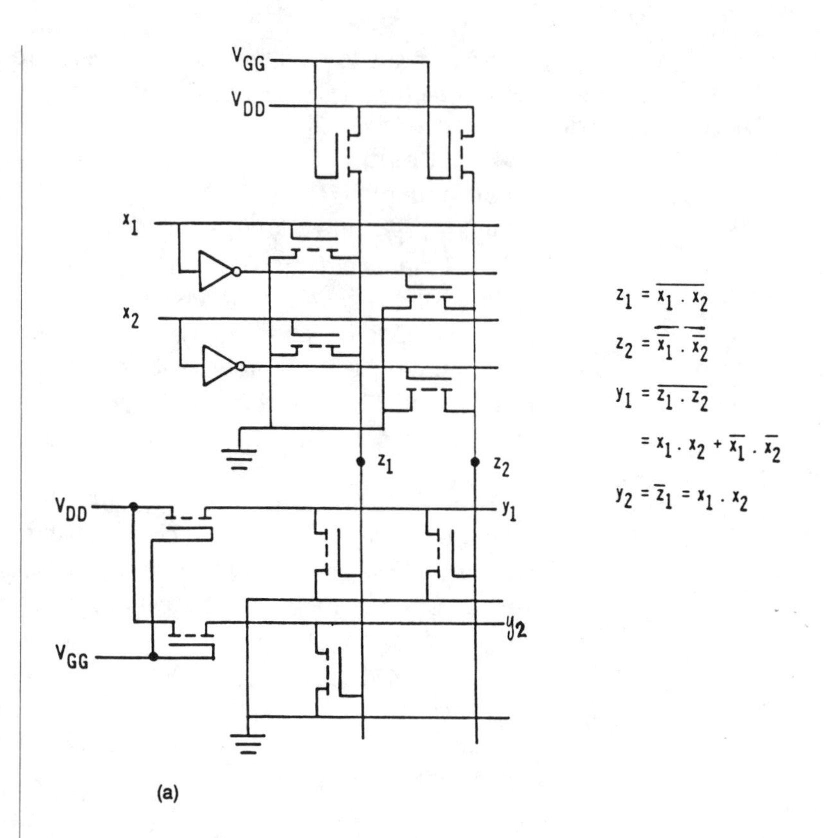

FIGURE 11.8 (a) Two-level logic array circuit configuration. (b, c, d, e) Derivation of PLA notation.

particular problem requires 16 memory spaces for ROM realization and only 12 memory spaces for PLA implementation without using the minimization process. Evidently, one may observe that given the number of switching functions, the number of product terms (minterms) determines the size of the logic array chip required. In comparing ROM with PLA implementations, there is another point worthy of mention. In implementing a sequential machine by means of ROM, one would have to provide external flip-flops or register, whereas with PLA one can use the flip-flops or register built inside the chip. Figure 11.11(e) shows the logic diagram of a typical PLA chip, the PLS105 built by Advanced Micro Devices (AMD), which is named the Programmable Logic Sequencer (PLS). For this device, notice that both the AND and OR matrices are programmable, and that it provides 14 built-in RS flip-flops.

c. Traffic Light Controller

Figure 11.11(f) shows the system block diagram for the traffic light controller as designed with the PLA device. Figure 11.11(g) shows the design in PLA notation with two built-in D flip-flops. It is a direct

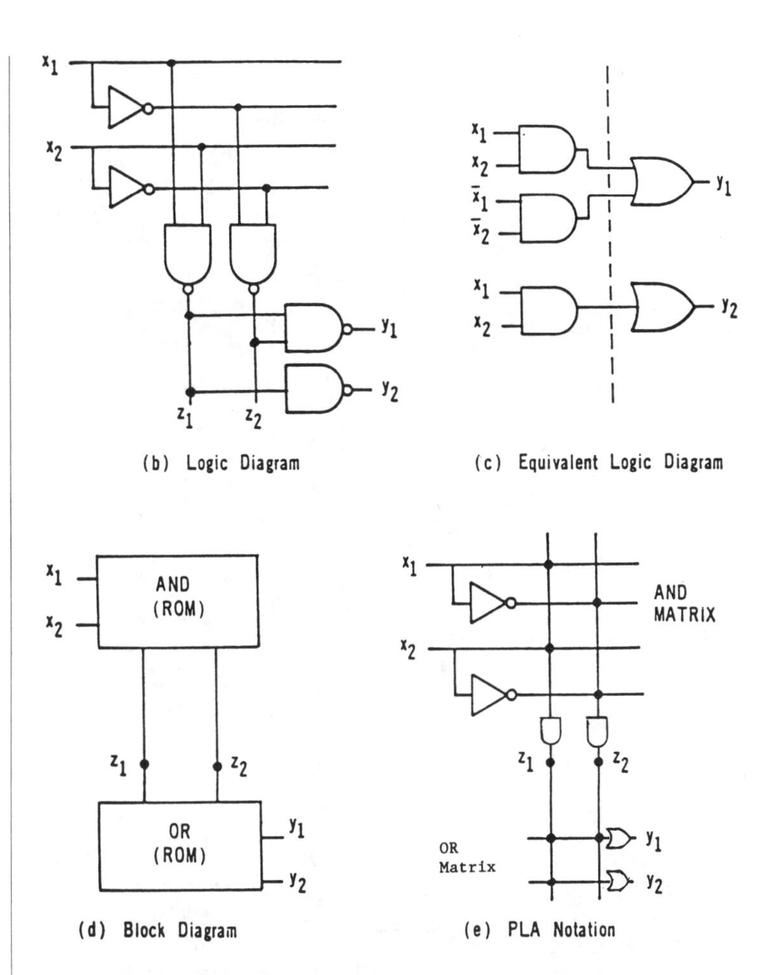

FIGURE 11.8 continued. (a) Two-level logic array circuit configuration. (b, c, d, e) Derivation of PLA notation.

translation of the design described in Steps 7 and 8 in Chapter 10, Section 10.4.2. The readers are urged to verify the contents of both AND and OR matrices shown in Figure 11.11(g) according to the result described in Chapter 10, Section 10.4.2 or Figure 10.3(e).

11.5. DESIGN WITH PAL

In this section, another useful and popular type of programmable logic device will be presented. It is called Programmable Array Logic, invented by Monolithic Memories, Inc. (MMI)/AMD. As is the case with PLA, PAL is also structured with AND and OR matrices with built-in register or flip-flops. However, in contrast to the other devices, in PAL the AND (product) matrix is programmable while the maximum number of OR-terms in the OR (sum) matrix is fixed and not user-programmable. Its property is just the opposite of ROM/PROM. Figure

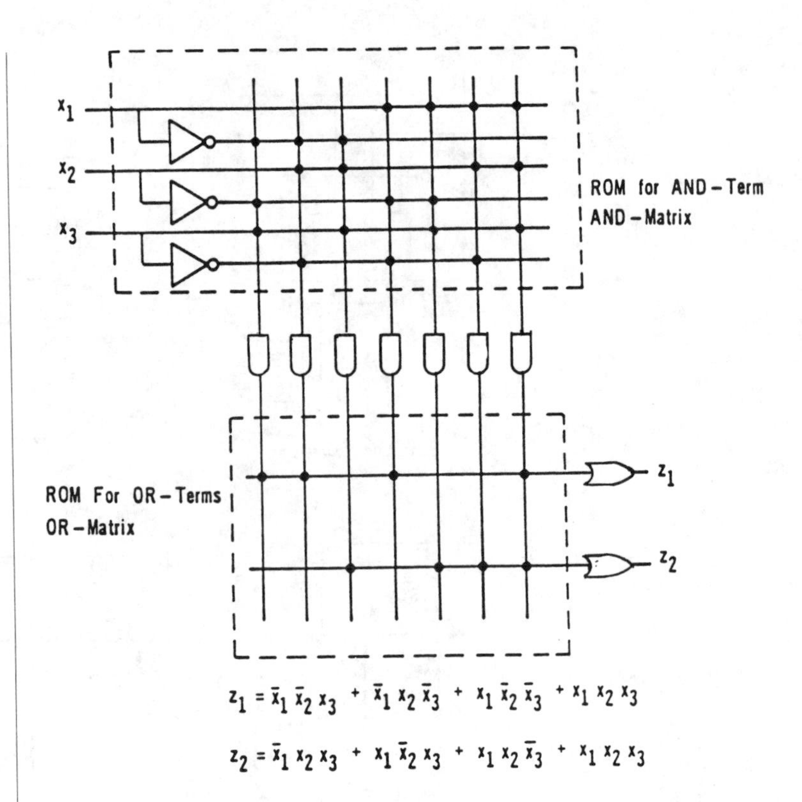

$$z_1 = \bar{x}_1 \bar{x}_2 x_3 + \bar{x}_1 x_2 \bar{x}_3 + x_1 \bar{x}_2 \bar{x}_3 + x_1 x_2 x_3$$

$$z_2 = \bar{x}_1 x_2 x_3 + x_1 \bar{x}_2 x_3 + x_1 x_2 \bar{x}_3 + x_1 x_2 x_3$$

FIGURE 11.9 PLA realization of a full-adder.

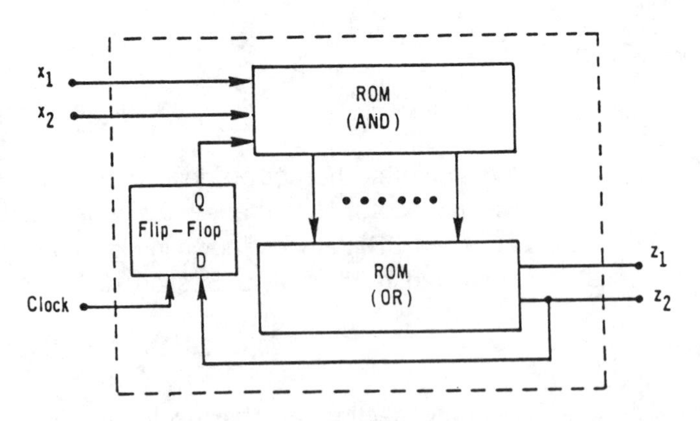

FIGURE 11.10 PLA realization of a serial-adder.

11.12(a) shows a logic diagram of a typical PAL, the TIBPAL16R4 by Texas Instruments, Inc. Note that the device provides eight inputs, four I/O pins that can be programmed to be either inputs or output ports, and four D flip-flops. Here we have 32 maximum possible product terms that are programmable. Among these 32 terms, there are 8 to 16 input switching variables (since there are eight pins which can be either inputs or outputs), 8 to 0 tri-state outputs, and 0 to 4 feedback inputs from the four D flip-flops. However, there is no OR matrix. Instead, it provides eight OR gates with eight FAN-INs. That is, the designer is allowed to realize a maximum of eight switching functions. Each function is limited to the SUM of a maximum of eight product terms. It is interesting to point out that although the device provides no programmable OR matrix, in practice the eight available OR gates can satisfy most of the design requirements. Reviewing the three design examples presented in the preceding sections, we note that none of their switching functions requires a sum of eight product terms. However, system designers quickly find that in many cases they do wish PALs to have flexible or programmable output logic ports available. Figure 11.12(b) shows the logic diagram of a more sophisticated PAL, the PALC18U8Q by AMD. Note that the rectangular blocks containing 1s and 0s at the outputs are programmable multiplexers (recall that multiplexers are electronic rotary switches). Thus, by programming the multiplexers, the output ports of the device can be the assertion/negation outputs of the OR gate or those of the flip-flops. In addition, one has the choice of having the feedback directly connected from the flip-flops or the tri-state outputs. This kind of block is sometimes called a *macro logic cell.*

11.6. STRUCTURAL COMPARISON OF PROM, PAL, AND PLA

As a summary of structural differences among PROM, PAL and PLA, Figures 11.13(a), (b), and (c), respectively, show the architecture of each of the three programmable devices, which are sometimes generically referred to as "Programmable Logic Devices," or PLDs. Notice in these figures that line intersections marked with dots are fixed or nonprogrammable, whereas intersections marked with "×" are programmable. Here, the "×" denotes that the linking fuse is intact. For example, the AND array marked with dots shown for the PROM is actually its built-in address decoder which yields the 16 minterms that are not programmable. As for the OR array marked with ×s, they denote the contents of the device, where one can store the desired information in the design phase. PROMs are relatively inexpensive and easy to program. Both PLA and PAL have built-in flip-flops, but the former is more flexible (and expensive) than the latter.

11.7. LOGIC CELL ARRAY

At present, the CMOS based Logic Cell Array (LCA) by Xilinx, Inc. of

Counting Sequence	Present State Binary Equivalent				Next State Binary Equivalent			
	Q_3	Q_2	Q_1	Q_0	d_3	d_2	d_1	d_0
0	0	0	0	0	0	0	1	1
3	0	0	1	1	0	1	0	1
5	0	1	0	1	0	1	1	1
7	0	1	1	1	1	0	0	1
9	1	0	0	1	0	0	0	1
1	0	0	0	1	0	0	1	0
2	0	0	1	0	0	1	0	0
4	0	1	0	0	1	0	0	0
8	1	0	0	0	1	0	1	0
10	1	0	1	0	1	1	1	0
14	1	1	1	0	1	1	0	0
12	1	1	0	0	0	0	0	0

(a)

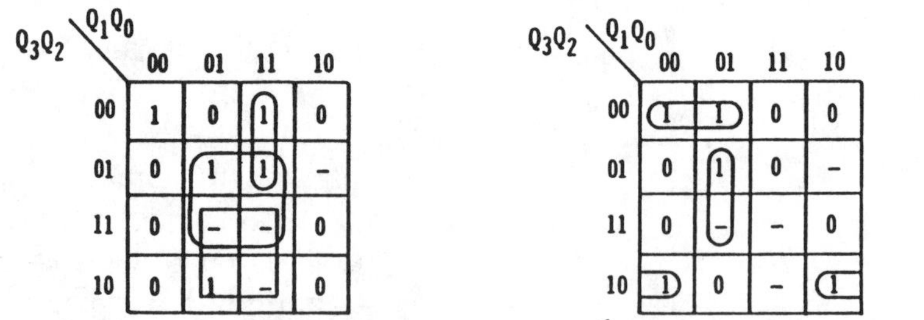

Q_3Q_2 \ Q_1Q_0	00	01	11	10
00	1	0	1	0
01	0	1	1	–
11	0	–	–	0
10	0	1	–	0

$d_0 = Q_0^{n+1} = Q_0Q_2 + Q_0Q_3 + Q_0\bar{Q}_1\bar{Q}_3 + \bar{Q}_0\bar{Q}_1\bar{Q}_2\bar{Q}_3$

Q_3Q_2 \ Q_1Q_0	00	01	11	10
00	1	1	0	0
01	0	1	0	–
11	0	–	–	0
10	1	0	–	1

$d_1 = Q_1^{n+1} = \bar{Q}_1\bar{Q}_2\bar{Q}_3 + Q_0\bar{Q}_1\bar{Q}_3 + \bar{Q}_0\bar{Q}_2Q_3$

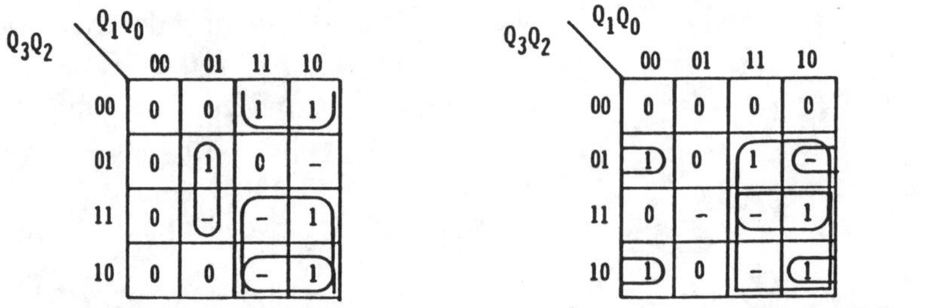

Q_3Q_2 \ Q_1Q_0	00	01	11	10
00	0	0	1	1
01	0	1	0	–
11	0	–	–	1
10	0	0	–	1

$d_2 = Q_2^{n+1} = Q_1Q_3 + Q_1\bar{Q}_2 + Q_0\bar{Q}_1Q_2$

Q_3Q_2 \ Q_1Q_0	00	01	11	10
00	0	0	0	0
01	1	0	1	–
11	0	–	–	1
10	1	0	–	1

$d_3 = Q_3^{n+1} = Q_1Q_2 + Q_1Q_3 + \bar{Q}_0Q_2\bar{Q}_3 + \bar{Q}_0\bar{Q}_2Q_3$

(b)

FIGURE 11.11 (a) Truth table of binary counter for any specific sequence. (b) Karnaugh maps. (c) PLA realization of a 4-bit counter with any specified sequence (with minimization). (d) PLA realization of a 4-bit counter with any specified sequence (before minimization). (e) Architecture of a PLS105 device (courtesy of AMD). (f) Block diagram of PLA realization of traffic light controller. (g) Schematic of PLA realization of traffic light controller.

FIGURE 11.11 continued. (a) Truth table of binary counter for any specific sequence. (b) Karnaugh maps. (c) PLA realization of a 4-bit counter with any specified sequence (with minimization). (d) PLA realization of a 4-bit counter with any specified sequence (before minimization). (e) Architecture of a PLS105 device (courtesy of AMD). (f) Block diagram of PLA realization of traffic light controller. (g) Schematic of PLA realization of traffic light controller.

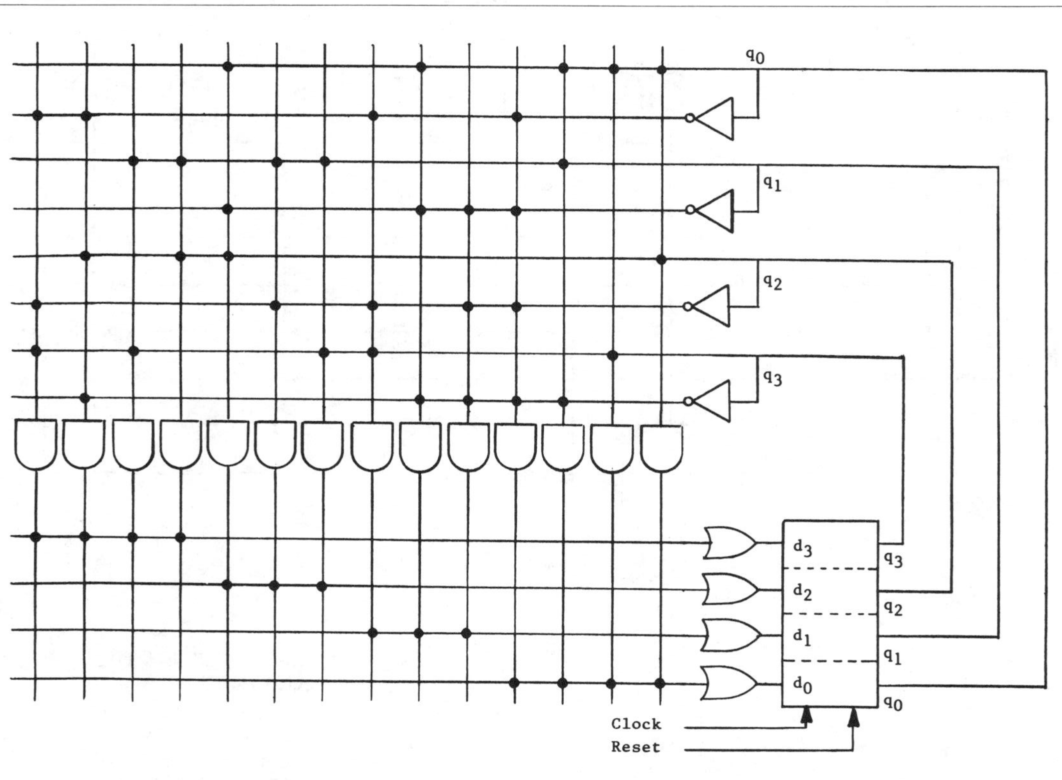

(c)

FIGURE 11.11 continued. (a) Truth table of binary counter for any specific sequence. (b) Karnaugh maps. (c) PLA realization of a 4-bit counter with any specified sequence (with minimization). (d) PLA realization of a 4-bit counter with any specified sequence (before minimization). (e) Architecture of a PLS105 device (courtesy of AMD). (f) Block diagram of PLA realization of traffic light controller. (g) Schematic of PLA realization of traffic light controller.

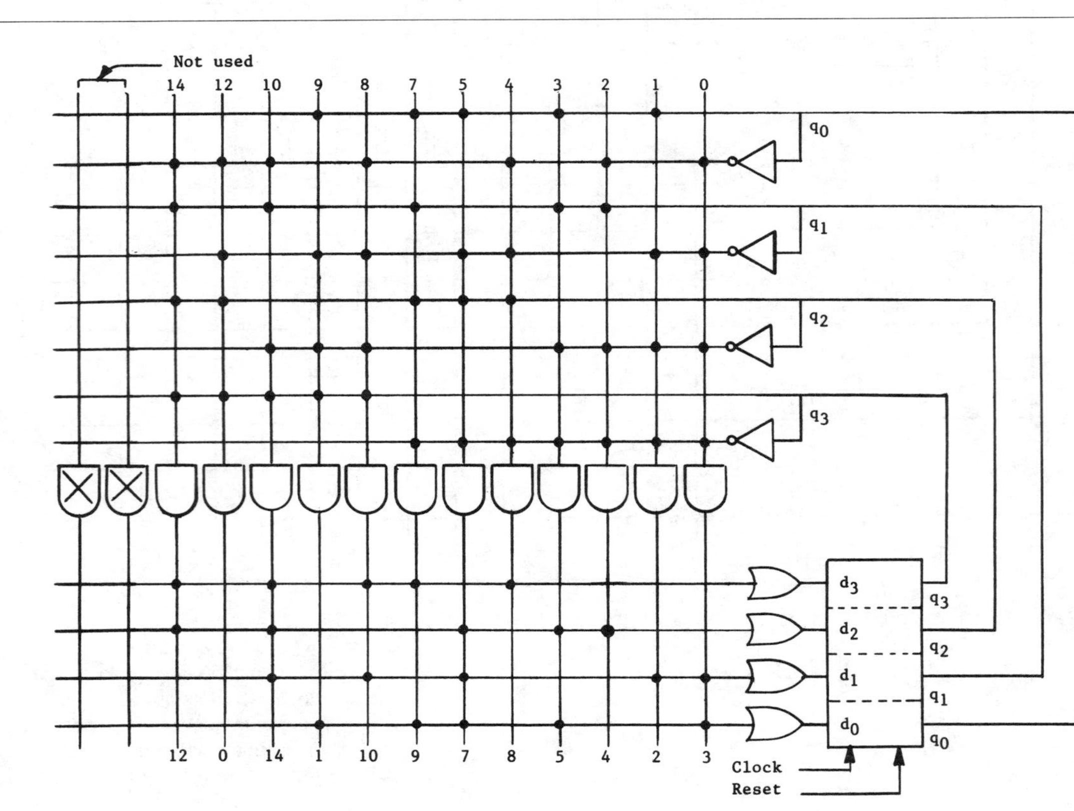

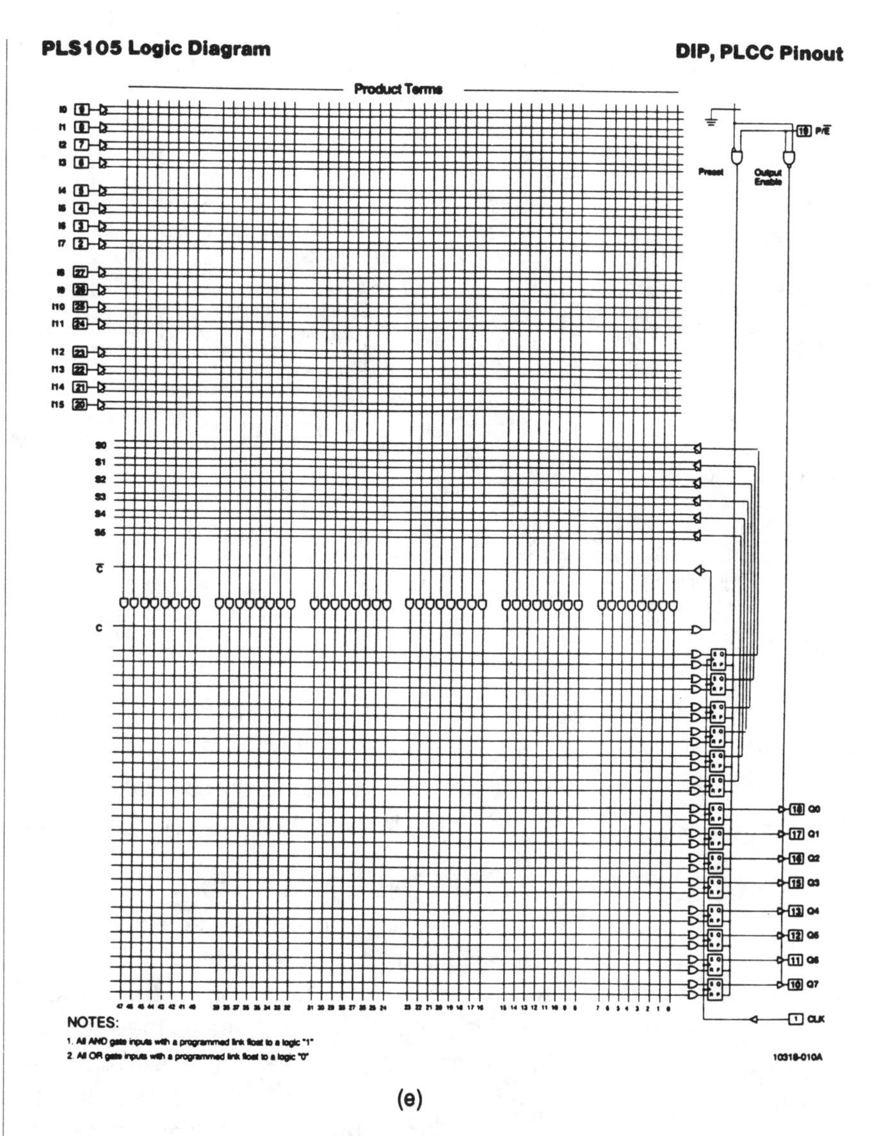

(e)

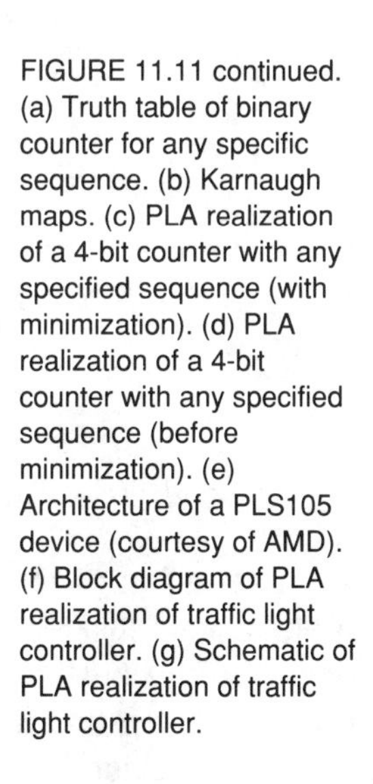
FIGURE 11.11 continued. (a) Truth table of binary counter for any specific sequence. (b) Karnaugh maps. (c) PLA realization of a 4-bit counter with any specified sequence (with minimization). (d) PLA realization of a 4-bit counter with any specified sequence (before minimization). (e) Architecture of a PLS105 device (courtesy of AMD). (f) Block diagram of PLA realization of traffic light controller. (g) Schematic of PLA realization of traffic light controller.

San Jose, CA, is considered the most powerful and flexible of all programmable devices. One of its salient features in comparison with the others is that it has a distributed cellular logic architecture. Unlike the PLA/PAL devices which contain only a limited number of flip-flops that are structured like a buffer register and placed near the output, the LCA has many flip-flops, each of which is supported by a block of programmable combinational logic to make up a so-called configurable logic block (CLB). These CLBs are then distributed over the chip in a cellular structure. The LCA is claimed to be a kind of chip that possesses the advantage of density found in gate-array and that of pro-

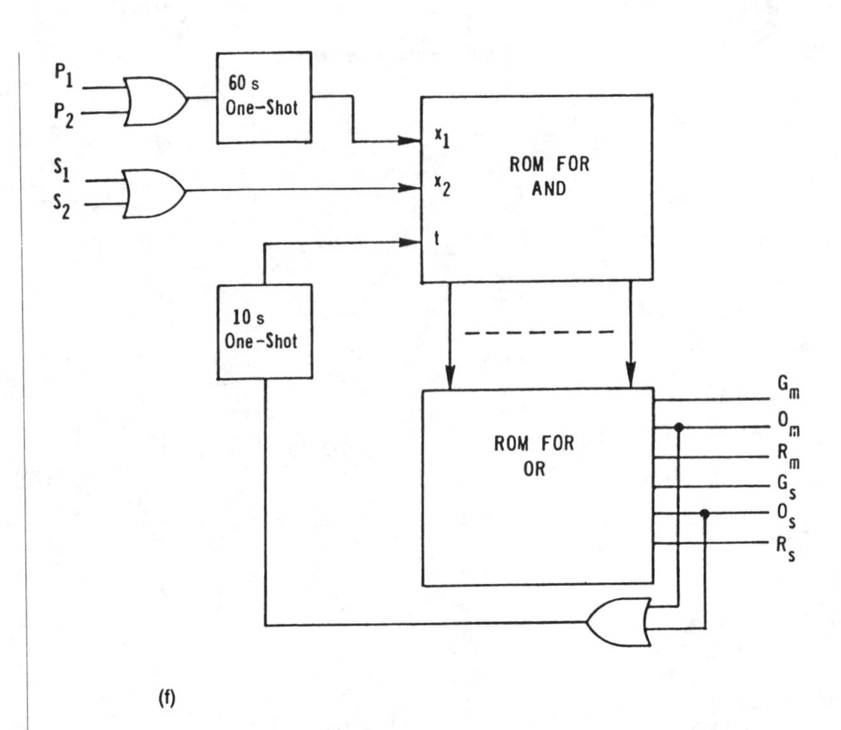

(f)

FIGURE 11.11 continued. (a) Truth table of binary counter for any specific sequence. (b) Karnaugh maps. (c) PLA realization of a 4-bit counter with any specified sequence (with minimization). (d) PLA realization of a 4-bit counter with any specified sequence (before minimization). (e) Architecture of a PLS105 device (courtesy of AMD). (f) Block diagram of PLA realization of traffic light controller. (g) Schematic of PLA realization of traffic light controller.

grammability found in PLA/PAL devices. LCAs appear to be laying the foundation for better design of Application-Specific Integrated Circuits (ASICs). A brief introduction to this powerful device follows.

11.7.1. System Structure

Figure 11.14(a) shows the simplified system structure of an LCA. Briefly, there are three kinds of building elements, namely, I/O block, configurable logic block (CLB), and interconnection area. As its name implies, the I/O block serves as the interface between the internal logic and the chip pins which are its "windows" or "doors" to the outside world. The CLB is the "brain cell" of the system. Its D flip-flop and a configurable/programmable combinational logic network provide the ingredients for being designed as a two-state sequential machine module. The interconnection areas are also user programmable, which allows the designer to customize the internal connections among the elements inside the chip in such a way that the whole chip can function as a complex and fast sequential machine. The reason that the CLBs can operate fast is that they can be configured to operate in parallel or serial to implement the functions designed by the system designer. In addition, the chip contains a thousand bits of high speed static RAM memory cells (not shown in Figure 11.14(a)) distributed on the chip to hold the system configuration designed by the user so that

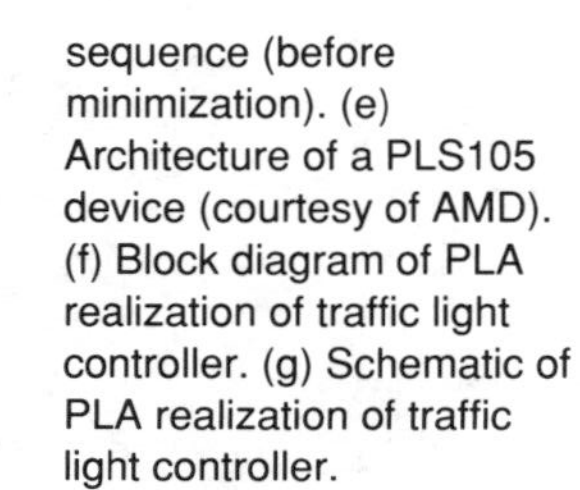

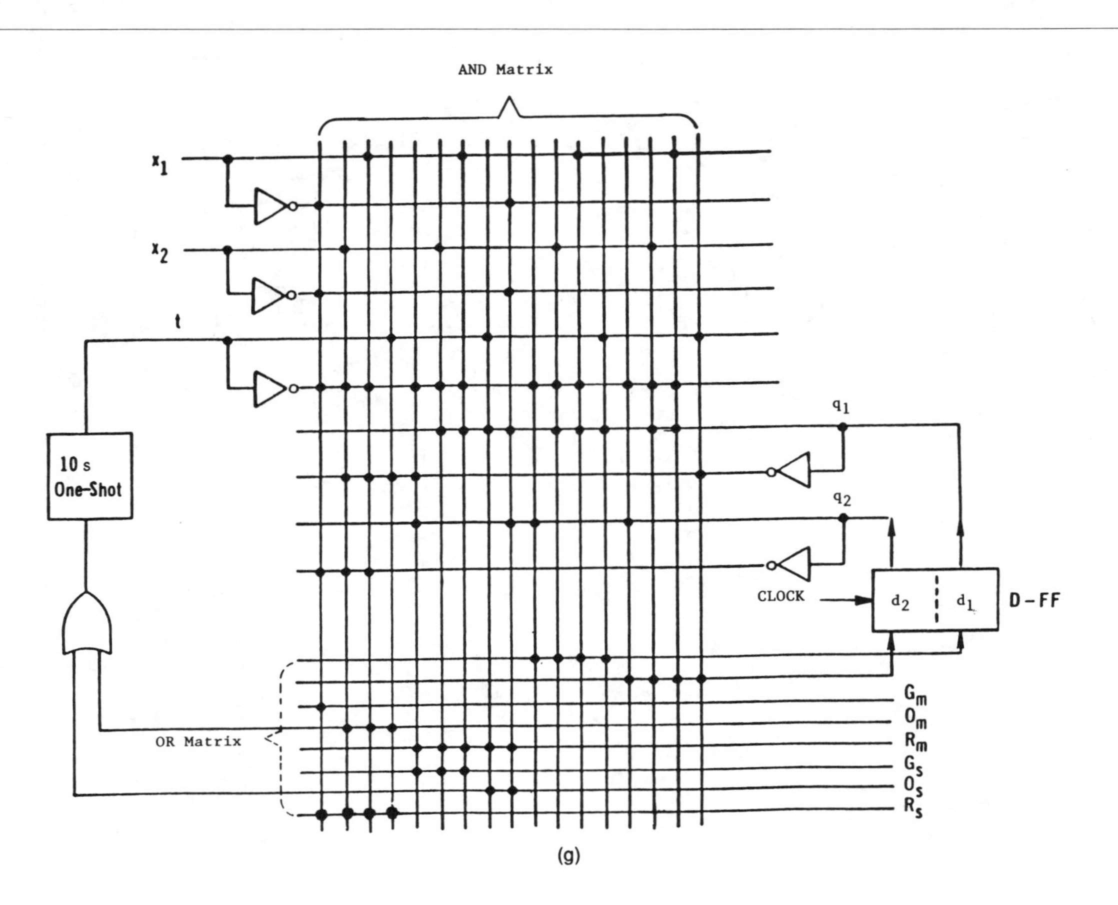

(g)

FIGURE 11.11 continued. (a) Truth table of binary counter for any specific sequence. (b) Karnaugh maps. (c) PLA realization of a 4-bit counter with any specified sequence (with minimization). (d) PLA realization of a 4-bit counter with any specified sequence (before minimization). (e) Architecture of a PLS105 device (courtesy of AMD). (f) Block diagram of PLA realization of traffic light controller. (g) Schematic of PLA realization of traffic light controller.

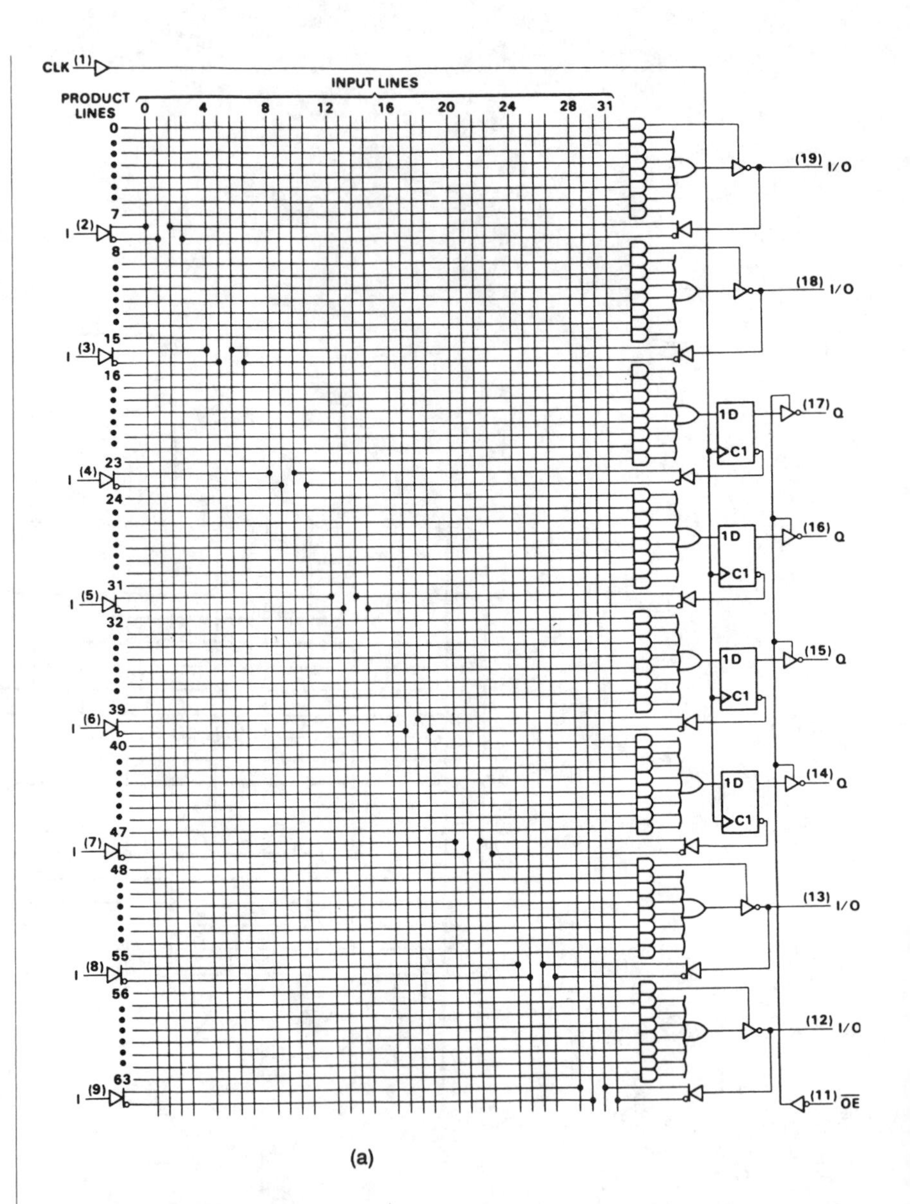

(a)

FIGURE 11.12 (a) TIBPAL12R4 logic diagram (courtesy of Texas Instruments, Inc.). (b) PALC18U8Q logic diagram (courtesy of AMD).

the chip will behave as the logic system that the designer wishes to implement. As a result, without changing the hardware architecture of the chip one can configure it to carry out any task, by reloading this memory with the system configuration information specifically designed for that task. Therefore, this memory is referred to as *configuration memory*. Figure 11.14(b) shows the circuit diagram of one configuration memory cell. It is simply a cross-coupled NAND flip-flop described in Chapter 5. More details of the three essential building blocks are given next.

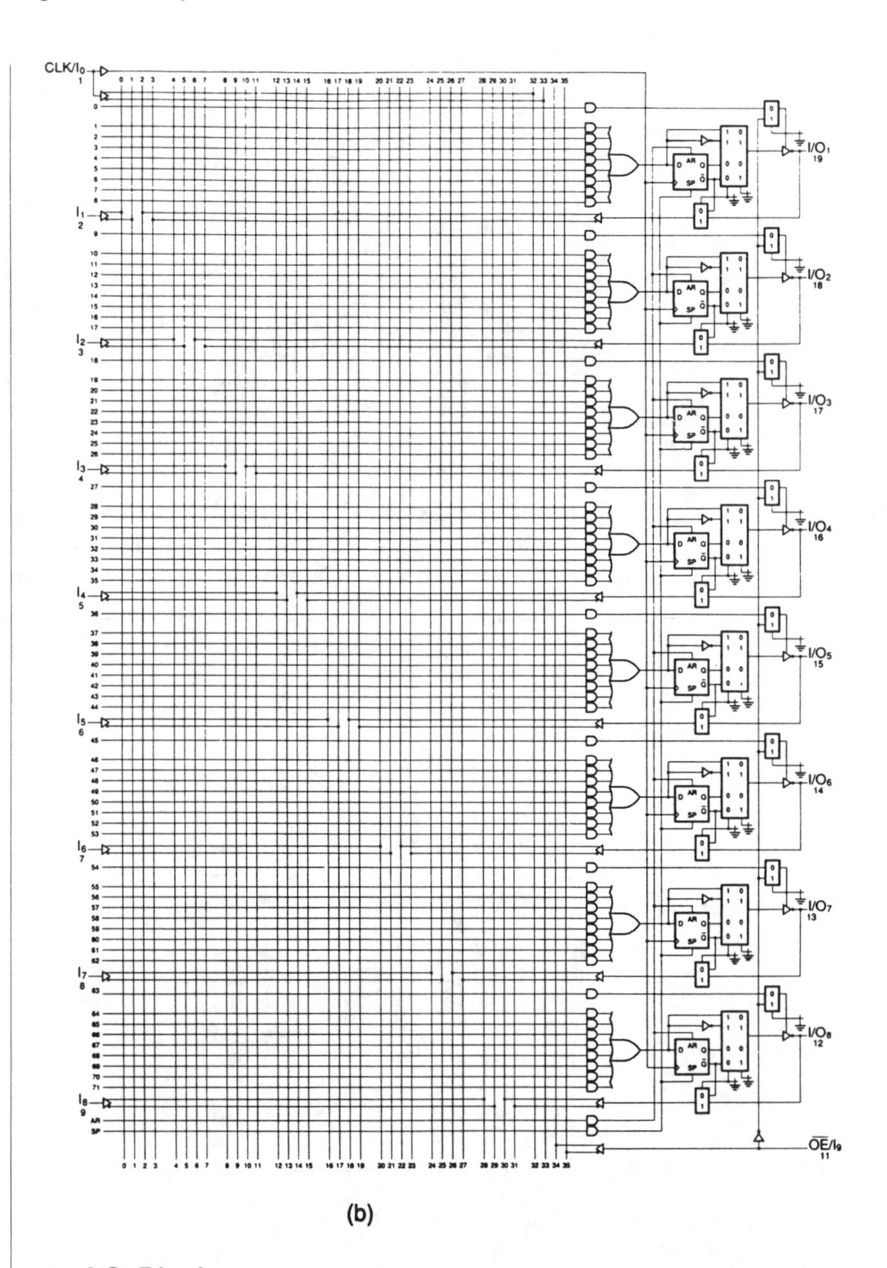

(b)

FIGURE 11.12 continued. (a) TIBPAL12R4 logic diagram (courtesy of Texas Instruments, Inc.). (b) PALC18U8Q logic diagram (courtesy of AMD).

a. I/O Block

Figure 11.14(c) shows the schematic diagram of an I/O block. Note that there are two program-controlled multiplexers, one D flip-flop, one tri-state gate, and one buffer gate. To configure the pin shown on the left of the figure as an input port, one could program the 3:1 multiplexer OFF to float the tri-state gate. (Recall that a multiplexer is equivalent to an electronic rotary switch.) Then, one may have the chance to

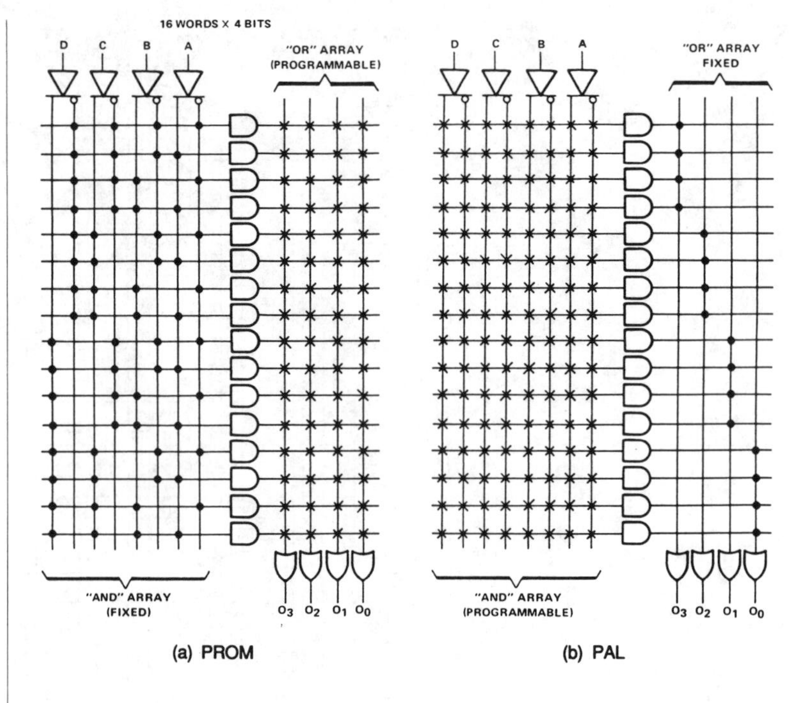

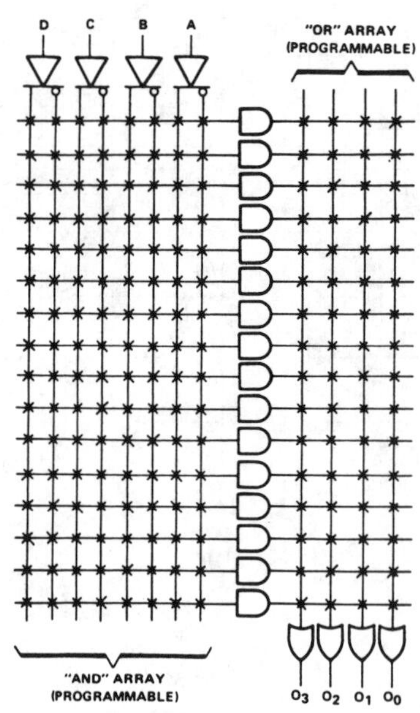

FIGURE 11.13 (a, b) Symbolic diagrams of PROM and PAL (courtesy of Texas Instruments, Inc.). (c) Symbolic diagram of PLA (courtesy of Texas Instruments, Inc.).

configure the I/O pin to be an input port with or without a latch by programming the 2:1 multiplexer. To configure that pin as an output port, program the 3:1 multiplexer to TS (tri-state) for passing the output-enable (low true) signal to the tri-state gate. Shown on the right are four lines. One of them is connected to the built-in I/O clock. The other three are to be connected to the internal CLB(s) by the configuration program designed by the system designer.

b. Configurable Logic Block (CLB)

As mentioned in the introduction to this section, the CLB is the brain cell of this powerful chip. As shown in Figure 11.14(d), there are six multiplexers and one D flip-flop with R-S inputs that overrule its clocked D input. That is, for the D flip-flop the R input clears the Q output to logic [0], while the S input sets Q to logic [1], regardless of what states the D and K (clock) inputs are. Therefore the R-S inputs are also called asynchronous inputs. In addition, there is a block (layer) of programmable combinational logic (some literature uses the term "combinatorial logic" instead). It is structured as follows.

(1) There are four inputs, one of which is shared by the independent input D and the feedback line from the Q output of the flip-flop. The rest of the inputs, A, B, and C can be independent input switching variables.
(2) G and F, the outputs of the block, can be the switching function of the input switching variables A, B, C, and D or Q.

This block may well consist of one of the kinds of programmable logic devices described in the previous sections of this chapter. It could be just a plain 16×1 PROM. Thus, it can provide the designer three options for grouping the inputs. For example, one can partition the memory space into two segments to yield two 8×1 PROM memory spaces. As a result, they can be used to implement two independent switching functions G and F as shown in the diagram. It is interesting to note that with this kind of CLB, we can easily implement a parallel full-adder or a serial full-adder by using this combinational logic block in conjunction with the D flip-flop.

Furthermore, examine the outputs X and Y and the multiplexers provided. We see that X and Y can be configured to be connected to G, F or Q. As a result, one can use one CLB to implement a two-state sequential machine of either the Mealy or Moore model, or a two-state sequential machine module. With a chip containing 64 CLBs one could build a complex sequential machine made of 64 modules. To do this,

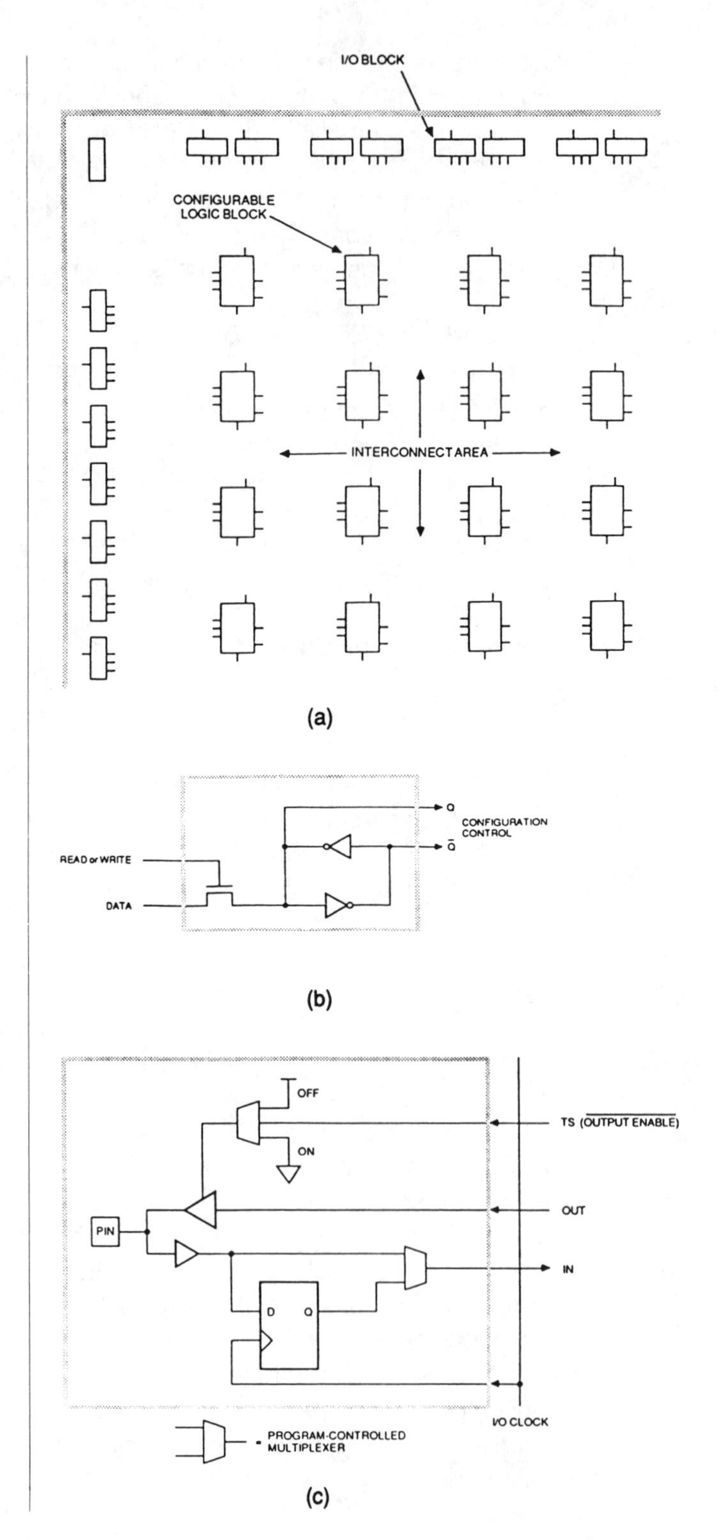

FIGURE 11.14 (a) Logic cell array structure (courtesy of Xilinx, Inc.). (b) Configuration memory cell (courtesy of Xilinx, Inc.). (c) I/O block (courtesy of Xilinx, Inc.). (d) Configurable logic block (courtesy of Xilinx, Inc.). (e) Long line interconnect (courtesy of Xilinx, Inc.).

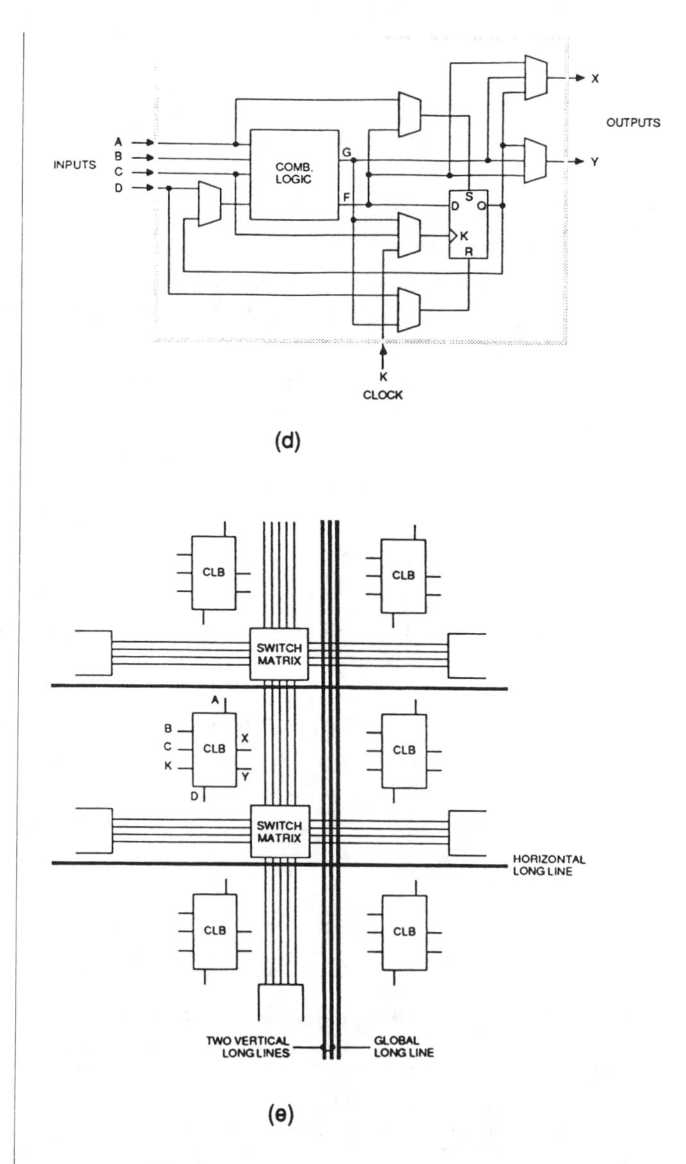

FIGURE 11.14 continued. (a) Logic cell array structure (courtesy of Xilinx, Inc.). (b) Configuration memory cell (courtesy of Xilinx, Inc.). (c) I/O block (courtesy of Xilinx, Inc.). (d) Configurable logic block (courtesy of Xilinx, Inc.). (e) Long line interconnect (courtesy of Xilinx, Inc.).

we would of course need some way to connect or "wire" these modules together for the final complex machine or system. The provision of the programmable interconnect will be described next.

c. Programmable Interconnect

Figure 11.14(e) highlights the structure of the programmable interconnect areas on the chip. They look just like our highway system. There are long lines just like our freeways, for remote communication purposes among the CLBs, and local lines with switch matrices for local communication purposes. All of these communication lines are user configurable.

11.7.2. Basic Operation

At this point you may ask, how do we make the whole system work, since we notice that all the elements, including the invisible distributed configuration memory, described above are volatile memory. In other words, when the power supply is off, all installed information will be gone; therefore, the chip needs some way to support its operations. Actually, there are two ways. (1) An external PROM or ROM chip(s) preprogrammed with the user's design (configuration) information can be installed on a PC board with the chip, and thus the configuration memory on the LCA chip can be loaded with its contents, by bit-serial or byte-serial fashion, during the chip or system initialization phase. (2) The configuration memory on the LCA chip can be initialized by a host microcomputer which can download a file containing the configuration information. One may ask why all this complication is needed; why not just make the LCA chip a PROM based chip? One reason is that with this arrangement, the LCA can be kept as a universal device and thus several kinds of standard chips can be made available on the market for designers to choose. The designers can then customize chips by supporting them with one or more ROM/PROM prestored with their proprietary configuration information. Currently, two LCA chips made by Xilinx, Inc. are available. These are the XC2064 and the XC2018. Their brief specifications are listed in Table 11.4.

TABLE 11.4
Specifications for Two Typical LCAs*

Part number	Logic capacity (usable gates)	Configurable logic blocks	User I/Os	Configuration program (bits)
XC2064	1200	64	58	12038
XC2018	1800	100	74	17878

* Courtesy of Xilinx, Inc.

Perhaps in the future, the configuration memory can be an EEPROM (electrical erasable programmable ROM). If so, the chip would be even more powerful than it is now. Another critical question one may ask is, how can one develop the configuration information? Fortunately, there is a whole set of development tools available from Xilinx. As time progresses, more different kinds of LCA chips and development tools should be available for designers to use. The tools (IBM PC/AT based) currently available to users from Xilinx include the following: In-Circuit Emulation and Simulation, XACT Design Editor, Logic Cell Array Evaluation Kit, Configuration PROM Programmer, etc. For more details the reader is referred to Reference 25.

11.8. MICROPROGRAMMABLE OR MICROPROGRAMMED CONTROL UNIT AND INSTRUCTION-BASED LOGIC SEQUENCER

In 1951, M. V. Wilkes proposed a model for a "microprogrammed device".[21] It is a device implementing his original concept on realization of random logic with programmable logic. Figure 11.15(a) shows Wilkes' model. This model was originally proposed for the design of the control unit of a digital computer. The control unit is basically a unit of sequential random logic network which is traditionally implemented with hard-wired logic. As one would expect, this unit is one of the most challenging tasks facing a computer designer, and it is also one of the most expensive units to be assembled and tested in the factory. Because of its randomness of circuit layout and timing sequence, the control unit is prone to be miswired and there could be errors buried deeply in its timing sequence. Thus, extensive and time-consuming troubleshooting would be required to correct the errors after it is built. For a conventional digital computer architecture, i.e., the Von Neumann model, the control unit in conjunction with the Arithmetic Logic Unit (ALU) constitute the Central Processing Unit (CPU) of the computer. The functions of the control unit are to fetch and decode the instructions in the main memory and to generate a set of control pulses for controlling the data paths and scheduling the operations within the CPU for executing the instructions.

What Wilkes proposed was that one could use ROM to replace the hard-wired random logic for implementing the control unit. Because of its symmetric structure and regular circuit layout characteristics, ROM is logically believed to be an ideal candidate for realization of random logic networks such as the control unit of the CPU. With ROM the problems caused by the randomness in realization of a control unit are minimized. Wilkes could never dream that the inexpensive ROM/PROM would become real 30 years later, since at that time building a ROM was an expensive task and to realize the control unit with ROM was almost an academic luxury. Thanks to rapid progress in solid-state technology, anyone can now enjoy the use of the concept proposed by Wilkes, not only in digital computer design but also in digital system design.

Let us turn to Wilkes' model again and see how it functions. First,

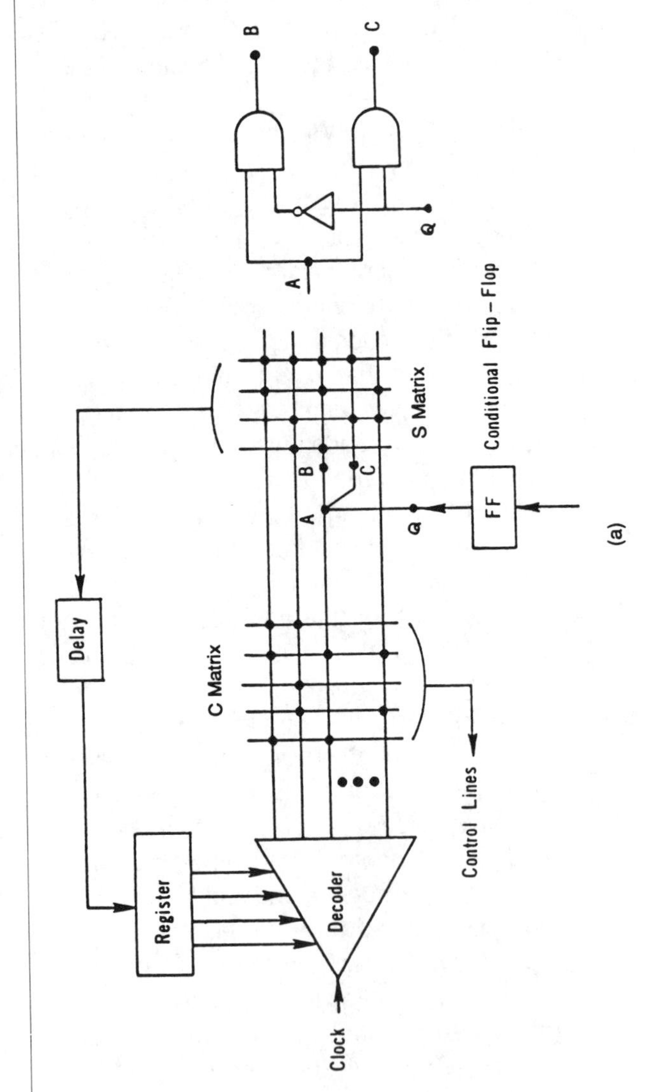

FIGURE 11.15 (a) Wilkes' model of microprogrammed device. (b) Simplified functional block diagram of a digital computer. (c) Functional block diagram of a control unit. (d) Am2910 microprogram sequencer (courtesy of AMD).

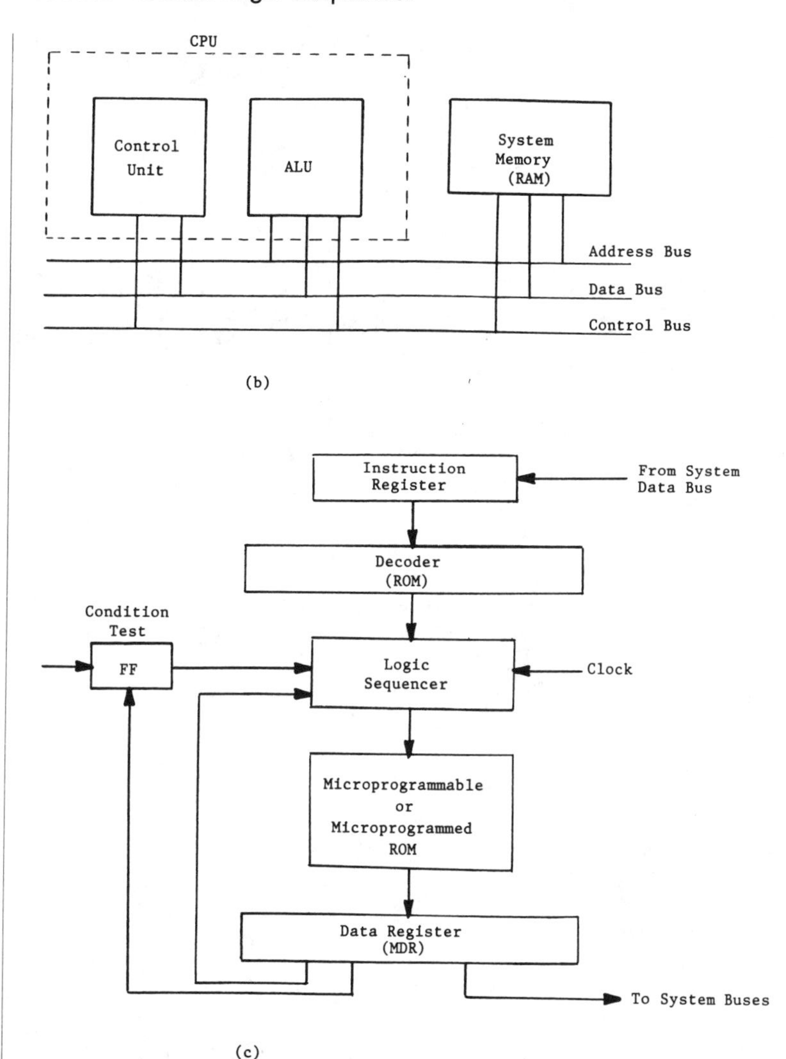

FIGURE 11.15 continued. (a) Wilkes' model of microprogrammed device. (b) Simplified functional block diagram of a digital computer. (c) Functional block diagram of a control unit. (d) Am2910 microprogram sequencer (courtesy of AMD).

we must understand that the model is indeed structured like a ROM subsystem. Here, both the C (control) matrix and the S (sequence) matrix make up the contents (9-bit word) of the ROM; the register and the decoder are nothing more than the address latch and address decoder in it. The delay block is made up of a bank of D flip-flops, or MDR. In the figure shown, consider the address of the ROM progressing downward, i.e., 0, 1, ..., and the data bits going from left to right, i.e., d_8, d_7, ..., d_0, with d_0 being the far right bit. The upper 5 bits are defined as the "control field," and the lower 4 bits as the "sequence field." At location 0, we see 10011 in the C matrix, and 0011 in the S matrix. Here, we can imagine that the control lines with 10011 are

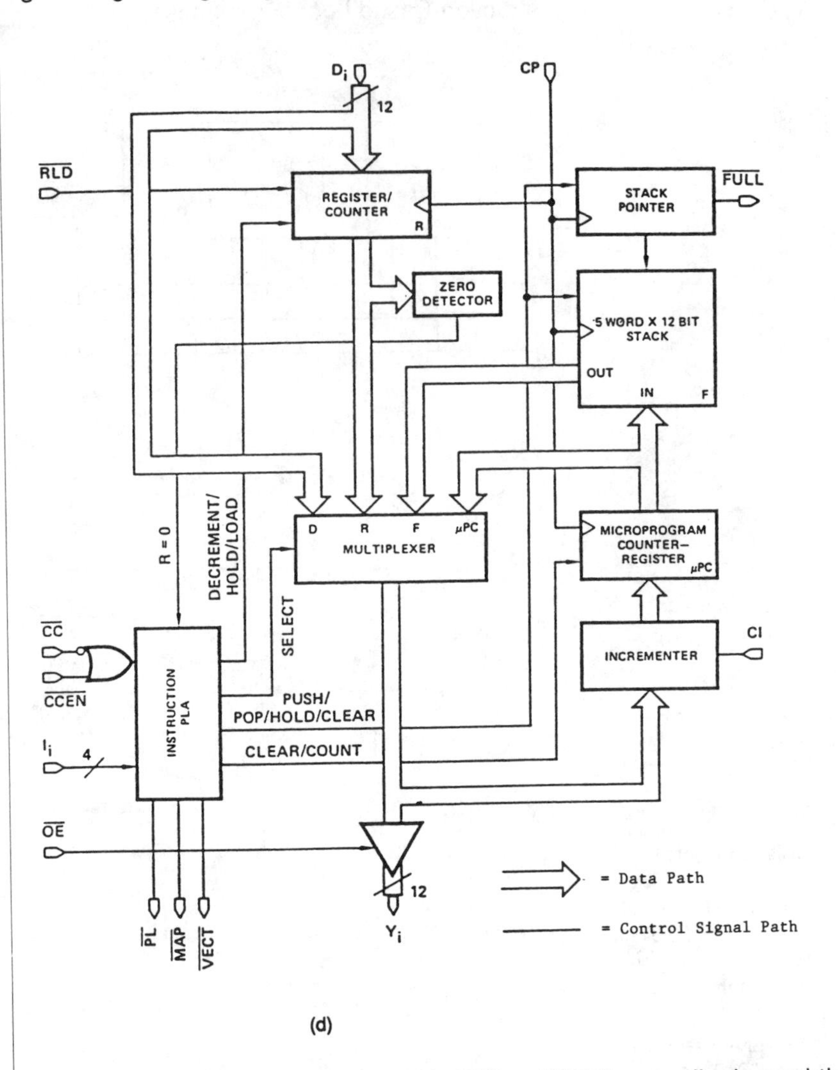

(d)

FIGURE 11.15 continued. (a) Wilkes' model of microprogrammed device. (b) Simplified functional block diagram of a digital computer. (c) Functional block diagram of a control unit. (d) Am2910 microprogram sequencer (courtesy of AMD).

turning five data paths in the ALU "ON" or "OFF" accordingly, and the 0011 is pointing at location 3 of the ROM. As the next clock pulse arrives, the contents of location 3 are read and processed accordingly. Here, the control paths are controlled by 10010, and the next address would be Branch B, or 1011, if Q of the conditional flip-flop (flag) is 0, otherwise the next address would be Branch C, or 0101. The conditional flip-flop is a flag which indicates a test result yielded from the ALU, or that yielded from some external sensor.

Consider the conditional flip-flop is used as the flag to indicate the current instruction execution result of the ALU. As an example, let the instruction being executed be "Subtract X, Y; if negative, then jump to ..., else continue ...". Then, if the result is negative, the conditional flip-flop would be set to logic [1] and the process would branch to some address specified by its sequence field, otherwise it would continue to the next address. Basically, the contents of the ROM here tells the CPU what to do through the control field and where to go through the sequence field.

Traditionally, the memory word is known as the "micro word," and the contents of the micro memory is called a microprogram. Thus, the process of determining what the contents stored in ROM should be is called "microprogramming," and the microprogram stored in the ROM is called "firmware." If a RAM is used as the micro memory, the unit is called a microprogrammable control unit, otherwise it is a microprogrammed control unit. Normally, the micro word format consists of more than the two fields described. Depending on the type of the ALU instruction, it may require one to several micro words to execute. Thus, a group of micro words that are designed to execute an instruction is called a "micro routine". As we can see, a control unit is nothing more than a sophisticated and complex special purpose sequential machine, and a micro routine is a modular sequential machine. A control unit thus consists of a number of micro routines that share the same hardware for execution. The reason we introduced the microprogrammed or microprogrammable control unit concept, originally proposed by Wilkes, is that the control units for most of the CISC (complex instruction set computing) computers, i.e., the 80486 microcomputer, are based on microprogram designs; thus, its complexity can be appreciated. We want to point out the striking similarity between the microprogram concept and the sequential machine design with PLA in digital systems. We can thus apply Wilkes' concept to simplify the design process for more complex digital systems.

Now, let us "walk an extra mile" and learn a little bit more about how the control unit of a digital computer is implemented with the microprogram concept. Figure 11.15(b) shows a simplified functional block diagram of a general purpose digital computer. We will focus our discussion on how the ALU and the microprogrammed control unit work together. Note that there are four blocks labeled with their functions. These blocks communicate with each other through the Address Bus and the Data Bus, while the Control Bus mostly carries control signals such as READ MEMORY, WRITE MEMORY, etc. from the control unit. A brief description of the operation of the control unit follows.

Assume that a program is already loaded in the system memory and the first instruction is at address 0. The control unit is to fetch that instruction and execute it accordingly. Having fetched the instruction, the control unit decodes it and locates the micro routine which is responsible for issuing the control sequence to the ALU to execute that instruction. Then the micro address pointer in the control unit points at the starting address of that micro routine and executes it accordingly. After finishing the execution, the micro address pointer would point at the micro routine which would fetch the next instruction from the system memory and the cycle continues. Since we are interested in the control unit, let us take a look at its structure.

Figure 11.15(c) is a simplified functional diagram of a control unit. Here, the blocks labeled "Logic Sequencer," "Microprogrammable /Microprogrammed ROM," "Data Register," and "Condition Test Flip-Flop" constitute the Wilkes model. Compare this structure with the sequential machine described in the previous chapter, and you will notice that the structure is a Mealy machine. The micro ROM functions as the input combinational logic layer, and its output provides the next state vector, D, of the unit. The output of the data register has three fields: one holds the present state vector, Q(i); one holds the output vector, Z(i), which is connected to the system buses; and one holds the condition test whose result will be sent as the partial components of the input vector to the input combinational logic layer via the Logic Sequencer. Here, the only block that we are not familiar with is the Logic Sequencer, which turns out to be the major contributor to the powerful control unit. It interfaces the input vector, Q, and the ALU instruction decoder, which is also implemented with a ROM, to the micro ROM. For a simple sequential machine, a straightforward ROM/PROM based combinational logic network without the sequencer would be sufficient. However, the control unit of a digital computer requires critical time sequence, branch/jump, and conditional branch/jump operations. Thus, a block dedicated to the listed functions, such as the Logic Sequencer, is needed.

Figure 11.15(d) shows the structure of a typical logic sequencer, the AM2910 by AMD. It is called an instruction-based microprogram sequencer. This is because of the fact that the four lines, I_i, at the lower left corner are the feedback lines from the data register (remember the feedback lines from the flip-flops in the Mealy machine). They provide the instruction codes for the sequencer. Therefore, the sequencer is able to make an intelligent decision for the control unit about what to do and where to go next. Note that the sequencer provides stack memory for micro routines, and a multiplexer for controlling the data flow within the unit and handling the branch/jump operations. By examining the functional label for each block, one should be able to tell what services the unit can provide. It is important to note here that a system designer can use this powerful instruction-based sequencer with ROM/PROM to design a fairly complex digital system with ease, since one would not need to repeatedly design the same or common details for the operations of test, branch, conditional branch, etc. for each system. Currently there are a number of different kinds of sequencer chips available for digital system designers to choose from. By selecting the proper chip, the designer can concentrate time and energy on the microprogram design.

11.9. CONCLUSION

In this chapter, we have described tools or programmable devices for implementing the logic design of combinational logic networks, sequential machines, and digital systems. We covered simple devices like multiplexers, memory families such as ROM, PLA, and PAL, and the powerful logic sequencer with its microprogramming technique. With tools or devices, one can design digital systems ranging from washing machines to complicated and intelligent missile controllers.

REFERENCES

1. Nichols, J. L., A logical next step for read-only-memories, *Electronics,* 40(12), 111—114, June 12, 1967.
2. Kvamme, F., Standard read-only-memory simplifies complex logic design, *Electronics,* 43(1), 88—95, January 5, 1970.
3. Hemel, A., Making small ROM's do math quickly, cheaply and easily, *Electronics,* 43(10), 104—111, May 11, 1970.
4. Reyling, G., PLAs enhance digital processor speed and cut component count, *Electronics,* 47(16), 109—114, August 8, 1974.
5. Fitchenbaum, M. L., Top-down design streamlines digital system projects, *Computer Design,* 15(9), 91—96, September 1976.
6. Carr, W. N. and Mize, J. P., *MOS/LSI Design and Application,* Sawyer and Miller, Eds., McGraw-Hill, New York, 1972.
7. Kenny, R., Microprogramming simplifies control system design, *Computer Design,* 14(2), 96—100, February 1975.
8. Fleisher, H. and Maissel, L. I., An introduction to array logic, *IBM J. Res. and Develop.,* 19(2), 98—109, March 1975.
9. Logue, J. C., Brickman, N. F., Howley, F., Jones, J. W., and Wu, W. W., Hardware implementation of a small system in PLA, *IBM J. Res. Develop.,* 19(2), 110—119, March 1975.
10. Jones, J. W., Array logic macros, *IBM J. Res. Develop.,* 19(2), 120—126, March 1975.
11. Mitchell, T. W., PLA's make simple controllers and decoders, *Electronic Design,* 24(15), 98—101, July 19, 1976.
12. Siebert, J. E., Digital multiplexers reduce chip count in logic design, *Electronics,* 50(9), 120—123, April 28, 1977.
13. Mageswaran, A., Digital multiplexers derive six-variable logic function, *Electronics,* 52(4), 148—151, February 15, 1979.

14. Cavlan, N. and Durham, S. J., Field-programmable arrays: powerful alternatives to random logic — Part I, *Electronics,* 52(14), 109—114, July 5, 1979.
15. Cavlan, N. and Durham, S. J., Sequencers and arrays transform truth tables into working systems — Part II, *Electronics,* 52(15), 132—144, July 19, 1979.
16. Peatman, J. B., *The Design of Digital Systems*, McGraw-Hill, New York, 1972.
17. Krieger, M., *Basic Switching Circuit Theory*, Macmillan, New York, 1967.
18. Friedman, A. D., *Logic Design of Digital Systems,* Computer Science Press, New York, 1975.
19. Blakeslee, T. T., *Digital Design with Standard MSI and LSI,* 2nd ed., John Wiley & Sons, New York, 1979.
20. Roth, C. H., Jr., *Fundamentals of Logic Design,* 3rd ed., West Publishing, St. Paul, MN, 1985.
21. Wilkes, M. V., The Best Way to Design an Automatic Calculating Machine, Manchester U. Computer Inaugural Conference, 1951.
22. Monolithic Memories/AMD, System Design Handbook, 2nd ed., 1985.
23. Monolithic Memories/AMD, PAL/PLE Device: Programmable Logic Array Handbook, 5th ed., 1986.
24. Texas Instruments, Inc., Programmable Logic, 1989.
25. Xilinx, Inc., The Programmable Gate Array Design Handbook, 1986.
26. Comer, D. J., *Digital Logic and State Machine Design,* Holt, Rinehart and Winston, 1984.
27. Mick, J. and Brick, J., *Bit-Slice Microprocessor Design,* McGraw-Hill, New York, 1980.
28. AMD, AM29203 Evaluation Board User's Guide, Advanced Micro Devices, Inc., 1985.
29. Husson, S. S., *Microprogramming — Principles and Practices,* Prentice-Hall, Englewood Cliffs, NJ, 1970.
30. Chu, Y., *Computer Organization and Microprogramming*, Prentice-Hall, Englewood Cliffs, NJ, 1962.
31. Lin, W. C., Ed., *Microprocessors: Fundamentals and Applications*, IEEE Press, 1977.
32. Agrawala, A. K. and Rauscher, T. G., *Foundations of Microprogramming,* Academic Press, New York, 1976.
33. Agrawala, A. K. and Rauscher, T. G., Microprogramming: perspective and status, *IEEE Trans. Computers,* C-23(8), 817—837, August 1974.

12 COMPUTER–AIDED DESIGN AND HARDWARE DEBUGGING

12.1. COMPUTER-AIDED DESIGN

In this part of the chapter, we briefly describe WHY and HOW a personal computer (PC) can be applied to assist a logic designer in designing digital systems. Specifically, we will present information on designing the PLD-based sequential machines with the assistance of a PC.

12.1.1. Why

In the preceding chapter, we presented the design of sequential machines with programmable devices, including multiplexers, ROM/PROMs, PLAs, and PALs. We have seen with design examples that realizing sequential machines with PLDs indeed reduces IC chip counts and labor costs, that increases system reliability and flexibility, and that a PLD-based system may easily be modified or upgraded by replacing or reprogramming the contents of those devices. For example, in the design of the traffic light controller, the hardware implementation may require ten or more discrete IC chips just for the sequential logic section; in contrast, implementation with PLD would require just one chip. Thus far, everything appears beautiful "on paper." However, let us consider the step of storing or writing the designed information onto a PLD device. For clarification, let us re-examine the design example of the binary counter with any specific sequence described in Chapter 11, Section 11.4.3(b), and for convenience, the final PLA realization of the design shown in Chapter 11, Figure 11.11(d) is reproduced in Figure 12.1(a). Note that we are using the second solution, or the one that realizes the design before logic minimization. For illustrating a different approach, let us now select a PAL chip instead of PLA to implement the design. For economic reasons, we choose a small but good enough standard PAL chip. Since our counter requires four flip-flops, we must select a chip which has at least this many. We choose the PAL16R4 chip described in Chapter 11, Section 11.5,

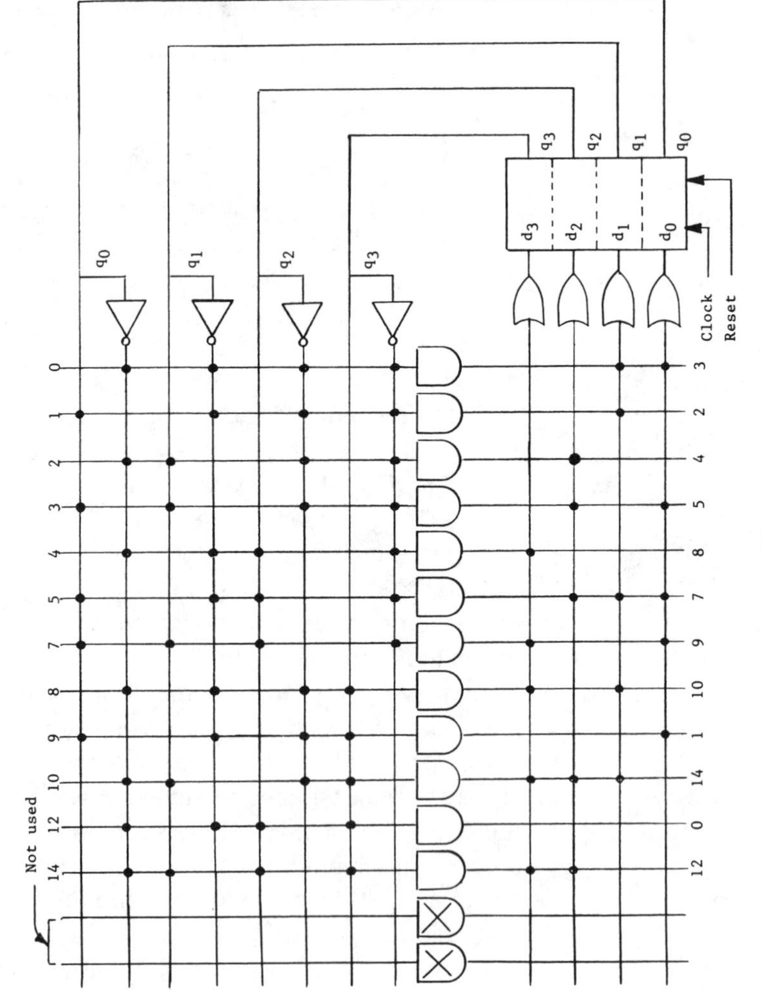

FIGURE 12.1 (a) PLA implementation of a 4-bit counter with any specified sequence. (b) PAL (TIBPAL16R4) implementation of a 4-bit counter with any specified sequence (courtesy of Texas Instruments, Inc.).

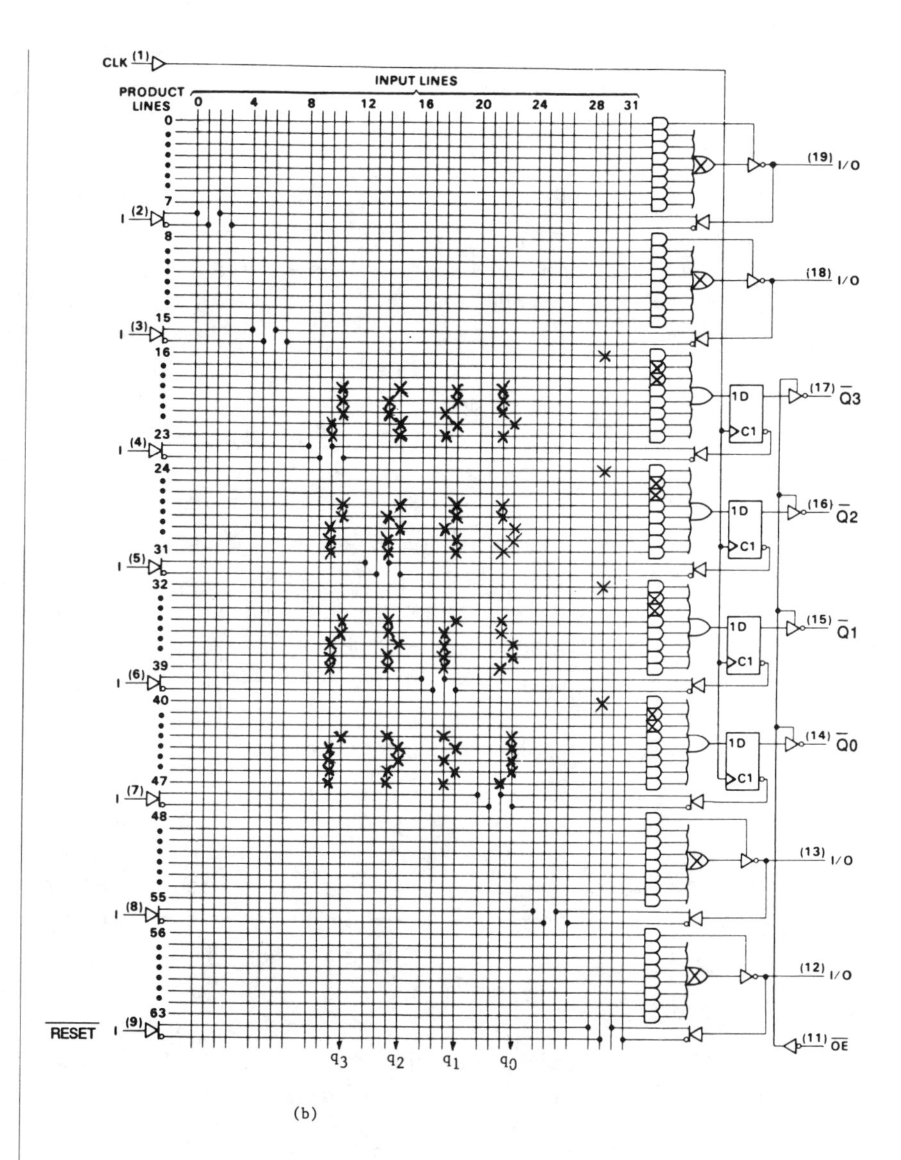

(b)

FIGURE 12.1 continued. (a) PLA implementation of a 4-bit counter with any specified sequence. (b) PAL (TIBPAL16R4) implementation of a 4-bit counter with any specified sequence (courtesy of Texas Instruments, Inc.).

since it has four flip-flops and it is a standard chip which we are familiar with. Let us "program" this chip manually by filling in the "sum of products", or the input combinational logic for the flip-flops with ×s on the schematic of the chip, according to the second solution described in Chapter 11, Section 11.4.3(b). Figure 12.1(b) shows the "manually reprogrammed" PAL16R4 chip, which is similar to the schematic shown in Figure 12.1(a), except that in the PAL implementation, the OR matrix becomes a set of fixed, eight fan-in OR gates. Note that the outputs of the flip-flops here are buffered and inverted by tri-state gates; thus, they can directly sink the LEDs for display purposes if so desired. However, the tri-state gates are controlled by Pin 11, which is labeled by /OE, where OE stands for OUTPUT-ENABLE, and " / " stands for LOW TRUE. Thus, when a logic 0 signal is applied at Pin 11, the chip outputs would be /Q0, /Q1, /Q2, /Q3 if we define Q0, Q1, Q2, and Q3 as the outputs of the flip-flops. The reader should verify the design by logically (but mentally) executing the circuit and seeing whether the counter will indeed count by the sequence 0-3-5-7-9-1-2-4-8-10-14-12-0-3... Although there are still many product (AND) terms and sum (OR) terms available on the chip which are not being used in this design, it is still economically justifiable since it is a commercially available standard PAL currently produced by MMI/AMD, Texas Instruments, and others.

At this point one may ask what would the designer do next? The task is obviously not yet completed, even though we have filled the contents with pencil on the PAL's logic diagram. One alternative is that we could send the logic design including the manually filled PAL logic diagram shown in Figure 12.1(b) to a chip manufacturer for factory fabrication. As mentioned previously, this is a good way if the design is to be produced in great quantity. However, we may still want to test, modify or upgrade the design before committing ourselves to the final product. What if there were some "bugs" in the design? Let us at this point implement the design with an erasable PAL. We will use the most primitive method to program it. That is, using "PROM programmer-like" equipment, we will actually program a PAL or blow the "fuses" on the chip by entering the design contents bit by bit onto the PAL programmer manually. With a design as simple as this example, plus patience (if any) and luck, we may be able to achieve the final goal manually with no mistakes. But what if we had a design that requires hundreds or more bits of contents to be loaded onto a PAL or PLA chip? Obviously we want a more civilized way to enter the design onto the chip than to do it manually. Furthermore, we might wish to simulate and test the design before actually storing it on the chip, without having to go through the tedious and mechanical minimization process using Karnaugh maps or other methods. Fortunately, all of these wishes are now fulfillable at a reasonable cost by using a type of software called computer-aided design application programs for logic design with PLDs. Most of the programs commercially available now can be installed on a PC with 640 kbytes of RAM and a floppy disk drive, or preferably a system with a hard disk drive of at least 20 Mbytes, such as the IBM PC/XT, or PC/AT, or equivalent.

12.1.2. How

a. Operation Principle of a PC

Before learning the details of using computer-aided design application software in our design, we must first understand the basic structure and operation principle of a Personal Computer (PC). Figure 12.2 illustrates this basic structure. As shown, the user communicates with the operating system of a PC through the keyboard and CRT. An operating system is a collection of system software routines that serves as the "middleman" between the user and the system hardware, so that the user need not know the details of how to use or manipulate the hardware attached to the system. That is, all the user has to know is how to issue the proper commands to the operating system; the details of the system hardware are not the user's concern. For example, to print a document from a PC, the user just issues a "print" command, followed by the name of the document, and the operating system does the rest. In addition to the responsibility just described, the operating system also provides other service software such as the line or screen editor, file management, command interpreter, etc. With the editor, the user can develop programs for his/her own particular applications; with file management, one can create or delete directories and

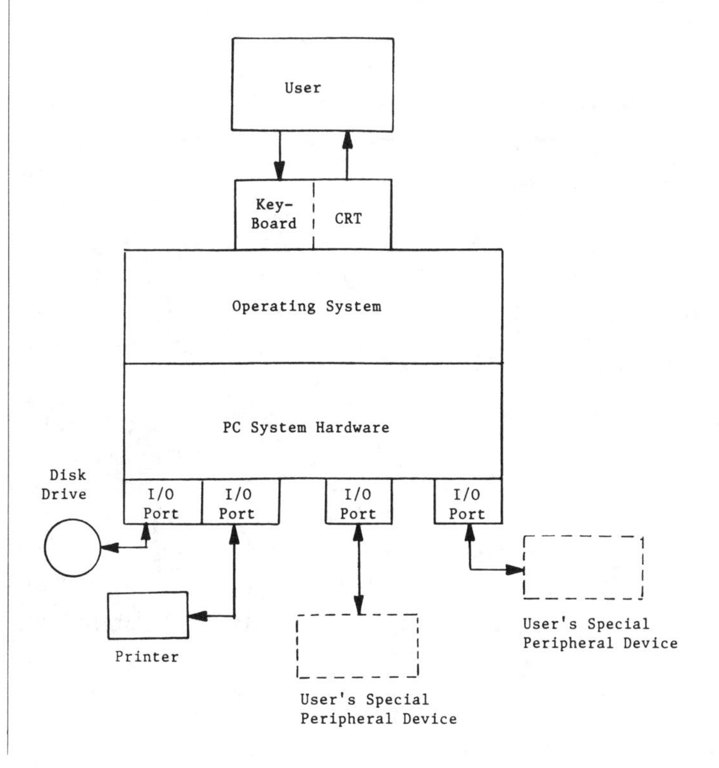

FIGURE 12.2 Basic structure of a personal computer.

files or install specific application software onto the system. Furthermore, as you may guess, the command interpreter accepts and executes commands typed by the user on the keyboard. Currently, the most popular operating system for PCs is MS-DOS or PC-DOS.

Referring to Figure 12.2 again, in addition to the two I/O ports for printer and diskdrive, the PC also provides two uncommitted I/O ports, one for parallel data transferring and one which is normally an RS232 formatted serial I/O port. These two ports are for the user to connect special peripheral devices, shown in the figure as dotted blocks, to the system. In our case, we will be using the RS232 serial port for our PAL programmer, assuming the programmer we purchased accepts RS232 serial inputs only. Having an understanding of these features, we can now proceed to learn how to use a PC to ease our design task. In summary, whenever we use a PC to assist our design we must first familiarize ourselves with its operating system and check whether there is any uncommitted I/O port, etc. available. It is not necessary to know details of how the operating system functions; mostly, it is sufficient to know the commands which are available.

b. Design Example

To illustrate what we have just described, let us follow up with a simple design example. Say we want to use a PC to help in the design of our familiar example, namely, the 4-bit binary counter described in Chapter 11, Section 11.4.3(a) and the design result as reproduced in Figure 12.1(a). Assume that we decide to use

(1) an IBM-PC/AT with 20 Mbyte hard disk as the host computer,
(2) PALASM2 (IBM-PC/AT version) by MMI/AMD as the computer-aided design software, and
(3) the University Lab Programmer, also by MMI/AMD, to program the PAL16R4 device shown in Figure 12.1(b).

Figure 12.3 depicts the system configuration for this example. We have an IBM-PC/AT (or equivalent) PC run by the operating system MS-DOS as the host computer, and the University-Lab-Programmer (product of AMD) connected to the uncommitted RS232 serial port which is provided. The programmer can program only a limited number of the most popular PALs, and the PAL16R4 is one of them. PALASM2 comes with a set of floppy disks which consists of the installation software. PALASM2 has the following features:

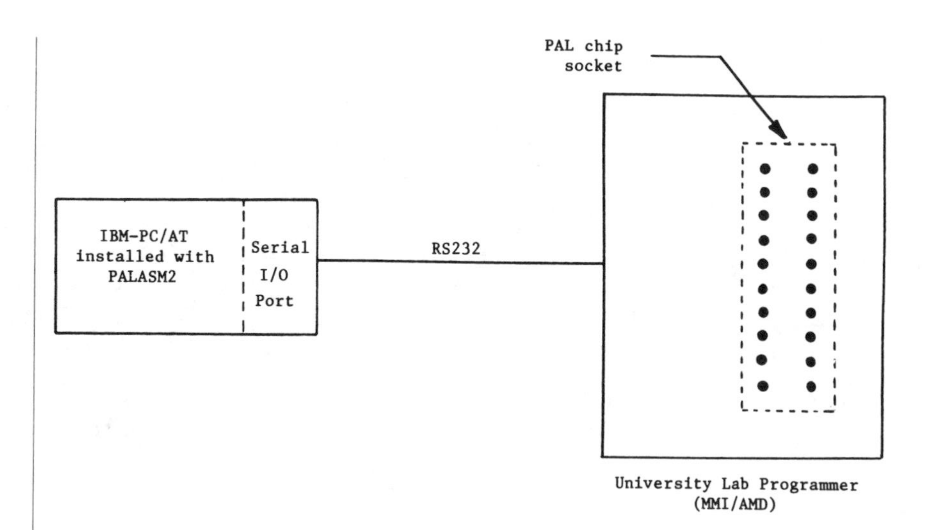

FIGURE 12.3 System configuration for programming PAL.

- Menu-driven software.
- Accepts user-developed or edited PAL device file in the format

  <filename>.pds

  which is structured as follows:

  (1) User-defined input/output variables
  (2) Input/output device pin assignments
  (3) Boolean or "sum of products" switching equations
  (4) Feedback flip-flop or register assignment

- Utility software: Logic Minimization, Waveform Display, PAL Device Programmer Driver

Having installed PALASM2, the user can use an editor available on the host computer to develop the user's file. The user's file for our counter example may be called "counter.pds" and can be listed as follows:

```
DEVICE TYPE = 16R4
PIN LIST NAMES  =
PIN NUMBER  =  1        PIN NAME = CLK
PIN NUMBER  =  2        PIN NAME = NC
PIN NUMBER  =  3        PIN NAME = NC
PIN NUMBER  =  4        PIN NAME = NC
PIN NUMBER  =  5        PIN NAME = NC
PIN NUMBER  =  6        PIN NAME = NC
PIN NUMBER  =  7        PIN NAME = NC
PIN NUMBER  =  8        PIN NAME = NC
PIN NUMBER  =  9        PIN NAME = /RESET
PIN NUMBER  = 10        PIN NAME = GND
PIN NUMBER  = 11        PIN NAME = /OE
PIN NUMBER  = 12        PIN NAME = NC
PIN NUMBER  = 13        PIN NAME = NC
PIN NUMBER  = 14        PIN NAME = /Q0
PIN NUMBER  = 15        PIN NAME = /Q1
PIN NUMBER  = 16        PIN NAME = /Q2
PIN NUMBER  = 17        PIN NAME = /Q3
PIN NUMBER  = 18        PIN NAME = NC
PIN NUMBER  = 19        PIN NAME = NC
PIN NUMBER  = 20        PIN NAME = VCC

[Note: NC stands for “no connection”]

EXPRESSIONS AND DESCRIPTION =
EXPRESSION [1] =
Q0 = /RESET  +  /Q0*/Q1*/Q2*/Q3  +  Q0*Q1*/Q2*/Q3  +
Q0*/Q1*/Q2*/Q3  +
Q0*Q1*Q2*/Q3  +  Q0*/Q1*/Q2*Q3

EXPRESSION [2] =
Q1 = /RESET  +  /Q0*/Q1*/Q2*/Q3  +  Q0*/Q1*/Q2*/Q3  +
Q0*/Q1*Q2*/Q3  +
/Q0*/Q1v/Q2*Q3  +  /Q0vQ1*/Q2*Q3

EXPRESSION [3] =
Q2 = /RESET  +  /Q0*Q1*/Q2*/Q3  +  Q0*Q1*/Q2*/Q3  +
Q0*/Q1*Q2*/Q3  +
/Q0*Q1*/Q2*Q3  +  /Q0*Q1*Q2*Q3

EXPRESSION [4] =
Q3 = /RESET  +  /Q0*/Q1*Q2*/Q3  +  Q0*Q1*Q2*/Q3  +
/Q0*/Q1*/Q2*Q3  +
/Q0*Q1*/Q2*Q3  +  /Q0*Q1*Q2*Q3

[Note: where “ / ” implies negation; i.e., /Q0 means
“negation of Q0”]
```

This is the user's file to be processed by PALASM2. Note that this file has three sections: (1) the name of the PAL chip selected, (2) the 20 pin mnemonic definitions of the chip as defined by the user, and (3) the four input equations of the four flip-flops, where the left hand side variables of the equations, Q0 – Q3, are the variables for coding the next state. Remember that these equations are derived directly from the state transition table, or the truth table of the counter (Chapter 11, Section 11.4.3(b)), and that we use the solution where the minimization step was skipped.

Once the user's file is complete, PALASM2 then takes over the process and programs the PAL16R4 (by "blowing" the fuses or the transistors at the intersections or keeping them intact) through the University Lab Programmer as shown in Figure 12.3, according to the four input equations of the flip-flops included in the user's file. However, in the process the user is able to see the so-called "Fuse Map" of the device as shown in Figure 12.4. This shows how the device would actually be programmed. Here, the fuses are intact at intersections marked by ×s, and the fuses are blown at those intersections marked by hyphens. Readers are urged to compare Figure 12.4 with Figure 12.1(b) and discover the similarity between them, and in Figure 12.4 to identify the parts that are equivalent to OR gates, the flip-flops Q0 – Q3, and the AND (product) terms in Figure 12.1(b).

c. Procedure for Programming a PAL Device

In the preceding section, we described how the PALASM2 software can perform the laborious task of constructing the device logic diagram as shown in Figure 12.1(b) with a fuse map and actually "blowing" the fuses on the device. In fact, PALASM2 can do more than we have just demonstrated, since it also includes other routines which perform simulation, check for syntax errors, etc. Figure 12.5 depicts the procedure for using PALASM2 to assist in designing sequential machines with PAL devices. The designer can simply enter the user's file which includes the Boolean equations he/she designed, and the software does all the "dirty work", up to having the chip programmed.

d. Other Available Computer-Aided Design Software and Programming Hardware

Having understood the basic concepts of computer-aided design and worked through a simple illustrative procedure for applying a typical software package to the design of a simple sequential machine, we should be able to quickly learn to use more sophisticated systems

```
0000 0000 0011 1111 1111 2222 2222 2233
0123 4567 8901 2345 6789 0123 4567 8901

xxxx xxxx xxxx xxxx xxxx xxxx xxxx xxxx -0 -
xxxx xxxx xxxx xxxx xxxx xxxx xxxx xxxx -1 -
xxxx xxxx xxxx xxxx xxxx xxxx xxxx xxxx -2 -
xxxx xxxx xxxx xxxx xxxx xxxx xxxx xxxx -3 -
xxxx xxxx xxxx xxxx xxxx xxxx xxxx xxxx -4 -
xxxx xxxx xxxx xxxx xxxx xxxx xxxx xxxx -5 -
xxxx xxxx xxxx xxxx xxxx xxxx xxxx xxxx -6 -
xxxx xxxx xxxx xxxx xxxx xxxx xxxx xxxx -7 -

xxxx xxxx xxxx xxxx xxxx xxxx xxxx xxxx -8 -
xxxx xxxx xxxx xxxx xxxx xxxx xxxx xxxx -9 -
xxxx xxxx xxxx xxxx xxxx xxxx xxxx xxxx 10 -
xxxx xxxx xxxx xxxx xxxx xxxx xxxx xxxx 11 -
xxxx xxxx xxxx xxxx xxxx xxxx xxxx xxxx 12 -
xxxx xxxx xxxx xxxx xxxx xxxx xxxx xxxx 13 -
xxxx xxxx xxxx xxxx xxxx xxxx xxxx xxxx 14 -
xxxx xxxx xxxx xxxx xxxx xxxx xxxx xxxx 15 -
Q3          =
---- ---- ---- ---- ---- ---- ---- -x-- 16 -
xxxx xxxx xxxx xxxx xxxx xxxx xxxx xxxx 17 -
xxxx xxxx xxxx xxxx xxxx xxxx xxxx xxxx 18 -
---- ---- ---x ---x ---x --x- ---- ---- 19 -
---- ---- ---x --x- ---x --x- ---- ---- 20 -
---- ---- ---x --x- --x- --x- ---- ---- 21 -
---- ---- --x- ---x ---x ---x ---- ---- 22 -
---- ---- --x- ---x --x- --x- ---- ---- 23 -
Q2          =
---- ---- ---- ---- ---- ---- ---- -x-- 24 -
xxxx xxxx xxxx xxxx xxxx xxxx xxxx xxxx 25 -
xxxx xxxx xxxx xxxx xxxx xxxx xxxx xxxx 26 -
---- ---- ---x ---x ---x --x- ---- ---- 27 -
---- ---- ---x --x- ---x --x- ---- ---- 28 -
---- ---- --x- ---x --x- ---x ---- ---- 29 -
---- ---- --x- --x- ---x ---x ---- ---- 30 -
---- ---- --x- --x- ---x --x- ---- ---- 31 -
Q1          =
---- ---- ---- ---- ---- ---- ---- -x-- 32 -
xxxx xxxx xxxx xxxx xxxx xxxx xxxx xxxx 33 -
xxxx xxxx xxxx xxxx xxxx xxxx xxxx xxxx 34 -
---- ---- ---x --x- ---x --x- ---- ---- 35 -
---- ---- ---x --x- --x- --x- ---- ---- 36 -
---- ---- --x- ---x --x- ---x ---- ---- 37 -
---- ---- --x- --x- --x- ---x ---- ---- 38 -
---- ---- --x- --x- --x- --x- ---- ---- 39 -
Q0          =
---- ---- ---- ---- ---- ---- ---- -x-- 40 -
xxxx xxxx xxxx xxxx xxxx xxxx xxxx xxxx 41 -
xxxx xxxx xxxx xxxx xxxx xxxx xxxx xxxx 42 -
---- ---- ---x --x- --x- ---x ---- ---- 43 -
---- ---- --x- ---x ---x ---x ---- ---- 44 -
---- ---- --x- ---x --x- ---x ---- ---- 45 -
---- ---- --x- --x- ---x ---x ---- ---- 47 -
---- ---- --x- --x- --x- --x- ---- ---- 48 -

xxxx xxxx xxxx xxxx xxxx xxxx xxxx xxxx 49 -
xxxx xxxx xxxx xxxx xxxx xxxx xxxx xxxx 50 -
xxxx xxxx xxxx xxxx xxxx xxxx xxxx xxxx 51 -
xxxx xxxx xxxx xxxx xxxx xxxx xxxx xxxx 52 -
xxxx xxxx xxxx xxxx xxxx xxxx xxxx xxxx 53 -
xxxx xxxx xxxx xxxx xxxx xxxx xxxx xxxx 54 -
xxxx xxxx xxxx xxxx xxxx xxxx xxxx xxxx 55 -
xxxx xxxx xxxx xxxx xxxx xxxx xxxx xxxx 56 -

xxxx xxxx xxxx xxxx xxxx xxxx xxxx xxxx 57 -
xxxx xxxx xxxx xxxx xxxx xxxx xxxx xxxx 58 -
xxxx xxxx xxxx xxxx xxxx xxxx xxxx xxxx 59 -
xxxx xxxx xxxx xxxx xxxx xxxx xxxx xxxx 60 -
xxxx xxxx xxxx xxxx xxxx xxxx xxxx xxxx 61 -
xxxx xxxx xxxx xxxx xxxx xxxx xxxx xxxx 62 -
xxxx xxxx xxxx xxxx xxxx xxxx xxxx xxxx 63 -
xxxx xxxx xxxx xxxx xxxx xxxx xxxx xxxx 63 -
```

FIGURE 12.4 PAL16R4 fuse map for the counter on any specific sequence.

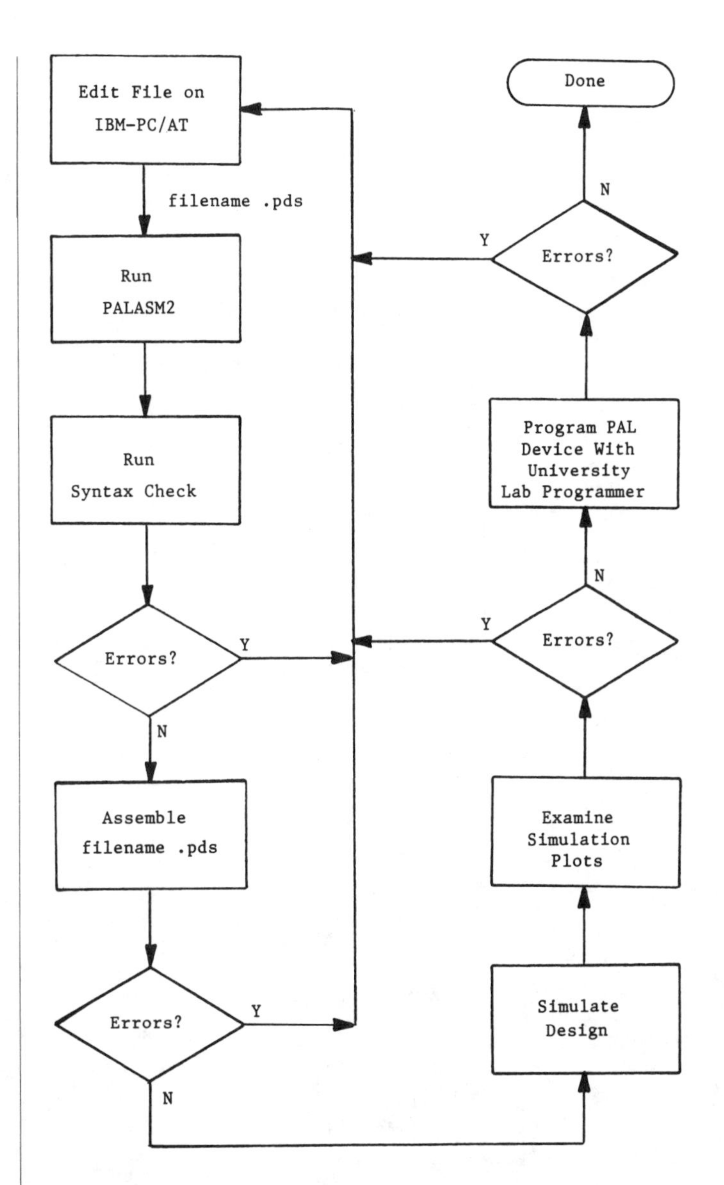

FIGURE 12.5 Procedure for using PALASM2 software.

without too much trouble. Figure 12.6 shows a general computer-aided design system for the design of PLD-based digital systems. Note that the system includes the software/hardware with which it can directly accept Boolean equations, or schematics, or state diagrams as the input files and yield the final programmed devices of different kinds of PLDs. The following is a partial list of software and hardware manufacturers in the U.S. who are currently active in this field.

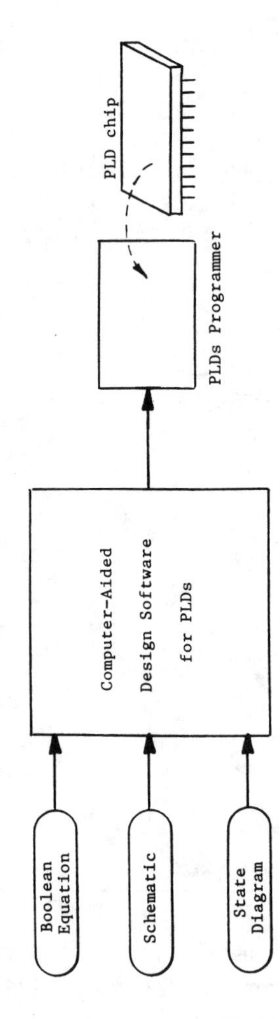

FIGURE 12.6 A general PLD programming system.

SOFTWARE: Assisted Technologies, Santa Clara, CA 95050
Data I/O, Redmond, WA 98052
MMI/Advanced Micro Devices, Sunnyvale, CA 94088
Intel, Inc., Santa Clara, CA 95052

HARDWARE: Citel, Santa Clara, CA 95050
Data I/O, Redmond, WA 98052
Digital Media, Costa Mesa, CA 92626
Kontron, Mountain View, CA 94039
MMI/Advanced Micro Devices, Sunnyvale, CA 94088
Stag Micro Systems, Sunnyvale, CA 94086
Storey Systems, Mesquite, TX 75150
Structured Design, Santa Clara, CA 95054
Sunrise Electronics, Glendora, CA 91740
Valley Data Sciences, Mountain View, CA 94043
Varix, Richardson, TX 75081
Wavetec/Digelec, Sunnyvale, CA 94089

12.1.3. Conclusion

The objective of presenting computer-aided design here is not to instantly convert the reader into an expert. Rather, it is to give an introduction to this powerful tool so that hopefully the reader will be willing to invest time and energy to pursue the subject further. An attempt has been made to simplify the subject to a beginner's level so that designers who are new to this field may first overcome whatever "computer phobia" they may have and discover that using a computer to assist their design work is not indeed a mysterious task. Briefly, one must first be familiar with the commands for operating the PC; next, install a PC-compatible software package following the manufacturer's directions; then, connect the proper programming hardware to the uncommitted I/O port of the PC. It is important to note that learning how to use a PC to aid the design process is not the major challenge; the challenges are (1) given a problem statement by a customer, the designer must intelligently interpret it and define the system specifications; (2) from system specifications, the designer must construct a proper system structure or functional system block diagram, whether Mealy or Moore model; (3) define the input/output vectors; (4) construct the state diagram (graph); (5) choose the proper PLD; and (6) develop a user's file for the specific computer-aided design software/hardware package and follow through the steps specified. Therefore, learning how to use computer-aided design packages to design digital systems is really similar to learning how to drive an automobile. The real task is learning how to use the automobile to help you achieve your goal as fast and efficiently as possible. Of course, you could still choose walking instead of going to the trouble of learning how to drive an automobile. For example, you could choose to walk from San Francisco to Los Angeles, CA, U.S. instead of driving; after all, there are only 425 miles between the two cities.

12.2. HARDWARE DEBUGGING

Having completed the computer-aided or paper design, one would normally assemble the hardware together by wire-wrapping or soldering to build a first prototype of the designed system. The prototype may consist of discrete logic: SSI (small-scale integrated) and MSI (medium-scale integrated) circuits only, or SSI and MSI plus LSI (large-scale integrated) circuits. And it must be thoroughly tested. Any designer who has ever built hardware will honestly say that an assembled prototype seldom functions as expected in the first time. Therefore, a process of troubleshooting or hardware debugging is always unavoidable in any design cycle. By design cycle, we mean the time period that begins with defining system specifications and ends with product delivery. Depending on how well it is designed, as well as how good the circuit layout is, not to mention its complexity, a prototype may require days or even months of troubleshooting. In this portion of the chapter, we intend to propose general guidelines for the designer to follow. These guidelines may not cure all problems encountered, but they help system designers to think clearly and take systematic steps to deal with whatever problems they may encounter. The guidelines are presented in three phases.

12.2.1. Preventive Phase

In this phase, we will emphasize two steps:

a. Modular Concept

First, the system designer must divide the system into modules. Ideally, assign one function to one module, and the number of signal lines for inter-module communication should be minimized. As an example, suppose we are to design a data acquisition system to collect an analog signal for a microcomputer. From Chapter 8, we know that the system would consist of an operational amplifier, a guard filter, a sample and hold circuit, a sampling clock, and an A/D converter. Thus, we may construct the system with the following modules: (1) operational amplifier, (2) guard filter, (3) sampling clock, (4) sample and hold circuit, and (5) A/D converter.

b. Circuit Layout

The circuit layout of the system must follow the rules described in Chapter 9. Consider the data acquisition example in part (a) of this section. The modules should be positioned on the circuit board from left to right as follows: amplifier — guard filter — sample and hold circuit — A/D converter, with the sampling clock in the vicinity of the sample and hold and A/D modules, but far from the amplifier. In addition, one must obey the rules for grounding, shielding, and decoupling described in Chapter 9.

12.2.2. Passive Test

Before applying power to the circuit board for the first time, one must do a simple passive test, i.e., using an ohmmeter to measure the resistance between all of the power lines entering the board. For example, consider the line between +5 V and ground. To start, connect the positive line of the ohmmeter to +5 and the negative to ground. Then reverse the connections and compare the readings. The two readings might not be equal but should not be zero or close to it. If it reads zero either way, one must check for a "dead short circuit" on the board.

12.2.3. Active Test

Active testing means that we connect the power supplies to the circuit board, turn them on, and do the static (dc) or dynamic (ac) tests accordingly. However, before performing the tests, one must observe that everything appears to be "quiet and peaceful" on the board after the power is turned on. For example, one should not see smoke or smell a burning odor. A few minutes later you may gently touch the chips and other circuit components. If any are hot enough to burn your fingers, the power must be turned off immediately and careful checking should be done to see if there was any miscalculation in the design phase or wrong connection in the assembly phase. If everything appears all right, you may proceed to perform the next two tests.

a. Static or dc Test

By static test, we mean that the circuit is in a quiescent condition. That is, the circuit is being tested with the condition that there is no real or artificial signal being applied to the system, and we can use an oscilloscope to observe the dc voltages at all critical points of each module. We must test one module at a time.

Analog Modules

As an example, let us conduct a static test on the amplifier module shown in Figure 12.7, where the pins of the amplifier chip are labeled with numbers and the circuit intersections on the P.C. (printed circuit) board and the device socket pins are labeled with letters. We may perform the following tests.

1. The f-g-j-k section denotes the power supply source on the board which may consist of a number of modules to be tested. Observe if the LED is ON; if not, check points f and k. If points f and k show the expected voltage, then either the LED is defective or the wiring connecting to it is loose. Use the ohmmeter to check for continuity. Of course, always turn the power off before using the ohmmeter for the circuit continuity test.
2. Check points d–1, p–i and 2–4 for expected voltage readings; otherwise, test for cold solder joints or circuit discontinuity with the ohmmeter. In many cases, one may find normal voltage across d–1 and no voltage across 2–4. This means a problem exists in the socket.
3. Check the voltage across a–n, b–m and 1–5; these should be zero. Similarly, the voltages across h–1 and 3–4 should also be zero. Otherwise, check for continuity of the feedback resistor network, R_1 and R_2.

Digital Modules

For digital module testing, the procedures for checking the power supply, ground, chip pins, socket pins, and circuit continuity are iden-

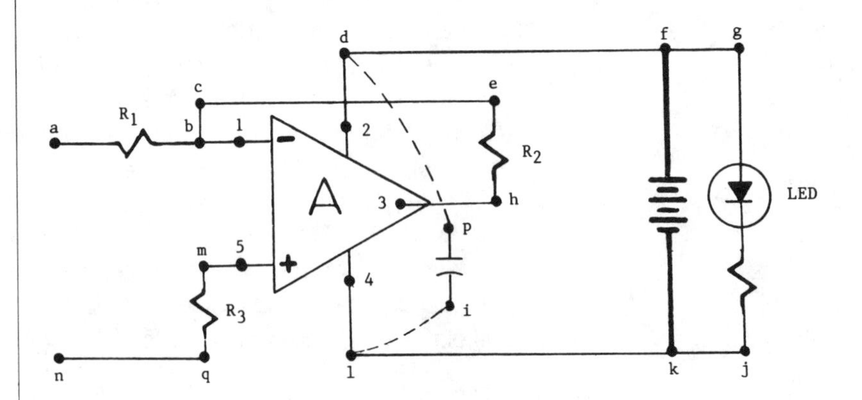

FIGURE 12.7 Hardware debugging of an analog module: amplifier.

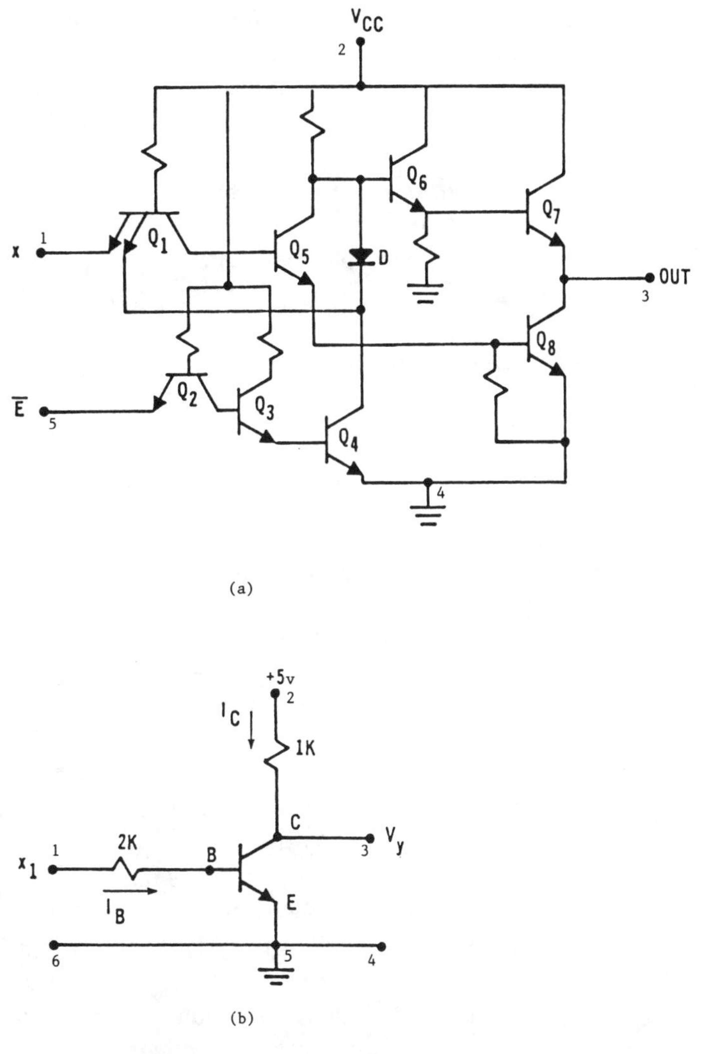

FIGURE 12.8 Hardware debugging of a digital circuit module.

tical to those for the analog modules. However, since the signals for digital circuits are binary, it is easier to simulate the signal for static testing. As a rule of thumb, one should examine the control lines immediately as soon as the test of the power and ground lines is completed. Figure 12.8(a) shows a simple digital module, a tri-state gate. This module has a control input $\overline{E}$ and data input X. Observe the output of the module with the control input = 0 then 1. See if the data at X can indeed be controlled by the control line or control input. The procedure of checking all control lines before examining any data lines is also valid for testing complex modules. Figure 12.8(b) is a single transistor circuit or module. It is used as a digital (not analog) switch.

Thus, we are interested in testing for saturation. For $X_1 = 1$, we expect the output to be 0.2 V and for $X_1 = 0$ it should be +5 V. For the saturation test, however, the voltage at C should be equal or less than that at B, otherwise the circuit is not saturated.

b. Dynamic or ac Test

By a dynamic test we mean one where the module is tested with a real or artificial input signal and observed for expected output responses.

Analog Modules

Referring to Figure 12.7, with an analog signal, E_i, applied at points a–n, we expect the output signal at points h–1 to be $E_o = -E_i \frac{R_2}{R_1}$. If it is not, inspect for continuity or the values of the feedback resistor network, or if we have a defective amplifier chip.

Digital Modules

If practical, apply a clock signal at the control input(s) to simulate the data responses to its inputs under logic LOW/HIGH control conditions. The responses must be a set of fairly clean waveforms. That is, they should not have excessive ringing, overshooting or undershooting, or any glitch which will yield a logical error. If any of these conditions are present, follow the rules described in Chapter 9. Mostly one finds that decoupling capacitors, grounding or transmission line problems are to blame. In addition, one should examine the power supplies during the dynamic test. For example, the power supply voltage for a TTL module should be within the range 4.75 to 5.25 V, and both the power and ground lines should be fairly clean. That is, there should not be any noise signal driving the voltage into uncertain logic levels.

For multiple input and output digital modules, it is very difficult to check for normal data responses with an oscilloscope. Fortunately, there is a very powerful device called a LOGIC ANALYZER which is available to the system designer. A logic designer may be considered as an oscilloscope that can measure binary data with probes having 8, 16, or more channels of testing hooks at the same time. In addition, it usually has a built-in RAM where a series of events can be stored for the user to examine. The sophisticated logic analyzers provide a set of control signals in addition to their signal sensing probes for the user to test digital modules. Some of them can "see" a glitch of one nanosecond or less in width. Most of the more recent logic analyzers are microprocessor or microcontroller based; they are menu-driven and user-friendly. Readers are urged to become familiar with these powerful devices for digital testing.

12.3. CONCLUSION

We have presented a general and logical guideline for the system designer to follow in hardware debugging. When a prototype device does not function as expected, its designer, likely a recently graduated novice, will sit or stand in front of his/her work bench staring at his/her hardware system and probably have difficulty coming to "enjoy" troubleshooting. But troubleshooting for a system designer is analogous to a detective trying to find out "whodunit". It is our hope that readers will not panic when the system they have painstakingly designed and assembled does not work the first time. Most troubles are due to miswiring, loose contacts, or careless circuit layout, not to design errors as inexperienced designers are prone to suspect. A piece of final advice is for designers to not get into a panicky state and start randomly testing their circuit. One must both design and test systems using a modular approach; circuits must be checked in logical sequence, not at random.

REFERENCES

1. Monolithic Memories, Inc., *System Design Handbook,* 2nd ed., 1985.
2. Monolithic Memories, Inc., *LSI Data Book,* 7th ed., 1986.
3. Monolithic Memories, Inc., *PAL/PLE Device Programmable Logic Array Handbook,* 5th ed., 1986.
4. Advanced Micro Devices, Inc., *PAL Device Data Book,* 1988.
5. Texas Instruments, Inc., *Programmable Logic Data Book,* 1989.
6. Monolithic Memories, Inc., *PALASM2 User Documentation,* Version 2, Revision 5C, 1987.
7. Xilinx, Inc., *The Programmable Gate Array Design Handbook,* 1st ed., 1987.

Explanation of Logic Symbols

by F.A. Mann

Contents

From "Explanation of New Logic Symbols" by F. A. Mann, in The TTL Data Book, Vol. 1, Texas Instruments, 1984, pp. 4—3 to 4—29. Reprinted by Permission of Texas Instruments.

List of Tables

List of Illustrations

Explanation of Logic Symbols†

1 Introduction

The International Electrotechnical Commission (IEC) has been developing a very powerful symbolic language that can show the relationship of each input of a digital logic circuit to each output without showing explicitly the internal logic. At the heart of the system is dependency notation, which will be explained in section 4.

The system was introduced in the USA in a rudimentary form in IEEE/ANSI Standard Y32.14-1973. Lacking at that time a complete development of dependency notation, it offered little more than a substitution of rectangular shapes for the familiar distinctive shapes for representing the basic functions of AND, OR, negation, etc. This is no longer the case.

Internationally, IEC Technical Committee TC-3 has approved a new document (Publication 617-12) that consolidates the original work started in the mid 1960s and published in 1972 (Publication 117-15) and the amendments and supplements that have followed. Similarly for the USA, IEEE Committee SCC 11.9 has revised the publication IEEE Std 91/ANSI Y32.14. Now numbered simply IEEE Std 91-1984, the IEEE standard contains all of the IEC work that has been approved, and also a small amount of material still under international consideration. Texas Instruments is participating in the work of both organizations, and this document introduces new logic symbols in accordance with the new standards. When changes are made as the standards develop, future editions will take those changes into account.

The following explanation of the new symbolic language is necessarily brief and greatly condensed from what the standards publications now contain. This is not intended to be sufficient for those people who will be developing symbols for new devices. It is primarily intended to make possible the understanding of the symbols used in various data books and the comparison of the symbols with logic diagrams, functional block diagrams, and/or function tables to further help that understanding.

2 Symbol Composition

A symbol comprises an outline or a combination of outlines together with one or more qualifying symbols. The shape of the symbol is not significant. As shown in Figure 1, general qualifying symbols are used to tell exactly what logical operation is performed by the elements. Table 1 shows general qualifying symbols defined in the new standards. Input lines are placed on the left, and output lines are placed on the right. When an exception is made to that convention, the direction of signal flow is indicated by an arrow as shown in Figure 11.

† Written by F. A. Mann.

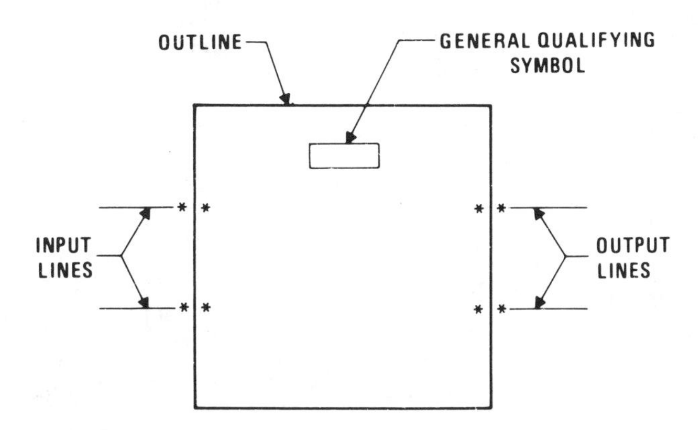

Figure 1. Symbol Composition

All outputs of a single, unsubdivided element always have identical internal logic states determined by the function of the element except when otherwise indicated by an associated qualifying symbol or label inside the element.

The outlines of elements may be abutted or embedded in which case the following conventions apply. There is no logic connection between the elements when the line common to their outlines is in the direction of signal flow. There is at least one logic connection between the elements when the line common to their outlines is perpendicular to the direction of signal flow. The number of logic connections between elements will be clarified by the use of qualifying symbols, and this is discussed further under that topic. If no indications are shown on either side of the common line, it is assumed there is only one connection.

When a circuit has one or more inputs that are common to more than one element of the circuit, the common-control block may be used. This is the only distinctively shaped outline used in the IEC system. Figure 2 shows that, unless otherwise qualified by dependency notation, an input to the common-control block is an input to each of the elements below the common-control block.

A common output depending on all elements of the array can be shown as the output of a common-output element. Its distinctive visual feature is the double line at its top. In addition, the common-output element may have other inputs as shown in Figure 3. The function of the common-output element must be shown by use of a general qualifying symbol.

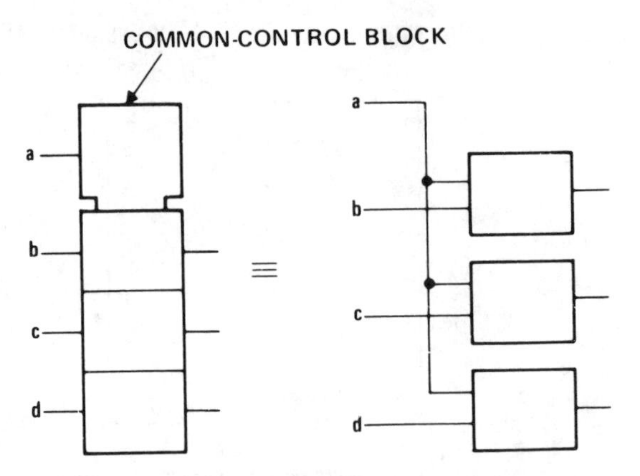

Figure 2. Common-Control Block

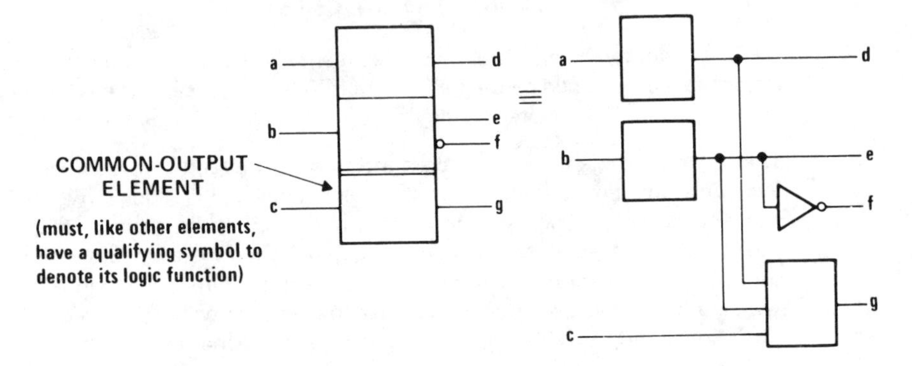

Figure 3. Common-Output Element

3 Qualifying Symbols

3.1 General Qualifying Symbols

Table 1 shows general qualifying symbols defined by IEEE Standard 91. These characters are placed near the top center or the geometric center of a symbol or symbol element to define the basic function of the device represented by the symbol or of the element.

Table 1. General Qualifying Symbols

SYMBOL	DESCRIPTION	CMOS EXAMPLE	TTL EXAMPLE
&	AND gate or function.	'HC00	SN7400
≥1	OR gate or function. The symbol was chosen to indicate that at least one active input is needed to activate the output.	'HC02	SN7402
=1	Exclusive OR. One and only one input must be active to activate the output.	'HC86	SN7486
=	Logic identity. All inputs must stand at the same state.	'HC86	SN74180
2k	An even number of inputs must be active.	'HC280	SN74180
2k + 1	An odd number of inputs must be active.	'HC86	SN74ALS86
1	The one input must be active.	'HC04	SN7404
▷ or ◁	A buffer or element with more than usual output capability (symbol is oriented in the direction of signal flow).	'HC240	SN74S436
⎍	Schmitt trigger; element with hysteresis.	'HC132	SN74LS18
X/Y	Coder, code converter (DEC/BCD, BIN/OCT, BIN/7-SEG, etc.).	'HC42	SN74LS347
MUX	Multiplexer/data selector.	'HC151	SN74150
DMUX or DX	Demultiplexer.	'HC138	SN74138
Σ	Adder.	'HC283	SN74LS385
P−Q	Subtracter.	†	SN74LS385
CPG	Look-ahead carry generator.	'HC182	SN74182
π	Multiplier.	†	SN74LS384
COMP	Magnitude comparator.	'HC85	SN74LS682
ALU	Arithmetic logic unit.	'HC181	SN74LS381
⎍	Retriggerable monostable.	'HC123	SN74LS422
1⎍	Nonretriggerable monostable (one-shot).	'HC221	SN74121
G ⎍⎍	Astable element. Showing waveform is optional.	†	SN74LS320
!G ⎍⎍	Synchronously starting astable.	†	SN74LS624
G! ⎍⎍	Astable element that stops with a completed pulse.	†	†
SRGm	Shift register. m = number of bits.	'HC164	SN74LS595
CTRm	Counter. m = number of bits; cycle length = 2^m.	'HC590	SN54LS590
CTR DIVm	Counter with cycle length = m.	'HC160	SN74LS668
RCTRm	Asynchronous (ripple-carry) counter; cycle length = 2^m.	'HC4020	†

† Not all of the general qualifying symbols have been used in TI's CMOS and TTL data books, but they are included here for the sake of completeness.

Table 1. General Qualifying Symbols (Continued)

SYMBOL	DESCRIPTION	CMOS EXAMPLE	TTL EXAMPLE
ROM	Read-only memory.	†	SN74187
RAM	Random-access read/write memory.	'HC189	SN74170
FIFO	First-in, first-out memory.	†	SN74LS222
I=0	Element powers up cleared to 0 state.	†	SN74AS877
I=1	Element powers up set to 1 state.	'HC7022	SN74AS877
Φ	Highly complex function; "gray box" symbol with limited detail shown under special rules.	'ACT2140	SN74LS608

†Not all of the general qualifying symbols have been used in TI's CMOS and TTL data books, but they are included here for the sake of completeness.

3.2 General Qualifying Symbols for Inputs and Outputs

Qualifying symbols for inputs and outputs are shown in Table 2, and many will be familiar to most users, a likely exception being the logic polarity symbol for directly indicating active-low inputs and outputs. The older logic negation indicator means that the external 0 state produces the internal 1 state. The internal 1 state means the active state. Logic negation may be used in pure logic diagrams; in order to tie the external 1 and 0 logic states to the levels H (high) and L (low), a statement of whether positive logic (1 = H, 0 = L) or negative logic (1 = L, 0 = H) is being used is required or must be assumed. Logic polarity indicators eliminate the need for calling out the logic convention and are used in various data books in the symbology for actual devices. The presence of the triangular polarity indicator indicates that the L logic level will produce the internal 1 state (the active state) or that, in the case of an output, the internal 1 state will produce the external L level. Note how the active direction of transition for a dynamic input is indicated in positive logic, negative logic, and with polarity indication.

The internal connections between logic elements abutted together in a symbol may be indicated by the symbols shown in Table 2. Each logic connection may be shown by the presence of qualifying symbols at one or both sides of the common line, and, if confusion can arise about the number of connections, use can be made of one of the internal connection symbols.

The internal (virtual) input is an input originating somewhere else in the circuit and is not connected directly to a terminal. The internal (virtual) output is likewise not connected directly to a terminal. The application of internal inputs and outputs requires an understanding of dependency notation, which is explained in section 4.

Table 2. Qualifying Symbols for Inputs and Outputs

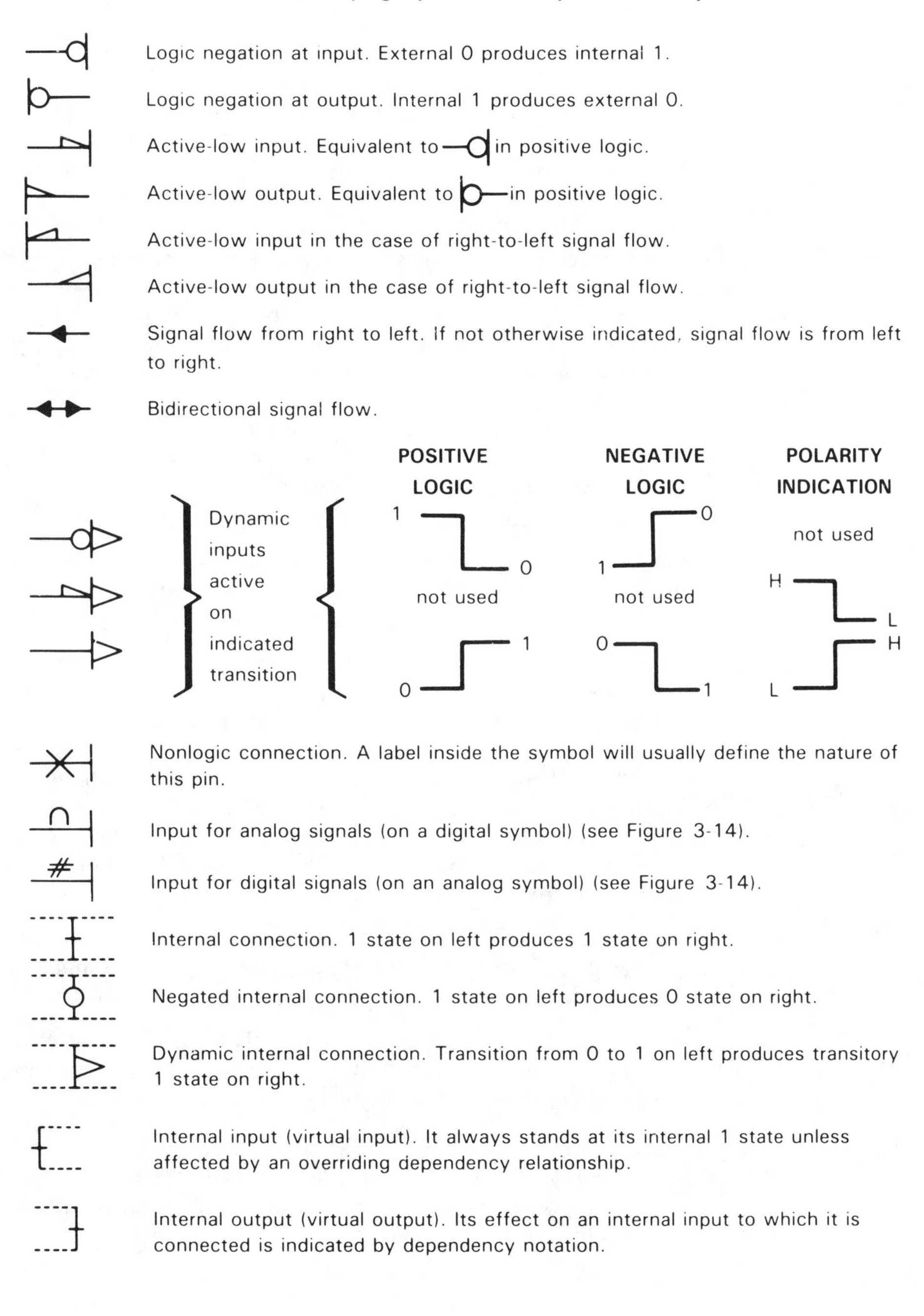

Logic negation at input. External 0 produces internal 1.

Logic negation at output. Internal 1 produces external 0.

Active-low input. Equivalent to in positive logic.

Active-low output. Equivalent to in positive logic.

Active-low input in the case of right-to-left signal flow.

Active-low output in the case of right-to-left signal flow.

Signal flow from right to left. If not otherwise indicated, signal flow is from left to right.

Bidirectional signal flow.

Nonlogic connection. A label inside the symbol will usually define the nature of this pin.

Input for analog signals (on a digital symbol) (see Figure 3-14).

Input for digital signals (on an analog symbol) (see Figure 3-14).

Internal connection. 1 state on left produces 1 state on right.

Negated internal connection. 1 state on left produces 0 state on right.

Dynamic internal connection. Transition from 0 to 1 on left produces transitory 1 state on right.

Internal input (virtual input). It always stands at its internal 1 state unless affected by an overriding dependency relationship.

Internal output (virtual output). Its effect on an internal input to which it is connected is indicated by dependency notation.

In an array of elements, if the same general qualifying symbol and the same qualifying symbols associated with inputs and outputs would appear inside each of the elements of the array, then these qualifying symbols are usually shown only in the first element. This is done to reduce clutter and to save time in recognition. Similarly, large identical elements that are subdivided into smaller elements may each be represented by an unsubdivided outline. The SN54HC242 or SN54LS440 symbol illustrates this principle.

3.3 Symbols Inside the Outline

Table 3 shows some symbols used inside the outline. Note particularly that open-collector (open-drain), open-emitter (open-source), and 3-state outputs have distinctive symbols. An EN input affects all the external outputs of the element in which it is placed, plus the external outputs of any elements shown to be influenced by that element. It has no effect on inputs. When an enable input affects only certain outputs, affects outputs located outside the indicated influence of the element in which the enable input is placed, and/or affects one or more inputs, a form of dependency notation will indicate this (see 4.10). The effects of the EN input on the various types of outputs are shown.

It is particularly important to note that a D input is always the data input of a storage element. At its internal 1 state, the D input sets the storage element to its 1 state, and at its internal 0 state, it resets the storage element to its 0 state.

The binary grouping symbol will be explained more fully in section 8. Binary-weighted inputs are arranged in order, and the binary weights of the least significant and the most significant lines are indicated by numbers. In this document, weights of input and output lines will usually be represented by powers of two only when the binary grouping symbol is used; otherwise, decimal numbers will be used. The grouped inputs generate an internal number on which a mathematical function can be performed or that can be an identifying number for dependency notation (Figure 31). A frequent use is in addresses for memories.

Reversed in direction, the binary grouping symbol can be used with outputs. The concept is analogous to that for the inputs, and the weighted outputs will indicate the internal number assumed to be developed within the circuit.

Other symbols are used inside the outlines in accordance with the IEC/IEEE standards but are not shown here. Generally, these are associated with arithmetic operations and are self-explanatory.

When nonstandardized information is shown inside an outline, it is usually enclosed in square brackets [like these]. The square brackets are omitted when associated with a nonlogic input, which is indicated by an X superimposed on the connection line outside the symbol.

Table 3. Symbols Inside the Outline

Postponed output (of a pulse-triggered flip-flop). The output changes when input initiating change (e.g., a C input) returns to its initial external state or level. See paragraph 5.

Bi-threshold input (input with hysteresis)

N-P-N open-collector or similar output that can supply a relatively low-impedance L level when not turned off. Requires external pull-up. Capable of positive-logic wired-AND connection.

Passive-pull-up output is similar to N-P-N open-collector output but is supplemented with a built-in passive pull-up.

N-P-N open-emitter or similar output that can supply a relatively low-impedance H level when not turned off. Requires external pull-down. Capable of positive-logic wired-OR connection.

Passive-pull-down output is similar to N-P-N open-emitter output but is supplemented with a built-in passive pull-down.

Three-state output

Output with more than usual output capability (symbol is oriented in the direction of signal flow).

EN — Enable input
When at its internal 1-state, all outputs are enabled.
When at its internal 0-state, open-collector and open-emitter outputs are off, three-state outputs are in the high-impedance state, and all other outputs (i.e., totem-poles) are at the internal 0-state.

J, K, R, S — Usual meanings associated with flip-flops (e.g., R = reset to 0, S = reset to 1).

T — Toggle input causes internal state of output to change to its complement.

D — Data input to a storage element equivalent to: S R

→m ←m — Shift right (left) inputs, m = 1, 2, 3, etc. If m = 1, it is usually not shown.

+m −m — Counting up (down) inputs, m = 1, 2, 3, etc. If m = 1, it is usually not shown.

0 … m — Binary grouping. m is highest power of 2.

Table 3. Symbols Inside the Outline (Continued)

Symbol	Description
CT = 15	The contents-setting input, when active, causes the content of a register to take on the indicated value.
CT = 9	The content output is active if the content of the register is as indicated.
	Input line grouping . . . indicates two or more terminals used to implement a single logic input. e.g., The paired expander inputs of SN7450. X, $\overline{X}$, E
"1"	Fixed-state output always stands at its internal 1 state. For example, see SN74185.

4 Dependency Notation

4.1 General Explanation

Dependency notation is the powerful tool that sets the IEC symbols apart from previous systems and makes compact, meaningful symbols possible. It provides the means of denoting the relationship between inputs, outputs, or inputs and outputs without actually showing all the elements and interconnections involved. The information provided by dependency notation supplements that provided by the qualifying symbols for an element's function.

In the convention for the dependency notation, use will be made of the terms "affecting" and "affected." In cases where it is not evident which inputs must be considered as being the affecting or the affected ones (e.g., if they stand in an AND relationship), the choice may be made in any convenient way.

So far, eleven types of dependency have been defined, and all of these are used in various TI data books. X dependency is used mainly with CMOS circuits. They are listed below in the order in which they are presented and are summarized in Table 4 at the end of section 4.

Section	Dependency Type or Other Subject
4.2	G, AND
4.3	General Rules for Dependency Notation
4.4	V, OR
4.5	N, Negate (Exclusive-OR)
4.6	Z, Interconnection
4.7	X, Transmission
4.8	C, Control
4.9	S, Set and R, Reset
4.10	EN, Enable
4.11	M, Mode
4.12	A, Address

4.2 G (AND) Dependency

A common relationship between two signals is to have them ANDed together. This has traditionally been shown by explicitly drawing an AND gate with the signals connected to the inputs of the gate. The 1972 IEC publication and the 1973 IEEE/ANSI standard showed several ways to show this AND relationship using dependency notation. While ten other forms of dependency have since been defined, the ways to invoke AND dependency are now reduced to one.

In Figure 4 input **b** is ANDed with input **a**, and the complement of **b** is ANDed with **c**. The letter G has been chosen to indicate AND relationships and is placed at input **b**, inside the symbol. A number considered appropriate by the symbol designer (1 has been used here) is placed after the letter G and also at each affected input. Note the bar over the 1 at input **c**.

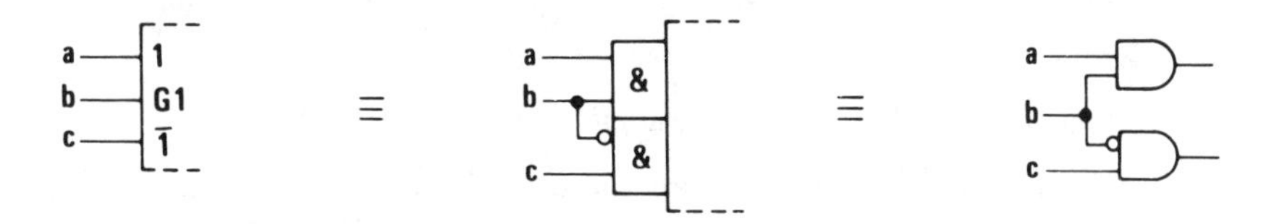

Figure 4. G Dependency Between Inputs

In Figure 5, output **b** affects input **a** with an AND relationship. The lower example shows that it is the internal logic state of **b**, unaffected by the negation sign, that is ANDed. Figure 6 shows input **a** to be ANDed with a dynamic input **b**.

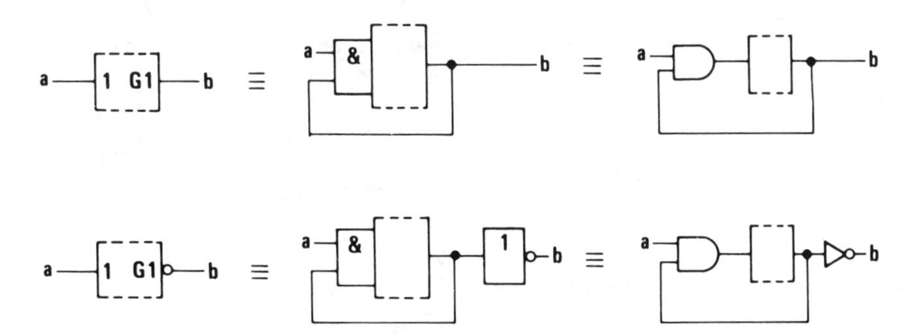

Figure 5. G Dependency Between Outputs and Inputs

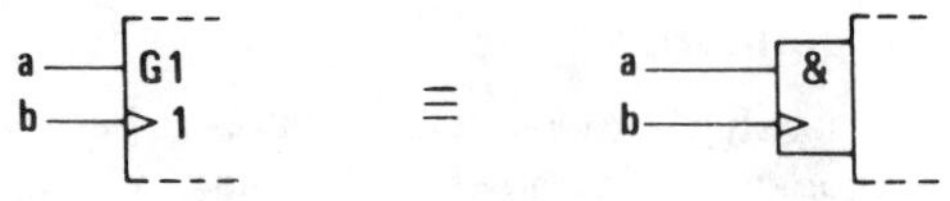

Figure 6. G Dependency with a Dynamic Input

The rules for G dependency can be summarized thus:

> When a G*m* input or output (*m* is a number) stands at its internal 1 state, all inputs and outputs affected by G*m* stand at their normally defined internal logic states. When the G*m* input or output stands at its 0 state, all inputs and outputs affected by G*m* stand at their internal 0 states.

4.3 Conventions for the Application of Dependency Notation in General

The rules for applying dependency relationships in general follow the same pattern as was illustrated for G dependency.

Application of dependency notation is accomplished by:

1) labeling the input (or output) *affecting* other inputs or outputs with the letter symbol indicating the relationship involved (e.g., G for AND) followed by an identifying number, appropriately chosen, and
2) labeling each input or output *affected* by that affecting input (or output) with that same number.

If it is the complement of the internal logic state of the affecting input or output that does the affecting, then a bar is placed over the identifying numbers at the affected inputs or outputs (Figure 4).

If two affecting inputs or outputs have the same letter and the same identifying number, they stand in an OR relationship to each other (Figure 7).

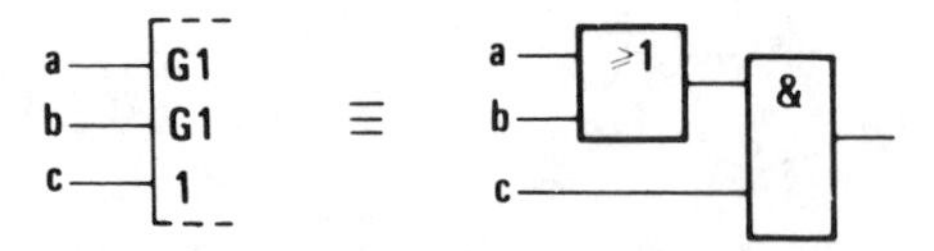

Figure 7. ORed Affecting Inputs

If the affected input or output requires a label to denote its function (e.g., "D"), this label will be *prefixed* by the identifying number of the affecting input (Figure 15).

If an input or output is affected by more than one affecting input, the identifying numbers of each of the affecting inputs will appear in the label of the affected one, separated by commas. The normal reading order of these numbers is the same as the sequence of the affecting relationships (Figure 15).

If the labels denoting the functions of affected inputs or outputs must be numbers (e.g., outputs of a coder), the identifying numbers to be associated with both affecting inputs and affected inputs or outputs may be replaced by another character selected to avoid ambiguity, e.g., Greek letters (Figure 8).

a — α a — 1
b — Gα ≡   b — G1
c — $\overline{\alpha}$ c — $\overline{1}$

Figure 8. Substitution for Numbers

4.4 V (OR) Dependency

The symbol denoting OR dependency is the letter V (Figure 9).

When a V*m* input or output stands at its internal 1 state, all inputs and outputs affected by V*m* stand at their internal 1 states. When the V*m* input or output stands at its internal 0 state, all inputs and outputs affected by V*m* stand at their normally defined internal logic states.

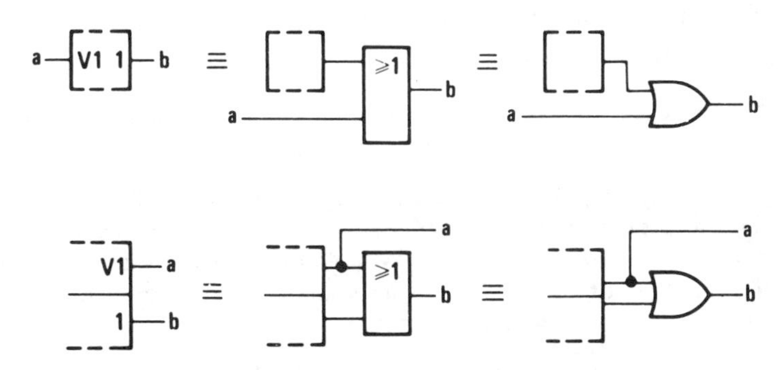

Figure 9. V (OR) Dependency

4.5 N (Negate) (Exclusive-OR) Dependency

The symbol denoting negate dependency is the letter N (Figure 10). Each input or output affected by an N*m* input or output stands in an Exclusive-OR relationship with the N*m* input or output.

When an N*m* input or output stands at its internal 1 state, the internal logic state of each input and each output affected by N*m* is the complement of what it would otherwise be. When an N*m* input or output stands at its internal 0 state, all inputs and outputs affected by N*m* stand at their normally defined internal logic states.

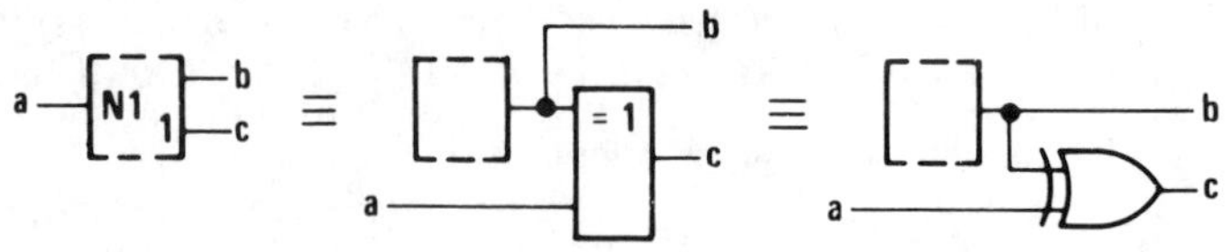

If $a = 0$, then $c = b$
If $a = 1$, then $c = \bar{b}$

Figure 10. N (Negate) (Exclusive-OR) Dependency

4.6 Z (Interconnection) Dependency

The symbol denoting interconnection dependency is the letter Z.

Interconnection dependency is used to indicate the existence of internal logic connections between inputs, outputs, internal inputs, and/or internal outputs.

The internal logic state of an input or output affected by a Z*m* input or output will be the same as the internal logic state of the Z*m* input or output, unless modified by additional dependency notation (Figure 11).

4.7 X (Transmission) Dependency

The symbol denoting transmission dependency is the letter X.

Transmission dependency is used to indicate controlled bidirectional connections between affected input/output ports (Figure 12).

When an X*m* input or output stands at its internal 1 state, all input-output ports affected by this X*m* input or output are bidirectionally connected together and stand at the same internal logic state or analog signal level. When an X*m* input or output stands at its internal 0 state, the connection associated with this set of dependency notation does not exist.

Although the transmission paths represented by X dependency are inherently bidirectional, use is not always made of this property. This is analogous to a piece of wire, which may be constrained to carry current in only one direction. If this is the case in a particular application, then the directional arrows shown in Figures 12, 13, and 14 are omitted.

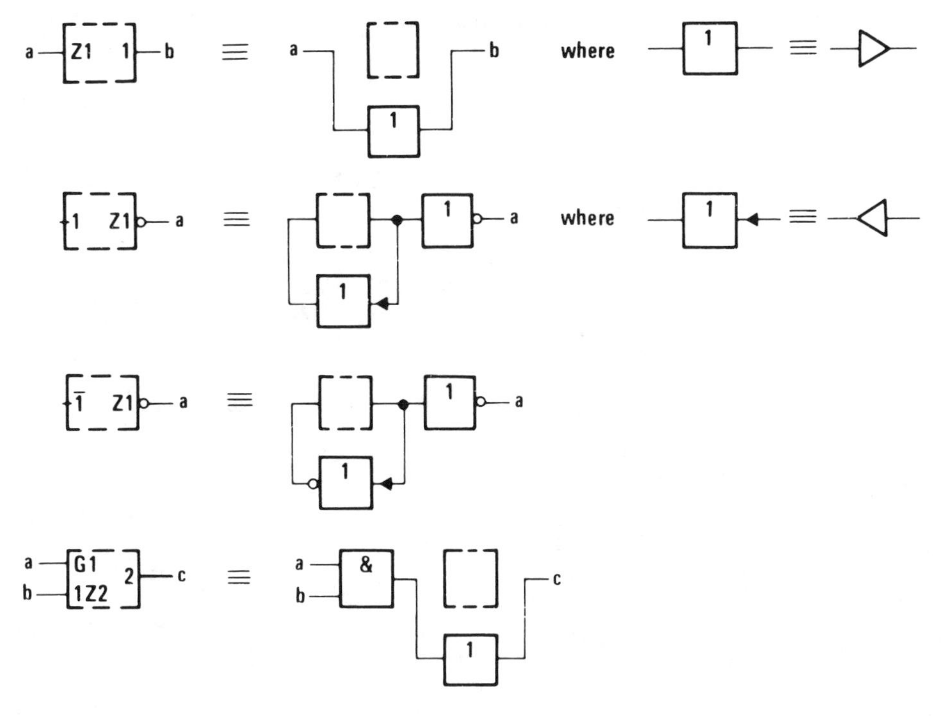

Figure 11. Z (Interconnection) Dependency

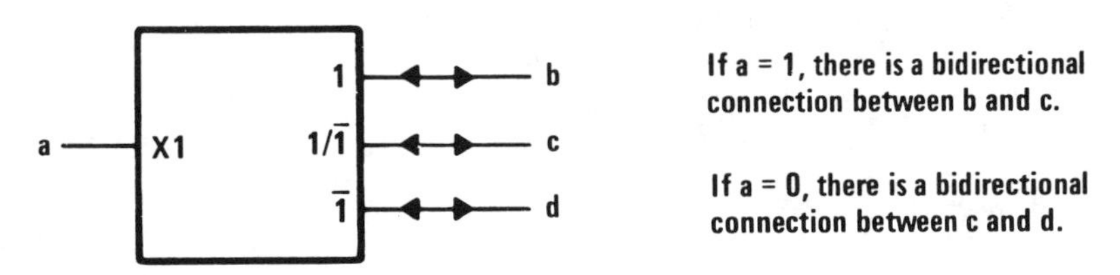

Figure 12. X (Transmission) Dependency

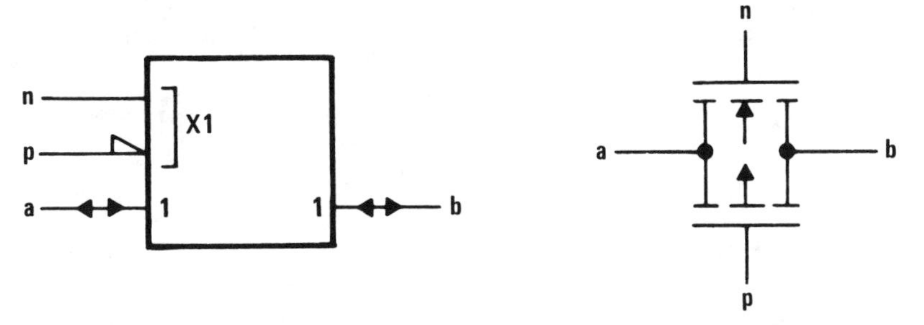

Figure 13. CMOS Transmission Gate Symbol and Schematic

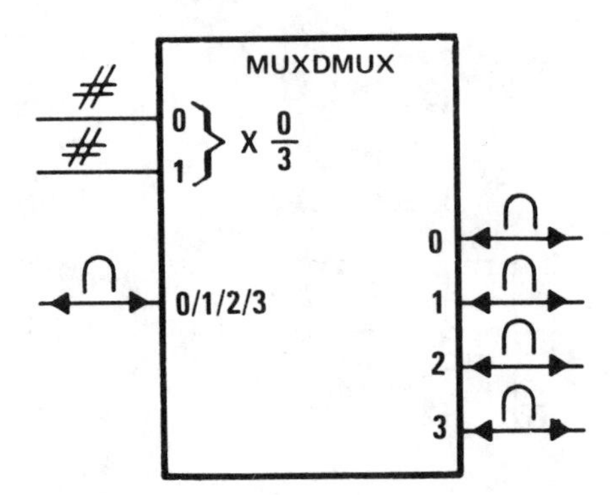

Figure 14. Analog Data Selector (Multiplexer/Demultiplexer)

4.8 C (Control) Dependency

The symbol denoting control dependency is the letter C.

Control inputs are usually used to enable or disable the data (D, J, K, R, or S) inputs of storage elements. They may take on their internal 1 states (be active) either statically or dynamically. In the latter case, the dynamic input symbol is used as shown in the third example of Figure 15.

When a C*m* input or output stands at its internal 1 state, the inputs affected by C*m* have their normally defined effect on the function of the element; i.e., these inputs are enabled. When a C*m* input or output stands at its internal 0 state, the inputs affected by C*m* are disabled and have no effect on the function of the element.

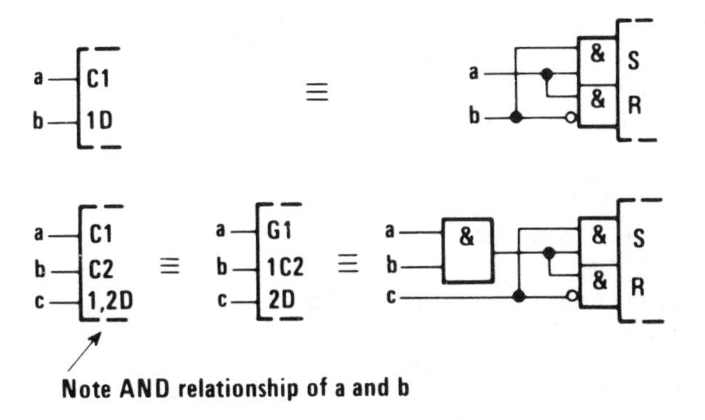

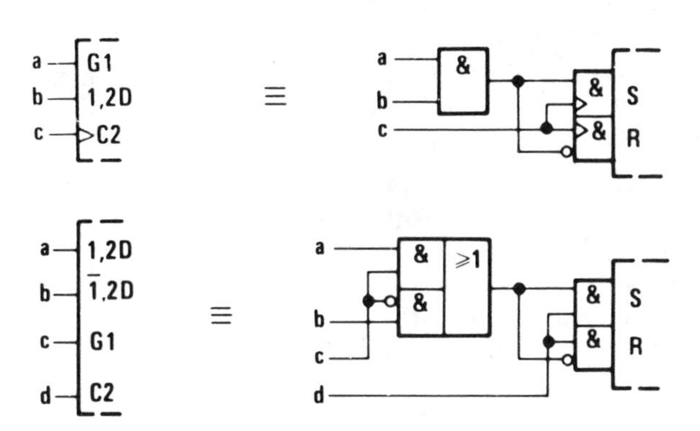

Input c selects which of a or b is stored when d goes low.

Figure 15. C (Control) Dependency

4.9 S (Set) and R (Reset) Dependencies

The symbol denoting set dependency is the letter S. The symbol denoting reset dependency is the letter R.

Set and reset dependencies are used if it is necessary to specify the effect of the combination R = S = 1 on a bistable element. Case 1 in Figure 16 does not use S or R dependency.

When an S*m* input is at its internal 1 state, outputs affected by the S*m* input will react, regardless of the state of an R input, as they normally would react to the combination S = 1, R = 0. See cases 2, 4, and 5 in Figure 16.

When an R*m* input is at its internal 1 state, outputs affected by the R*m* input will react, regardless of the state of an S input, as they normally would react to the combination S = 0, R = 1. See cases 3, 4, and 5 in Figure 16.

When an S*m* or R*m* input is at its internal 0 state, it has no effect.

Note that the noncomplementary output patterns in cases 4 and 5 are only pseudo stable. The simultaneous return of the inputs to S = R = 0 produces an unforeseeable stable and complementary output pattern.

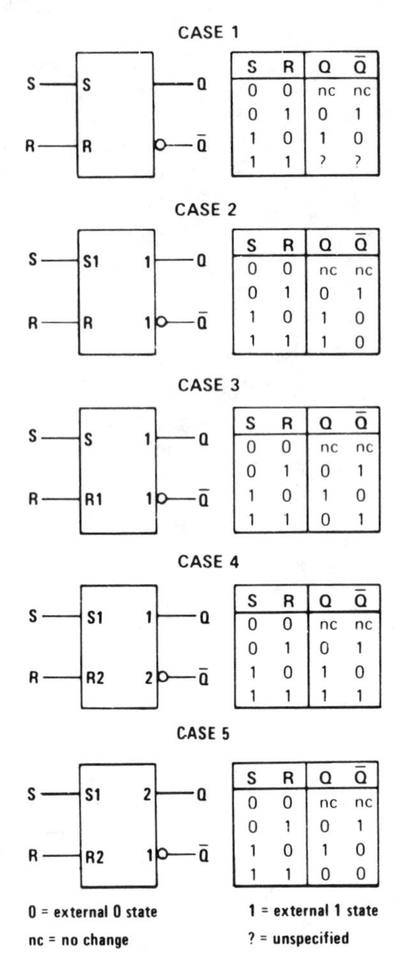

Figure 16. S (Set) and R (Reset) Dependencies

4.10 EN (Enable) Dependency

The symbol denoting enable dependency is the combination of letters EN.

An EN*m* input has the same effect on outputs as an EN input, see 3.3, but it affects only those outputs labeled with the identifying number *m*. It also affects those inputs labeled with the identifying number *m*. By contrast, an EN input affects all outputs and no inputs. The effect of an EN*m* input on an affected input is identical to that of a C*m* input (Figure 17).

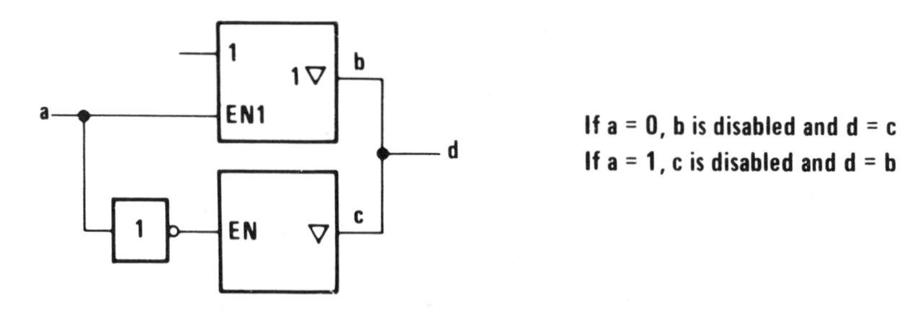

Figure 17. EN (Enable) Dependency

When an EN*m* input stands at its internal 1 state, the inputs affected by EN*m* have their normally defined effect on the function of the element, and the outputs affected by this input stand at their normally defined internal logic states; i.e., these inputs and outputs are enabled.

When an EN*m* input stands at its internal 0 state, the inputs affected by EN*m* are disabled and have no effect on the function of the element, and the outputs affected by EN*m* are also disabled. Open-collector outputs are turned off, three-state outputs stand at their normally defined internal logic states but externally exhibit high impedance, and all other outputs (e.g., totem-pole outputs) stand at their internal 0 states.

4.11 M (MODE) Dependency

The symbol denoting mode dependency is the letter M.

Mode dependency is used to indicate that the effects of particular inputs and outputs of an element depend on the mode in which the element is operating.

If an input or output has the same effect in different modes of operation, the identifying numbers of the relevant affecting M*m* inputs will appear in the label of that affected input or output between parentheses and separated by solidi (Figure 22).

4.11.1 M Dependency Affecting Inputs

M dependency affects inputs the same as C dependency. When an M*m* input or M*m* output stands at its internal 1 state, the inputs affected by this M*m* input or M*m* output have their normally defined effect on the function of the element; i.e., the inputs are enabled.

When an M*m* input or M*m* output stands at its internal 0 state, the inputs affected by this M*m* input or M*m* output have no effect on the function of the element. When an affected input has several sets of labels separated by solidi (e.g., C4/2→/3+), any set in which the identifying number of the M*m* input or M*m* output appears has no effect and is to be ignored. This represents disabling of some of the functions of a multifunction input.

The circuit in Figure 18 has two inputs, **b** and **c**, that control which one of four modes (0, 1, 2, or 3) will exist at any time. Inputs **d, e,** and **f** are D inputs subject to dynamic control (clocking) by the **a** input. The numbers 1 and 2 are in the series chosen to indicate the modes so inputs **e** and **f** are only enabled in mode 1 (for parallel loading), and input **d** is only enabled in mode 2 (for serial loading). Note that input **a** has three functions. It is the clock for entering data. In mode 2, it causes right shifting of data, which means a shift away from the control block. In mode 3, it causes the contents of the register to be incremented by one count.

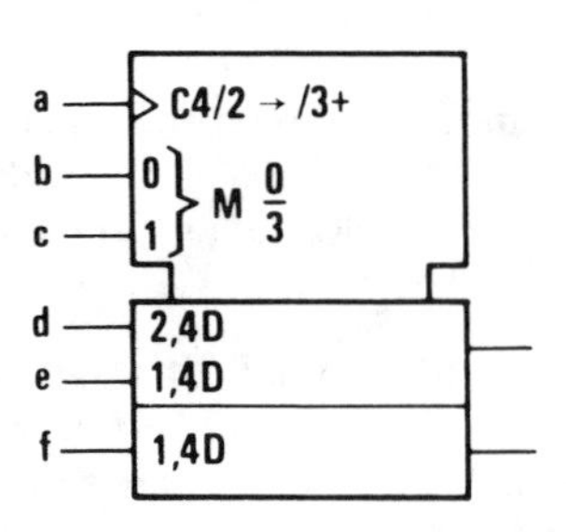

Note that all operations are synchronous.

In MODE 0 (b = 0, c = 0), the outputs remain at their existing states as none of the inputs has an effect.

In MODE 1 (b = 1, c = 0), parallel loading takes place thru inputs e and f.

In MODE 2 (b = 0, c = 1), shifting down and serial loading thru input d take place.

In MODE 3 (b = c = 1), counting up by increment of 1 per clock pulse takes place.

Figure 18. M (Mode) Dependency Affecting Inputs

4.11.2 M Dependency Affecting Outputs

When an M*m* input or M*m* output stands at its internal 1 state, the affected outputs stand at their normally defined internal logic states; i.e., the outputs are enabled.

When an M*m* input or M*m* output stands at its internal 0 state, at each affected output any set of labels containing the identifying number of that M*m* input or M*m* output has no effect and is to be ignored. When an output has several different sets of labels separated by solidi (e.g., 2,4/3,5), only those sets in which the identifying number of this M*m* input or M*m* output appears are to be ignored.

Figure 19 shows a symbol for a device whose output can behave as either a 3-state output or an open-collector output depending on the signal applied to input **a**. Mode 1 exists when input **a** stands at its internal 1 state, and, in that case, the three-state symbol applies, and the open-element symbol has no effect. When **a** = 0, mode 1 does not exist so the three-state symbol has no effect, and the open-element symbol applies.

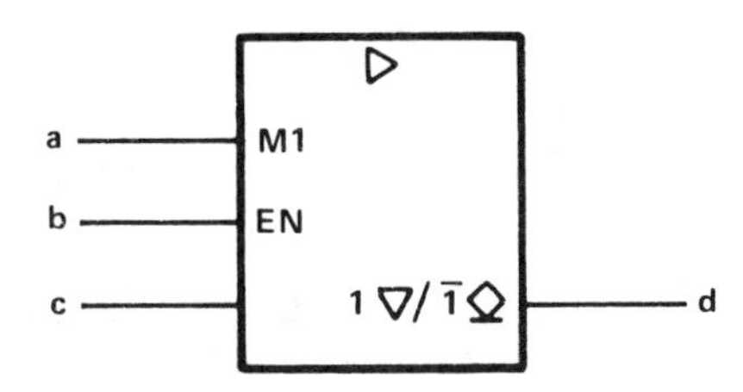

Figure 19. Type of Output Determined by Mode

In Figure 20, if input **a** stands at its internal 1 state establishing mode 1, output **b** will stand at its internal 1 state only when the content of the register equals 9. Since output **b** is located in the common-control block with no defined function outside of mode 1, the state of this output outside of mode 1 is not defined by the symbol.

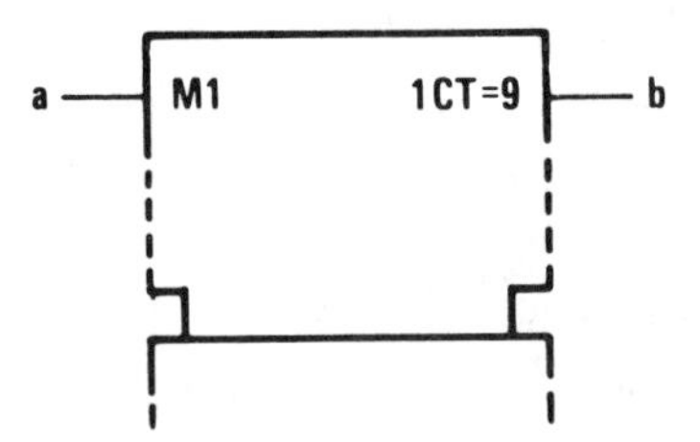

Figure 20. An Output of the Common-Control Block

In Figure 21, if input **a** stands at its internal 1 state establishing mode 1, output **b** will stand at its internal 1 state only when the content of the register equals 15. If input **a** stands at its internal 0 state, output **b** will stand at its internal 1 state only when the content of the register equals 0.

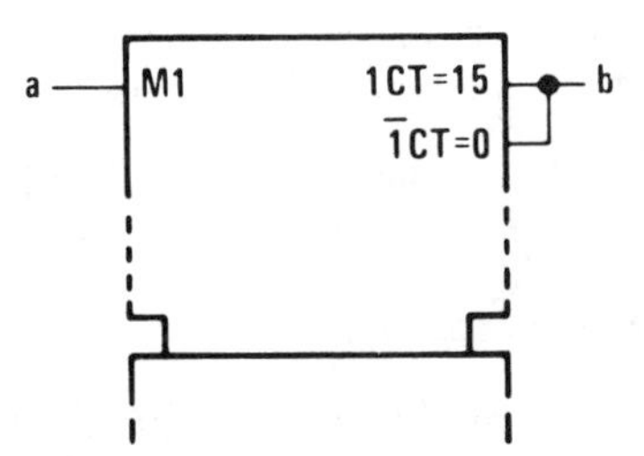

Figure 21. Determining an Output's Function

In Figure 22, inputs **a** and **b** are binary weighted to generate the numbers 0, 1, 2, or 3. This determines which one of the four modes exists.

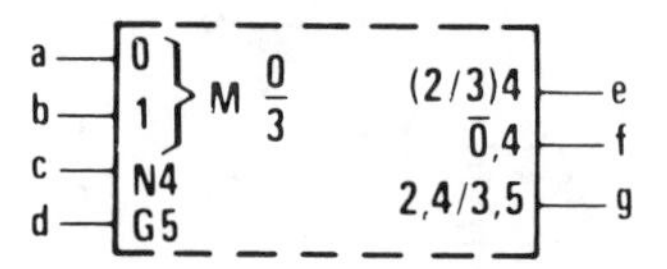

Figure 22. Dependent Relationships Affected by Mode

At output **e**, the label set causing negation (if **c** = 1) is effective only in modes 2 and 3. In modes 0 and 1, this output stands at its normally defined state as if it had no labels. At output **f**, the label set has effect when the mode is not 0 so output **e** is negated (if **c** = 1) in modes 1, 2, and 3. In mode 0, the label set has no effect so the output stands at its normally defined state. In this example, $\overline{0}$,4 is equivalent to (1/2/3)4. At output **g**, there are two label sets: the first set, causing negation (if **c** = 1), is effective only in mode 2; the second set, subjecting **g** to AND dependency on **d**, has effect only in mode 3.

Note that in mode 0 none of the dependency relationships has any effect on the outputs, so **e**, **f**, and **g** will all stand at the same state.

4.12 A (Address) Dependency

The symbol denoting address dependency is the letter A.

Address dependency provides a clear representation of those elements, particularly memories, that use address control inputs to select specified sections of a multidimensional array. Such a section of a memory array is usually called a word. The purpose of address dependency is to allow a symbolic presentation of the entire array. An input of the array shown at a particular element of this general section is common to the corresponding elements of all selected sections of the array. An output of the array shown at a particular element of this general section is the result of the OR function of the outputs of the corresponding elements of selected sections.

Inputs that are not affected by any affecting address input have their normally defined effect on all sections of the array, whereas inputs affected by an address input have their normally defined effect only on the section selected by that address input.

An affecting address input is labeled with the letter A followed by an identifying number that corresponds with the address of the particular section of the array selected by this input. Within the general section presented by the symbol, inputs and outputs affected by an A*m* input are labeled with the letter A, which stands for the identifying numbers, i.e., the addresses, of the particular sections.

Figure 23 shows a 3-word by 2-bit memory having a separate address line for each word and uses EN dependency to explain the operation. To select word 1, input **a** is taken to its 1 state, which establishes mode 1. Data can now be clocked into the inputs marked "1,4D." Unless words 2 and 3 are also selected, data cannot be clocked in at the inputs marked "2,4D" and "3,4D." The outputs will be the OR functions of the selected outputs; i.e., only those enabled by the active EN functions.

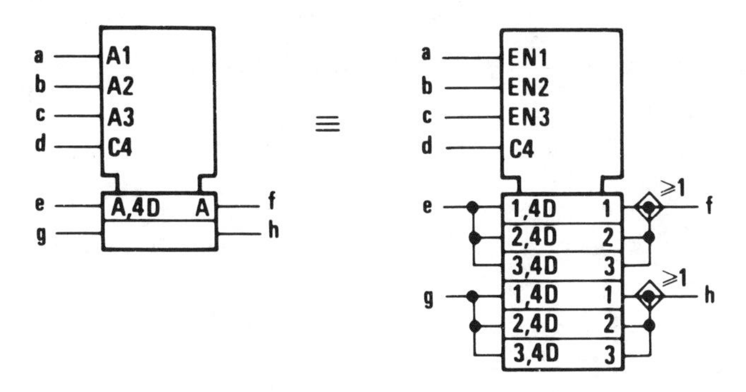

Figure 23. A (Address) Dependency

The identifying numbers of affecting address inputs correspond with the addresses of the sections selected by these inputs. They need not necessarily differ from those of other affecting dependency inputs (e.g., G, V, N, . . .), because, in the general section presented by the symbol, they are replaced by the letter A.

If there are several sets of affecting A*m* inputs for the purpose of independent and possibly simultaneous access to sections of the array, then the letter A is modified to 1A, 2A, ... Since they have access to the same sections of the array, these sets of A inputs may have the same identifying numbers. The symbols for 'HC170 or SN74LS170 make use of this.

Figure 24 is another illustration of the concept.

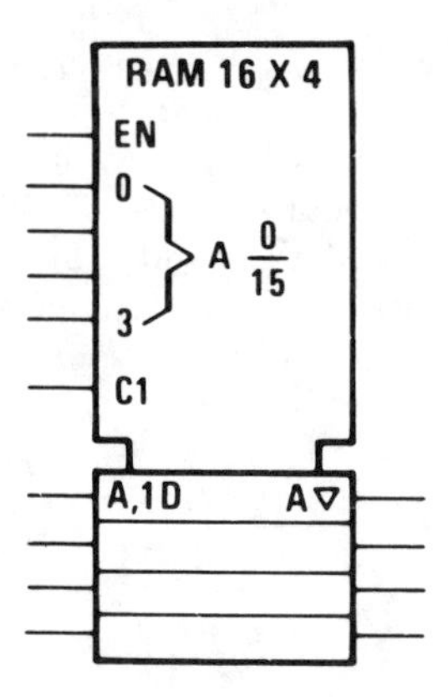

Figure 24. Array of 16 Sections of Four Transparent Latches with State Outputs Comprising a 16-Word × 4-Bit Random-Access Memory

Table 4. Summary of Dependency Notation

TYPE OF DEPENDENCY	LETTER SYMBOL*	AFFECTING INPUT AT ITS 1-STATE	AFFECTING INPUT AT ITS 0-STATE
Address	A	Permits action (address selected)	Prevents action (address not selected)
Control	C	Permits action	Prevents action
Enable	EN	Permits action	Prevents action of inputs ◇outputs off ▽outputs at external high impedance, no change in internal logic state Other outputs at internal 0 state
AND	G	Permits action	Imposes 0 state
Mode	M	Permits action (mode selected)	Prevents action (mode not selected)
Negate (Ex-OR)	N	Complements state	No effect
Reset	R	Affected output reacts as it would to S = 0, R = 1	No effect
Set	S	Affected output reacts as it would to S = 1, R = 0	No effect
OR	V	Imposes 1 state	Permits action
Transmission	X	Bidirectional connection exists	Bidirectional connection does not exist
Interconnection	Z	Imposes 1 state	Imposes 0 state

* These letter symbols appear at the AFFECTING input (or output) and are followed by a number. Each input (or output) AFFECTED by that input is labeled with that same number. When the labels EN, R, and S appear at inputs without the following numbers, the descriptions above do not apply. The action of these inputs is described under "Symbols Inside the Outline," see 3.3.

5 Bistable Elements

The dynamic input symbol, the postponed output symbol, and dependency notation provide the tools to differentiate four main types of bistable elements and make synchronous and asynchronous inputs easily recognizable (Figure 25). The first column shows the essential distinguishing features; the other columns show examples.

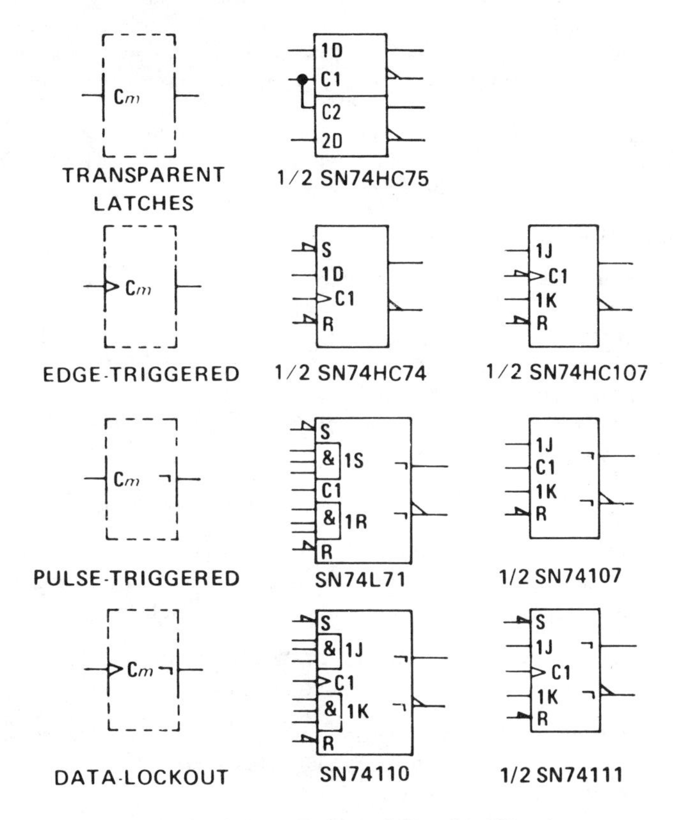

Figure 25. Four Types of Bistable Circuits

Transparent latches have a level-operated control input. The D input is active as long as the C input is at its internal 1 state. The outputs respond immediately. Edge-triggered elements accept data from D, J, K, R, or S inputs on the active transition of C. Pulse-triggered elements require the setup of data before the start of the control pulse; the C input is considered static since the data must be maintained as long as C is at its 1 state. The output is postponed until C returns to its 0 state. The data-lockout element is similar to the pulse-triggered version except that the C input is considered dynamic in that, shortly after C goes through its active transition, the data inputs are disabled, and data does not have to be held. However, the output is still postponed until the C input returns to its initial external level.

Notice that synchronous inputs can be readily recognized by their dependency labels (1D, 1J, 1K, 1S, 1R) compared to the asynchronous inputs (S, R), which are not dependent on the C inputs.

6 Coders

The general symbol for a coder or code converter is shown in Figure 26. X and Y may be replaced by appropriate indications of the code used to represent the information at the inputs and at the outputs, respectively.

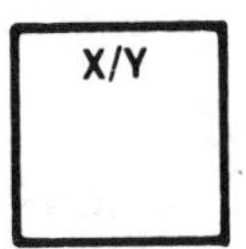

Figure 26. Coder General Symbol

Indication of code conversion is based on the following rule:

> Depending on the input code, the internal logic states of the inputs determine an internal value. This value is reproduced by the internal logic states of the outputs, depending on the output code.

The indication of the relationships between the internal logic states of the inputs and the internal value is accomplished by:

1) labeling the inputs with numbers. In this case, the internal value equals the sum of the weights associated with those inputs that stand at their internal 1-state, or by
2) replacing X by an appropriate indication of the input code and labeling the inputs with characters that refer to this code.

The relationships between the internal value and the internal logic states of the outputs are indicated by:

1) labeling each output with a list of numbers representing those internal values that lead to the internal 1-state of that output. These numbers shall be separated by solidi as in Figure 27. This labeling may also be applied when Y is replaced by a letter denoting a type of dependency

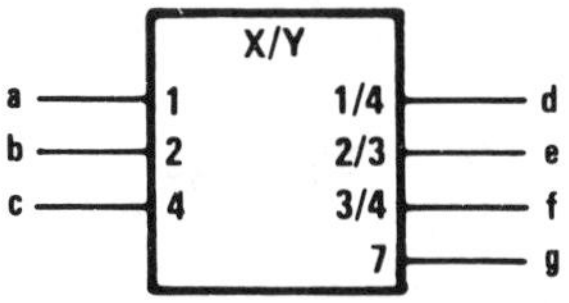

TRUTH TABLE

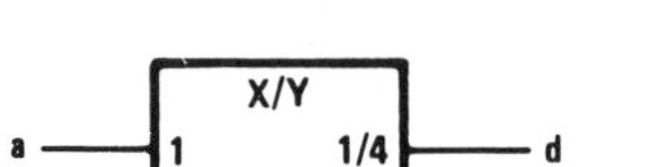

INPUTS			OUTPUTS			
c	b	a	g	f	e	d
0	0	0	0	0	0	0
0	0	1	0	0	0	1
0	1	0	0	0	1	0
0	1	1	0	1	1	0
1	0	0	0	1	0	1
1	0	1	0	0	0	0
1	1	0	0	0	0	0
1	1	1	1	0	0	0

Figure 27. An X/Y Code Converter

(see section 7). If a continuous range of internal values produces the internal 1 state of an output, this can be indicated by two numbers that are inclusively the beginning and the end of the range, with these two numbers separated by three dots (e.g., 4 . . . 9 = 4/5/6/7/8/9) or by

2) replacing Y by an appropriate indiction of the output code and labeling the outputs with characters that refer to this code as in Figure 28.

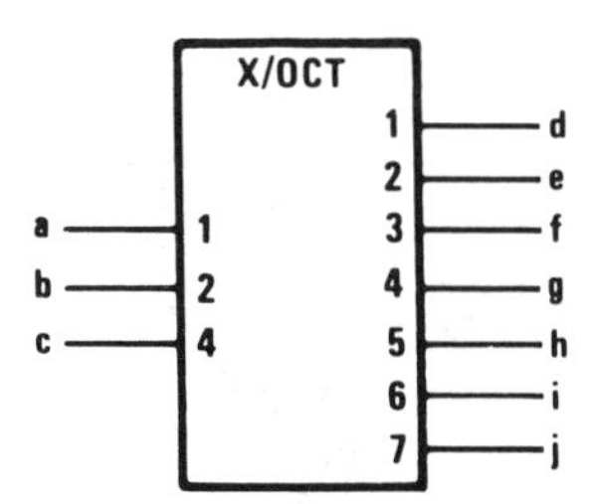

TRUTH TABLE

INPUTS			OUTPUTS						
c	b	a	j	i	h	g	f	e	d
0	0	0	0	0	0	0	0	0	0
0	0	1	0	0	0	0	0	0	1
0	1	0	0	0	0	0	0	1	0
0	1	1	0	0	0	0	1	0	0
1	0	0	0	0	0	1	0	0	0
1	0	1	0	0	1	0	0	0	0
1	1	0	0	1	0	0	0	0	0
1	1	1	1	0	0	0	0	0	0

Figure 28. An X/Octal Code Converter

Alternatively, the general symbol may be used together with an appropriate reference to a table in which the relationship between the inputs and outputs is indicated. This is a recommended way to symbolize a PROM after it has been programmed.

7 Use of a Coder to Produce Affecting Inputs

It often occurs that a set of affecting inputs for dependency notation is produced by decoding the signals on certain inputs to an element. In such a case, use can be made of the symbol for a coder as an embedded symbol (Figure 29).

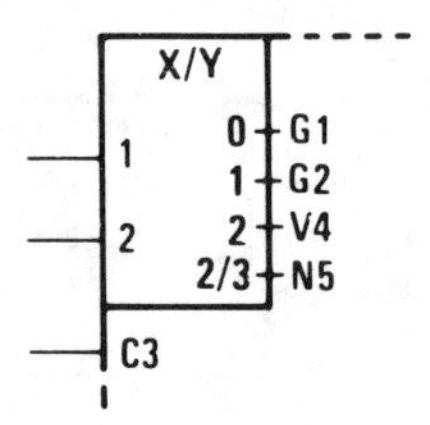

Figure 29. Producing Various Types of Dependencies

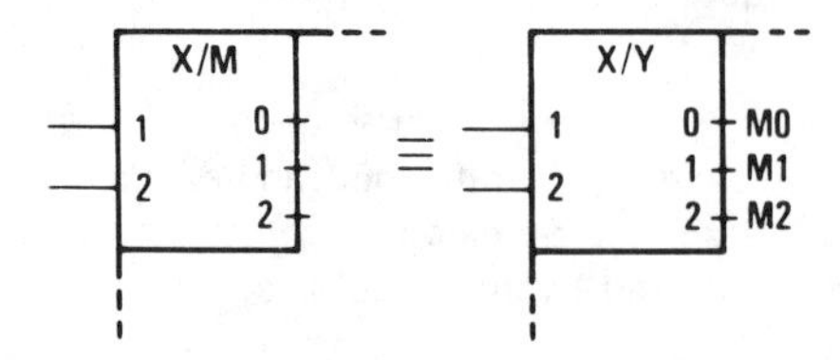

Figure 30. Producing One Type of Dependency

If all affecting inputs produced by a coder are of the same type as their identifying numbers shown at the outputs of the coder, Y (in the qualifying symbol X/Y) may be replaced by the letter denoting the type of dependency. The indications of the affecting inputs should then be omitted (Figure 30).

8 Use of Binary Grouping to Produce Affecting Inputs

If all affecting inputs produced by a coder are of the same type and have consecutive identifying numbers not necessarily corresponding with the numbers that would have been shown at the outputs of the coder, use can be made of the binary grouping symbol. k external lines effectively generate 2^k internal inputs. The bracket is followed by the letter denoting the type of dependency followed by m1/m2. The m1 is to be replaced by the smallest identifying number and the m2 by the largest one, as shown in Figure 31.

9 Sequence of Input Labels

If an input having a single functional effect is affected by other inputs, the qualifying symbol (if there is any) for that functional effect is preceded by the labels corresponding to the affecting inputs. The left-to-right order of these preceding labels is the order in which the effects or modifications must be applied. The affected input has no functional effect on the element if the logic state of any one of the affecting inputs, considered separately, would cause the affected input to have no effect, regardless of the logic states of other affecting inputs.

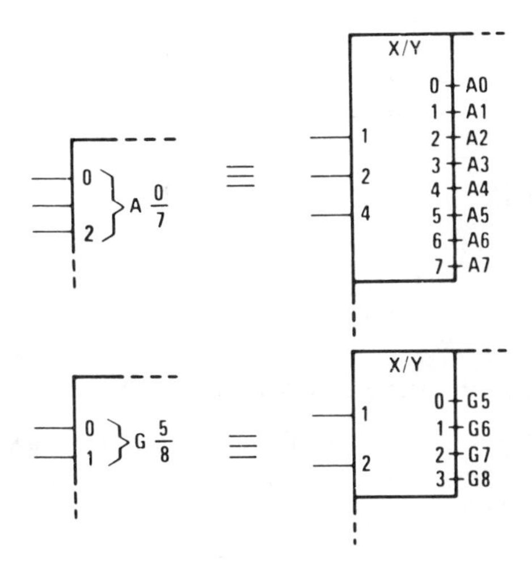

Figure 31. Use of the Binary Grouping Symbol

If an input has several different functional effects or has several different sets of affecting inputs, depending on the mode of action, the input may be shown as often as required. However, there are cases in which this method of presentation is not advantageous. In those cases, the input may be shown once with the different sets of labels separated by solidi (Figure 32). No meaning is attached to the order of these sets of labels. If one of the functional effects of an input is that of an unlabeled input to the element, a solidus will precede the first set of labels shown.

If all inputs of a combinational element are disabled (caused to have no effect on the function of the element), the internal logic states of the outputs of the element are not specified by the symbol. If all inputs of a sequential element are disabled, the content of this element is not changed, and the outputs remain at their existing internal logic states.

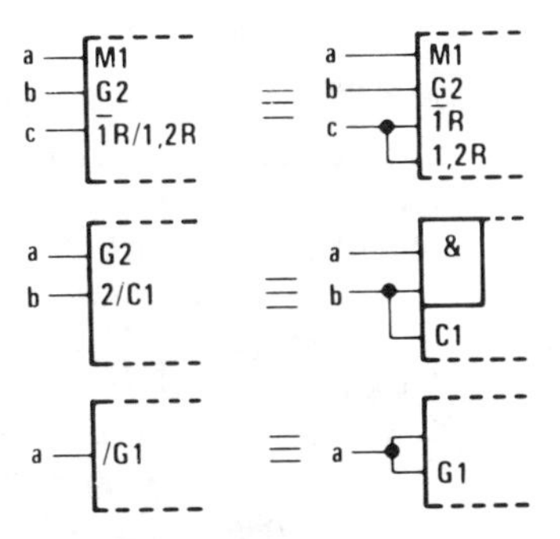

Figure 32. Input Labels

Labels may be factored using algebraic techniques (Figure 33).

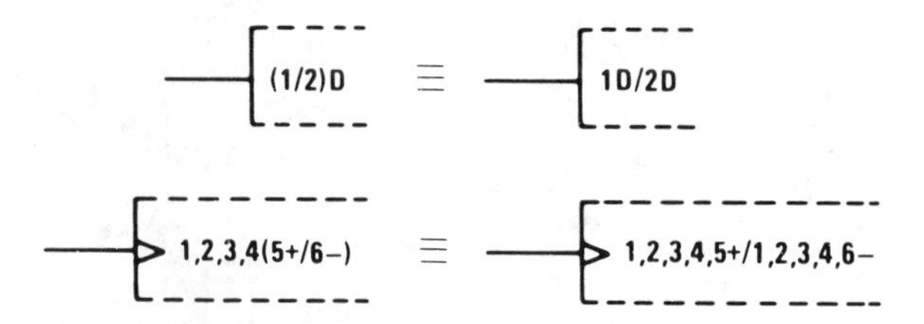

Figure 33. Factoring Input Labels

10 Sequence of Output Labels

If an output has a number of different labels, regardless of whether they are identifying numbers of affecting inputs or outputs or not, these labels are shown in the following order:

1) If the postponed output symbol has to be shown, this comes first, if necessary preceded by the indications of the inputs to which it must be applied
2) Followed by the labels indicating modifications of the internal logic state of the output, such that the left-to-right order of these labels corresponds with the order in which their effects must be applied
3) Followed by the label indicating the effect of the output on inputs and other outputs of the element.

Symbols for open-circuit or 3-state outputs, where applicable, are placed just inside the outside boundary of the symbol adjacent to the output line (Figure 34).

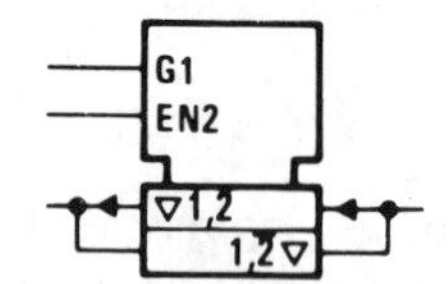

Figure 34. Placement of 3-State Symbols

If an output needs several different sets of labels that represent alternative functions (e.g., depending on the mode of action), these sets may be shown on different output lines that must be connected outside the outline. However, there are cases in which this method of presentation is not advantageous. In those cases, the output may be shown once with the different sets of labels separated by solidi (Figure 35).

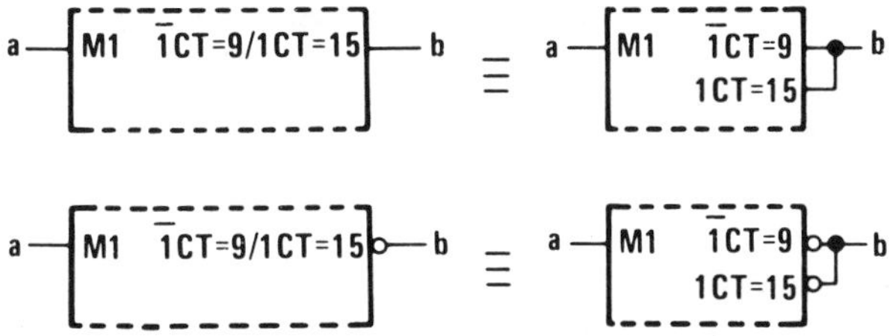

Figure 35. Output Labels

Two adjacent identifying numbers of affecting inputs in a set of labels that are not already separated by a nonnumeric character should be separated by a comma.

If a set of labels of an output not containing a solidus contains the identifying number of an affecting M*m* input standing at its internal 0 state, this set of labels has no effect on that output.

Labels may be factored using algebraic techniques (Figure 36).

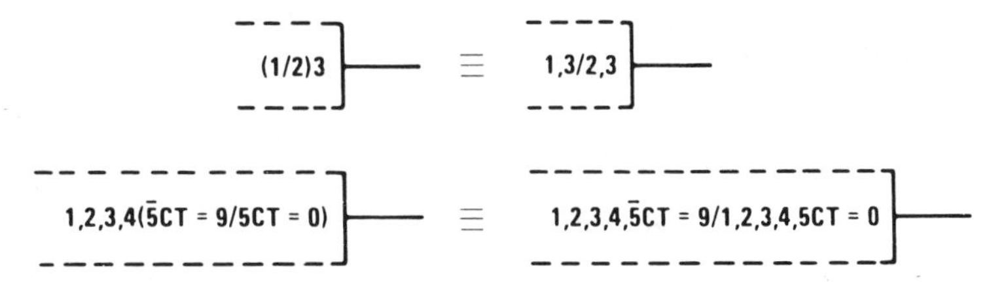

Figure 36. Factoring Output Labels

If you have questions on this Explantion of Logic Symbols, please contact:

Texas Instruments Incorporated
F.A. Mann, MS 3684
P.O. Box 655303
Dallas, Texas 75265

Telephone (214) 997-2489

IEEE Standards may be purchased from:

Institute of Electrical and Electronic Engineers, Inc.
IEEE Standards Office
445 Hoes Lane
P.O. Box 1331
Piscataway, N.J. 08855-1331

International Electrotechnical Commission (IEC) publications my be purchased from:

American National Standards Institute, Inc.
1430 Broadway
New York, N.Y. 10018

INDEX

A

B

H

I

J

K

L

S

W

X

Z